T0180344

Lecture Notes in Computer Science 12885

Advanced Research in Computing and Software Science

Subline of Lecture Notes in Computer Science

More information about this subseries at http://www.springer.com/series/7409

Ioannis Caragiannis ·
Kristoffer Arnsfelt Hansen (Eds.)

Algorithmic Game Theory

14th International Symposium, SAGT 2021
Aarhus, Denmark, September 21–24, 2021
Proceedings

Editors
Ioannis Caragiannis
Aarhus University
Aarhus, Denmark

Kristoffer Arnsfelt Hansen
Aarhus University
Aarhus, Denmark

ISSN 0302-9743 ISSN 1611-3349 (electronic)
Lecture Notes in Computer Science
ISBN 978-3-030-85946-6 ISBN 978-3-030-85947-3 (eBook)
https://doi.org/10.1007/978-3-030-85947-3

LNCS Sublibrary: SL3 – Information Systems and Applications, incl. Internet/Web, and HCI

This Springer imprint is published by the registered company Springer Nature Switzerland AG
The registered company address is: Gewerbestrasse 11, 6330 Cham, Switzerland

Preface

This volume contains the papers and extended abstracts presented at the 14th International Symposium on Algorithmic Game Theory (SAGT 2021), held during September 21–24, 2021, at Aarhus University, Denmark.

The purpose of SAGT is to bring together researchers from Computer Science, Economics, Mathematics, Operations Research, Psychology, Physics, and Biology to present and discuss original research at the intersection of Algorithms and Game Theory.

This year, we received a record number of 73 submissions, which were all rigorously peer-reviewed by the Program Committee (PC). Each paper was reviewed by at least 3 PC members, and evaluated on the basis of originality, significance, and exposition. The PC eventually decided to accept 30 papers to be presented at the conference.

The works accepted for publication in this volume cover most of the major aspects of Algorithmic Game Theory, including auction theory, mechanism design, markets and matchings, computational aspects of games, resource allocation problems, and computational social choice. To accommodate the publishing traditions of different fields, authors of accepted papers could ask that only a one-page abstract of the paper appeared in the proceedings. Among the 30 accepted papers, the authors of 4 papers selected this option.

Furthermore, due to the generous support by Springer, we were able to provide a Best Paper Award. The PC decided to give the award to the paper "Descending the Stable Matching Lattice: How Many Strategic Agents are Required to Turn Pessimality to Optimality?" authored by Ndiamé Ndiaye, Sergey Norin, and Adrian Vetta.

The program also included three invited talks by distinguished researchers in Algorithmic Game Theory, namely Yiling Chen (Harvard University, USA), Elias Koutsoupias (University of Oxford, UK), and Rahul Savani (University of Liverpool, UK). In addition, SAGT 2021 featured tutorial talks given by Vasilis Gkatzelis (Drexel University, USA) and Martin Hoefer (Goethe University Frankfurt, Germany).

We would like to thank all the authors for their interest in submitting their work to SAGT 2021, as well as the PC members and the external reviewers for their great work in evaluating the submissions. We also want to thank Springer and the COST Action GAMENET (CA16228) for their generous financial support. We are grateful to the Aarhus Institute of Advanced Studies for hosting the conference. Finally, we would also like to thank Anna Kramer at Springer for helping with the proceedings, and the EasyChair conference management system for facilitating the peer-review process.

July 2021

Ioannis Caragiannis
Kristoffer Arnsfelt Hansen

Organization

Program Committee

Haris Aziz	UNSW Sydney, Australia
Siddharth Barman	Indian Institute of Science, India
Xiaohui Bei	Nanyang Technological University, Singapore
Simina Brânzei	Purdue University, USA
Ioannis Caragiannis (co-chair)	Aarhus University, Denmark
Jing Chen	Stony Brook University, USA
Ágnes Cseh	Hungarian Academy of Sciences, Hungary
Argyrios Deligkas	Royal Holloway University of London, UK
Kousha Etessami	University of Edinburgh, UK
Aris Filos-Ratsikas	University of Liverpool, UK
Felix Fischer	Queen Mary University of London, UK
Michele Flammini	Gran Sasso Science Institute, Italy
Paul Goldberg	University of Oxford, UK
Nick Gravin	Shanghai University of Finance and Economics, China
Kristoffer Arnsfelt Hansen (co-chair)	Aarhus University, Denmark
Ayumi Igarashi	National Institute of Informatics, Japan
Thomas Kesselheim	University of Bonn, Germany
Max Klimm	TU Berlin, Germany
Maria Kyropoulou	University of Essex, UK
Pascal Lenzner	Hasso Plattner Institute, Germany
Stefano Leonardi	Sapienza University of Rome, Italy
Pinyan Lu	Shanghai University of Finance and Economics, China
Troels Bjerre Lund	IT University of Copenhagen, Denmark
Swaprava Nath	IIT Kanpur, India
Britta Peis	RWTH Aachen University, Germany
Maria Polukarov	King's College London, UK
Emmanouil Pountourakis	Drexel University, USA
Marco Scarsini	Luiss Guido Carli, Italy
Nisarg Shah	University of Toronto, Canada
Eric Sodomka	Facebook, USA
Adrian Vetta	McGill University, Canada

Organizing Committee

Ioannis Caragiannis	Aarhus University, Denmark
Kristoffer Arnsfelt Hansen	Aarhus University, Denmark
Signe L. Jensen	Aarhus University, Denmark

Steering Committee

Elias Koutsoupias	University of Oxford, UK
Marios Mavronicolas	University of Cyprus, Cyprus
Dov Monderer	Technion, Israel
Burkhard Monien	University of Paderborn, Germany
Christos Papadimitriou	Columbia University, USA
Giuseppe Persiano	University of Salerno, Italy
Paul Spirakis (Chair)	University of Liverpool, UK

Additional Reviewers

Andreas Abels
Ben Abramowitz
Mete Şeref Ahunbay
Michele Aleandri
Alessandro Aloisio
Mikel Álvarez-Mozos
Georgios Amanatidis
Piyush Bagad
Márton Benedek
Umang Bhaskar
Rangeet Bhattacharyya
Georgios Birmpas
Alexander Braun
Hau Chan
Federico Corò
Andrés Cristi
Josu Doncel
Soroush Ebadian
Eduard Eiben
Thomas Erlebach
Tomer Ezra
John Fearnley
Simone Fioravanti
Federico Fusco
Erick Galinkin
Ganesh Ghalme
Hugo Gilbert
Kira Goldner
Daniel Halpern
Tesshu Hanaka
Zhiyi Huang
Yaonan Jin
Naoyuki Kamiyama
Panagiotis Kanellopoulos
Bojana Kodric
Frederic Koessler
Anand Krishna
Anilesh Kollagunta Krishnaswamy
Christian Kroer
Alex Lam
Philip Lazos
Bo Li
Yingkai Li
Shengxin Liu
Zhengyang Liu
Xinhang Lu
Junjie Luo
Simon Mauras
Themistoklis Melissourgos
Evi Micha
Shivika Narang
Vishnu Narayan
Tim Oosterwijk
Adèle Pass-Lanneau
Neel Patel
Daniel Paulusma
Nidhi Rathi
Rebecca Reiffenhäuser
Niklas Rieken
Daniel Schmand
Marc Schroder
Steffen Schuldenzucker
Paolo Serafino
Garima Shakya
Oskar Skibski
Tyrone Strangway

Warut Suksompong
Mashbat Suzuki
Zhihao Gavin Tang
Laura Vargas Koch
Xavier Venel
Paritosh Verma
Cosimo Vinci
Alexandros Voudouris
Jasmin Wachter
Kangning Wang
Zihe Wang
Jinzhao Wu
Haifeng Xu
Xiang Yan
Chunxue Yang
Tom van der Zanden
Yuhao Zhang

Invited Talks

Mechanisms for Selling Information

Yiling Chen

Harvard University, USA

Abstract. Different from traditional goods, information (private signals) can be sold in more flexible ways. A salient feature of information as goods is that it can be revealed partially. This not only means that one can sell partial information at a granularity of his choice but also suggests that partial information revelation can be used to advertise the value of the remaining information. Hence, the space of mechanisms for selling information is rich. In this talk, I will discuss designing optimal mechanisms for a revenue-driven, monopoly information holder to sell his information to information buyers in a few scenarios.

Biography: Yiling Chen is a Gordon McKay Professor of Computer Science at Harvard University. She received her Ph.D. in Information Sciences and Technology from the Pennsylvania State University. Prior to working at Harvard, she spent two years at Yahoo! Research in New York City. Her research lies in the intersection of computer science, economics and other social sciences, with a focus on social aspects of computational systems. She was a recipient of The Penn State Alumni Association Early Career Award, and was selected by IEEE Intelligent Systems as one of "AI's 10 to Watch" early in her career. Her work received best paper awards at ACM EC, AAMAS, ACM FAT* (now ACM FAccT) and ACM CSCW conferences. She has co-chaired the 2013 Conference on Web and Internet Economics (WINE'13), the 2016 ACM Conference on Economics and Computation (EC'16) and the 2018 AAAI Conference on Human Computation and Crowdsourcing (HCOMP18) and has served as an associate editor for several journals.

On the Nisan-Ronen Conjecture for Graphs

Elias Koutsoupias

University of Oxford, UK

Abstract. The Nisan-Ronen conjecture states that no truthful mechanism for makespan-minimization when allocating a set of tasks to n unrelated machines can have approximation ratio less than n. Over more than two decades since its formulation, little progress has been made in resolving it. In this talk, I will discuss recent progress towards validating the conjecture by showing a lower bound of $1 + \sqrt{n-1}$. The lower bound is based on studying an interesting class of instances that can be represented by multi-graphs in which vertices represent machines and edges represent tasks, and each task should be allocated to one of its two incident machines.

Biography: Elias Koutsoupias is a professor of computer science at the University of Oxford. His research interests include algorithmic aspects of game theory, economics and networks, online algorithms, decision-making under uncertainty, distributed algorithms, design and analysis of algorithms, and computational complexity. He previously held faculty positions at the University of California, Los Angeles (UCLA) and the University of Athens. He studied at the National Technical University of Athens (BSc in electrical engineering) and the University of California, San Diego (PhD in computer science). He received the Goedel Prize of theoretical computer science for his work on the Price of Anarchy, in reference to laying the foundations of algorithmic game theory.

The Complexity of Gradient Descent

Rahul Savani

University of Liverpool, UK

Abstract. PPAD and PLS are successful classes that each capture the complexity of important game-theoretic problems: finding a mixed Nash equilibrium in a bimatrix game is PPAD-complete; and finding a pure Nash equilibrium in a congestion game is PLS-complete. Many important problems, such as solving a Simple Stochastic Game or finding a mixed Nash equilibrium of a congestion game, lie in both classes. However, it was strongly believed that their intersection does not have natural complete problems. We show that it does: any problem that lies in both classes can be reduced in polynomial time to the problem of finding a stationary point of a function. Our result has been used to show that computing a mixed equilibrium of a congestion game is also complete for the intersection of PPAD and PLS.

This is joint work with John Fearnley, Paul Goldberg, and Alexandros Hollender.

Biography: Rahul Savani is a Professor of Economics and Computation at the University of Liverpool. He has worked extensively on the computation of equilibria in game-theoretic models. The paper that he will present won a Best Paper Award at STOC'21.

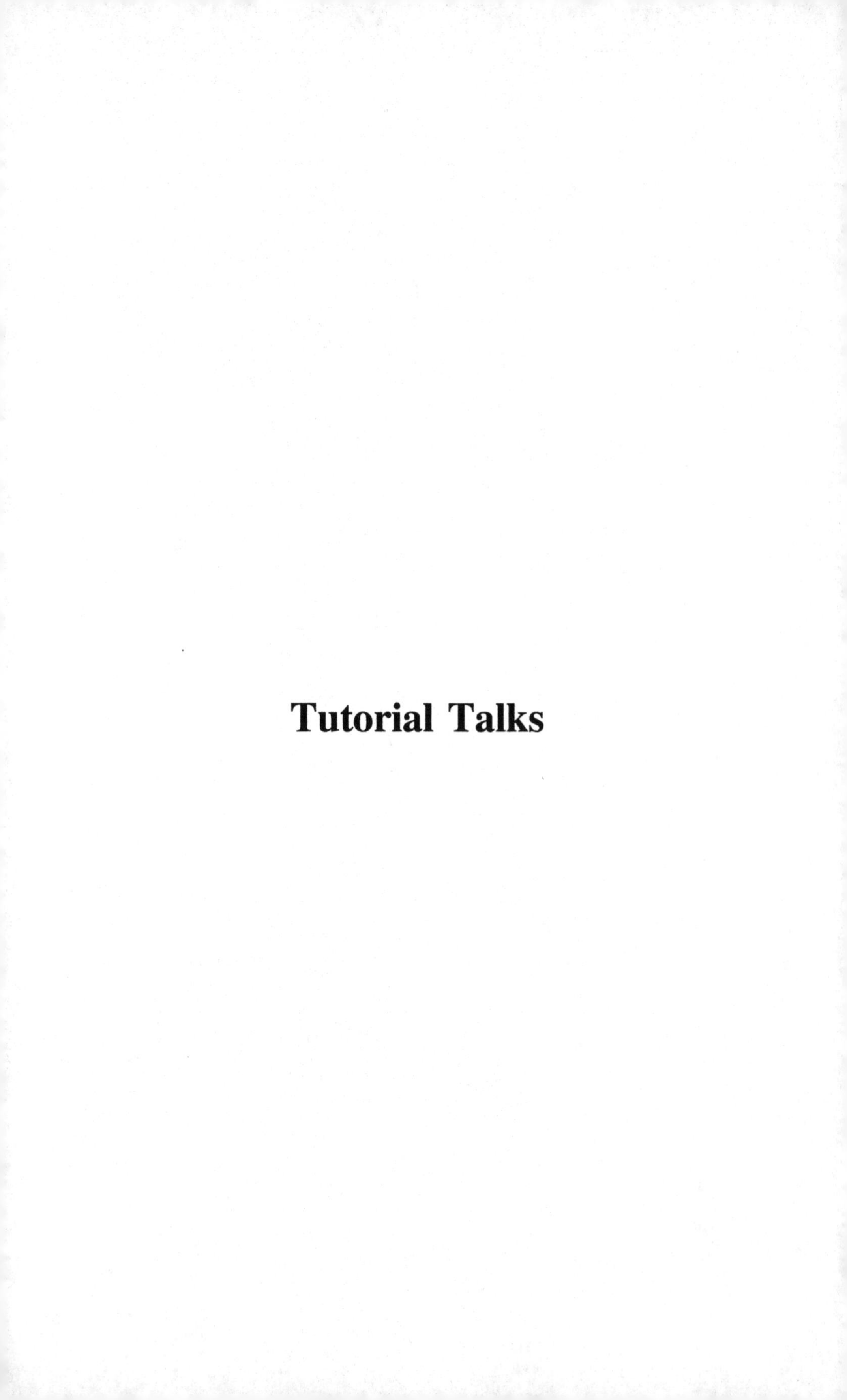

Tutorial Talks

Understanding the Power and Limitations of Clock Auctions

Vasilis Gkatzelis

Drexel University, USA

Abstract. In this tutorial, we will be focusing on the class of (deferred-acceptance) clock auctions, introduced by economists Paul Milgrom and Ilya Segal. Clock auctions satisfy a sequence of impressive properties: i) they are obviously strategyproof, which implies that it is very easy for the participating bidders to identify their optimal strategy, ii) they are weakly group-strategyproof, which guarantees that even if the bidders collude, they cannot all benefit from manipulating the auction, iii) they are transparent and do not require that the bidders trust the auctioneer, and iv) they satisfy unconditional winner privacy, which means that the winners of the auction do not need to reveal their true value. This unique combination of benefits that clock auctions provide make them ideal for real-world problems, since they require very little from the participating bidders. Our presentation will first discuss these properties in detail, it will then study the extent to which clock auctions can match the state-of-the-art performance guarantees of previously known auctions (proving both positive and negative results) and will conclude with a discussion of some open problems and future directions.

Biography: Vasilis Gkatzelis is an assistant professor in computer science at Drexel University and his research focuses on problems in algorithmic game theory and approximation algorithms. He is a recipient of the NSF CAREER award. Prior to joining Drexel University, he held positions as a postdoctoral scholar at the computer science departments of UC Berkeley and Stanford University, and as a research fellow at the Simons Institute for the Theory of Computing. He received his PhD from the Courant Institute of New York University.

Algorithmic Challenges in Information Design

Martin Hoefer

Goethe University Frankfurt, Germany

Abstract. Information is a crucial resource in modern economy. Collecting and sharing information strategically is central to the business strategy of many major companies, including search engines, recommendation engines, and two-sided market platforms.

In all these domains, there is an informed "sender" (often a company or platform) who shares information in order to motivate an uninformed "receiver" (e.g., a potential customer) to take actions that are beneficial to the sender. Information design, alternatively known also as Bayesian persuasion, studies how the sender can disclose information optimally while accounting for the incentives that govern the behavior of the receiver.

Over the last decade, ideas from information design have found many applications in economics, and the area offers interesting challenges for algorithmic work. In this tutorial, we concetrate on algorithms for optimization problems arising in basic pesuasion problems. We also touch upon recent work on extensions to restricted communication, dynamic arrival, multiple senders and receivers, learning, and more. Along the way, we mention open problems and opportunities for future research.

Biography: Martin Hoefer is a professor in Computer Science at Goethe University Frankfurt. He received a PhD in Computer Science from Konstanz University in 2007. Subsequently, he was a postdoc at Stanford University and a junior professor at RWTH Aachen University. In 2012 he joined MPI Informatik as a senior researcher, and in 2016 Goethe University as full professor. His research investigates algorithms for problems at the intersection of computer science and game theory in a broad sense, with a recent focus on information design.

Contents

Markets and Matchings

Social Choice and Cooperative Games

Abstracts

Auctions and Mechanism Design

Improved Two Sample Revenue Guarantees via Mixed-Integer Linear Programming

Mete Şeref Ahunbay[1(✉)] and Adrian Vetta[2]

[1] Department of Mathematics and Statistics, McGill University, Montréal, QC, Canada
mete.ahunbay@mail.mcgill.ca

[2] Department of Mathematics and Statistics and School of Computer Science, McGill University, Montréal, QC, Canada
adrian.vetta@mcgill.ca

Abstract. We study the performance of the Empirical Revenue Maximizing (ERM) mechanism in a single-item, single-seller, single-buyer setting. We assume the buyer's valuation is drawn from a regular distribution F and that the seller has access to *two* independently drawn samples from F. By solving a family of mixed-integer linear programs (MILPs), the ERM mechanism is proven to guarantee at least .5914 times the optimal revenue in expectation. Using solutions to these MILPs, we also show that the worst-case efficiency of the ERM mechanism is at most .61035 times the optimal revenue. These guarantees improve upon the best known lower and upper bounds of .558 and .642, respectively, of Daskalakis and Zampetakis [4].

1 Introduction

We study a primitive setting in revenue maximization: there is a single seller wishing to sell a single item to a single buyer, where the buyer's valuation for the item is drawn from a *regular* distribution F on $[0, \infty)$. Further, we incorporate the now widespread supposition that the valuation distribution F is *unknown to the seller.* Specifically, we present quantitative expected revenue guarantees when the seller is allowed access to **two** random, independently drawn sample valuations from F before she selects a mechanism by which to sell the item.

When F is known to the seller, Myerson [12] showed that the optimal mechanism the seller can implement is a *posted price mechanism.* In a posted price mechanism, the seller chooses a price p and the buyer decides to either buy the item or not. Of course, under the implementation of such a mechanism, the buyer would purchase the item if and only if his valuation for the item is greater than p. Given this, the seller simply picks a price p which maximizes her expected revenue. Formally, denoting the probability she sells the item for price p' as $1 - F(p')$, the seller picks a price $p \in \max_{p' \in \mathbb{R}_+} p' \cdot (1 - F(p'))$.

I. Caragiannis and K. A. Hansen (Eds.): SAGT 2021, LNCS 12885, pp. 3–17, 2021.
https://doi.org/10.1007/978-3-030-85947-3_1

But what about the case when F is unknown to the seller? When the seller has *sample access* to F, the natural approach is for the seller to assume the buyer's valuation distribution is given by the *empirical distribution* $\hat{F}$ induced by the set of samples; she may then simply implement the optimal mechanism of Myerson [12] using the empirical distribution. This method, called the *Empirical Revenue Maximising* (ERM) mechanism, provides surprisingly good performance guarantees even in the case of a single sample. Specifically, Dhangwatnotai et al. [6] showed that for the ERM mechanism just one sample suffices to give a $\frac{1}{2}$-approximation to the optimal revenue. Huang et al. [11] showed that this factor $\frac{1}{2}$ bound is tight for any *deterministic* mechanism. In contrast, Fu et al. [7] gave a *probabilistic* mechanism obtaining at least $\frac{1}{2} + 5 \cdot 10^{-9}$ times the optimal revenue using a single sample.

On the other hand, another line of work studies the performance of the ERM mechanism with respect to *sample complexity*. This asks how many samples are necessary and/or sufficient to obtain a $(1-\epsilon)$-approximation of the optimal revenue, in expectation or with high probability. Dhangwatnotai et al. [6] noted that even in our simple setting, the ERM mechanism does *not* provide distribution independent polynomial sample complexity bounds; however, a *guarded* variant of the ERM mechanism which ignores an ϵ fraction of the largest samples does produce a $(1-\epsilon)$-approximate reserve price with probability $(1-\delta)$ given $\Omega\left(\epsilon^{-3} \cdot \ln\left(\frac{1}{\epsilon\delta}\right)\right)$ samples. Later, Huang et al. [11] showed that any pricing algorithm that obtains a $(1-\epsilon)$-approximation of the optimal revenue requires $\Omega(\epsilon^{-3})$ samples, implying the factor ϵ^{-3} in the sample complexity result of [6] is tight. For more on the sample complexity of the ERM mechanism and its variants, see [1,3,5,8–10,13].

Motivated by the gap in our knowledge on sample complexity between the cases of a large number of samples and a single sample, Babaioff et al. [2] asked for revenue guarantees (in expectation) for a fixed number of samples ≥ 2. Through a very rigorous case analysis they proved that, for *two* samples, the ERM mechanism breaks the factor $\frac{1}{2}$ barrier, guaranteeing at least .509 times the optimal revenue in expectation. Significant improvements in revenue guarantees were then provided by Daskalakis and Zampetakis [4], who showed that with two samples a *rounded* version of the ERM mechanism obtains in expectation at least .558 times the optimal revenue. To achieve this they constructed a family of SDPs whose solutions provide lower bounds on the performance of the rounded ERM mechanism. Furthermore, through their primal solutions, they also showed that there exists a distribution of the buyer's valuation for which, with two samples, the ERM mechanism obtains in expectation at most .642 times the optimal revenue.

In this paper, we study the ERM mechanism with two samples by building upon the optimization perspective of Daskalakis and Zampetakis [4], using an MILP-based framework to inspect the performance of the ERM mechanism in our setting. Our key technical contribution is to present an MILP to bound the performance of the ERM mechanism and which, despite the presence of $\geq$1000 binary variables, can be approximately solved in a reasonable amount of time with *provably* small error guarantees. This allows us to prove the ERM mechanism obtains at least .5914 times the optimal revenue. Furthermore, primal solutions to our MILPs show that there is a distribution F for the buyer such that the ERM mechanism obtains at most .61035 times the optimal revenue.

2 Preliminaries

There are two agents: a seller and a buyer. The seller wishes to sell a single item to the buyer, whose valuation v is drawn from a distribution F. To do so, the seller runs a posted price mechanism – the seller commits to a price p, and the buyer can either take it or leave it. The buyer is utility maximizing, and his utility is quasilinear in payment. In particular, the buyer purchases the item if and only if $v \geq p$. Further, we make the standard assumption that the distribution of the buyer's valuation, F, is *regular*. A distribution F on $\mathbb{R}_+$, given by its cumulative distribution function $F : \mathbb{R}_+ \to [0, 1]$, is called regular if its revenue curve $R(q) = (1 - q) \cdot F^{-1}(q)$ is concave on $(0, 1)$. The objective of the seller is to maximize her revenue, but the distribution F is unknown to her. Instead she must select the posted price based upon (two) independently drawn samples from F.

We assume the seller does the obvious and implements the *Empirical Revenue Maximizing* (ERM) mechanism. That is, she simply posts a price which maximizes her expected revenue with respect to the *empirical distribution* $\bar{F}$ she obtains via her two samples $t \leq s$:

$$\bar{F}(p) = \begin{cases} 0 & p < t \\ 1/2 & t \leq p < s \\ 1 & s \leq p \end{cases}$$

Thus the seller sets price $p = t$ if $s < 2t$, and sets price $p = s$ if $s > 2t$. If $s = 2t$, since we are interested in worst case revenue, we may assume that the seller picks $p \in \{s, t\}$ which minimizes $p \cdot (1 - F(p))$. Denote the expected revenue from posting price p by $r(p) = p \cdot (1 - F(p))$. Next, let the *bisample expected revenue* $\psi_F(\cdot, \cdot)$ be defined as follows. When $s \geq t$, set

$$\psi_F(s,t) = r(s) \cdot \mathbb{I}(s > 2t) + r(t) \cdot \mathbb{I}(s < 2t) + \min\{r(s), r(t)\} \cdot \mathbb{I}(s = 2t), \quad (1)$$

and, when $s < t$, set $\psi_F(s, t) = \psi_F(t, s)$. Then the seller's revenue for implementing the ERM mechanism is exactly:

$$\bar{r}_F = \int_{(s,t)\in\mathbb{R}_+^2} \psi_F(s,t) \cdot dF(s) \times dF(t) \quad (2)$$

In turn, the optimal revenue for distribution F is given by $r_F = \max_{p\in\mathbb{R}_+} p \cdot (1 - F(p))$. In this paper, we are interested in providing lower and upper bounds for the relative performance of the ERM mechanism and the optimal mechanism, $\alpha = \inf_{F|F \text{ is regular}} \bar{r}_F / r_F$.

Following Daskalakis and Zampetakis [4], we will be deriving our bounds on α via a reduction to a set of optimization programs for which we compute solutions. However, we make a different choice of variables, working in the *quantile space* (i.e. with the revenue curve) rather than working in the *price space* (i.e. with the PDF/CDF of the distribution directly). Towards this end, note first that if R is the revenue curve of the distribution F then $R(q) = (1-q) \cdot F^{-1}(q)$ for any $q \in [0,1]$. Therefore, for each $q \in [0,1]$, the revenue curve provides the *price inverse* of q:

$$F^{-1}(q) = \begin{cases} R(q)/(1-q) & q \in [0,1) \\ \lim_{q' \to 1^-} R(q')/(1-q') & q = 1 \end{cases}$$

Via the price inverse, we may define the *bisample revenue function* $\phi_R(\cdot,\cdot)$ on $[0,1]^2$. To do this, if $(x,y) \in [0,1]^2$ and $x \geq y$, set

$$\begin{aligned} \phi_R(x,y) \;=\; & R(x) \cdot \mathbb{I}[F^{-1}(x) > 2\,F^{-1}(y)] \;+\; R(y) \cdot \mathbb{I}[F^{-1}(x) < 2\,F^{-1}(y)] \\ & +\; \min\{R(x), R(y)\} \cdot \mathbb{I}[F^{-1}(x) = 2\,F^{-1}(y)] \end{aligned} \tag{3}$$

If instead $x < y$, we symmetrically extend the function by setting $\phi_R(x,y) = \phi_R(y,x)$. We then write $\bar{r}_F$ as a double integral on $[0,1]^2$, which by (2) has the form:

$$\bar{r}_R = \int_{(x,y)\in[0,1]^2} \phi_R(x,y) \cdot d(x,y) \tag{4}$$

For any concave $R : [0,1] \to \mathbb{R}_+$ the bisample revenue function $\phi_R(x,y)$ is Riemann integrable, which implies that

$$\alpha = \inf_{F | F \text{ is regular}} \bar{r}_F / r_F = \inf_{R | R:[0,1]\to\mathbb{R}_+ \text{ is concave}} \bar{r}_R / r_R, \tag{5}$$

where $r_R = \max_{q\in[0,1]} R(q)$.

3 Approximation Programs

The Riemann integrability of ϕ_R on $[0,1]^2$ also suggests a possible optimization formulation for our problem. Given a gauge, we can try to find a concave and non-negative function R on $[0,1]$, suitably constrained, such that an approximation of $\bar{r}_R$ is minimized. To do so, we first need to define a gauge on $[0,1]^2$. We opt for the natural approach, defining a gauge on $[0,1]^2$ by considering product intervals arising from a gauge on $[0,1]$. Suppose we divide the interval $[0,1]$ into subintervals of the form $I(i) = [q_i, q_{i+1}]$ for $1 \leq i \leq n$, where $q_1 = 0$, $q_{n+1} = 1$, and $q_{i+1} > q_i$ for any $1 \leq i \leq n$. Also denote by $I(i,j)$ the product interval $[q_i, q_{i+1}] \times [q_j, q_{j+1}]$. Then we may rewrite integral (4) as:

$$\bar{r}_R = \sum_{1\leq i\leq n} \int_{(x,y)\in I(i,i)} \phi_R(x,y) \cdot d(x,y) + 2 \cdot \sum_{1\leq j<i\leq n} \int_{(x,y)\in I(i,j)} \phi_R(x,y) \cdot d(x,y) \tag{6}$$

We want primal solutions to our problems to describe approximately minimal value distributions for the buyer. One way to do so is to include variables that correspond to the values the revenue curve attains. Specifically, for $(q_i)_{1\leq i\leq n+1}$, we will include variables $R(q_i)$. For notational convenience later on, let $\vec{R}$ denote the vector containing all $R(q_i)$. Then each $R(q_i)$ corresponds to a value attained by a non-negative, concave function. This implies that the following constraints must hold:

$$R(q_i) \geq \frac{R(q_{i+1})(q_i - q_{i-1}) + R(q_{i-1})(q_{i+1} - q_i)}{(q_{i+1} - q_{i-1})} \qquad \forall\, 1 < i < n+1 \tag{7}$$

$$R(q_i) \geq 0 \qquad \forall\, 1 \leq i \leq n+1 \tag{8}$$

Furthermore, we want R to be normalized such that $\max_{q\in[0,1]} R(q) = 1$. Unfortunately, this is non-trivial to implement linearly. So, instead, we constrain the set of revenue curves so that there exists $1 \leq OPT \leq n$ and $q^* \in [q_{OPT}, q_{OPT+1}]$ such that $R(q^*) = 1$. By the concavity and non-negativity of R, this implies that:

$$R(q_{OPT}) \geq q_{OPT}/q_{OPT+1} \tag{9}$$

$$R(q_{OPT+1}) \geq (1 - q_{OPT+1})/(1 - q_{OPT}) \tag{10}$$

Furthermore, by concavity, $R(\cdot)$ should be weakly increasing before q_{OPT} and weakly decreasing beyond q_{OPT+1}:

$$\begin{aligned} R(q_{i+1}) - R(q_i) &\leq 0 \qquad \forall 1 \leq i < q_{OPT} \\ -R(q_{i+1}) + R(q_i) &\leq 0 \qquad \forall q_{OPT+1} \leq i < n \end{aligned} \tag{11}$$

We also model the indicator functions in (3) as binary variables:

Lemma 1. *For any $(x, y) \in [0, 1)$ such that $x > y$,*

$$\phi_R(x,y) = \min_{w(x,y)\in\{0,1\}} \quad R(x)\cdot w(x,y) + R(y)\cdot(1 - w(x,y))$$

$$\text{subject to} \quad w(x,y)\cdot[R(x)\cdot(1-y) - 2R(y)\cdot(1-x)] \geq 0 \tag{12}$$

$$(1 - w(x,y))\cdot[R(x)\cdot(1-y) - 2R(y)\cdot(1-x)] \leq 0 \tag{13}$$

To compute a Riemann sum, we evaluate w on a set T of points in $[0, 1]^2$ such that for each non-diagonal area element $I(i, j)$ for $1 \leq j < i \leq n$ contains a point where we evaluate w. Formally,

$$\forall 1 \leq j < i \leq n, \quad \exists(\bar{q}_i, \bar{q}_j) \in T, \ (\bar{q}_i, \bar{q}_j) \in [q_i, q_{i+1}] \times [q_j, q_{j+1}].$$

We include variables for the value R attains on endpoints of intervals, but w may be evaluated (in principle) anywhere on $I(i, j)$. Then for $(x, y) \in T$, to be able to impose constraints of the form (12) and (13) on $w(x, y)$, we find $R(x)$ and $R(y)$ by linear interpolation on $\vec{R}$. In particular, if $x \in [q_i, q_{i+1}]$, then:

$$R(x) \cdot (q_{i+1} - q_i) = R(q_i) \cdot (q_{i+1} - x) + R(q_{i+1}) \cdot (x - q_i),$$

and likewise for $R(y)$. So setting $\vec{w}$ to be the vector containing all $w(x_\ell, y_\ell)$, for each individual summand in (6) we may approximate

$$\int_{(x,y)\in I(i,j)} \phi_R(x, y) \cdot d(x, y) \simeq A(i, j) \cdot f_{ij}(\vec{R}, \vec{w})$$

where $A(i, j) = (q_{i+1} - q_i)(q_{j+1} - q_j)$ is the area of $I(i, j)$, for $1 \leq j \leq i \leq n$, and f_{ij} is some function determined by our approximation scheme, homogeneous of degree one in $\vec{R}$.

This provides the form of our most general optimization formulation: we consider a set of gauges indexed by a set J, $(\vec{q}^k)_{k\in J}$, such that $\cup_{k\in J}[q^k_{OPT^k}, q^k_{OPT^k+1}] = [0, 1]$, and find R that minimizes our approximation of $\bar{r}_R$ by computing:

$$\begin{aligned} \min_{k\in J} \min_{\vec{R},\vec{w}} \quad & \sum_{1\leq i\leq n} A_k(i, i) \cdot f_{ii}(\vec{R}, \vec{w}) + 2 \cdot \sum_{1\leq j<i\leq n} A_k(i, j) \cdot f_{ij}(\vec{R}, \vec{w}) \qquad (14) \\ \text{subject to} \quad & (7), (9), (10), (11), (12), (13) \\ & \vec{R} \in [0, 1]^{n+1} \\ & \vec{w} \in \{0, 1\}^T \end{aligned}$$

3.1 Upper Bound: A Quadratic Formulation

To derive an upper bound, we will need to find an approximately-minimal revenue curve. We consider a straightforward implementation of (14) to do this. For $n \in \mathbb{N}$, we take the uniform gauge given by $q_i = (i - 1)/n$ for $1 \leq i \leq n + 1$, and consider each case when the peak of the revenue curve is in $[q_k, q_{k+1}]$ for $1 \leq k \leq n$. To evaluate the Riemann sum, mark the midpoint of each interval, $\bar{q}_i = (q_i + q_{i+1})/2$. Then to approximate our Riemann integral, for each $I(i, j)$ we will evaluate the function ϕ_R at $(\bar{q}_i, \bar{q}_j)$. So we set:

$$f_{ii} = \frac{R(q_i) + R(q_{i+1})}{2} \qquad \forall 1 \leq i \leq n$$

$$f_{ij} = \frac{R(q_i) + R(q_{i+1})}{2} w(\bar{q}_i, \bar{q}_j) + \frac{R(q_j) + R(q_{j+1})}{2} (1 - w(\bar{q}_i, \bar{q}_j)) \qquad \forall 1 \leq j < i \leq n$$

If we evaluate the resulting optimization problem, the constraints (9) and (10) tend to "chip off" the peak of the revenue curve in the primal solutions. This is unlikely to be a feature of an actual minimal revenue curve, so we will convert the constraints (9) and (10) into a single equality constraint, at the cost of increasing the size of the index set J by one. First observe that

$$\max\{R(q_{OPT}), R(q_{OPT+1})\} \geq \max\{q_{OPT}/q_{OPT+1}, (1 - q_{OPT+1})/(1 - q_{OPT})\}$$

for any feasible solution $(\vec{R}, \vec{w})$. Let $\vec{R}^* = \max\{R(q_{OPT}), R(q_{OPT+1})\}^{-1}\vec{R}$. Then for any $1 \le j \le i \le n$, by the homogeneity of f_{ij} in $\vec{R}$ it can be shown that

$$A(i,j) \cdot f_{ij}(\vec{R}, \vec{w}) \ge \frac{n-1}{n+1} \cdot A(i,j) \cdot f_{ij}(\vec{R}^*, \vec{w}). \tag{15}$$

Now, $\vec{R}^*$ has either $R(q_{OPT}) = 1$ or $R(q_{OPT+1}) = 1$. So we consider imposing such an equality constraint in our optimization programs to normalise the maximum of the revenue curve, dropping the optimality constraints (9) and (10) from our optimization program. We are also able to drop the constraint (11), since it is implied by $R(q_k) = 1$, $R \le 1$, and the concavity constraints (7).

Finally, note that with the uniform gauge, $A(i,j) = 1/n^2$, for any $1 \le j \le i \le n$. Thus, our Riemann sum minimization program is:

$$\begin{aligned} \hat{\alpha}(n) = \min_{1 \le k \le n+1} \min_{\vec{R}, \vec{w}} \quad & \sum_{1 \le i \le n} \frac{1}{n^2} \cdot f_{ii}(\vec{R}, \vec{w}) + 2 \cdot \sum_{1 \le j < i \le n} \frac{1}{n^2} \cdot f_{ij}(\vec{R}, \vec{w}) \\ \text{subject to} \quad & (7), (12), (13) \\ & R(q_k) = 1 \\ & \vec{R} \in [0,1]^{n+1} \\ & \vec{w} \in \{0,1\}^{\binom{n}{2}} \end{aligned} \tag{16}$$

Intuitively, since the factor $(n-1)/(n+1)$ in (15) goes to 1 as n grows large, this program should be able to approximate α. To be able to prove this, we emphasize an important *monotonicity* property of $w(x, y)$: it is non-decreasing in the first argument and non-increasing in the second argument:

Lemma 2. *Suppose that R is concave and non-negative on $[0, 1]$, and w is determined as in Lemma 1. Then for any $x, y \in [0, 1]^2$ such that $x > y$:*

(i) If $x' > x$, then $w(x', y) \ge w(x, y)$.
(ii) If $x > y' > y$, then $w(x, y') \le w(x, y)$.

These monotonicity properties of w imply that only few $w(\bar{q}_i, \bar{q}_j)$'s may be "misspecified". In particular, for some revenue curve R, the objective contributions $A(i,j) \cdot f_{ij}(\vec{R}, \vec{w})$ all underestimate their corresponding terms in 6 except for a vanishing fraction of product intervals $I(i, j)$:

Lemma 3. *Let $(\vec{R}, \vec{w})$ be a feasible solution of (16), and let R be a revenue curve agreeing with $\vec{R}$ on the gauge $(q_i)_{1 \le i \le n+1}$. Then for at least $\binom{n}{2} - 2n + 3$ many pairs (i, j) such that $1 \le j < i \le n$, w is a constant function on $I(i, j)$. In particular, for such pairs (i, j):*

$$\int_{(x,y) \in I(i,j)} \phi_R(x, y) \cdot d(x, y) \ge A(i,j) \cdot f_{ij}(\vec{R}, \vec{w}),$$

with equality if R is the linear interpolation of $\vec{R}$.

To prove Lemma 3 we define a notion of *constantness* for w on any $I(i,j)$ with $j < i$. We will say that the pair (i,j) is 1-*definite* if $w(q_{i+1}, q_j) = w(q_i, q_{j+1}) = 1$, and 0-*definite* if $w(q_{i+1}, q_j) = w(q_i, q_{j+1}) = 1$. Else, by the monotonicity of w, it must be that $w(q_{i+1}, q_j) = 1$ and $w(q_i, q_{j+1}) = 0$; we call such a pair (i,j) *indefinite*. Then, by the monotonicity of w, it holds that:

1. If (i,j) is 1-definite then $(i+1, j-1)$ is 1-definite.
2. If (i,j) is 0-definite and $j+1 < i-1$, then $(i-1, j+1)$ is 0-definite.
3. If (i,j) is indefinite then $(i+1, j-1)$ is 1-definite, and if also $j+1 < i-1$, then $(i-1, j+1)$ is 0-definite.

Thus at most $\binom{n}{2} - 2n + 3$ pairs (i,j) such that $1 \le j < i \le n$ may be indefinite, from which the proof follows. Consideration of definiteness also allows us to provide an explicit convergence result for $\hat{\alpha}(n)$:

Theorem 1. *For each* $n \ge 1$, $\hat{\alpha}(n) - \frac{2}{n-1} - \frac{5n-6}{n^2} \le \alpha \le \hat{\alpha}(n) + \frac{5n-6}{n^2}$.

The convergence result suggests a natural optimization scheme to find an approximately minimal distribution – we linearize the terms of the form $R(q_\ell) \cdot w(\bar{q}_i, \bar{q}_j)$ in the objective and the constraints, adding in the constraints from the second-order Sherali-Adams lift of (16) that include such terms. In particular, we add in the constraints

$$\begin{aligned} Rw(\ell,i,j) &\ge 0 \\ w(\bar{q}_i,\bar{q}_j) - Rw(\ell,i,j) &\ge 0 \\ R(\bar{q}_\ell) - Rw(\ell,i,j) &\ge 0 \\ -R(\bar{q}_\ell) - w(\bar{q}_i,\bar{q}_j) + Rw(\ell,i,j) &\ge -1 \qquad \forall 1 \le j < i \le n,\ \ell \in \{i,j\}, \end{aligned} \tag{17}$$

where $Rw(\ell,i,j)$ is a variable representing the product $R(\bar{q}_\ell) \cdot w(\bar{q}_i, \bar{q}_j)$. We then replace the product terms in constraints (12), (13) and in the objective with the corresponding linearized variable. Note that these constraints imply that $Rw(\ell,i,j) = R(\bar{q}_\ell) \cdot w(\bar{q}_i, \bar{q}_j)$ whenever $w(\bar{q}_i, \bar{q}_j)$ is $\{0,1\}$-valued. Therefore, the mixed-integer LP formulation is exact.

Finally, we impose the monotonicity constraints implied by Lemma 2. Even though these constraints are *redundant* for our formulation, we have found that the inclusion of monotonicity constraints improves the performance of the solver. Then our approximate MILP has the following form:

$$\begin{aligned} \min_{1 \le k \le n+1} \min_{\vec{R}, \vec{w}, Rw} \quad & \sum_{1 \le i \le n} \frac{1}{n^2} \cdot R(\bar{q}_i) + \sum_{1 \le j < i \le n} \frac{2}{n^2} \cdot (Rw(i,i,j) + R(\bar{q}_j) - Rw(j,i,j)) \\ \text{subject to} \quad & (7), (12), (13), 17 \\ & w(\bar{q}_i, \bar{q}_j) \le w(\bar{q}_{i+1}, \bar{q}_j)\ \forall 1 \le j < i < n \\ & w(\bar{q}_i, \bar{q}_j) \ge w(\bar{q}_i, \bar{q}_{j+1})\ \forall 1 \le j < i+1 \le n \\ & R(q_k) = 1 \\ & \vec{R} \in [0,1]^{n+1} \\ & \vec{w} \in \{0,1\}^{\binom{n}{2}} \\ & Rw \in [0,1]^{2 \times \binom{n}{2}} \end{aligned} \tag{18}$$

3.2 Lower Bound: A Cubic Formulation

While the MILP (18) does provide certifiable lower bounds for α by Theorem 1, the exponential nature of the problem kicks in before we can certify any improvement on the lower bound of $\simeq .558$ provided Daskalakis and Zampetakis [4]. We work around this problem by considering a cubic program which, given a gauge, lower bounds the contribution of any area element. We first fix our gauge: for the general formulation of the problem (14), we find a set of gauges $(\vec{q}^k)_{k \in I}$ with prescribed *optimal intervals* $[q^k_{OPT^k}, q^k_{OPT^k+1}]$ such that $\cup_{k\in I}[q^k_{OPT^k}, q^k_{OPT^k+1}] = [0,1]$. As evidenced by (15), we will want the freedom to pick $q^k_{OPT^k+1} - q^k_{OPT^k}$ small for each gauge $\vec{q}^k$ to minimize the loss from relaxing the optimality constraint to (9) and (10). To this end, for some $N \in \mathbb{N}$ "significantly larger" than n, we will set $J = \{1, 2, ..., N\}$ and

$$q^k_{OPT^k} = \frac{k-1}{N}, \quad q^k_{OPT^k+1} = \frac{k}{N}.$$

Then by (15), we expect degredations on the quality of the lower bound caused by the optimality constraints to be of order $\sim 1/N$ as we impose larger N. Note that this only comes at a linear cost of having to compute N MILPs.

Next, we need to decide on where to evaluate each $w(q_i, q_j)$. By Lemma 2, to decide on the definiteness of an area element $[q_i, q_{i+1}] \times [q_j, q_{j+1}]$ for $1 \le j < i \le n$, we need to check $w(q_{i+1}, q_j)$ and $w(q_i, q_{j+1})$. Due to this constraint, we also need to assign a value to w on $(q_i, q_i)_{1 \le i \le n+1}$. The defining constraints (12) and (13) become degenerate on such points. Instead we will opt to always fix $w(q_i, q_i) = 0$, as such an assignment respects monotonicity and we wish to avoid adding even more binary variables.

We now derive lower bounds on the contribution of each area element. Lower bounding the contribution of diagonal area elements is straightforward:

Lemma 4. *Suppose $R : [0,1] \to \mathbb{R}_+$ is concave and R attains its maximum in $I^k(OPT^k)$. Then the following hold:*

(i) If $i < OPT^k$, then $\int_{(x,y)\in I^k(i,i)} \phi_R(x,y) \cdot d(x,y) \ge A_k(i,i)\left(\frac{2R(q^k_i)}{3} + \frac{R(q^k_{i+1})}{3}\right)$.

(ii) If $i = OPT^k$, then $\int_{(x,y)\in I^k(i,i)} \phi_R(x,y) \cdot d(x,y) \ge 0$.

(iii) If $i > OPT^k$, then $\int_{(x,y)\in I^k(i,i)} \phi_R(x,y) \cdot d(x,y) \ge A_k(i,i)\left(\frac{R(q^k_i)}{3} + \frac{2R(q^k_{i+1})}{3}\right)$.

Next, we provide lower bounds on off-diagonal area elements conditional on their *definiteness*:

Lemma 5. *Suppose $R : [0,1] \to \mathbb{R}_+$ is concave, $1 \leq j < i \leq n$, and R attains its maximum in $I^k(OPT^k)$. Then the following hold:*

(a) If the pair (i,j) is 1-definite, or if the pair (i,j) is indefinite and $j > OPT^k$, then:

$$\int_{(x,y)\in I(i,j)} \phi_R(x,y) \cdot d(x,y) \geq A_k(i,j) \cdot \frac{R(q_i^k) + R(q_{i+1}^k)}{2}.$$

(b) If the pair (i,j) is 0-definite, or if the pair (i,j) is indefinite and $i < OPT^k$, then:

$$\int_{(x,y)\in I(i,j)} \phi_R(x,y) \cdot d(x,y) \geq A_k(i,j) \cdot \frac{R(q_j^k) + R(q_{j+1}^k)}{2}.$$

(c) For any pair (i,j) – in particular if (a) and (b) do not hold – we have

$$\int_{(x,y)\in I(i,j)} \phi_R(x,y) \cdot d(x,y) \geq A_k(i,j) \cdot \mathbb{E}[\min\{\underline{R}(x), \underline{R}(y)\} | (x,y) \in I(i,j)],$$

where $\underline{R}$ is the minimum concave and non-negative function on $[0,1]$ satisfying (9) and (10).

This allows us to write a cubic expression which lower bounds the contribution from an off-diagonal area element to the revenue:

Corollary 1. *Suppose $R : [0,1] \to \mathbb{R}_+$ is concave, $1 \leq j < i \leq n$, and and R attains its maximum in $I^k(OPT^k)$. Let $f_{1ij}(\vec{R}), f_{0ij}(\vec{R}), f_{\iota ij}(\vec{R})$ be respectively the lower bounds on the revenue contribution from the area element $I(i,j)$, conditional respectively on the pair (i,j) being 1-definite, 0-definite or indefinite as in Lemma 5. Then:*

$$\begin{aligned}\int_{(x,y)\in I(i,j)} \phi_R(x,y) \cdot d(x,y) \geq\ & A_k(i,j) f_{1ij}(\vec{R}) w(q_{i+1}^k, q_j^k) w(q_i^k, q_{j+1}^k) \\ &+ A_k(i,j) f_{0ij}(\vec{R})(1 - w(q_{i+1}^k, q_j^k))(1 - w(q_i^k, q_{j+1}^k)) \\ &+ A_k(i,j) f_{\iota ij}(\vec{R}) w(q_{i+1}^k, q_j^k))(1 - w(q_i^k, q_{j+1}^k)) \\ &+ A_k(i,j) f_{\iota ij}(\vec{R})(1 - w(q_{i+1}^k, q_j^k)) w(q_i^k, q_{j+1}^k)\end{aligned}$$

Note that the fourth term of the lower bound in Corollary 1 is *redundant* – it will equal zero for any integral solution for $\vec{w}$ by monotonicity. Still, the term allows us to gain some more strength in the LP relaxation of the program, so we retain it in our final formulation.

Given a gauge $(\vec{q}^k)$, a lower bound function $f_{ij}(\vec{R}, \vec{w})$ for each $1 \leq j \leq i \leq n$ is then provided by Lemma 4 and Corollary 1. To linearize the objective function, we again consider incorporating the relevant variables from the degree 3 Sherali-Adams lift of the problem, with their defining inequalities.

For the objective, we consider variables:

1. w^2 corresponding to terms of type $w(q_{i+1}, q_j) \cdot w(q_i, q_{j+1})$,
2. Rw corresponding to terms of type $R(q_\ell) \cdot w(q_{i+1}, q_j)$ or $R(q_\ell) \cdot w(q_i, q_{j+1})$, and
3. Rw^2 corresponding to terms of type $R(q_\ell) \cdot w(q_{i+1}, q_j) \cdot w(q_i, q_{j+1})$.

For w^2, the Sherali-Adams inequalities are then:

$$\begin{aligned} \forall 1 \le j < i \le n, \\ -w(q_{i+1}, q_j) + w^2(i,j) \le 0 \\ -w(q_i, q_{j+1}) + w^2(i,j) \le 0 \\ w(q_{i+1}, q_j) + w(q_i, q_{j+1}) - w^2(i,j) \le 1 \end{aligned} \tag{19}$$

In turn, for Rw, the Sherali-Adams inequalities are given:

$$\begin{aligned} \forall 1 \le j < i \le n, \forall \ell \in \{i, i+1, j, j+1\}, \forall (s,t) \in \{(i+1, j), (i, j+1)\}, \\ -R(q_\ell) + Rw(\ell, s, t) \le 0 \\ -w(q_s, q_t) + Rw(\ell, s, t) \le 0 \\ R(q_\ell) + w(q_s, q_t) - Rw(\ell, s, t) \le 1 \end{aligned} \tag{20}$$

Finally, we have the Sherali-Adams inequalities for Rw^2:

$$\begin{aligned} \forall 1 \le j < i \le n, \forall \ell \in \{i, i+1, j, j+1\}, \\ -w^2(i,j) + Rw^2(\ell,i,j) \le 0 \\ -Rw(\ell, i+1, j) + Rw^2(\ell,i,j) \le 0 \\ -Rw(\ell, i, j+1) + Rw^2(\ell,i,j) \le 0 \\ -w(q_{i+1}, q_j) + w^2(i,j) + Rw(\ell, i+1, j) - Rw^2(\ell,i,j) \le 0 \\ -w(q_i, q_{j+1}) + w^2(i,j) + Rw(\ell, i, j+1) - Rw^2(\ell,i,j) \le 0 \\ -R(q_\ell) + Rw(\ell, i+1, j) + Rw(\ell, i, j+1) - Rw^2(\ell,i,j) \le 0 \\ R(q_\ell) + w(q_{i+1}, q_j) + w(q_i, q_{j+1}) ... \\ ... - Rw(\ell, i+1, j) - Rw(\ell, i, j+1) - w^2(i,j) + Rw^2(\ell,i,j) \le 1 \end{aligned} \tag{21}$$

For the defining constraints for w, (12) and (13), we linearize terms of the form $R(q_\ell) \cdot w(\bar{q}_i, \bar{q}_j)$ to $Rw(\ell, i, j)$, coinciding with the previously defined Rw term whenever necessary. These terms have defining inequalities:

$$\begin{aligned} \forall 1 \le j < i \le n, \forall \ell \in \{i, j\}, \\ -R(q_\ell) + Rw(\ell, i, j) \le 0 \\ -w(q_i, q_j) + Rw(\ell, i, j) \le 0 \\ R(q_\ell) + w(q_i, q_j) - Rw(\ell, i, j) \le 1 \end{aligned} \tag{22}$$

Finally, we again impose the monotonicity constraints for w to improve the performance of our solver. This implies that our lower bounding MILP has the following form:

$$\min_{k\in\{1,2,\ldots,N\}} \min_{\vec{R},\vec{w},w^2,Rw,Rw^2} \sum_{1\leq i\leq n} A_k(i,i)\cdot f_{ii}(\vec{R},\vec{w}) + 2\cdot \sum_{1\leq j<i\leq n} A_k(i,j)\cdot f_{ij}(\vec{R},\vec{w}) \tag{23}$$

$$\begin{aligned}
\text{subject to}\quad & (7),(9),(10),(11),(12),(13),(19),(20),(21),(22)\\
& w(\bar{q}_i,\bar{q}_j)\leq w(\bar{q}_{i+1},\bar{q}_j)\ \forall 1\leq j<i<n\\
& w(\bar{q}_i,\bar{q}_j)\geq w(\bar{q}_i,\bar{q}_{j+1})\ \forall 1\leq j<i+1\leq n\\
& w(q_i,q_i)=0\ \forall 1\leq i\leq n\\
& \vec{R}\in[0,1]^{n+1}\\
& \vec{w}\in\{0,1\}^{\binom{n+1}{2}}\\
& w^2,Rw,Rw^2\geq 0
\end{aligned}$$

We still need to explicitly pick a gauge for (23) for each $k\in\{1,2,...,N\}$. An immediate candidate is the "approximately uniform" gauge. For such a gauge, when $k=1$, we divide $[1/N,1]$ into $n-1$ equal size intervals. Likewise, when $k=N$, we divide $[0,1-1/N]$ into $n-1$ equal size intervals. If instead $1<k<N$, we choose m such that:

$$m\in\arg\min_{1<\mu<n-1}\left|\frac{k-1}{N\cdot\mu}-\frac{N-k}{N\cdot(n-\mu-1)}\right|$$

We then divide $[0,(k-1)/N]$ into m equal size intervals, and $[k/N,1]$ into $n-m-1$ equal size intervals.

While straightforward, this choice of gauge is problematic. In particular, the approximately uniform gauge results in "jagged" behaviour for the objective values of (23) parametrised by q_{OPT} when (approximately) $q_{OPT}\in[0,.2]$. The upwards kinks occur roughly when $k\to k+1$ causes $m\to m+1$. This implies that, for some initial segment of $[0,1]$, the quality of our lower bounds improve when we add more intervals in the segment $[0,(k-1)/N]$. So we consider a modification of the approximately uniform gauge, *square weighing* the gauge on $[0,1/2]$. In particular, for $k<N/2$, we instead choose m such that

$$m\in\arg\min_{1<\mu<n-1}\left|\frac{k-1}{N\cdot\mu^2}-\frac{N-k}{N\cdot(n-\mu-1)^2}\right|.$$

Unfortunately, using this square-weighted gauge results in considerable slowdown of computations, when $k\lesssim N/10$. For this reason, we lower the relative efficiency guarantees of our solver when $k\leq N/10$. This results in a jump "discontinuity" in our computed revenue *guarantees*, but the derived lower bounds are smoothed on the initial segment of $[0,1]$ by the weighing and the quality of the lower bounds we obtain increase. The reason for why such a weighing works is unknown to us; indeed, we found the square-weighing rule by trial-and-error.

4 Results: Lower and Upper Bounds

We are now ready to present lower and upper bounds on the performance of the ERM mechanism with two samples. We compute (18) and (23) using MATLAB

+ CPLEX as our solver of choice[1]. We compute (18) for $n = 80$, obtaining an approximate *conditional*[2] *minimum expected revenue curve*. Each computation for $k \in \{1, ..., 81\}$ also provides us with an approximately minimal distribution; given primal solution (R, w, Rw) to (18) for k, we consider the minimum concave function R_k such that $R_k(j/n) = R(j/n)$ for any $j = \{0, 1, ..., 80\}$. By numerically evaluating the integral (4) in Mathematica for each such R_k, we obtain upper bounds on the performance of the ERM mechanism.

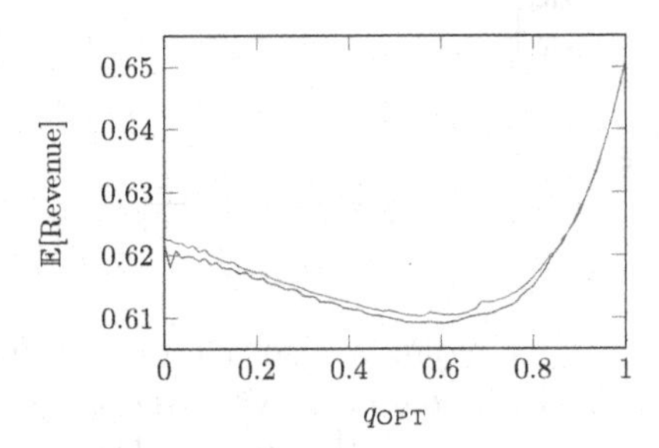

Fig. 1. Results of computation of (18) for $n = 80$. The objective values of (18) are shown in blue, while upper bounds obtained from primal solutions are shown in orange. (Color figure online)

The results of this computation is shown in Fig. 1. Numerically computing the integral (4) for each primal solution we obtain, the internal error estimates provided by Mathematica are $\leq 10^{-6}$ for each integral approximation. We find that our primal solution for $n = 80$ and $q_{\text{OPT}} = 44/80$ provides a regular revenue curve for which the ERM mechanism obtains $\leq .61035$ times the optimal revenue. Also as seen in Fig. 2, our primal solutions show that minimal distributions are closely approximated by piecewise linear functions on at most three intervals (3-piecewise linear functions).

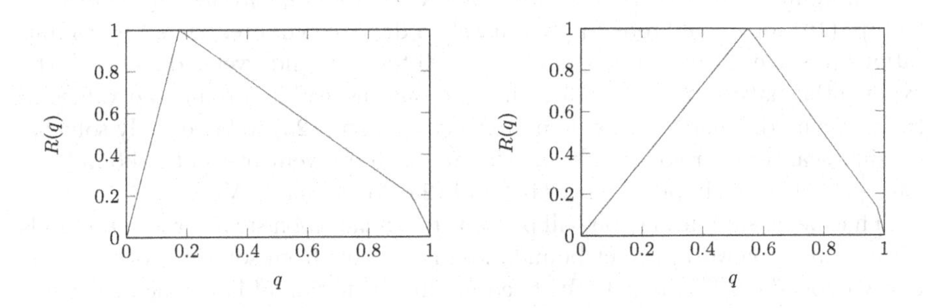

Fig. 2. Revenue curves of approximately minimal distributions (conditional on q_{OPT}) obtained from (18) for $n = 80$ and $k = 15, 45$.

[1] Our code is available at meteahunbay.com/files/code-twoSampleMILP.zip.
[2] On $\arg\max_{q \in [0,1]} R(q)$.

Finally, we compute (23) for $n = 50$ and $N = 500$, running our solver at 99.8% relative tolerance for $k > 50$ and 99% relative tolerance for $k \leq 50$. The results of the computation are shown in Fig. 3. Our results show that the ERM mechanism guarantees an expected revenue $\geq .5914$ times the optimal revenue.

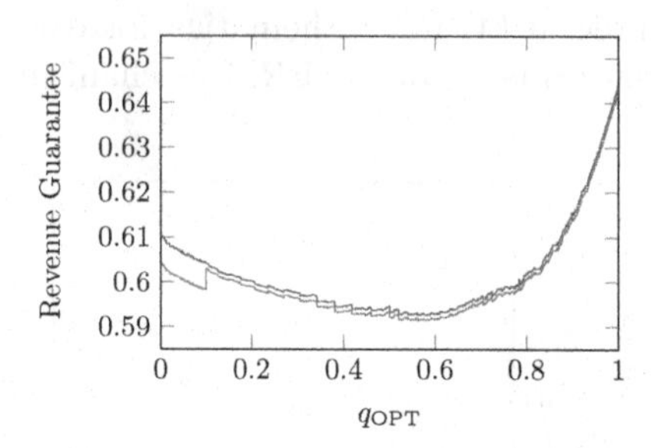

Fig. 3. Results of computation of (23) for $n = 50$ and $N = 500$. The blue line shows the value of the objective for primal solutions found, while the orange line shows conditional lower bounds on the expected revenue of the auction. These results are corrected for tolerance of the MILP solver. (Color figure online)

5 Conclusion

In this paper, we presented an MILP formulation to inspect the expected revenue of the ERM mechanism in the single item, single buyer, two sample setting. Working within this formulation has allowed us to greatly improve upon the known upper and lower bounds of the expected revenue guarantees of the ERM mechanism with two samples, and provided us with insights on what minimum revenue distributions may look like.

Despite the sheer number of binary variables involved, computations to certify our bounds were relatively *cheap* – on a ASUS ROG Zephyrus M (GU502GV) laptop, (18) for $n = 80$ took approximately a day to compute, while the computations to solve (23) for $n = 50$ and $N = 500$ took around twelve days. Still, the exponential nature of the problem had become noticable around the values of (n, N) we used. Therefore, we do not expect (18) and (23) to be feasibly solvable for significantly finer gauges, disallowing major improvements on the bounds we have provided by simply solving (18) and (23) for larger n, N.

That being said, it may be still possible to extract even stronger lower bounds within our framework. Lower bounds we may derive from solutions of (18) currently depend on Theorem 1. For fixed n, our estimation of how much the value of (18) overestimates α is $2/(n-1) + (5n-6)/n^2$. For $n = 80$, this error estimate is $\lesssim .0869$, which means that our computations for (18) can only certify a lower bound of .5210. However, Fig. 1 hints that the actual error might be much smaller than our estimate. Improving this estimate could then help certify stronger lower bounds on α.

Finally, we note that our formulation should extend naturally to the setting with ≥3 samples. However, in such an extension, the number of binary variables would blow up exponentially as the number of samples increases for fixed number of intervals, n. This implies that the extension of (18) and (23) to a setting with ≥3 samples might not be feasibly solvable. Still, for settings in which the performance of solvers do not depreciate too much, our techniques should be readily applicable.

References

1. Alon, N., Babaioff, M., Gonczarowski, Y.A., Mansour, Y., Moran, S., Yehudayoff, A.: Submultiplicative Glivenko-Cantelli and uniform convergence of revenues. In: Proceedings of the 31st International Conference on Neural Information Processing Systems, NIPS 2017, pp. 1655–1664 (2017)
2. Babaioff, M., Gonczarowski, Y.A., Mansour, Y., Moran, S.: Are two (samples) really better than one? In: Proceedings of the 2018 ACM Conference on Economics and Computation, EC 2018, p. 175 (2018)
3. Cole, R., Roughgarden, T.: The sample complexity of revenue maximization. In: Proceedings of the 46th Annual ACM Symposium on Theory of Computing, STOC 2014, pp. 243–252 (2014)
4. Daskalakis, C., Zampetakis, M.: More revenue from two samples via factor revealing SDPs. In: Proceedings of the 21st ACM Conference on Economics and Computation, EC 2020, pp. 257–272 (2020)
5. Devanur, N.R., Huang, Z., Psomas, C.A.: The sample complexity of auctions with side information. In: Proceedings of the 48th Annual ACM Symposium on Theory of Computing, STOC 2016, pp. 426–439 (2016)
6. Dhangwatnotai, P., Roughgarden, T., Yan, Q.: Revenue maximization with a single sample. In: Proceedings of the 11th ACM Conference on Electronic Commerce, EC 2010, pp. 129–138 (2010)
7. Fu, H., Immorlica, N., Lucier, B., Strack, P.: Randomization beats second price as a prior-independent auction. In: Proceedings of the 16th ACM Conference on Economics and Computation, EC 2015, p. 323 (2015)
8. Gonczarowski, Y.A., Nisan, N.: Efficient empirical revenue maximization in single-parameter auction environments. In: Proceedings of the 49th Annual ACM SIGACT Symposium on Theory of Computing, STOC 2017, pp. 856–868 (2017)
9. Gonczarowski, Y.A., Weinberg, S.M.: The sample complexity of up-to-ϵ multi-dimensional revenue maximization. In: IEEE 59th Annual Symposium on Foundations of Computer Science, FOCS 2018, pp. 416–426 (2018)
10. Guo, C., Huang, Z., Zhang, X.: Settling the sample complexity of single-parameter revenue maximization. In: Proceedings of the 51st Annual ACM SIGACT Symposium on Theory of Computing, STOC 2019, pp. 662–673 (2019)
11. Huang, Z., Mansour, Y., Roughgarden, T.: Making the most of your samples. In: Proceedings of the 16th ACM Conference on Economics and Computation, EC 2015, pp. 45–60 (2015)
12. Myerson, R.B.: Optimal auction design. Math. Oper. Res. **6**(1), 1–158 (1981)
13. Roughgarden, T., Schrijvers, O.: Ironing in the dark. In: Proceedings of the 2016 ACM Conference on Economics and Computation, EC 2016, pp. 1–18 (2016)

The Price of Stability of Envy-Free Equilibria in Multi-buyer Sequential Auctions

Mete Şeref Ahunbay[1](✉), Brendan Lucier[2], and Adrian Vetta[1]

[1] McGill University, Montreal, Canada
mete.ahunbay@mail.mcgill.ca, adrian.vetta@mcgill.ca
[2] Microsoft Research New England, Cambridge, USA
brlucier@microsoft.com

Abstract. Motivated by auctions for carbon emission permits, we examine the social welfare of *multi-buyer sequential auctions* for identical items. Assuming the buyers have weakly-decreasing incremental (concave) valuation functions, we study subgame perfect equilibria of a repeated second-price auction with n buyers and T time periods. We show that these auctions admit envy-free subgame-perfect equilibria that $(1 - 1/e)$-approximate the optimal welfare. The equilibria we construct have a natural interpretation: each bidder guarantees for herself the best outcome she could obtain if all other bidders were non-strategic. Without the envy-freeness condition, the price of anarchy can be as bad as $\Theta(1/T)$ even when restricting to equilibria that satisfy a no-overbidding condition. We also consider the restricted class of envy-free subgame perfect equilibria that survive iterated deletion of weakly undominated strategies. For this class of equilibria we prove constant bounds on the price of anarchy for three settings with differing levels of market competitiveness, based on the number of buyers with oligopsony power (market power).

1 Introduction

We study sequential auctions with identical items. There are T time periods, and in each period a single item is sold to n bidders via a second-price (or first-price) auction. The same set of bidders participates in each round and a bidder can win multiple items. Our motivation is that sequential auctions with identical items form the basis of cap-and-trade systems and other emission license markets. Thus, to assess the potential effectiveness of these system in combating climate change, a central task is to quantify the quality of outcomes in sequential auctions. In particular, our aim in this paper is to evaluate the structure and efficiency of equilibria when the buyers have weakly decreasing incremental (concave) valuation functions.

I. Caragiannis and K. A. Hansen (Eds.): SAGT 2021, LNCS 12885, pp. 18–33, 2021.
https://doi.org/10.1007/978-3-030-85947-3_2

1.1 Motivation: Carbon Pricing

One motivation for this work is the use of carbon pricing in reducing climate change. For example, in a cap-and-trade system carbon permits are sold by auction repeatedly over time. Ergo, to understand such mechanisms we must understand sequential auctions with identical items (permits). Indeed, the auction mechanism studied in this paper forms the basis of both the *Regional Greenhouse Gas Initiative (RGGI)*, consisting of Eastern states in the United States, and the *Western Climate Initiative (WCI)*, including California and Québec. Still, knowledge of carbon pricing markets is not required for the game-theoretic auction analyses presented in this paper.

Formally, we want to evaluate the price of stability and the price of anarchy when the number of bidders n and the number of time periods T gets large.[1] There are multiple factors that may cause inefficiencies in sequential auctions of carbon licenses. The first is that the carbon market involves a relatively small number of very large participants. In such oligopolistic markets, the participants have *price-making* abilities via strategic bidding. The extent of these abilities and the magnitude of any resultant deleterious effects varies with the specific market. Second, carbon permit auctions take place every quarter and this has potentially serious consequences with respect to strategic behaviours. For example, the repeated nature of cap-and-trade auction allows firms to experiment with strategies. This opens up the possibility of *learning* bidding strategies that benefit it but worsen the performance of the mechanism. Potentially, this could mean that a cap-and-trade system initially works well but its performance degrades as the firms learn to play the mechanism. In addition, it is well-known that repeated auctions are more vulnerable to other problematic behaviours, such as signaling and implicit collusion. To better understand these effects and their welfare impact, we study classes of subgame perfect equilibria in sequential auctions.

1.2 Background on Sequential Auctions

Sequential auctions with identical goods have been long-studied in the economics community. Early experimental and empirical studies of such sequential auctions were given by Pitchik and Schotter [15] and by Ashtenfelter [3] who was motivated by wine and art auctions. These papers highlighted the additional complexity that arises in multi-period (sequential) auctions compared to one-period (static) auctions. Indeed, Ashtenfelter begins with the memorable line "At the first wine auction I ever attended, I saw the repeal of the law of one price." Specifically, the selling price of identical lots of wine differed over the course of just "a matter of seconds"; in fact, the prices decreased over time. This surprising property, known as the *declining price anomaly* [11], has since been observed in numerous sequential auctions [4,7,8,14,17,18]. Furthermore, Gale

[1] The *price of stability* is the worst case ratio over all instances between the welfare of the *best* equilibrium (for a specified class of equilibria) and the optimal welfare, and the *price of anarchy* is the worst case ratio over all instances between the welfare of the *worst* equilibrium (for a specified class of equilibria) and the optimal welfare.

and Stegeman [10] presented a model of sequential auctions (which we use in this paper) and proved that, for the case of two buyers, the observational results above were not coincidental: at any subgame perfect equilibrium that survives the iterative deletion of weakly dominated strategies, the price weakly declines over time.

There has also been a great deal of interest in the computer science community in sequential auctions for both identical [5,6] and non-identical goods [9,13]. Of particular relevance, Ahunbay et al. [1,2] study the *price of anarchy* for 2-buyer sequential auctions for identical goods. Under the model of Gale and Stegeman [10], they prove that the price of anarchy is at most $1 - \frac{1}{e}$, assuming weakly-decreasing incremental (concave) valuation functions.

In contrast, little is known about sequential auctions with more than two buyers. The reader may ask: what is special about the two-buyer setting? It turns out the 2-buyer is much simpler to analyze because, from the perspective of each buyer, at any time period there are only two outcomes: win or lose. Thus, at each node in the game tree for the auction, we can associate a specific *marginal value* for winning to each buyer. Thus, each subgame corresponds to a standard auction with independent values [10,13]. Consequently, applying backwards induction on the game tree then allows us to compute an equilibrium. For example, in the second-price setting, the unique subgame perfect equilibrium that survives the iterative deletion of weakly dominated strategies corresponds to each buyer bidding their marginal value at each game node. Unfortunately, for more than two buyers this argument falls apart. From the perspective of any agent, there are no longer just two outcomes in each round: losing to agent i is not the same as losing to agent j, since the winner's identity might change what happens in future rounds. Agents can therefore implicitly impose externalities on each other, and the marginal value of winning is not well-defined for any buyer. In particular, each subgame now corresponds to a auction with inter-dependent values [12,13]. This makes analyzing the multi-buyer setting much more complicated. Indeed, unlike 2-buyer auctions, prices need not weakly decrease over time in sequential auctions with three or more buyers; see Narayan et al. [12].

1.3 Our Results

In Sect. 2, we present the formal model of sequential auctions with identical goods where each buyer has a weakly concave valuation function.

Our first result is a negative one. In Sect. 2, we prove the *price of anarchy* is $\Theta(1/T)$ for second-price sequential auctions with T time periods. The example we construct has the property that agents do not overbid their incremental values in any round, so imposing a no-overbidding assumption is not enough to bypass these bad equilibria. This motivates the study of envy-free second-price equilibria. These correspond to equilibria of both first-price and second-price sequential auctions [13]. Our main result, given in Sect. 4, is that the *price of stability* for envy-free equilibria in sequential auctions for identical goods is at least $1 - \frac{1}{e}$.

In fact, the envy-free equilibrium we construct has a natural economic interpretation. We suppose that each buyer considers a hypothetical scenario where all other participants bid truthfully every round, and then decides on an optimal response. This response might involve *strategic demand reduction* (from the theory of multi-unit auctions), where a buyer uses their oligopsony power (i.e., market power) to influence their price. In our envy-free equilibrium, each agent *simultaneously* obtains at least the utility they would obtain from this hypothetical scenario. To show this we derive structural results and strategic concepts of importance in analyzing envy-free equilibria, and find that agents can use greedy bidding strategies to implement their oligopsonist outcome starting from any round of the auction, and that this results in at least a $1 - \frac{1}{e}$ factor of the optimal welfare (and this bound is tight). The main result follows by constructing an envy-free equilibrium related to these greedy bidding strategies.

In Sect. 5, we consider a different class of subgame perfect equilibria that survive the iterated deletion of weakly dominated strategies and satisfy no-incremental-overbidding. There might not be a unique such equilibrium when there are three or more bidders. We therefore present bounds on the *price of anarchy* for equilibria in this class. Specifically we present results for distinct cases related to the number of buyers with oligopsony power. First, we show the price of anarchy is 1 if the auction is suitably "competitive" and none of the buyers have oligopsony power. Second, we prove that if there is a unique buyer with oligopsony power then the price of anarchy is at least $1 - \frac{1}{e}$. Third, we present constant price of anarchy bounds for a specific family of valuation profiles where multiple buyers have oligopsony power.

2 Preliminaries

In this section, we present the complete information model for sequential multiunit auctions, due to Gale and Stegeman [10], and our concept of equilibrium. In a sequential multiunit auction, there is a set $[T]$ of $T \geq 1$ identical items to be sold and a set $[n]$ of $n \geq 1$ buyers. Each buyer $i \in [n]$ has a *valuation function* $V_i : [T] \cup \{0\} \rightarrow \mathbb{R}_+$, where $V_i(k)$ is buyer i's value for obtaining k items. We assume that valuation functions are non-decreasing (free-disposal) and normalized so that $V_i(0) = 0$. We also assume that each V_i is weakly concave, i.e. buyers' valuations exhibit diminishing incremental returns. We define buyer i's incremental valuation function as $v_i(k) = V_i(k) - V_i(k-1)$, denoting the value buyer i has for obtaining a k'th additional item. Note that, since V_i is concave, v_i is weakly decreasing.

The items are sold in a sequential auction over T rounds. In each round a single item is allocated via a sealed-bid auction. More specifically, in each round $t \in [T]$, each buyer i submits a real-valued bid b_{it}. An auction rule is then applied to determine which buyer obtains the item and at what price. We will consider both first-price and second-price variations of the auction.

Example 1. Consider a two-buyer auction with two items, where the incremental valuations are $(v_1(1), v_1(2)) = (11, 9)$ and $(v_2(1), v_2(2)) = (7, 3)$. The optimal outcome is for buyer 1 to receive both copies of the item for a social welfare of $11 + 9 = 20$. Interestingly, economic theory predicts a very different equilibrium outcome. Specifically, using either a first-price or a second-price auction in each time period, buyer 2 should win the first item sold for a price of 5, and buyer 1 should win the second item sold for a price of 3. To see this, imagine that buyer 1 wins the first item. Then in the second period she will have to pay 7 to beat buyer 2 for the second item. Given this, buyer 2 will also be willing to pay up to 7 to win the first item. Thus, buyer 1 will win both permits for 7 each and obtains a *utility* (profit) of $20 - 14 = 6$. On the other hand, imagine that buyer 2 wins the first item. Now in the second period, buyer 1 will only need to pay 3 to beat buyer 2 for the second item, giving her a profit of $11 - 3 = 8$. Consequently, by bidding 5 in the first period, buyer 1 can guarantee herself a profit of 8 regardless of whether or not she wins in the first period. Given this bid, buyer 2 will maximize his own utility by winning the first item for 5. The claimed equilibrium follows. In fact, this is the unique subgame perfect equilibrium surviving the iterative deletion of weakly dominated strategies (see Sect. 2.2). Note that this equilibrium outcome gives a suboptimal social welfare of $11 + 7 = 18$. ◇

2.1 Bidding Strategies

To investigate equilibria we must further formalise the sequential auction model. A *history* describes the bid profiles and auction outcomes for a prefix of the auction rounds. We write $\mathbb{H}$ for the set of all histories. We think of a history as describing the public information revealed over the course of (a prefix of) the auction. We say a history h is an *outcome* if it describes all T auction rounds, otherwise we say it is a *decision node.* We write $\mathbb{O}$ for the set of outcomes and $\mathbb{D}$ for the set of all decision nodes.

A *bidding strategy* $b_i \colon \mathbb{D} \to \mathbb{R}_+$ maps each decision node to a bid. A profile of bidding strategies b describes a bidding strategy for each bidder, where we think of $b_i(h)$ as the bid placed by agent i following history h. We write $\mathbb{B}$ for the set of all profiles of bidding strategies. A *tie-breaking rule* is a function $\pi : [n] \times \mathbb{D} \times \mathbb{B} \to [0, 1]$, where $\pi_i(h|b)$ is the probability that bidder i is allocated the item in the round at decision node h, given that bidders follow the strategies in b. Note that for all $b \in \mathbb{B}$ and $h \in \mathbb{D}$, we have $\sum_{i\in[n]} \pi_i(h|b) = 1$.[2] A tie-breaking rule is deterministic precisely when $\pi_i(b|h) \in \{0, 1\}$ for all i, b, and h. A *payment rule* is a function $p : \mathbb{D} \times \mathbb{B} \to \mathbb{R}_+$, where $p(h|b)$ describes the payment to be made by a buyer at decision node h conditional on being allocated an item that round.

A history contains information about all past bids, as well as the winners and payments of previous auction rounds. However in all of the bidding strategies we consider, we will use only the number of items allocated to each buyer in prior rounds. We will therefore sometimes abuse notation and describe a history h as

[2] In alternate settings, such as when reserve prices are incorporated, it may be the case that $\sum_{i\in[n]} \pi_i(h|b) < 1$.

a vector in $\mathbb{Z}_+^n$, where the i'th component of the vector denotes the number of items allocated to buyer i following history h. As we deal with integer vectors representing item allocations, we will denote histories with x, y, z as needed. For notational convenience, we let e_i be the vector whose ith component equals 1 and all other components equal 0. So for example, $x + e_i$ denotes the history that proceeds according to x, followed by bidder i winning the subsequent round. Finally, we let $t(x) = T - \sum_i x_i$ denote the number of rounds remaining after history x; i.e., the length of sequential subauction that begins following history x.

In this paper we focus on sequential first-price and second-price auctions. In both of these auction types, the allocation rule π is such that, for all $i \in [n]$, $b \in \mathbb{B}$, and $x \in \mathbb{D}$, we have that $\pi_i(x|b) > 0$ implies $b_i(x) \geq b_j(x)$ for all $j \in [n]$. In other words, a bidder with maximum bid will win in each round, but tie-breaking can be arbitrary and possibly randomized. In the first-price auction, the payment rule is $p(x|b) = \max_i b_i(x)$, while in the second-price auction, $p(x|b) = \min_{i\in[n]} \max_{j\in[n]\setminus\{i\}} b_j(x)$.

Given bidding strategies b and a tie-breaking rule π, we may compute forward utilities of buyers at decision node x through induction on $t(x)$. When $t(x) = 0$, the auction has ended and for each $i \in [n], u_i(x|b) = 0$. If $t(x) > 0$, for each buyer i we have

$$u_i(x|b) = \pi_i(x|b) \cdot [v_i(x_i + 1) - p(x|b)] + \sum_{j\in[n]} \pi_j(x|b) \cdot u_i(x + e_j|b).$$

We write $V(k|x)$ for the *total valuation function* at x, denoting the value of the global optimal assignment of k items beginning at decision node x. That is, $V(k|x)$ is the maximum of $\sum_{i\in[n]} V_i(x_i + k_i)$, over all profiles $(k_1, \ldots, k_n)$ with $\sum_{j\in[n]} k_j = k$. We note that $V(\cdot|x)$ is non-decreasing and weakly concave for every x. We will also write $V_{-i}(k|x)$ and $v_{-i}(k|x)$ for the corresponding global optimal assignment and marginal values when we exclude buyer i. That is, $V_{-i}(k|x)$ is the maximum of $\sum_{j\in[n]} V_j(x_j + k_j)$ over profiles $(k_1, \ldots, k_n)$ with $\sum_{j\in[n]} k_j = k$ and $k_i = 0$, and $v_{-i}(k|x) = V_{-i}(k|x) - V_{-i}(k-1|x)$. We refer to $v_{-i}(\cdot|x)$ as the *opposing incremental value function* of buyer i at x.

2.2 Equilibria in Sequential Auctions

We now discuss our equilibrium concept. Bidding strategies b paired with tie-breaking rule π constitutes a **subgame perfect equilibrium** of the sequential multiunit auction if

$$u_i(x|(\beta_i, b_{-i})) \leq u_i(x|b) \qquad \forall i \in [n], \forall \beta_i \in \mathbb{B}_i, \forall x \in \mathbb{D}.$$

Subgame perfection requires that each agent is choosing a utility-optimizing action at each decision node, given the strategies of other agents (including how other agents will respond in future rounds).

In fact, the equilibrium concept we study here is that of an *envy-free* subgame perfect equilibrium with *no-overbidding*. A bidding profile b satisfies **no-(incremental) overbidding** if $\forall x \in \mathbb{D}$, $b_i(x) \leq v_i(x_i + 1)$. That is, no agent bids more than their incremental value for winning an addition item, in any round. We say that bidding strategies b (for tie-breaking rule π) constitute an **envy-free equilibrium** if for every decision node x and buyer i,

$$u_i(x|b) \geq v_i(x_i + 1) - p(x|b) + u_i(x + e_i|b).$$

That is, in an envy-free equilibrium, no bidder who loses in a given round would strictly prefer to win at the price paid by the winner.

3 Inefficiency of Non-Envy-Free Equilibria

It is well-known that subgame perfect equilibria may be arbitrarily inefficient, even for single-item auctions, if buyers may overbid for an item.[3] Overbidding is extremely risky in practice, so we focus on equilibria that satisfy the aforementioned standard no-overbidding property – at each decision node buyers do not bid more than their incremental value for winning an additional item. But we note that the simple prohibition of overbidding does not ensure that equilibria have good social welfare. Whilst the price of anarchy is bounded away from zero (Theorem 1), it can still be negligible (Example 2).

Theorem 1. *At any equilibrium in the sequential second price auction with no-incremental overbidding, the social welfare is at least* $\frac{OPT}{T}$.

The crude bound of $\frac{1}{T}$ in Theorem 1 is tight for SPE of second-price sequential multiunit auctions. This is shown by the following example.

Example 2. We consider a sequential auction of $T \geq 3$ identical items with three buyers. Buyer 1 and buyer 2 have value 1 for every item they win. In turn, buyer 3 has value ϵ for the each item he wins up to $T-2$ items, and has value 0 for any extra items he may win. Consider the following bidding strategy profile: in the first $T-2$ rounds, if buyer 3 has won every previous item, buyer 1 and buyer 2 both bid 0 and buyer 3 bids ϵ. In round $T-1$ if buyer 3 has won every previous item, buyer 2 bids 1 while buyer 1 and buyer 3 both bid 0. Finally, in round T if buyer 3 has won $T-2$ items and buyer 2 has won one item, buyer 1 bids 1 while buyer 2 and 3 both bid 0. At any other decision node, all buyers bid their valuation for an additional item; in particular both buyer 1 and buyer 2 bid 1 while buyer 3 bids $\leq \epsilon$. This bidding profile is a subgame perfect equilibrium, and on the equilibrium path buyer 3 wins $T-2$ items while buyer 1 and buyer 2 win only one item each afterwards. Therefore, the social welfare of the outcome is $2 + (T-2) \cdot \epsilon$. On the other hand, a welfare-optimal allocation gives each item to either buyer 1 or buyer 2, for a total value of T. Thus by taking $\epsilon > 0$ sufficiently small, we obtain a price of anarchy of $\Theta(1/T)$. ◇

[3] For example, take a single-item second-price auction with two buyers. Suppose $v_1(1) = 1$ and $v_2(1) = \epsilon$. The bids $b_1(0) = 0$ and $b_2(0) = 2$ form an equilibrium with social welfare ϵ, but the optimal welfare is 1.

The problem inherent in Example 2 is that subgame perfect equilibria allow for *signalling behaviour* even if overbidding is prohibited. Specifically, buyer 2 lets buyer 1 win at price 0 in the final round in response to buyer 1 letting buyer 2 win in the penultimate round. But this is arguably unsatisfying: the final round is a second-price auction over a single item, and buyer 2's strategy in the final round is weakly dominated by him bidding 1 instead. Such bidding behavior cannot arise in an envy-free equilibrium, since no losing buyer would strictly prefer to win the item *at the price paid by the winning buyer*. This is exactly the condition which eliminates the possibility of a buyer overbidding to the point it deters other buyers from partaking in the auction.

4 An Envy-Free Equilibrium

In this section, we show the price of stability of envy-free equilibria is $(1 - \frac{1}{e})$. The subgame perfect equilibrium we construct also has the property that bidders do not incrementally overbid.[4] To do this, we first study an elementary bidding scheme for buyers with oligopsony power. The bidding scheme is simply *strategic demand reduction*, which is a restricted form of greedy bidding whereby each agent aims at obtaining a certain "safe" payoff. This generalizes the Residual Monopsonist procedure of Rodriguez [16] to arbitrary valuations, and the greedy bidding strategies in Ahunbay et al. [1] to arbitrarily many buyers. We show that if the buyers with oligopsony power simultaneously apply strategic demand reduction then the outcome is guaranteed to provide at least a $(1-\frac{1}{e})$ factor of the optimal social welfare. We then show how to construct an envy-free equilibrium with no-incremental overbidding based upon the greedy strategies.

4.1 A Greedy Bidding Strategy

To start we define quantities that help compare how high a buyer's valuation is compared to other buyers. The **strong oligopsony factor** $f_i(x)$ and the **weak oligopsony factor** $g_i(x)$ of buyer i at decision node x are respectively given by:

$$f_i(x) = \max[\{0\} \cup \{1 \leq k \leq t(x) | v_i(x_i + k) > v_{-i}(t(x) - k + 1|x)\}] \quad (1)$$
$$g_i(x) = \max[\{0\} \cup \{1 \leq k \leq t(x) | v_i(x_i + k) \geq v_{-i}(t(x) - k + 1|x)\}] \quad (2)$$

So $f_i(x)$ is the minimum number of items that buyer i can obtain in a welfare-optimal allocation (starting from node x) and $g_i(x)$ is the maximum. We say that buyer i has **oligopsony power** at decision node x if $f_i(x) > 0$. Note that if there are no ties in incremental values then $f_i(x) = g_i(x)$. If $g_i(x) = t(x)$ then it is welfare-optimal to allocate all the remaining items to buyer i. Moreover, if $f_i(x) = t(x)$ then this optimal allocation is unique. We will say that buyer

[4] We note that this is an equilibrium refinement rather than a restriction of the action space. Bidders are still able to consider deviations in which they overbid.

i is a **monopsonist** at x if $g_i(x) = t(x)$ and a **strict monopsonist** at x if $f_i(x) = t(x)$. Let

$$\lambda_{ij}(x) = \min\{k \geq 0 | v_j(x_j + 1) = v_{-i}(k|x)\}. \tag{3}$$

Then $\lambda_{ij}(x)$ measures the *position* of buyer j's value of an additional item at x in the opposing incremental valuation function $v_{-i}(x)$. For instance, suppose that buyer 1's opposing incremental valuation at 0 is as shown in Fig. 1, and buyer 2 has value 4 for the first item he obtains. Then $\lambda_{12}(0) = 2$, as buyer 2's first valuation equals the height of the second bin in Fig. 1.

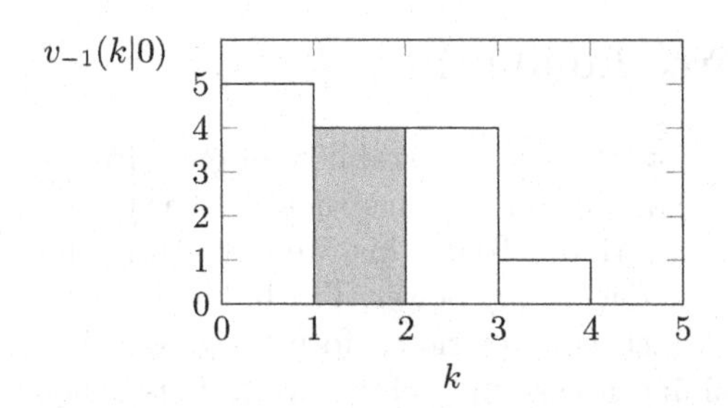

Fig. 1. Example opposing valuations for buyer 1. If buyer 2 has $v_2(1) = 4$ then buyer 2's incremental valuation is *at the second position earliest* in the histogrammatic display of $v_{-1}(\cdot|0)$, hence $\lambda_{12}(0) = 2$.

Given any bidding profile b for buyers, in the subauction starting from decision node x, any buyer i may consider deviating to a simple bidding strategy targeting the purchase of $0 \leq k \leq f_i(x)$ items. At decision node $y \geq x$, suppose buyer i has purchased less than k items and that the highest bidding buyer j has $v_j(y_j+1) \leq v_{-i}(t(x)-k+1|x)$. Then buyer i can deviate to bidding $v_j(y_j+1)+\epsilon$ for small $\epsilon > 0$; else buyer i can "pass" on winning the item by bidding below other buyers. Then for each $0 \leq k \leq f_i(x)$ buyer i may guarantee for herself at x a forward utility equal to $\bar{\mu}_i(k|x) = \sum_{\ell=1}^{k} v_i(x_i + \ell) - k \cdot v_{-i}(t(x) - k + 1|x)$. We then denote the **greedy utility** of buyer i at decision node x to be

$$\mu_i(x) = \max_{0 \leq k \leq t(x)} \bar{\mu}_i(k|x). \tag{4}$$

Buyer i's **greedy demand** is the minimum number of items buyer i may target the purchase of with these simple bidding strategies to attain their greedy utility,

$$\kappa_i(x) = \min \arg\max_{0 \leq k \leq t(x)} \bar{\mu}_i(k|x). \tag{5}$$

We ask if there exists an equilibrium where *every* buyer obtains their greedy utility at every decision node x. If this is the case, at decision node x no buyer i should let a buyer $j \neq i$ win at price less than $\rho_{ij}(x) = v_i(x_i+1) + \mu_i(x+e_i) - \mu_i(x+e_j)$. We call this price the **threshold price** of buyer i **against** buyer j at x. If buyer i is to obtain their greedy utility at x, it should also be the case that buyer i does not win an item at a price strictly below

$$\rho_i(x) = v_i(x_i+1) + \mu_i(x+e_i) - \mu_i(x). \tag{6}$$

This price is called the **threshold price** of buyer i at x, and it can be shown that if buyer i has oligopsony power ($f_i(x) > 0$) and does not demand the entire supply (i.e. $\kappa_i(x) < t(x)$), then for every buyer j with $\lambda_{ij}(x) \leq t(x) - \kappa_i(x)$ we have $\rho_{ij}(x) = \rho_i(x)$. Finally, via the simple bidding strategies we consider and by the no-overbidding constraint, no buyer i should let a buyer j with incremental value $v_j(x_j+1) < v_{-i}(t(x) - \kappa_i(x) + 1|x)$ win. We thus define the **baseline price** of buyer i as

$$\beta_i(x) = \begin{cases} v_{-i}(t(x) - \kappa_i(x) + 1|x) & \kappa_i(x) > 0 \\ v_i(x_i+1) & \kappa_i(x) = 0 \end{cases} \tag{7}$$

Example 3. Consider again a two-item auction with two buyers and incremental valuations $(v_1(1), v_1(2)) = (11, 9)$ and $(v_2(1), v_2(2)) = (7, 3)$. In the first round, buyer 1 has a greedy utility of 8 (buying one item at price $v_2(2) = 3$) while buyer 2 has a greedy utility of 0 (as she cannot guarantee herself any items). In the second round, buyer 2 has a greedy utility of 0 whether he has won in the first round or not. Buyer 1, on the other hand, has a greedy utility of 2 after winning an item (by obtaining the second item at price $v_2(1) = 7$), and 8 if she lets buyer 2 win the first item. This implies that buyer 1 has a threshold price of $11 + 2 - 8 = 5$ in the first round. Since buyer 1 may attain its greedy utility by purchasing a single item at price 3, buyer 1 has a baseline price of 3. Buyer 2 has $f_2(0,0) = 0$ so her threshold and baseline price both equal $v_2(1) = 7$. ◇

We analyze the behaviour of threshold and baseline prices via an extension of the arguments in [1]. Suppose that at decision node x, a buyer i with oligopsony power does not let any buyer j win at price $\rho_{ij}(x)$. Then if buyer $j \neq i$ wins at x, ρ_i weakly decreases and β_i remains constant. If instead buyer i wins at x and $f_i(x+e_i) > 0$ then ρ_i and β_i both weakly increase. Moreover, if buyer i does not demand the entire supply then there exists a buyer j such that $v_i(x_i+1) > v_j(x_j+1) \geq \rho_{ij}(x) = \rho_i(x)$. If buyer i does demand the entire supply at x then $\rho_i(x) = v_{-i}(1|x)$, while $\rho_{ij}(x) > v_j(x_j+1)$ for all $j \neq i$. We conclude that buyer i may obtain their greedy utility by outbidding every other buyer.

4.2 Efficiency Under Strategic Demand Reduction

Before we construct our equilibrium, we first inspect the evolution of the auction when buyers implement *strategic demand reduction*. At each decision node x, each buyer i bids their threshold price $\rho_i(x)$. This may induce ties, but there is a way to select the winner so that each buyer obtains at least their greedy payoff in the auction. Specifically, we can choose a winning buyer i such that for any other buyer j, $\rho_{ji}(x) = \rho_j(x)$. These strategies might not form an equilibrium as-is, but in Sect. 4.3 we will use them as the basis of an envy-free subgame perfect equilibrium.

At decision node x, if some buyer i demands the entire supply then $\rho_{ij}(x) > v_{-i}(1|x)$ for any buyer j. Also for a buyer $j \neq i$ we have $\rho_{ji}(x) = \rho_j(x) = v_j(x_j + 1)$. Then buyer i wins an item at price $v_{-i}(1|x)$. At $x + e_i$ buyer i still demands the entire supply and $v_{-i}(1|x+e_i) = v_{-i}(1|x)$, so buyer i keeps purchasing every item at price $v_{-i}(1|x)$, earning a payoff equal to $\bar{\mu}_i(\kappa_i(x)|x) = \mu_i(x)$.

Suppose instead that no buyer demands the entire supply. If some buyer i with $g_i(x) > 0$ wins an item, then $\kappa_i(x + e_i) \geq \kappa_i(x) - 1$. If such a buyer i does not win the item, then the winning buyer j has $\lambda_{ij}(x) \leq t(x) - \kappa_i(x)$ and buyer i's demand is unchanged. This implies that every buyer i wins at least $\kappa_i(x)$ items, and that there can be at most one buyer j who wins fewer than $f_j(x)$ items. If such a buyer j exists, he must earn payoff exactly equal to $\mu_i(x)$ in the subauction starting from x.

That each buyer earns their greedy utility and that there may only be one buyer j who wins $< f_i(x)$ items allows us to lower bound the welfare of outcomes. If there exists some buyer i with $g_i(x) > 0$ who wins fewer than $g_i(x)$ items, the social welfare of the outcome reached in equilibrium, $\text{SW}(x)$, is bounded below:

$$\text{SW}(x) \geq \sum_{\ell=1}^{\kappa_i(x)} v_i(x_i + \ell) + \sum_{\ell=1}^{t(x)-\kappa_i(x)} v_{-i}(\ell|x).$$

Example 4. Suppose in a six-item auction with five buyers that buyer 1 has value 30 for each item she wins, except for a sixth item for which she has value 20. In turn buyers $2, 3, 4, 5$ are unit-demand, with valuations of $24, 18, 15$ and 10 respectfully. Then buyer 1's incremental valuations and opposing incremental valuations at the beginning of the auction can be displayed as in Fig. 2. When buyers implement their greedy bidding strategies and if ties are never broken in favor of buyer 1, buyer 1 obtains $\kappa_1(0) = 2$ items and each other buyer obtains a single item, so the welfare of the auction outcome equals the area of the shaded region. It is immediate that among outcomes that award at least $\kappa_1(0) = 2$ items to buyer 1 and $f_2(0) = 1$ item to buyer 2, the shaded area equals the lowest possible social welfare.

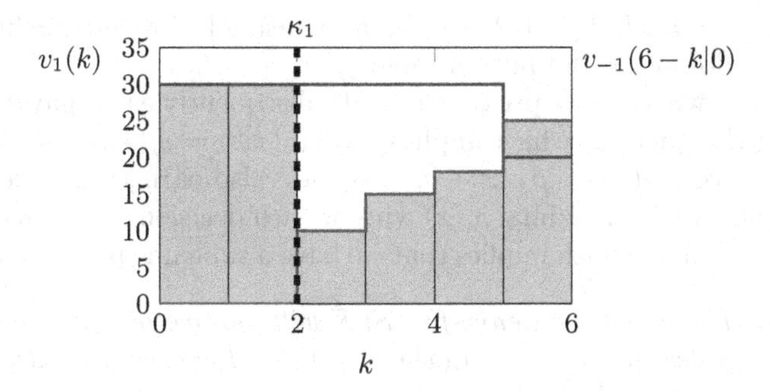

Fig. 2. A histogrammatic display of buyers' valuations in Example 4. The red curve shows buyer 1's incremental valuations, while the blue curve right-to-left shows buyer 1's opposing incremental valuation function in the beginning of the auction. The shaded area equals the welfare of the auction outcome when buyers bid their threshold prices. (Color figure online)

Thus social welfare is lower bounded by the outcome of a two-buyer auction, where we contract all buyers other than buyer i into a single buyer $-i$ winning $t(x) - \kappa_i(x)$ items. Efficiency bounds in this setting therefore *reduce* to the two buyer setting with no-overbidding considered in [1], when the two buyers implement their greedy bidding strategies.

Theorem 2. *If all buyers implement strategic demand reduction, the outcome of the auction provides a* $(1 - 1/e)$*-approximation of the optimal social welfare.*

From a practical perspective it is important to note that the implementation of these greedy bidding strategies does *not* require complete information. Strategic demand reduction can be implemented by an agent i knowing only the aggregate market demand v_{-i}. In practice (such as in emission license auctions), it is reasonable to assume that such aggregate demand, or a good approximation to it, is known to the bidders.

4.3 Price of Stability of Envy-Free Equilibria

We obtain our price of stability result by modifying the bidding strategies that result from strategic demand reduction to obtain an envy-free equilibrium with no-overbidding in which every buyer i earns exactly their greedy payoff. If some buyer i demands the entire supply at x, we let buyer i bid $\min\{v_i(x_i+1), \rho_i(x)\} > v_{-i}(1|x)$ and leave other buyers' bids unchanged. In this case, buyer i again wins all items at price $v_{-i}(1|x)$ without overbidding.

If no buyer demands the entire supply, suppose buyer i wins when buyers implement demand reduction. If buyer i has $f_i(x) > 0$ then there exists a buyer j with $v_j(x_j + 1) \geq \rho_i(x)$. We then increase the bid of buyer j to $\rho_i(x)$. Thus buyer i and j both bid $\rho_i(x)$ at x, and we break the tie[5] at x in favor of buyer i.

[5] Alternately, we can let buyer i make an infinitesimally greater bid than $\rho_i(x)$. Such bids are considered in [12,13] for equilibria of first-price sequential auctions.

Other buyers $j' \neq i, j$ bid $\rho_{j'i}(x) = \rho_j(x)$ at x instead – by maximality of $\rho_i(x)$ such buyers j' do not outbid buyers i and j.

Then buyer i wins at x at price $\rho_i(x)$ and subsequently earns payoff $\mu_i(x+e_i)$ in the rest of the auction, which implies that $u_i(x|b) = \rho_i(x)$. Any other buyer $j' \neq i$ must have $\mu_{j'}(x) = \mu_{j'}(x + e_i)$, so they also earn their greedy utility. Finally, as a buyer with maximal $\rho_i(x)$ wins at each decision node x no buyer has a profitable deviation, which implies that we have a subgame perfect equilibrium.

Theorem 3. *There exists an envy-free SPE with no-overbidding where for any buyer i and any decision node x, $u_i(x|b) = \mu_i(x)$. Moreover the outcome of this equilibrium is supported by buyers implementing strategic demand reduction, and if buyer i wins at decision node x she purchases the item at price $\rho_i(x)$.*

In particular, the outcome of this envy-free equilibrium with no-overbidding attains a $(1 - 1/e)$-approximation of the optimal social welfare. Moreover, this approximation factor is tight for our construction: a sequence of two-buyer valuations profiles such that this equilibrium has efficiency converging to $(1 - 1/e)$.

Corollary 1. *The price of stability of envy-free equilibria in sequential multiunit auctions, under the no-incremental overbidding constraint, is at least* $(1 - 1/e)$.

5 Equilibria with Iterated Elimination of Weakly Dominated Strategies

In this section we study a different class of equilibria that survive the iterated elimination of weakly dominated strategies (IEWDS). Recall, from Theorem 1 and Example 2, that the price of anarchy is $\frac{1}{T}$. But, as discussed, Example 2 relied on the inherent use of signalling. Implicitly, the corresponding equilibrium class encompassed by Theorem 1 is that of subgame perfect equilibria *with signalling*. Can we say anything about less permissive (and more natural) classes of subgame perfect equilibria? Indeed we can for the fundamental class of subgame perfect equilibria that survive the iterative elimination of weakly dominated strategies.

5.1 An Ascending Price Auction Mechanism

Take any decision node in the sequential auction and consider the following ascending-price mechanism. Starting at $p = 0$ continuously raise the price p. At price p, buyer i remains in the auction as long as there is at least one buyer j *still in the auction* who buyer i is willing to pay a price p to beat. The last buyer to drop out wins at the corresponding price. This procedure produces a unique *dropout bid* τ_i for each buyer i for the decision node.

This mechanism induces exactly the set of bids that survives IEWDS for both first-price sequential auctions [13] and second-price sequential auctions; see [13] and [12], respectively, for details. To wit, the ascending-price mechanism outputs an envy-free equilibrium, albeit a different one than that of Sect. 4.

Moreover, there are strong reasons to believe that equilibria that survive the iterative elimination of weakly dominated strategies form *the natural class* of equilibrium for sequential auctions [12,13]. Consequently, for the price of anarchy results that follow we restrict attention to this class of equilibria, that is, those equilibrium induced by the ascending-price mechanism.

5.2 The Price of Anarchy Under Competition

We now study the price of anarchy for equilibria that survive IEWDS, under a no-incremental-overbidding assumption on the set of strategies. This class of equilibria naturally extends the analysis of [1], which considered such no-overbidding equilibria for the case of $n = 2$ bidders.

Specifically, we present constant price of anarchy bounds for three settings with varying degrees of oligopsony power. The first result concerns the case where no buyer has oligopsony power, that is, for every buyer i we have $f_i(0) = 0$. Example 2 is of this type. In this setting there is a unique equilibrium that survives IEWDS and the auction attains full efficiency.

Theorem 4. *Suppose that at decision node x, for any buyer i we have $f_i(x) = 0$. Then for any buyer i, $u_i(x) = 0$, and prices equal $v(1|x)$ in every subsequent round of the auction. In particular, the price of anarchy is* 1.

Second, consider the case where there is exactly one buyer i with oligopsony power. Example 1 is of this type. In this setting, we prove a multi-buyer result paralleling Theorem 3 of [1] for the 2-buyer case – at every decision node x of the auction, buyer i obtains her greedy utility $\mu_i(x)$. This result is driven by the fact that buyer i is the unique price-setter throughout the auction as long as she retains oligopsony power. Because all the other buyers profit off of buyer i's demand reduction and since buyer i's threshold price increases after she wins an item, the buyers are incentivised to outbid buyer i while buyer i holds oligopsony power. This causes buyer i to be constrained to her greedy utility, which in turn induces her to win a number of items no less than her greedy demand.

Theorem 5. *Suppose that there exists a unique buyer i such that $f_i(0) > 0$. Then for every decision node x, $u_i(x) = \mu_i(x)$. Moreover, the price of anarchy is at least* $1 - 1/e$.

Finally, consider the case where multiple buyers have oligopsony power. To obtain constant price of anarchy bounds for this setting we make an additional restriction on the valuation profiles. We say the buyers' valuations are **flat-optimal** if $v(1|0) = v(T|0)$. In this case, for every pair of buyers i and j and for any k, ℓ such that $1 \leq k \leq g_i(x)$, $1 \leq \ell \leq g_j(x)$, we have $v_i(k) = v_j(\ell)$. We note that this family of valuations includes the worst-case efficiency instances for two buyers [1,2].

To bound the efficiency of equilibria in this setting, we invoke Theorem 4 and Theorem 5 and use a counting argument. Suppose that $(0, x_1, x_2, ..., x_T)$ is an equilibrium path, and $(i_1, i_2, ..., i_T)$ the sequence of winners on this equilibrium path. We say that round $1 \leq t \leq T$ is a *loss round* if $v_{i_t}(x_{t-1}+1) < v(1|0)$. Note that if round t is a loss round, then for every buyer i we have $f_i(x_t) = \max\{f_i(x_{t-1}) - 1, 0\}$. We count the number of loss rounds until the subauction we reach in equilibrium satisfies either the conditions of Theorem 4 or Theorem 5. This can equal at most the second-highest $f_i(0)$. Thus there are at most $\lfloor T/2 \rfloor$ loss rounds before we reach a subauction in which efficiency is at least $(1 - 1/e)$. However, the non-loss rounds before do not hurt efficiency, which provides a lower bound on the efficiency of the equilibrium outcome.

Theorem 6. *If the valuations are flat-optimal then the price of anarchy is at least* $\frac{1}{2} \cdot (1 - 1/e)$.

One interpretation of Theorem 6 is that, for a certain family of valuations, even if a large number of buyers hold oligopsony power and ties are broken adversarially, the efficiency of the auction does not depreciate too much. This suggests that having many buyers with market power is not necessarily enough to allow coordination on an inefficient outcome, for the natural class of equilibria given by the ascending-price mechanism.

Acknowledgements. The authors are very grateful for advice from and discussions with Craig Golding, Tom Johnson and Alex Wood of the former Ontario Ministry of the Environment and Climate Change, and from Christopher Regan, Dale Beugin and Jason Dion of Canada's Ecofiscal Commission.

References

1. Ahunbay, M.Ş, Lucier, B., Vetta, A.: Two-buyer sequential multiunit auctions with no overbidding. In: Harks, T., Klimm, M. (eds.) SAGT 2020. LNCS, vol. 12283, pp. 3–16. Springer, Cham (2020). https://doi.org/10.1007/978-3-030-57980-7_1
2. Ahunbay, M.Ş, Vetta, A.: The price of anarchy of two-buyer sequential multiunit auctions. In: Chen, X., Gravin, N., Hoefer, M., Mehta, R. (eds.) WINE 2020. LNCS, vol. 12495, pp. 147–161. Springer, Cham (2020). https://doi.org/10.1007/978-3-030-64946-3_11
3. Ashenfelter, O.: How auctions work for wine and art. J. Econ. Perspect. **3**(3), 23–36 (1989)
4. Ashenfelter, O., Genesove, D.: Legal negotiation and settlement. Am. Econ. Rev. **82**, 501–505 (1992)
5. Bae, J., Beigman, E., Berry, R., Honig, M., Vohra, R.: Sequential bandwidth and power auctions for distributed spectrum sharing. IEEE J. Sel. Areas Commun. **26**(7), 1193–1203 (2008)
6. Bae, J., Beigman, E., Berry, R., Honig, M., Vohra, R.: On the efficiency of sequential auctions for spectrum sharing. In: Proceedings of the 1st Conference on Game Theory for Networks (GameNets), pp. 199–205 (2009)
7. Buccola, S.: Price trends at livestock auctions. Am. J. Agr. Econ. **64**, 63–69 (1982)
8. Chanel, O., Gérard-Varet, L., Vincent, S.: Auction theory and practice: evidence from the market for jewellery. In: Ginsburgh, V., Menger, P. (eds.) Economics of the Arts: Selected Essays, pp. 135–149. North-Holland, Amsterdam (1996)

9. Feldman, M., Lucier, B., Syrgkanis, V.: Limits of efficiency in sequential auctions. In: Chen, Y., Immorlica, N. (eds.) WINE 2013. LNCS, vol. 8289, pp. 160–173. Springer, Heidelberg (2013). https://doi.org/10.1007/978-3-642-45046-4_14
10. Gale, I., Stegeman, M.: Sequential auctions of endogenously valued objects. Games Econom. Behav. **36**(1), 74–103 (2001)
11. McAfee, P., Vincent, D.: The declining price anomaly. J. Econ. Theory **60**, 191–212 (1993)
12. Narayan, V.V., Prebet, E., Vetta, A.: The declining price anomaly is not universal in multi-buyer sequential auctions (but almost is). In: Fotakis, D., Markakis, E. (eds.) SAGT 2019. LNCS, vol. 11801, pp. 109–122. Springer, Cham (2019). https://doi.org/10.1007/978-3-030-30473-7_8
13. Paes Leme, R., Syrgkanis, V., Tardos, E.: Sequential auctions and externalities. In: Proceedings of 23rd Symposium on Discrete Algorithms (SODA), pp. 869–886 (2012)
14. Pesando, J., Shum, P.: Price anomalies at auction: evidence from the market for modern prints. In: Ginsburgh, V., Menger, P. (eds.) Economics of the Arts: Selected Essays, pp. 113–134. North-Holland, Amsterdam (1996)
15. Pitchik, C., Schotter, A.: Perfect equilibria in budget-constrained sequential auctions: an experimental study. RAND J. Econ. **19**, 363–388 (1988)
16. Rodriguez, G.: Sequential auctions with multi-unit demands. B.E. J. Theor. Econ. **9**(1), 1–35 (2009)
17. Salladare, F., Guilloteau, P., Loisel, P., Ollivier, P.: The declining price anomaly in sequential auctions of identical commodities with asymmetric bidders: empirical evidence from the Nephrops Norvegicus market in France. Agric. Econ. **48**, 731–741 (2017)
18. Thiel, S., Petry, G.: Bidding behaviour in second-price auctions: rare stamp sales, 1923–1937. Appl. Econ. **27**(1), 11–16 (1995)

Auctions with Interdependence and SOS: Improved Approximation

Ameer Amer(✉) and Inbal Talgam-Cohen(✉)

Faculty of Computer Science, Technion – Israel Institute of Technology, Haifa, Israel
{ameeramer,italgam}@cs.technion.ac.il

Abstract. Interdependent values make basic auction design tasks – in particular maximizing welfare truthfully in single-item auctions – quite challenging. Eden et al. recently established that if bidders' valuation functions are submodular over their signals (a.k.a. SOS), a truthful 4-approximation to the optimal welfare exists. We show existence of a mechanism that is truthful and achieves a tight 2-approximation to the optimal welfare when signals are binary. Our mechanism is randomized and assigns bidders only 0 or $\frac{1}{2}$ probabilities of winning the item. Our results utilize properties of submodular set functions, and extend to matroid settings.

Keywords: Mechanism design · Welfare maximization · Submodularity

1 Introduction

One of the greatest contributions of Robert Wilson and Paul Milgrom, the 2020 Nobel Laureates in economics, is their formulation of a framework for auction design with *interdependent* values [14]. Up to their work, the standard assumption underlying auction design theory was that each bidder fully knows her value for the item being auctioned, because this value depends only on her own private information. This assumption is, however, far from reality in very important settings – for example, when the auction is for drilling rights, the information one bidder has about whether or not there is oil to be found is extremely relevant to how another bidder evaluates the rights being auctioned. Works like [18] and [12] lay the foundation for rigorous mathematical research of such settings, yet many key questions still remain unanswered.

For concreteness, consider an auction with a single item for sale (our main setting of interest). In the interdependent values model, every bidder $i \in [n]$ has a privately-known signal s_i, and her value v_i is a (publicly-known) function of all the signals, i.e., $v_i = v_i(s_1, s_2, ..., s_n)$. Thus, in this model, not only the auctioneer is in the dark regarding a bidder's willingness to pay for the item being auctioned; so is the bidder herself (who knows s_i and $v_i(\cdot)$ but not s_{-i})!

This stark difference from the standard, independent private values (IPV) model creates a big gap in our ability to perform seemingly-simple auction design

I. Caragiannis and K. A. Hansen (Eds.): SAGT 2021, LNCS 12885, pp. 34–48, 2021.
https://doi.org/10.1007/978-3-030-85947-3_3

tasks. Arguably the most fundamental such task is *truthful welfare maximization*. For IPV, the truthful welfare-maximizing Vickrey auction [17] is a pillar of mechanism design (e.g., it has many practical applications and is usually the first auction taught in a mechanism design course). But with interdependence, welfare and truthfulness are no longer perfectly compatible: Consider two bidders reporting their signals s_1, s_2 to the auction, which allocates the item to the highest-value bidder according to these reports; if the valuation functions v_1, v_2 are such that bidder 1 wins when $s_1 = 0$ but loses when $s_1 = 1$, this natural generalization of Vickrey to interdependence is non-monotone and thus non-truthful. This is the case, for example, if $v_1 = 1 + s_1$ and $v_2 = H \cdot s_1$ for $H > 2$ (see [5, Example 1.2]).

The classic economics literature addressed this challenge by introducing a somewhat stringent condition on the valuation functions called "single-crossing", which ensures truthfulness of the natural generalization of Vickrey (in particular, single-crossing is violated by $v_1 = 1 + s_1, v_2 = H \cdot s_1$). Recently, a breakthrough result of Eden et al. [5] took a different approach: For simplicity consider *binary* signals – e.g., "oil" or "no oil" in an auction for drilling rights. Formally, $s_i \in \{0, 1\}$ (we focus on the binary case throughout the paper). The valuations are now simply *set functions* over the signals, objects for which a rich mathematical theory exists. Eden et al. applied a *submodularity* assumption to these set functions (in particular, submodularity holds for $v_1 = 1+s_1$ and $v_2 = H \cdot s_1$). Under such submodularity over the signals (*SOS*), they shifted focus from maximizing welfare to *approximating* the optimal welfare. While they showed that no truthful mechanism can achieve a better approximation factor than 2 (guaranteeing more than half the optimal welfare), they constructed a truthful randomized mechanism that achieves a 4-approximation (guaranteeing at least a quarter of the optimal welfare). The gap between 2 and 4 was left as an open problem.

Our Results and Organization. In this work we resolve the above open problem of [5] for binary signals. More precisely, we show that in the binary signal case there exists a truthful randomized mechanism that achieves a 2-approximation to the optimal welfare (for a formal statement see Theorem 1). Our result holds for any number n of bidders, and is constructive – that is, we give an algorithm that gets the n valuation functions as input, and returns the mechanism as output.[1]

The fact that our mechanism is randomized is unsurprising given another result of Eden et al. [5], who show that a deterministic mechanism cannot achieve a constant approximation to the optimal welfare even with SOS. This result is in fact proved with the above example of $v_1 = 1 + s_1, v_2 = H \cdot s_1$ and $s_i \in \{0, 1\}$. An interesting corollary of our construction is that a 2-approximation is achievable by a mechanism that is only "slightly" randomized – the only allocation probabilities it uses are 0 and $\frac{1}{2}$.

---

[1] The algorithm runs in time polynomial in its input size, which consists of set functions over n elements and so is exponential in n.

Our algorithm is arguably quite simple and streamlined – for every signal profile it searches for a feasible pair of bidders whose aggregate value exceeds that of the highest bidder, and randomly allocates the item among these two (this explains the factor of 2 in the approximation guarantee). Only if no such pair exists, the item is randomly either allocated to the highest bidder or left unallocated. To maintain monotonicity, the algorithm *propagates* allocation probabilities to neighboring signal profiles. Despite its relative simplicity, the algorithm requires careful analysis, which in particular relies on new properties of collections of submodular functions (Sect. 4.2). The main technical challenge is in showing that the 2-approximation guarantee holds despite the propagations.

Example. To illustrate our method, consider again the above example of $v_1 = 1 + s_1$ and $v_2 = H \cdot s_1$ where $s_i \in \{0, 1\}$. Our algorithm returns a randomized allocation rule that gives the item to bidder 1 with probability $\frac{1}{2}$ if $s_1 = 0$, and randomly allocates it to one of the two bidders if $s_1 = 1$.[2] This allocation rule is monotone (unlike the natural generalization of Vickrey), and leads to a truthful mechanism with a 2-approximation guarantee.

Extensions. In the full version of the paper we extend our main result to beyond single-item settings, namely to general single-parameter settings in which the set of winning bidders must satisfy a *matroid* constraint [15]. As in [5], we can also extend our positive results from welfare to *revenue* maximization using a reduction of [4].

Organization. After presenting the preliminaries in Sect. 2, we state our main theorem and give an overview of our algorithm in Sect. 3. The analysis appears in Sect. 4. Section 6 summarizes with future directions. The full version of the paper includes the pseudo-code and running time, additional details of the analysis, the extension to matroids and our results for non-binary signals.

Additional Related Work. Interdependent values have been extensively studied in the economic literature (see, e.g., [1,3,8,11]). In computer science, most works to date focus on the objective of maximizing revenue [2,4,10,16]. The work of [4] considers welfare maximization with a relaxed *c-single-crossing* assumption, where parameter $c \geq 1$ measures how close the valuations are to satisfying classic single-crossing. This work achieves a c-approximation for settings with binary signals. Their mechanisms also use propagations but otherwise are quite different than ours. The work of [7] also focuses on welfare but does not assume single-crossing; instead it partitions the bidders into ℓ "expertise groups" based on how their signal can impact the values for the good, and using clock auctions achieves approximation results parameterized by ℓ (and by the number of possible signals). The main paper our work is inspired by is [5]. It introduces

[2] Our algorithm has two iterations: At $s_1 = 0$, an appropriate pair is not found and so the highest bidder (bidder 1) wins the item with probability $\frac{1}{2}$, which is propagated forward to this bidder at $s_1 = 1$. At $s_1 = 1$, an appropriate pair is again not found and so the highest bidder (bidder 2) wins the item with probability $\frac{1}{2}$.

the *Random Sampling Vickrey auction*, which by excluding roughly half of the bidders achieves a 4-approximation to the optimal welfare for single-parameter, downward-closed SOS environments. The authors also show positive results for combinatorial SOS environments under various natural constraints. Finally, [6] also study welfare maximization in single- and multi-parameter environments but by simple, non-truthful parallel auctions.

2 Setting

Signals. As our main setting of interest we consider single-item auctions with n bidders. Every bidder i has a binary *signal* $s_i \in \{0,1\}$, which encompasses the bidder's private bit of information about the item. It is convenient to identify a signal profile $s = (s_1, s_2, ..., s_n)$ with its corresponding set $S = \{i \mid s_i = 1\}$ (by treating s_i as an indicator of whether $i \in S$).

Values. The bidders have *interdependent values* for the item being auctioned: Every bidder i's value v_i is a non-negative function of all bidders' signals, i.e., $v_i = v_i(s_1, s_2, ..., s_n) \geq 0$. We adopt the standard assumption that the valuation function v_i is weakly increasing in each coordinate and strongly increasing in s_i. Using the set notation we also write $v_i = v_i(S)$.[3] This makes $v_i(\cdot)$ a monotone *set function* over subsets of $[n]$.

Who Knows What. A setting is summarized by the valuation functions $v_1, \ldots, v_n$, which are publicly known (as is the signal domain $\{0,1\}$). The instantiation of the signals is private knowledge, that is, signal s_i is known only to bidder i.

SOS Valuations. The term *SOS valuations* was coined by Eden et al. [5] to describe interdependent valuation functions that are *submodular over the signals* (see also [2,4,13]).[4] With binary signals, valuations are SOS if $v_i(\cdot)$ is a submodular set function for every $i \in [n]$.

Definition 1 (Submodular set function). *A set function $v_i : 2^{[n]} \rightarrow \mathbb{R}$ is submodular if for every $S, T \subseteq [n]$ such that $S \subseteq T$ and $i \in [n] \backslash T$ it holds that $v_i(S \cup \{i\}) - v_i(S) \geq v_i(T \cup \{i\}) - v_i(T)$.*

A weaker definition that will also be useful for us is subadditivity. Every submodular set function is subadditive, but not vice versa.

Definition 2 (Subadditive set function). *A set function $v_i : 2^{[n]} \rightarrow \mathbb{R}$ is subadditive if for every $S, T \subseteq [n]$ is holds that $f(S) + f(T) \geq f(S \cup T)$.*

---

[3] This notation is not to be confused with the value for a *set of items* S; in our model there is a *single* item, and a bidder's interdependent value for it is determined by the *set of signals*, i.e., which subset of signals is "on".

[4] As mentioned above, submodularity over signals is not to be confused with submodularity over items in combinatorial auctions.

Given a set function v_i and a subset $S \subseteq [n]$, we use $v_i(\cdot \mid S)$ to denote the following set function: $v_i(T \mid S) = v_i(T \cup S) - v_i(S)$ for every $T \subseteq [n]$. It is known that submodularity of v_i implies subadditivity of $v_i(\cdot \mid S)$:

Proposition 1 (e.g., Lemma 1 of [9]). *If v_i is submodular then $v_i(\cdot \mid S)$ is subadditive for every subset $S \subseteq [n]$.*

We refer to an ordering $S_1, S_2, \ldots, S_{2^n}$ of all subsets of the ground set of elements $[n]$ as *inclusion-compatible* if for every pair S_k, S_ℓ such that $S_k \subset S_\ell$, it holds that $k < \ell$ (the included set is before the including one in the ordering).

Notation. Consider a set function v_i and two elements $j, k \in [n]$. For brevity we often write $v_i(j)$ for $v_i(\{j\})$, and $v_i(jk)$ for $v_i(\{j,k\})$.

2.1 Auctions with Interdependence

Randomized Mechanisms. Due to strong impossibility results for deterministic mechanisms [5], we focus on randomized mechanisms as follows: A randomized mechanism $M = (x, p)$ for interdependent values is a pair of allocation rule x and payment rule p. The mechanism solicits signal reports from the bidders, and maps a reported signal profile s to non-negative allocations $x = (x_1, ..., x_n)$ and expected payments $p = (p_1, ..., p_n)$, such that the item is *feasibly* allocated to bidder i with probability x_i (feasibility means $\sum_{i=1}^{n} x_i(s) \leq 1$).

Truthfulness. With interdependence, it is well-established that the appropriate notion of truthfulness is *ex post IC* (incentive compatibility) and *IR* (individual rationality). Mechanism M is ex post IC-IR if the following holds for every bidder i, true signal profile s and reported signal s'_i: Consider bidder i's expected utility when the others truthfully report s_{-i}:

$$x_i(s_{-i}, s'_i)v_i(s) - p_i(s_{-i}, s'_i);$$

then this expected utility is non-negative and maximized by truthfully reporting $s'_i = s_i$.[5]

Similarly to independent private values, the literature on interdependent values provides a characterization of ex post IC-IR mechanisms – as the class of mechanisms with a *monotone* allocation rule x. Allocation rule x satisfies monotonicity if for every signal profile s, bidder i and $\delta \geq 0$, increasing i's signal report by δ while holding other signals fixed increases i's allocation probability:

$$x_i(s_{-i}, s_i) \leq x_i(s_{-i}, s_i + \delta).$$

The characterization also gives a payment formula which, coupled with the monotone allocation rule, results in an ex post IC-IR mechanism. In more detail, the expected payment of bidder i is achieved by finding her critical signal report

[5] Note the difference from *dominant-strategy* IC, in which this guarantee should hold no matter how other bidders report.

and plugging it into her valuation function while holding others' signals fixed (see [16] for a comprehensive derivation of the payments).

Welfare Maximization. Our objective in this work is to design ex post IC-IR mechanisms for interdependent values that maximize social welfare. For a given setting and true signal profile s, the optimal welfare $\mathrm{OPT}(s)$ is achieved by giving the item to the bidder with the highest value, i.e., $\mathrm{OPT}(s) = \max_i\{v_i(s)\}$. Given a randomized ex post IC-IR mechanism $M = (x, p)$ for this setting, $\mathrm{ALG}(s)$ is its welfare in expectation over the internal randomness, i.e., $\mathrm{ALG}(s) = \sum_{i=1}^{n} x_i(s)v_i(s)$. We say mechanism M achieves a c-approximation to the optimal welfare for a given setting if for every signal profile s, $\mathrm{ALG}(s) \geq \frac{1}{c}\,\mathrm{OPT}(s)$ (note that the required approximation guarantee here is "universal", i.e., should hold individually for every s). Since Eden et al. [5] devise a setting for which no randomized ex post IC-IR mechanism can achieve better than a 2-approximation, we aim to design mechanisms that achieve a c-approximation to the optimal welfare where $c \geq 2$ (the closer to 2 the better).

3 Main Result and Construction Overview

Our main result is the following:

Theorem 1. *For every single-item auction setting with n bidders, binary signals and interdependent SOS valuations, there exists an ex post IC-IR mechanism that achieves a 2-approximation to the optimal welfare.*

Our proof of Theorem 1 is constructive – we design an algorithm that gets as input the valuation functions $v_1, \ldots, v_n$, and outputs an allocation rule x. Note that the main goal of the algorithm is to establish existence. Rule x is guaranteed to be both feasible and monotone. Thus, coupled with the appropriate expected payments p (based on critical signal reports), it constitutes an ex post IC-IR mechanism $M = (x, p)$. The main technical challenge is in showing that mechanism M has the following welfare guarantee: for every signal profile s, $\mathrm{ALG}(s) \geq \frac{1}{2}\,\mathrm{OPT}(s)$. We prove this approximation ratio and establish x's other properties like monotonicity in Sect. 4. We now give an overview of the algorithm (for the pseudo-code see the full version).

3.1 Construction Overview

In this section we give an overview of our main algorithm. The algorithm maintains an "allocation table" with rows corresponding to the n bidders, and columns corresponding to subsets $S \subseteq [n]$. At termination, column S will represent the allocation rule $x(S)$, with entry (i, S) encoding $x_i(S)$. For clarity of presentation the encoding is via colors: At initiation, all entries of the table are colored white to indicate they have not yet been processed. During its run, the algorithm colors each entry (i, S) of the table either red or black. Once a cell has been colored red or black, its color remains invariant until termination. The colors represent allocation probabilities as follows:

- red = bidder i gets the item with probability $\frac{1}{2}$ at signal profile S;
- black = i does not get the item at S (i.e., gets the item with probability 0).

As an interesting consequence, the allocation rule achieving the 2-approximation guarantee of Theorem 1 uses only two allocation probabilities, namely $\frac{1}{2}$ and 0. Note that for feasibility, no more than two entries in a single column should be colored red, and the remaining entries should be colored black.

We now explain roughly how the colors are determined by the algorithm. Consider an inclusion-compatible ordering of all subsets of $[n]$. The algorithm iterates over the ordered subsets, with S representing the current subset. We say bidder i *can be colored red at iteration* S if at the beginning of the iteration, (i, S) is colored either white or red. We define a notion of *favored bidder(s) at iteration* S – these are the ones the algorithm "favors" as winners of the item given signal profile S, and so will color them red at S. First, if there is a pair of bidders $i \neq j$ for which the following conditions all hold, we say they are favored at iteration S with **Priority 1**:

1. Bidder i and j's signals both belong to S (i.e., $s_i = s_j = 1$);
2. Bidders i and j can both be colored red at iteration S;
3. No other bidder $k \neq i, j$ is colored red at the beginning of iteration S;
4. The sum of values $v_i(S) + v_j(S)$ is at least $\mathrm{OPT}(S)$ (recall that $\mathrm{OPT}(S)$ is the highest value of any bidder for the item given signal profile S).

If such a pair does not exist, but there exists a bidder i who satisfies the following alternative conditions, we say i is favored at iteration S with **Priority 2**:

1. Bidder i can be colored red at iteration S;
2. The value $v_i(S)$ equals $\mathrm{OPT}(S)$.

Our main technical result in the analysis of the algorithm is to show that, unless at the beginning of iteration S two bidders are already colored red, then one of the two cases above must hold. That is, in every iteration S with no two reds, there is always either a favored pair with **Priority 1**, or a single favored bidder with **Priority 2**. Assuming this holds, the algorithm proceeds as follows. At iteration S it checks whether two bidders are already red, and if so continues to the next iteration. Otherwise, it colors the favored bidder(s) red by priority, and all other bidders black. The algorithm then performs *propagation* to other subsets S' in order to maintain monotonicity of the allocation rule (the term propagation was introduced in our context by [5]):

- If bidder $i \notin S$ is colored red at subset S, then red is propagated *forward* to bidder i at subset $S' = S \cup \{i\}$.
- If bidder $i \in S$ is colored black at subset S, then black is propagated *backward* to bidder i at subset $S' = S \setminus \{i\}$.

This completes the overview of our construction.

4 Proof of Theorem 1

We begin with a simple but useful observation:

Observation 1. *Consider a subset $S \subsetneq [n]$ and $i \notin S$. If during iteration S bidder i is colored red then $v_i(S) = \text{OPT}(S)$.*

Proof. The algorithm colors a bidder with a low signal red only if this bidder has **Priority 2**, and in this case her value must be highest among all bidders.

We now prove our main theorem, up to three lemmas that appear in Sects. 4.1, 4.2 and 4.3, respectively. Section 4.2 also develops a necessary tool for the proof in Sect. 4.3.

Proof (Theorem 1). We show that our algorithm returns an allocation rule that is feasible, monotone, and achieves a 2-approximation to the optimal welfare. For such an allocation rule there exist payments that result in an ex post IC-IR mechanism (see Sect. 2.1), establishing the theorem.

Let x be the allocation rule returned by the algorithm. We first show x is *feasible.* That is, for every $S \subseteq [b]$, the algorithm colors (i, S) either red or black for every bidder i, and at most two bidders are colored red in column S. To show this we invoke Lemma 1, by which the algorithm never reaches one of its error lines. Given that there are no errors, observe that the algorithm goes over all subsets, and for every subset $S \subseteq [n]$ either (i) skips to the next subset (if two bidders are already red), or (ii) finds a **Priority 1** pair or **Priority 2** bidder and colors them red. Indeed, by Lemma 4, if (i) does not occur then (ii) is necessarily successful. Once a **Priority 1** pair or **Priority 2** bidder is found, the rest of the column is colored black. Furthermore, once any two bidders in a column are colored red, the rest of the column is colored black. This establishes feasibility.

We now show x is *monotone.* Since the only allocation probabilities x assigns are $\frac{1}{2}$ and 0 (and one of these is always assigned), it is sufficient to show that for every $S \subseteq [n]$ and $i \notin S$, if $x(i, S) = \frac{1}{2}$ then $x(i, S \cup \{i\}) = \frac{1}{2}$. This holds since every time the algorithm calls ColorRed to color (i, S), it propagates the color red forward to $(i, S \cup \{i\})$ as well.

It remains to show that x achieves a 2-approximation to the optimal welfare. By definition of **Priority 1** and **Priority 2**, if such bidders are colored red then a 2-approximation is achieved for the corresponding signal profiles. It remains to consider signal subsets S for which at the beginning of iteration S, two cells i, j in the column are already colored red. These reds propagated forward from $v_i(S \backslash \{i\})$ and $v_j(S \backslash \{j\})$. Let $v_k(S)$ be the highest value at S. By Observation 1, $v_i(S \backslash \{i\})$ and $v_j(S \backslash \{j\})$ are highest at $S \backslash \{i\}$ and $S \backslash \{j\}$, respectively:

$$v_i(S \backslash \{i\}) \geq v_k(S \backslash \{i\});$$
$$v_j(S \backslash \{j\}) \geq v_k(S \backslash \{j\}).$$

Applying Lemma 2 to the above inequalities, it cannot simultaneously hold that $v_k(S) > v_i(S) + v_j(S)$. So $v_i(S) + v_j(S) \geq v_k(S)$, and the approximation guarantee holds, completing the proof. □

4.1 No Errors

Lemma 1 (No errors). *The algorithm runs without producing an error.*

The proof of Lemma 1 appears in the full version.

4.2 Properties of SOS Valuations

In this section we state and prove two lemmas for SOS valuations. The first is used in the proof of Theorem 1, and the second is the workhorse driving the proof in Sect. 4.3 that either **Priority 1** or **Priority 2** always hold. Very roughly, the first lemma (Lemma 2) states that if bidders i, j have higher values than bidder k when their own signals are low, then bidder k's value cannot exceed their sum when their signals are high. The second lemma (Lemma 3) is more complex, and to give intuition for what it states we provide a visualization in Figs. 1, 2 and 3 (we use a similar visualization to sketch our main proof in Sect. 4.3). The proof is by induction and is deferred to the full version of the paper.

Lemma 2. *Consider a subset $S' \subseteq [n]$ and three bidders i, j, k (not necessarily distinct) with SOS valuations over binary signals; let $S^* = S' \cup \{i, j\}$. If the following three inequalities hold simultaneously then they all hold with equality:*

- $v_i(S^* \backslash \{i\}) \geq v_k(S^* \backslash \{i\})$;
- $v_j(S^* \backslash \{j\}) \geq v_k(S^* \backslash \{j\})$;
- $v_k(S^*) \geq v_i(S^*) + v_j(S^*)$.

Proof. Define $u_i(\cdot) = v_i(\cdot \mid S')$ for every $i \in [n]$. Using this notation, to prove the lemma we need to show that if the following three inequalities hold simultaneously, they must all hold with equality:

- $u_i(j) + v_i(S') \geq u_k(j) + v_k(S')$;
- $u_j(i) + v_j(S') \geq u_k(i) + v_k(S')$;
- $u_k(ij) + v_k(S') \geq u_i(ij) + v_i(S') + u_j(ij) + v_j(S')$.

Assume the inequalities hold. Summing them and simplifying we get

$$u_i(j) + u_j(i) + u_k(ij) \geq u_k(j) + u_k(i) + v_k(S') + u_i(ij) + u_j(ij). \quad (1)$$

We now use the fact that $u_k(\cdot)$ is subadditive (Proposition 1), so $u_k(ij) \leq u_k(j) + u_k(i)$. Thus by Inequality (1),

$$u_i(j) + u_j(i) \geq v_k(S') + u_i(ij) + u_j(ij).$$

By monotonicity of set functions u_i and u_j, $u_i(j) \leq u_i(ij)$ and $u_j(i) \leq u_j(ij)$. We conclude that $v_k(S') \leq 0$, which can hold only with equality. But this equality would be violated if one of the three inequalities was strict, completing the proof. □

Lemma 3. *Consider a subset $S' \subseteq [n]$ and $3 + \ell_1 + \ell_2 + \ell_3$ bidders $E = \{i, j, k, t_1, \ldots, t_{\ell_1}, t'_1, \ldots, t'_{\ell_2}, t''_1, \ldots, t''_{\ell_3}\}$ (not necessarily distinct) with SOS valuations over binary signals. Let $t_{\ell_1+1} = k$, $t'_{\ell_2+1} = k$, $t''_{\ell_3+1} = i$, and $S^* = S' \cup E$. If the following $3 + \ell_1 + \ell_2 + \ell_3$ inequalities hold simultaneously then they all hold with equality:*

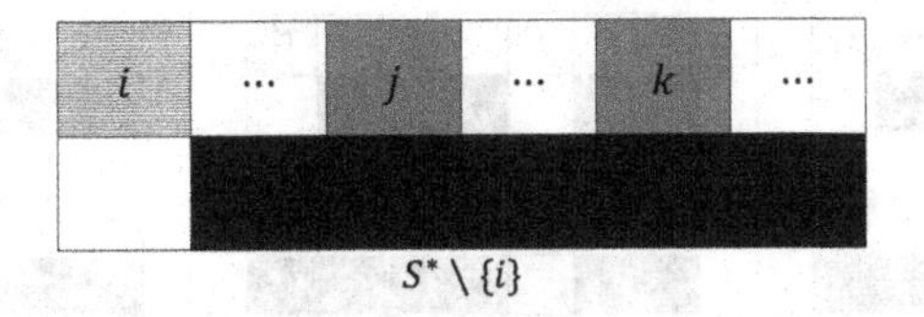

Fig. 1. Visualization of inequality (2).

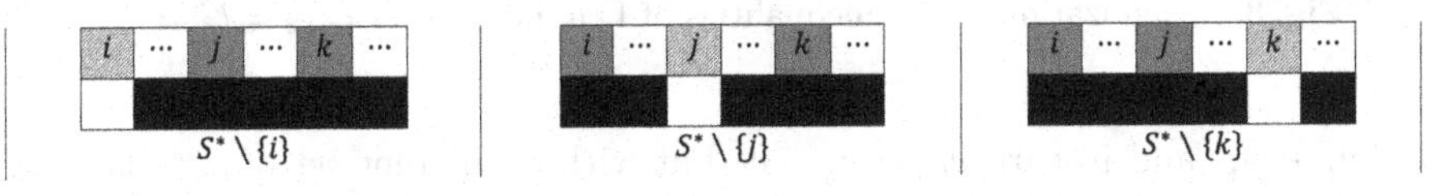

Fig. 2. Visualization of the inequalities of Lemma 3 for $\ell_1 = \ell_2 = \ell_3 = 0$.

- $v_i(S^*\backslash\{i\}) \geq v_{t_1}(S^*\backslash\{i\}) + v_j(S^*\backslash\{i\})$;
- $\forall h \in [\ell_1] : v_{t_h}(S^*\backslash\{i, t_1, ..., t_h\}) \geq v_{t_{h+1}}(S^*\backslash\{i, t_1, ..., t_h\}) + v_j(S^*\backslash\{i, t_1, ..., t_h\})$;
- $v_j(S^*\backslash\{j\}) \geq v_{t'_1}(S^*\backslash\{j\}) + v_i(S^*\backslash\{j\})$;
- $\forall h \in [\ell_2] : v_{t'_h}(S^*\backslash\{j, t'_1, ..., t'_h\}) \geq v_{t'_{h+1}}(S^*\backslash\{j, t'_1, ..., t'_h\}) + v_i(S^*\backslash\{j, t'_1, ..., t'_h\})$;
- $v_k(S^*\backslash\{k\}) \geq v_{t''_1}(S^*\backslash\{k\}) + v_j(S^*\backslash\{k\})$;
- $\forall h \in [\ell_3] : v_{t''_h}(S^*\backslash\{k, t''_1, ..., t''_h\}) \geq v_{t''_{h+1}}(S^*\backslash\{k, t''_1, ..., t''_h\}) + v_j(S^*\backslash\{k, t''_1, ..., t''_h\})$.

Visualization of Lemma 3

Consider the case $\ell_1 = \ell_2 = \ell_3 = 0$. The first inequality of Lemma 3 in this case, using that $t_1 = t_{\ell_1+1}$ (which equals k by definition), is:

$$v_i(S^*\backslash\{i\}) \geq v_k(S^*\backslash\{i\}) + v_j(S^*\backslash\{i\}). \tag{2}$$

We introduce a visualization of Inequality (2) as the $2 \times n$ "block" shown in Fig. 1. The columns of the block correspond to the bidders. The first row of the block represents which bidders participate in the inequality (in this case i, j, k), with the bidder on the greater (left) side of the inequality depicted in striped red (in this case i); the second row represents the signal set in the inequality (in this case $S^*\backslash i$), with the signals not in the set depicted in white (in this case i).

We can use the above visualization to depict all inequalities of Lemma 3. For the case that $\ell_1 = \ell_2 = \ell_3 = 0$, these are shown in Fig. 2. Consider now the case $\ell_1 = \ell_2 = \ell_3 = 1$, with the following inequalities (among others):

$$v_i(S^*\backslash\{i\}) \geq v_j(S^*\backslash\{i\}) + v_{t_1}(S^*\backslash\{i\});$$
$$v_{t_1}(S^*\backslash\{i, t_1\}) \geq v_j(S^*\backslash\{i, t_1\}) + v_k(S^*\backslash\{i, t_1\}).$$

In this case we have a fourth bidder t_1 who "bridges" between i, j, k. Instead of an inequality requiring that $v_i \geq v_j + v_k$ directly, here it is required that

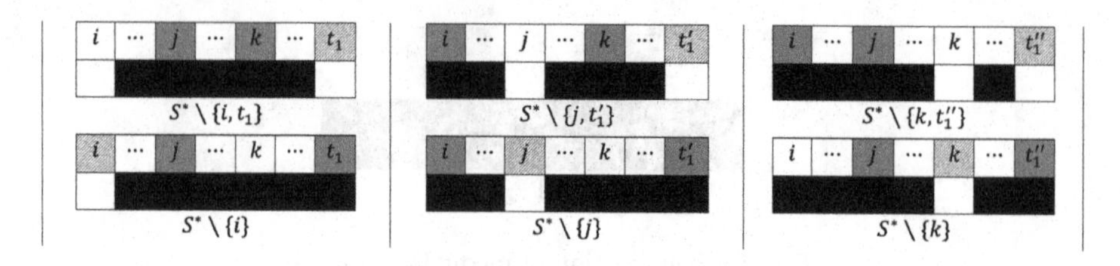

Fig. 3. Visualization of the inequalities of Lemma 3 for $\ell_1 = \ell_2 = \ell_3 = 1$.

$v_i \geq v_j + v_{t_1}$, and in turn $v_{t_1} \geq v_j + v_k$ but with a different set of signals. The full system of inequalities for this case (with 6 inequalities) appears in Fig. 3.

More generally, Lemma 3 holds for any number of "bridge" bidders. The general case is shown in the full version of the paper.

4.3 When Are Priorities 1 or 2 Guaranteed

In this section we prove the following lemma (some details of the proof are deferred to the full version of the paper):

Lemma 4. *Assume the algorithm runs on n bidders with SOS valuations over binary signals. Then for every $S \subseteq [n]$, if at the beginning of iteration S there are less than two red bidders, either* **Priority 1** *or* **Priority 2** *must hold.*

We begin with two observations that will be useful in the proof of Lemma 4.

Observation 2. *If at the beginning of iteration S no bidder is colored red, and during the iteration bidder i whose signal is low ($i \notin S$) is colored red, then for every pair $j, k \in S$, $v_i(S) \geq v_j(S) + v_k(S)$.*

Proof. Assume for contradiction that $v_j(S)+v_k(S) > v_i(S)$, then since j, k both have high signals and can be colored red at iteration S, they have **Priority 1** and should be colored in place of bidder i, contradiction. □

Observation 3. *If at the beginning of iteration S only bidder t whose signal is high ($t \in S$) is colored red, and during the iteration bidder i whose signal is low ($i \notin S$) is colored red, then for every $j \in S$, $v_i(S) \geq v_j(S) + v_t(S)$.*

Proof. Assume for contradiction that $v_j(S) + v_t(S) > v_i(S)$, then since j, t both have high signals and can be colored red at iteration S (t is already red and j can be colored red since there are no other reds besides t), they have **Priority 1** and should be colored in place of bidder i, contradiction. □

We can now prove our main lemma; missing details appear in the full version of the paper.

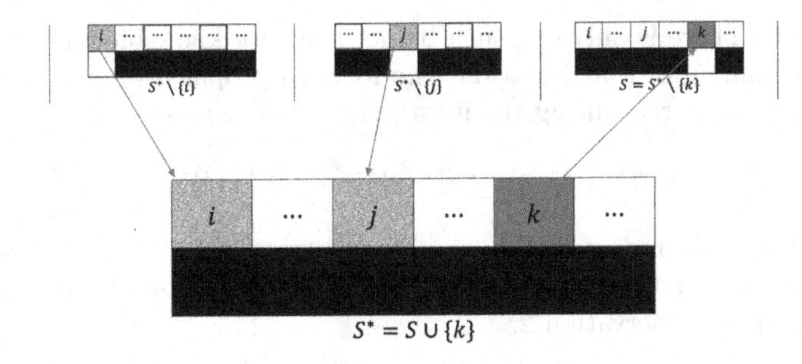

Fig. 4. Colorings at S, S^*, $S^*\backslash\{i\}$ and $S^*\backslash\{j\}$ given that (k, S) is black.

Proof (Lemma 4, sketch). Fix an iteration S with < 2 red bidders at its beginning. By *highest bidder* we mean the bidder whose value at S equals $\mathrm{OPT}(S)$. We split the analysis into cases; the most challenging cases technically are when the highest bidder is colored black in column S, and there are either no red cells or a single red cell in this column at the beginning of the iteration. Here we focus on the first among these cases and remark at the end how to treat the second, showing in both why a **Priority 1** pair exists in column S. The remaining cases are addressed in the full version of the paper.

Case 1: No Red Cells. Assume that at the beginning of iteration S, the highest bidder is colored black and there are no red cells in column S. Denote the highest bidder by k and observe that its color must have propagated backward from $(k, S \cup \{k\})$; let $S^* = S \cup \{k\}$. In column S^* there must therefore be two red bidders, whom we refer to as i and j, due to which k is colored black in this column. Red must have propagated forward to column S^* from $S^*\backslash\{i\}$ and $S^*\backslash\{j\}$. Figure 4 shows the allocation status of the relevant bidders at the beginning of iteration S for subsets $S, S^*, S^*\backslash\{i\}, S^*\backslash\{j\}$ – we use the same visualization as in Sect. 4.2, but with colors in the first row representing those set by the algorithm and arrows representing propagations.

Towards establishing existence of a **Priority 1** pair in column S, consider first the case in which the following two conditions hold:

1. At the beginning of iteration $S^*\backslash\{i\}$, no cells in that column are red;
2. At the beginning of iteration $S^*\backslash\{j\}$, no cells in that column are red.

By Observation 2,

$$v_i(S^*\backslash\{i\}) \geq v_j(S^*\backslash\{i\}) + v_k(S^*\backslash\{i\}); \tag{3}$$

$$v_j(S^*\backslash\{j\}) \geq v_i(S^*\backslash\{j\}) + v_k(S^*\backslash\{j\}). \tag{4}$$

If both (3) and (4) hold, by Lemma 3 with $\ell_1 = \ell_2 = \ell_3 = 0$ it cannot simultaneously hold that $v_k(S) > v_i(S) + v_j(S)$ (see Fig. 2 and related text). Thus $v_i(S) + v_j(S) \geq v_k(S)$, and pair i, j has **Priority 1**.

Now consider the case in which one of the two conditions does not hold, w.l.o.g. Condition (1). That is, at the beginning of iteration $S^*\backslash\{i\}$, a bidder t_1

is colored red. (There is only one such bidder since we know bidder i cannot be red at the beginning of that iteration – no forward propagation as $i \notin S^*\backslash\{i\}$ – and that i is colored red during the iteration). By Observation 3,

$$v_i(S^*\backslash\{i\}) \geq v_j(S^*\backslash\{i\}) + v_{t_1}(S^*\backslash\{i\}). \tag{5}$$

Since $(t_1, S^*\backslash\{i\})$ is red at the beginning of iteration $S^*\backslash\{i\}$, the color red necessarily propagated forward from $S^*\backslash\{i, t_1\}$. If Condition (1) now holds for $S^*\backslash\{i, t_1\}$ then by Observation 2,

$$v_{t_1}(S^*\backslash\{i, t_1\}) \geq v_j(S^*\backslash\{i, t_1\}) + v_k(S^*\backslash\{i, t_1\}) \tag{6}$$

If Inequalities (4)–(6) hold simultaneously, then by Lemma 3 with $\ell_1 = 1, \ell_2 = \ell_3 = 0$ we can again conclude that $v_i(S) + v_j(S) \geq v_k(S)$, and pair i, j has **Priority 1**. Notice that t_1 is the "bridge" bidder we discussed in Sect. 4.2. For visualization we note that Inequality (4) is the one depicted in Fig. 2 (middle) while Inequalities (5)–(6) are shown in Fig. 3 (left); these are the inequalities that correspond to $\ell_1 = 1$ and $\ell_2 = 0$.

If Condition (1) does not hold for subset $S^*\backslash\{i, t_1\}$, then there is an additional "bridge" bidder t_2 that is colored red at $S^*\backslash\{i, t_1\}$, and we can possibly apply Lemma 3 with $\ell_1 = 2$. If not, we continue in this way until either Condition (1) holds or only j, k remain in the subset. In either case, denote the final number of "bridge" bidders by ℓ_1. In the latter case, either Observation 2 or Observation 3 hold, and so

$$v_{t_{\ell_1}}(\{j, k\}) \geq v_j(\{j, k\}) + v_k(\{j, k\}).$$

By applying Lemma 3 with $\ell_1 > 0$ (and $\ell_2 = \ell_3 = 0$) we conclude that pair i, j has **Priority 1**.

Observe that the same analysis holds if both Condition (1) and Condition (2) are relaxed. In this case Lemma 3 applies with $\ell_1 > 0, \ell_2 > 0$ (and $\ell_3 = 0$).

Case 2: Single Red Cell. Finally, we address the case in which there exists a red bidder t''_1 in column S at the beginning of iteration S. We can write S as $S^*\backslash\{k\}$; the color red of t''_1 necessarily propagated forward from $S^*\backslash\{k, t''_1\}$. Assume the following third condition holds:

3. At the beginning of iteration $S^*\backslash\{k, t''_1\}$, no cells in that column are red.

By Observation 2,

$$v_{t''_1}(S^*\backslash\{k, t''_1\}) \geq v_j(S^*\backslash\{k, t''_1\}) + v_i(S^*\backslash\{k, t''_1\}). \tag{7}$$

If Inequality (7) holds (alongside previous inequalities) then Lemma 3 applies with $\ell_1 > 0, \ell_2 > 0, \ell_3 = 1$. If Condition (3) does not hold for subset $S^*\backslash\{k, t''_1\}$, we continue as above, denoting the final number of "bridge" bidders by ℓ_3. By applying Lemma 3 with $\ell_1 > 0, \ell_2 > 0$ and $\ell_3 > 0$, we conclude that pair i, t''_1 has **Priority 1**. □

5 Matroid Auction Settings

In this section we establish, by reduction to the single item case, the existence of a truthful mechanism for matroid settings that achieves a 2-approximation to the optimal welfare. A *matroid auction setting* is defined by a matroid $([n], \mathcal{I})$, where $\mathcal{I}$ contains all *independent sets* of bidders, i.e., subsets of bidders who can simultaneously *win* in the auction. We also refer to sets in $\mathcal{I}$ as *feasible.* For example, if there are k units of the item to allocate in the auction, $\mathcal{I}$ can be all possible subsets of k bidders. The rest of the setting is as before, i.e., every bidder i has a signal s_i and a valuation $v_i(s_1, \ldots, s_n)$ for winning. It is well known that in matroid settings *with given values*, welfare maximization is achieved by greedily adding bidders to the winner set W while keeping W feasible.

Proposition 2. *If for single item auctions there exists a monotone feasible allocation rule that achieves a 2-approximation to the optimal welfare, then there exists such an allocation rule for matroid auction settings as well. Specifically, Algorithm 1 is an allocation mechanism for matroid settings that achieves a 2-approximation to the optimal welfare.*

Algorithm 1. Matroid Settings

1: **function** MATROIDMECHANISM($(E,\mathcal{I}), k, S = \{s_i \mid s_i = 1\}, V = (v_1, \ldots, v_n)$)
2: $W = \emptyset$ ▷ Current winning bidders
3: **for** $j \in [k]$ **do** ▷ Run k times where k is the matroid rank
4: $x =$ ALLOCATE(E,V) ▷ ALLOCATE: the algorithm for the single item case
5: $b =$ randomly choose a bidder such that every bidder $b' \in E$ is chosen with probability $x(b', S)$
6: $W = W \cup \{b\}$ ▷ Add b to the winning set
7: $E =$ all bidders b' such that $W \cup \{b'\} \in \mathcal{I}$ ▷ Remove bidders who can no longer feasibly win
8: **end for**
9: **end function**

6 Summary and Future Directions

Tension between optimization and truthfulness is an important theme of algorithmic game theory. With interdependent values, this tension appears even without computational considerations. Since with interdependence there is an inherent clash between welfare maximization and truthfulness, the approximation toolbox comes in handy. We apply it to arguably the simplest possible setting (single-item auctions with binary signals), and get a tight understanding of the tradeoff (i.e., what fraction of the optimal welfare can be guaranteed by a truthful mechanism). Our results extend beyond single items.

Two promising future directions are: (i) generalizing our results beyond binary signals, and (ii) designing an "on the fly" tractable version of the 2-approximation truthful mechanism (i.e., a version that gets signal reports and returns an allocation only for the reported signal profile). For the former direction, non-binary signals pose additional challenges since two priorities are no longer sufficient in the algorithm, and additionally the propagation is more complex. In the full version of the paper we present progress towards resolving these challenges (in particular, we provide an extension of Lemma 3 to functions over general integer signals).

References

1. Ausubel, L.: A Generalized Vickrey auction. In: Econometric Society World Congress 2000 Contributed Papers 1257, Econometric Society (2000)
2. Chawla, S., Fu, H., Karlin, A.R.: Approximate revenue maximization in interdependent value settings. In: Proceedings of the 15th ACM Conference on Economics and Computation, EC, pp. 277–294 (2014)
3. Dasgupta, P., Maskin, E.: Efficient auctions. Q. J. Econ. **115**(2), 341–388 (2000)
4. Eden, A., Feldman, M., Fiat, A., Goldner, K.: Interdependent values without single-crossing. In: Proceedings of the 19th ACM Conference on Economics and Computation, EC, p. 369 (2018)
5. Eden, A., Feldman, M., Fiat, A., Goldner, K., Karlin, A.R.: Combinatorial auctions with interdependent valuations: SOS to the rescue. In: Proceedings of the 20th ACM Conference on Economics and Computation, EC, pp. 19–20 (2019)
6. Eden, A., Feldman, M., Talgam-Cohen, I., Zviran, O.: PoA of simple auctions with interdependent values. In: The 35th AAAI Conference on Artificial Intelligence (2021)
7. Gkatzelis, V., Patel, R., Pountourakis, E., Schoepflin, D.: Prior-free clock auctions for bidders with interdependent values. In: 14th International Symposium on Algorithmic Game Theory (2021)
8. Jehiel, P., Moldovanu, B.: Efficient design with interdependent valuations. Econometrica **69**(5), 1237–1259 (2001)
9. Lehmann, B., Lehmann, D., Nisan, N.: Combinatorial auctions with decreasing marginal utilities. Games Econ. Behav. **55**(2), 270–296 (2006)
10. Li, Y.: Approximation in mechanism design with interdependent values. Games Econ. Behav. **103**(C), 225–253 (2017)
11. Maskin, E.: Auctions and privatization. In: Siebert, H. (ed.) Privatization, pp. 115–136. J.C.B. Mohr Publisher (1992)
12. Milgrom, P.R., Weber, R.J.: A theory of auctions and competitive bidding. Econometrica **50**(5), 1089–1122 (1982)
13. Niazadeh, R., Roughgarden, T., Wang, J.R.: Optimal algorithms for continuous non-monotone submodular and DR-submodular maximization. J. Mach. Learn. Res. **21**, 125:1–125:31 (2020)
14. Nobel Committee for the Prize in Economic Sciences: Scientific background: Improvements to auction theory and inventions of new auction formats. The Royal Swedish Academy of Sciences Press Release (2020)
15. Oxley, J.G.: Matroid Theory (Oxford Graduate Texts in Mathematics). Oxford University Press, Inc., USA (2006)
16. Roughgarden, T., Talgam-Cohen, I.: Optimal and robust mechanism design with interdependent values. ACM Trans. Econ. Comput. **4**(3), 18:1–18:34 (2016)
17. Vickrey, W.: Counterspeculation, auctions, and competitive sealed tenders. J. Financ. **16**(1), 8–37 (1961)
18. Wilson, R.: A bidding model of perfect competition. Rev. Econ. Stud. **44**(3), 511–518 (1977)

Approximate Mechanism Design for Distributed Facility Location

Aris Filos-Ratsikas[1] and Alexandros A. Voudouris[2](✉)

[1] Department of Computer Science, University of Liverpool, Liverpool, UK
aris.filos-ratsikas@liverpool.ac.uk

[2] School of Computer Science and Electronic Engineering, University of Essex, Colchester, UK
alexandros.voudouris@essex.ac.uk

Abstract. We consider a single-facility location problem, where agents are positioned on the real line and are partitioned into multiple disjoint districts. The goal is to choose a location (where a public facility is to be built) so as to minimize the total distance of the agents from it. This process is distributed: the positions of the agents in each district are first aggregated into a representative location for the district, and then one of the district representatives is chosen as the facility location. This indirect access to the positions of the agents inevitably leads to inefficiency, which is captured by the notion of *distortion*. We study the discrete version of the problem, where the set of alternative locations is finite, as well as the continuous one, where every point of the line is an alternative, and paint an almost complete picture of the distortion landscape of both general and strategyproof distributed mechanisms.

Keywords: Facility location · Mechanism design · Distortion

1 Introduction

Social choice theory deals with the aggregation of different, often contrasting opinions into a common decision. There are many applications where the nature of the aggregation process is distributed, in the sense that it is performed in the following two steps: smaller groups of people first reach a consensus, and then their representative choices are aggregated into a final collective decision. This can be due to multiple reasons, such as scalability (local decisions are much easier to coordinate when dealing with a large number of individuals), or the inherent roles of the participants (for example, being member states in the European Union or electoral bodies in different regional districts). However, although often necessary, this distributed nature is known to lead to outcomes that do not accurately reflect the views of society. A prominent example of this fact is the 2016 US presidential election, where Donald Trump won despite receiving only 46.1% of the popular vote, as opposed to Hillary Clinton's 48.2%.

To quantify the inefficiency that arises in distributed social choice settings, recently Filos-Ratsikas *et al.* [20] adopted and extended the notion of *distortion*,

I. Caragiannis and K. A. Hansen (Eds.): SAGT 2021, LNCS 12885, pp. 49–63, 2021.
https://doi.org/10.1007/978-3-030-85947-3_4

which is broadly used in social choice theory to measure the deterioration of an aggregate objective (typically the utilitarian social welfare) due to the lack of complete information, and thus provides a systematic way of comparing different mechanisms. In their work, Filos-Ratsikas *et al.* considered a very general social choice scenario with unrestricted agent preferences, and showed asymptotically tight upper and lower bounds on the distortion of plurality-based mechanisms. We follow a similar approach in this paper for a fundamental structured domain of agent preferences, the well-known *facility location* problem on the line of real numbers.

The facility location problem is one of the most important in social choice, and has been considered in both the economics and the computer science literature. It is a special case of the *single-peaked preferences* domain [9,26] equipped with linear agent cost functions. Furthermore, it is the most prominent setting where the agents have *metric preferences*, and as such it has been studied extensively in the related distortion literature for centralized settings [3,6]. Finally, facility location was the paradigm used by Procaccia and Tennenholtz [29] to put forward their agenda of *approximate mechanism design without money*, which resulted in a plethora of works in computer science ever since.

In the agenda of Procaccia and Tennenholtz, the goal is to design mechanisms that are *strategyproof* (that is, they do not provide incentives to the agents to lie about their true preferences) and have good performance in terms of some aggregate objective, as measured by having low *approximation ratio*. The need for approximation now comes from the strategyproofness requirement, rather than the lack of information. In fact, the distortion and the approximation ratio are essentially two sides of the same coin, differentiated by the reason for the loss in efficiency. We will be concerned with *distributed mechanisms*, both strategyproof and not, in a quest to quantify the effect of distributed decision making on facility location, both independently and in conjunction with strategyproofness. Hence, our work follows the agendas of both approximate mechanism design [29] and of distributed distortion [20], and can be cast as *approximate mechanism design for distributed facility location.*

Our Setting and Contribution. We study the distributed facility location problem on the real line $\mathbb{R}$. As in the standard centralized problem, there is a set of agents with ideal positions and a set of alternative locations where the facility can be built. We consider both the *discrete* setting, where the set of alternatives is some finite subset of $\mathbb{R}$, as well as the *continuous* setting, where the set of alternatives is the whole $\mathbb{R}$. In the distributed version, the agents are partitioned into *districts*, and the aggregation of their positions into a single facility location is performed in two steps: In the first step, the agents of each district select a *representative* location for their district, and in the second step, *one of the representatives* is chosen as the final facility location; in Sect. 6, we discuss how our results extend to the case of *proxy voting*, where the location can be chosen from the set of all alternatives.

Our goal is to find the mechanism with the smallest possible *distortion*, which is defined as the worst-case ratio (over all instances of the problem) between the social cost of the location chosen by the mechanism and the minimum social cost over all locations; the social cost of a location is the total distance between the agent positions and the location. Note that the optimal location is calculated as if the agents are *not* partitioned into districts, and thus the distortion accurately measures the effect of selecting the facility location in a distributed manner to the efficiency of the system. We are also interested in *strategyproof mechanisms*, for which the distortion quantifies the loss in performance both due to lack of information and due to requiring strategyproofness. We mainly focus on the case of *symmetric* districts, which have equal size; in Sect. 6 we also discuss the case of *asymmetric* districts and other extensions. Our results are as follows:

- For the discrete setting, the best possible distortion by any mechanism is 3, and the best possible distortion by any strategyproof mechanism is 7.
- For the continuous setting, the best possible distortion by any mechanism is between 2 and 3, and the best possible distortion by any strategyproof mechanism is 3.

The mechanisms we design are adaptations of well-known mechanisms for the centralized facility location problem. In the discrete setting, the mechanism with the best possible distortion of 3 selects the representative of each district to be the location that minimizes the social cost of the agents therein, and then chooses the median representative as the facility location; we refer to this mechanism as MinimizeMedian. By modifying the first step so as to select the representative of a district to be the location that is the closest to the median agent in the district, we obtain the DistributedMedian mechanism, which is the best possible strategyproof mechanism with distortion 7. When we move to the continuous setting, selecting the median agent within each district minimizes the social cost of the agents therein, and thus DistributedMedian is an implementation of MinimizeMedian. The proofs of our upper bounds in Sects. 3 and 5 rely on a characterization of the structure of worst-case instances (in terms of distortion) for each of these mechanisms, which is obtained by carefully modifying the positions of some agents without decreasing the distortion.

For the lower bounds, we employ the following main idea. We construct instances of the problem for which any mechanism with low distortion (depending on the bound we are aiming for) must satisfy some constraints about the representative y it can choose for a particular district, namely, either that y is some specific location (in the discrete setting), or that it must lie in some specific interval (in the continuous setting). Then, because of the distributed nature of the mechanism, we can exploit the fact that y must represent this district in any instance that contains it, and use such instances to either argue about the distortion of the mechanism, or to impose constraints on the representatives of other districts. This idea is used repeatedly and inductively, and in conjunction with strategyproofness arguments when necessary.

Related Work. The notion of distortion was first introduced by Procaccia and Tennenholtz [28], who considered a setting where the agents have *normalized* cardinal valuations and the objective is to choose a single alternative. In its original definition, the distortion measured the performance of ordinal social choice mechanisms in terms of a cardinal objective, namely the utilitarian social welfare (the total utility of the agents for the chosen outcome). However, if one interprets the lack of information as the reason for the loss in efficiency, the distortion actually captures much wider scenarios, like the distributed social choice setting studied by Filos-Ratsikas *et al.* [20].

Although the number of papers dealing with (variants of) the aforementioned normalized setting is substantial (e.g., see [2,10,11,25]) the literature on the distortion flourished after Anshelevich *et al.* [3] and Anshelevich and Postl [6] studied settings in which the agents have *metric* preferences. Such preferences are constrained by the fact that the utility (or cost in the particular case) of every agent for different alternatives must satisfy the triangle inequality, which effectively results in the distortion bounds being small constants, rather than asymptotic bounds depending on the number of agents and alternatives. Similar investigations have given rise to a plethora of papers on this topic (e.g., see [1,7, 18,27]). For a comprehensive introduction to the distortion literature, we refer the reader to the recent survey of Anshelevich *et al.* [4].

As already mentioned, the facility location problem plays an important role in the literature at the intersection of computer science and economics. It became extremely popular in the economics and computation community after Procaccia and Tennenholtz [29] used it to put forward their agenda of *approximate mechanism design without money*, and has been studied for different objectives [17,19], multiple facilities [23,24], different domains [30], and several variants of the problem [13–16,21,22,31,32]; See also the recent survey of Chan *et al.* [12].

The most related setting to our work is an extension studied by Procaccia and Tennenholtz [29] with super-agents controlling multiple locations, whose cost is the total distance between their locations and the facility. They showed that the mechanism that first selects the median location of each super-agent and then the median of those is strategyproof and 3-approximate for the social cost. This implies an upper bound of 3 on the distortion of strategyproof mechanisms in our continuous setting, by interpreting the super-agents as district representatives; we show that this bound can be obtained by simple extensions of our techniques for the discrete setting. Procaccia and Tennenholtz also showed a matching lower bound, which however requires the super-agents to be truthful, and thus does not have any implications for our setting. This model was later extended by Babaioff *et al.* [8] to a setting where the locations are themselves strategic agents, and the agents of the higher level are strategic *mediators*.

2 Preliminaries

We consider the following distributed facility location problem. There is a set $\mathcal{N}$ of n *agents* positioned on the line of real numbers; let $x_i \in \mathbb{R}$ denote the *position* of agent $i \in \mathcal{N}$, and denote by $\mathbf{x} = (x_i)_{i \in \mathcal{N}}$ the *position profile* of all agents. The agents are partitioned into k *districts*; let $\mathcal{D}$ be the set of districts. We denote by $d(i)$ the district containing agent i, and by $\mathcal{N}_d$ the set of agents that belong to district $d \in \mathcal{D}$. In the main part of our paper, we focus on the case of *symmetric* districts such that $|N_d| = \frac{n}{k} = \lambda$; the case of *asymmetric* districts is discussed in Sect. 6. We will use the notation $\mathbf{x}_d = (x_i)_{i \in \mathcal{N}_d}$ for the restriction of $\mathbf{x}$ to the positions of the agents in district d, and we will refer to $\mathbf{x}_d$ as a *district position profile*. We say that two districts d and d' are *identical* if $\mathbf{x}_d = \mathbf{x}_{d'}$.

For two points $x, y \in \mathbb{R}$, let $\delta(x, y) = \delta(y, x) = |x - y|$ denote their *distance*. Given a position profile $\mathbf{x}$, the *social cost* of point $z \in \mathbb{R}$ is the total distance of the agents from z:

$$\text{SC}(z|\mathbf{x}) = \sum_{i \in \mathcal{N}} \delta(x_i, z)$$

Our goal is to select a location z^* from a set of *alternative locations* $\mathcal{Z} \subseteq \mathbb{R}$ to minimize the social cost: $z^* \in \arg\min_{z \in \mathcal{Z}} \text{SC}(z|\mathbf{x})$. In the *discrete* setting, the set of alternative locations is finite and denoted by $\mathcal{A}$, whereas, in the *continuous* setting, the set of alternative locations is the whole $\mathbb{R}$. Hence, $\mathcal{Z} = \mathcal{A}$ in the discrete version, or $\mathcal{Z} = \mathbb{R}$ in the continuous version.

We will use the term *instance* to refer to a tuple $\mathcal{I} = (\mathbf{x}, \mathcal{D}, \mathcal{Z})$ consisting of a position profile $\mathbf{x}$, a set of districts $\mathcal{D}$, and a set of alternative locations $\mathcal{Z}$; we omit the set of agents $\mathcal{N}$ as it is implied by $\mathbf{x}$. In the continuous setting, since the set of alternative locations is clear, we will simplify our notation further and use a pair $(\mathbf{x}, \mathcal{D})$ to denote an instance.

If we had access to the positions of all the agents, it would be easy to select the optimal location in both versions of the problem. However, in our setting the positions are assumed to be *locally* known, within each district. To decide the facility location we deploy *distributed mechanisms* (or, simply, *mechanisms*). A mechanism $\mathcal{M}$ consists of the following two steps of aggregation:

1. For every district $d \in \mathcal{D}$, the positions of the agents in d are aggregated into the *representative* location $z_d \in \mathcal{Z}$ of d. This step is *local*: z_d is a result of the corresponding district profile $\mathbf{x}_d$ only. Formally, for any two instances that contain two identical districts d_1 and d_2, we have that $z_{d_1} = z_{d_2} \in \mathcal{Z}$.
2. The district representatives are aggregated into a single *facility location*. That is, the facility location $\mathcal{M}(\mathcal{I})$ chosen by $\mathcal{M}$ when given as input the instance $\mathcal{I}$ is selected from the set of representatives.

The Distortion of Mechanisms. Due to lack of global information, the facility location chosen by a mechanism will inevitably be suboptimal. To quantify this inefficiency, we adopt and extend the notion of distortion to our setting. The *distortion* of an instance $\mathcal{I} = (\mathbf{x}, \mathcal{D}, \mathcal{Z})$ subject to using a mechanism $\mathcal{M}$ is the ratio between the social cost of the location $\mathcal{M}(\mathcal{I})$ and the social cost of the optimal location $\text{OPT}(\mathcal{I}) = \arg\min_{z \in \mathcal{Z}} \text{SC}(z|\mathbf{x})$ for $\mathcal{I}$:

$$\texttt{dist}(\mathcal{I}|\mathcal{M}) = \frac{\text{SC}(\mathcal{M}(\mathcal{I})|\mathbf{x})}{\text{SC}(\text{OPT}(\mathcal{I})|\mathbf{x})}.$$

Then, the distortion of $\mathcal{M}$ is the worst-case distortion over all possible instances:

$$\texttt{dist}(\mathcal{M}) = \sup_{\mathcal{I}} \texttt{dist}(\mathcal{I}|\mathcal{M}).$$

We say that a mechanism $\mathcal{M}$ is *unanimous*, if it chooses the representative of a district to be $z \in \mathcal{Z}$, whenever all agents in the district are positioned at z. The following lemma shows that it is without loss of generality to focus on unanimous mechanisms. Due to lack of space, the proof of the lemma as well as of other results are omitted.

Lemma 1. *Any mechanism with finite distortion is unanimous.*

Strategyproofness. Besides achieving low distortion, we are also interested in mechanisms which ensure that the agents report their positions *truthfully*, that is, they have no incentive to misreport hoping to change the outcome of the mechanism to a location that is closer to their position. Formally, let $\mathcal{I} = (\mathbf{x}, \mathcal{D}, \mathcal{Z})$ be an instance, where $\mathbf{x}$ is the *true* position profile of the agents, and let $\mathcal{J} = (\mathbf{y}, \mathcal{D}, \mathcal{Z})$ be any instance with position profile $\mathbf{y} = (y_i, \mathbf{x}_{-i})$, in which agent i reports y_i and all other agents report their positions according to $\mathbf{x}$. A mechanism $\mathcal{M}$ is *strategyproof* if the location chosen by $\mathcal{M}$ when given as input $\mathcal{I}$ is closer to the position x_i of any agent i than the location chosen by $\mathcal{M}$ when given as input $\mathcal{J}$. In other words, for every agent i and $y_i \in \mathbb{R}$, it must hold that

$$\delta(x_i, \mathcal{M}(\mathbf{x}, \mathcal{D}, \mathcal{Z})) \leq \delta(x_i, \mathcal{M}((y_i, \mathbf{x}_{-i}), \mathcal{D}, \mathcal{Z})).$$

This added requirement of strategyproofness imposes further restrictions, and potentially impacts the achievable distortion as well. Hence, our goal is to design strategyproof mechanisms with as low distortion as possible.

We now define the class of mechanisms that are strategyproof within districts. Intuitively, such mechanisms prevent the agents from misreporting in hopes of changing the representative of their district to a location closer to them. Observe that a strategyproof mechanism could in principle allow such a local manipulation, only to eliminate it in the second step (for example, by completely ignoring the representatives and choosing an arbitrary fixed facility location). We show that for mechanisms with a finite distortion, this is impossible.

Formally, a mechanism $\mathcal{M}$ is *strategyproof within districts* if for any district $d \in \mathcal{D}$, the representative of d on input $\mathcal{I} = (\mathbf{x}, \mathcal{D}, \mathcal{Z})$ is closer to the true position x_i of every agent i than the representative of d on input $\mathcal{J} = ((y_i, \mathbf{x}_{-i}), \mathcal{D}, \mathcal{Z})$. We can now show the following useful property of stratefyproof mechanisms.

Lemma 2. *Any strategyproof mechanism with finite distortion is strategyproof within districts.*

3 Mechanisms for the Discrete Setting

We begin the exposition of our results from the discrete setting. We consider the mechanisms MinimizeMedian (MM) and DistributedMedian (DM), which operate as follows. Given the representatives of the districts, both mechanisms select the facility location to be the median representative. The main difference between the two mechanisms is on how they select the representatives of the

Algorithm 1: MINIMIZEMEDIAN and DISTRIBUTEDMEDIAN

Mechanism MINIMIZEMEDIAN($\mathbf{x}, \mathcal{D}, \mathcal{A}$)
- **for** *each district* $d \in \mathcal{D}$ **do**
  - $z_d \leftarrow$ left-most location in $\arg\min_{z \in \mathcal{A}} \sum_{i \in \mathcal{N}_d} \delta(x_i, z)$
- **return** MEDIAN($\{z_d\}_{d \in \mathcal{D}}$)

Mechanism DISTRIBUTEDMEDIAN($\mathbf{x}, \mathcal{D}, \mathcal{A}$)
- **for** *each district* $d \in \mathcal{D}$ **do**
  - $z_d \leftarrow \arg\min_{z \in \mathcal{A}} \delta(Median(\mathbf{x}_d), z)$
- **return** MEDIAN($\{z_d\}_{d \in \mathcal{D}}$)

Rule MEDIAN($\mathbf{y}$)
- $\eta \leftarrow |\mathbf{y}|$
- sort $\mathbf{y} = (y_1, ..., y_\eta)$ in non-decreasing order
- **return** $\mathbf{y}_{\lfloor \eta/2 \rfloor}$

districts: MM selects the representative of each district to be the alternative location that minimizes the social cost of the agents within the district, while DM selects the representative of each district to be the location which is closest to the median agent in the district. In case there are at least two median representatives or at least two locations minimizing the social cost within some district, the left-most such option is chosen. See Algorithm 1 for a description of both mechanisms using pseudocode.

As one might expect, the fact that MM minimizes the social cost within the districts may give the opportunity to some agents to misreport their positions hoping to affect the outcome. On the other hand, by choosing the median location both within and over the districts, DM does not allow such manipulations. We have the following statement.

Theorem 1. MM *is not strategyproof, whereas* DM *is strategyproof.*

We now focus on bounding the distortion of these mechanisms. To this end, we first show in Sect. 3.1 that the instances achieving the worst-case distortion have a specific structure, which is common for both mechanisms. We then exploit this structure in Sect. 3.2 to show an upper bound of 3 on the distortion of MM and an upper bound of 7 on the distortion of DM.

3.1 Worst-Case Instances

For any mechanism $\mathcal{M} \in \{\text{MM}, \text{DM}\}$, let $\text{wc}(\mathcal{M})$ be the class of instances $\mathcal{I} = (\mathbf{x}, \mathcal{D}, \mathcal{A})$ such that:

(P1) For every agent $i \in \mathcal{N}$, $x_i \geq \mathcal{M}(\mathcal{I})$ if $\mathcal{M}(\mathcal{I}) < \text{OPT}(\mathcal{I})$, or $x_i \leq \mathcal{M}(\mathcal{I})$ if $\mathcal{M}(\mathcal{I}) > \text{OPT}(\mathcal{I})$.

(P2) For every location $z \in \mathcal{A}$, which is representative for a set of districts $\mathcal{D}_z \neq \varnothing$, the agents in those districts are positioned in the interval defined by z and $\text{OPT}(\mathcal{I})$.

Lemma 3. *The distortion of* $\mathcal{M} \in \{\text{MM}, \text{DM}\}$ *is equal to*

$$\sup_{\mathcal{I} \in \text{wc}(\mathcal{M})} \texttt{dist}(\mathcal{I}|\mathcal{M}).$$

Proof (sketch). It suffices to show that for every instance $\mathcal{J} \notin \text{wc}(\mathcal{M})$, there is an instance $\mathcal{I} \in \text{wc}(\mathcal{M})$, such that $\texttt{dist}(\mathcal{J}|\mathcal{M}) \leq \texttt{dist}(\mathcal{I}|\mathcal{M})$. Due to symmetry, assume that $\mathcal{M}(\mathcal{J}) = w < o = \text{OPT}(\mathcal{J})$. We transform $\mathcal{J}$ into $\mathcal{I}$:

(T1) Every agent i with position $x_i < w$ is moved to w.

(T2) For every location z which is representative for a set of districts $\mathcal{D}_z \neq \varnothing$ in $\mathcal{J}$, every agent in $\mathcal{D}_z$ whose position does not lie in the interval defined by z and o is moved to the boundaries of this interval.

Observe that, because (T1) is performed before (T2), an agent i with position $x_i < w < z < o$ who belongs to a district in $\mathcal{D}_z$ can be moved twice: once from x_i to w, and then again to z. These movements define a sequence of intermediate instances with the same districts and alternative locations, but different position profiles. We show that these instances preserve the following three properties, which are sufficient to show by induction that the distortion does not decrease as we go from $\mathcal{J}$ to $\mathcal{I}$: (a) The facility location chosen by the mechanism is always w; (b) The optimal location is always o; (c) For any two consecutive intermediate instances with position profiles $\mathbf{x}$ and $\mathbf{y}$, $\frac{\text{SC}(w|\mathbf{x})}{\text{SC}(o|\mathbf{x})} \leq \frac{\text{SC}(w|\mathbf{y})}{\text{SC}(o|\mathbf{y})}$. □

3.2 Bounding the Distortion

Given Lemma 3, we are ready to bound the distortion of both mechanisms.

Theorem 2. *The distortion of* MINIMIZEMEDIAN *is at most* 3.

Proof. Let $\mathcal{I} = (\mathbf{x}, \mathcal{D}, \mathcal{A}) \in \text{wc}(\mathcal{M})$ be an instance such that $\mathcal{M}(\mathcal{I}) = w < o = \text{OPT}(\mathcal{I})$, without loss of generality. For every $z \in \mathcal{A}$, let $\mathcal{D}_z$ be the set of districts for which z is the representative, and define the set $Z = \{z \in \mathcal{A} : \mathcal{D}_z \neq \varnothing\}$. Since the location w is selected by the mechanism, it must be $w \in Z$. For every $z \in Z$ and $y \in \mathcal{A}$, let

$$\text{SC}_z(y|\mathbf{x}) = \sum_{d \in \mathcal{D}_z} \sum_{i \in \mathcal{N}_d} \delta(x_i, y)$$

be the total distance of the agents in the districts of $\mathcal{D}_z$ from y. Also, recall that each district contains exactly λ agents. We make the following observations:

- Consider a location $z \in Z$. By property (P2), for every district $d \in \mathcal{D}_z$, we have that $\delta(z, o) = \delta(x_i, z) + \delta(x_i, o)$ for every $i \in \mathcal{N}_d$, and hence $\text{SC}_z(z|\mathbf{x}) + \text{SC}_z(o|\mathbf{x}) = \delta(z, o) \cdot \lambda |\mathcal{D}_z|$. Also, since $z_d = z$ for every $d \in \mathcal{D}_z$, z minimizes the total distance of the agents in d, and thus $\text{SC}_z(z|\mathbf{x}) \leq \text{SC}_z(o|\mathbf{x})$. Combining these, we obtain

$$\text{SC}_z(z|\mathbf{x}) \leq \frac{1}{2}\delta(z,o) \cdot \lambda|\mathcal{D}_z|; \tag{1}$$

$$\text{SC}_z(o|\mathbf{x}) \geq \frac{1}{2}\delta(z,o) \cdot \lambda|\mathcal{D}_z|. \tag{2}$$

- Consider a location $z \in Z\setminus\{w\}$. By (P1), w is the left-most representative, and thus $z > w$. By (P2), every agent i in a district of $\mathcal{D}_z$ lies in the interval defined by z and o, which means that $\delta(x_i, w) \leq \delta(w, o)$ if $z \leq o$, and $\delta(x_i, w) \leq \delta(w, z) = \delta(w, o) + \delta(z, o)$ if $z > o$. Since $\delta(z, o) \geq 0$, by summing over all the agents in the districts of $\mathcal{D}_z$, we obtain that

$$\text{SC}_z(w|\mathbf{x}) \leq \Big(\delta(w,o) + \delta(z,o)\Big) \cdot \lambda|\mathcal{D}_z|. \tag{3}$$

- Since w is the left-most representative (due to (P1)) and the median among all representatives (since it is chosen by the mechanism), it must be the case that w is the representative of more than half of the districts, and thus

$$|\mathcal{D}_w| \geq \sum_{z\in Z\setminus\{w\}} |\mathcal{D}_z|. \tag{4}$$

Given the above observations, we will now upper-bound the social cost of w and lower-bound the social cost of o. By the definition of $\text{SC}(w|\mathbf{x})$, and by applying (1) for $z = w$, (3) for $z \neq w$, and (4), we obtain

$$\begin{aligned}
\text{SC}(w|\mathbf{x}) &= \text{SC}_w(w|\mathbf{x}) + \sum_{z\in Z\setminus\{w\}} \text{SC}_z(w|\mathbf{x}) \\
&\leq \delta(w,o)\lambda\Big(\frac{1}{2}|\mathcal{D}_w| + \sum_{z\in Z\setminus\{w\}} |\mathcal{D}_z|\Big) + \sum_{z\in Z\setminus\{w\}} \delta(z,o)\lambda|\mathcal{D}_z| \\
&\leq \frac{3}{2}\delta(w,o)\lambda|\mathcal{D}_w| + \sum_{z\in Z\setminus\{w\}} \delta(z,o)\lambda|\mathcal{D}_z| \\
&\leq \frac{3}{2}\sum_{z\in Z} \delta(z,o)\lambda|\mathcal{D}_z|.
\end{aligned} \tag{5}$$

By the definition of $\text{SC}(o|\mathbf{x})$ and by applying (2), we obtain

$$\text{SC}(o|\mathbf{x}) = \sum_{z\in Z} \text{SC}_z(o|\mathbf{x}) \geq \frac{1}{2}\sum_{z\in Z} \delta(z,o)\lambda|\mathcal{D}_z|. \tag{6}$$

Therefore, by (5) and (6), the distortion of $\mathcal{I}$ subject to MM is at most 3, and since $\mathcal{I}$ is an arbitrary (up to symmetry) instance of wc(DM), Lemma 3 implies that $\texttt{dist}(\text{DM}) \leq 3$. □

The bound on the distortion of DM follows by a similar proof.

Theorem 3. *The distortion of* DISTRIBUTEDMEDIAN *is at most* 7.

4 Lower Bounds for the Discrete Setting

In this section, we present our lower bounds for the discrete setting. Specifically, we show that for any $\varepsilon > 0$, the distortion of any mechanism is at least $3 - \varepsilon$, and the distortion of any strategyproof mechanism is at least $7 - \varepsilon$. These lower bounds match the upper bounds of Sect. 3, and thus show that MM and DM are the best possible general and strategyproof mechanisms, respectively.

Without loss of generality, we assume that when given as input any instance with two districts each of which has a different representative, any mechanism chooses the left-most representative as the facility location; otherwise, the same bounds follow by symmetric arguments. Also, throughout this section, we write $\mathrm{SC}(z)$ instead of $\mathrm{SC}(z|\mathbf{x})$ for the social cost of z; $\mathbf{x}$ will be clear from context.

We start with general mechanisms.

Lemma 4. *Let $\mathcal{M}$ be a mechanism with distortion strictly less than 3. Let $\mathcal{I}$ be an instance with set of alternative locations $\mathcal{A} = \{0, 1\}$, and $k = 2\mu + 1$ districts such that 0 is the representative of μ districts and 1 is the representative of $\mu + 1$ districts, for every integer $\mu \geq 1$. Then, (i) $\mathcal{M}(\mathcal{I}) = 1$, and (ii) the representative of any district d for which all agents are positioned at $\frac{2\mu+1}{4(\mu+1)}$ is $z_d = 0$.*

Proof. We prove the statement by induction on μ.

Base Case: $\mu = 1$. For (i), assume there exists an instance $\mathcal{I}$ such that 0 is the representative of one district and 1 is the representative of two districts, but $\mathcal{M}(\mathcal{I}) = 0$. In particular, let $\mathcal{I}$ be as follows: In the first district, all agents are positioned at $1/4$; the representative must be 0, as otherwise the distortion would be 3. In the other two districts, all agents are positioned at 1; since $\mathcal{M}$ is unanimous, the representative of these districts is 1. Since $\mathrm{SC}(0) = \lambda/4 + 2\lambda = 9\lambda/4$ and $\mathrm{SC}(1) = 3\lambda/4$, we have that $\texttt{dist}(\mathcal{M}) \geq \texttt{dist}(\mathcal{I}|\mathcal{M}) = 3$, a contradiction.

For (ii), assume that the representative of the district in which all agents are positioned at $\frac{2\mu+1}{4(\mu+1)} = 3/8$ is 1. Let $\mathcal{J}$ be the following instance: In the first district, all agents are positioned at 0, and thus its representative is 0. In the other two districts, all agents are positioned at $3/8$, and thus their representative is 1. Since $\mathcal{J}$ satisfies the properties of the lemma, by (i), it must be $\mathcal{M}(\mathcal{J}) = 1$. However, since $\mathrm{SC}(0) = 3\lambda/4$ and $\mathrm{SC}(1) = \lambda + 10\lambda/8 = 9\lambda/4$, we again have that $\texttt{dist}(\mathcal{M}) \geq \texttt{dist}(\mathcal{J}|\mathcal{M}) = 3$, a contradiction.

Induction Step: We assume the lemma is true for $\mu = \ell - 1$, and will show that it is also true for $\mu = \ell$. For (i), consider the following instance $\mathcal{I}$ with $2\ell + 1$ districts: In the first ℓ districts, all agents are positioned at $\frac{2\ell-1}{4\ell} = \frac{2(\ell-1)+1}{4((\ell-1)+1)}$; by part (ii) for $\mu = \ell - 1$, we have that the representative of all these districts is 0. In the remaining $\ell + 1$ districts, all agents are positioned at 1, and thus their representative is 1. We have that

$$\mathrm{SC}(0) = \ell \cdot \frac{(2\ell - 1)\lambda}{4\ell} + (\ell + 1) \cdot \lambda = \frac{3(2\ell + 1)\lambda}{4}$$

and

$$\text{SC}(1) = \ell \cdot \frac{(2\ell+1)\lambda}{4\ell} = \frac{(2\ell+1)\lambda}{4}.$$

If $\mathcal{M}(\mathcal{I}) = 0$ then $\texttt{dist}(\mathcal{M}) \geq \texttt{dist}(\mathcal{I}|\mathcal{M}) = 3$. Therefore, for the mechanism to achieve distortion strictly less than 3, it must be the case that $\mathcal{M}(\mathcal{I}) = 1$.

For (ii), assume that the representative of a district in which all agents are positioned at $\frac{2\ell+1}{4(\ell+1)}$ is 1 instead. Let $\mathcal{J}$ be the following instance with $2\ell+1$ districts: In the first ℓ districts, all agents are at 0, and thus their representative is 0. In the remaining $\ell+1$ districts, all agents are positioned at $\frac{2\ell+1}{4(\ell+1)}$, and their representative is 1, by assumption. Since (i) holds for $\mu = \ell$, it must be $\mathcal{M}(\mathcal{J}) = 1$. However, since

$$\text{SC}(0) = (\ell+1) \cdot \frac{(2\ell+1)\lambda}{4(\ell+1)} = \frac{\lambda(2\ell+1)}{4}$$

$$\text{SC}(1) = \ell \cdot \lambda + (\ell+1) \cdot \frac{(2\ell+3)\lambda}{4(\ell+1)} = \frac{3\lambda(2\ell+1)}{4},$$

we have that $\texttt{dist}(\mathcal{M}) \geq \texttt{dist}(\mathcal{J}|\mathcal{M}) = 3$. □

We are now ready to prove the main theorem.

Theorem 4. *In the discrete setting, the distortion of any mechanism is at least $3 - \varepsilon$, for any $\varepsilon > 0$.*

Proof. Let $\mathcal{M}$ be any mechanism with distortion less than $3-\varepsilon$, for any $\varepsilon > 0$. We consider instances with set of alternative locations $\mathcal{A} = \{0, 1\}$. We will establish that $\mathcal{M}$ must choose 1 as the representative of any district in which all the agents are positioned at 1/2. Assume otherwise, and consider the following instance $\mathcal{I}$ with two districts: In the first district, all agents are positioned at 1/2; hence, the representative is 0. In the second district, all agents are positioned at 1; by unanimity, the representative of the second district is 1. Since there are two districts, one with representative 0 and one with representative 1, $\mathcal{M}$ selects the left-most district representative as the facility location, that is, $\mathcal{M}(\mathcal{I}) = 0$. However, since $\text{SC}(0) = \lambda/2 + \lambda = 3\lambda/2$ and $\text{SC}(1) = \lambda/2$, this decision leads to $\texttt{dist}(\mathcal{M}) \geq \texttt{dist}(\mathcal{I}|\mathcal{M}) = 3$, a contradiction.

Finally, consider the following instance $\mathcal{J}$ with $k = 2\mu + 1$ districts: In the first μ districts, all agents are positioned at 0; by unanimity, the representative of all these districts is 0. In the remaining $\mu+1$ districts, all agents are positioned at 1/2; by the above discussion, the representative of these districts is 1. By (i) of Lemma 4, we have that $\mathcal{M}(\mathcal{J}) = 1$. Since $\text{SC}(0) = (\mu+1) \cdot \frac{\lambda}{2}$ and $\text{SC}(1) = \mu \cdot \lambda + (\mu+1) \cdot \frac{\lambda}{2} = \frac{(3\mu+1)\lambda}{2}$, we have that $\texttt{dist}(\mathcal{J}|\mathcal{M}) = \frac{3\mu+1}{\mu+1}$. The theorem follows by choosing μ to be sufficiently large. □

Next, we show the following lower bound for stategyproof mechanisms.

Theorem 5. *In the discrete setting, the distortion of any strategyproof mechanism is at least $7 - \varepsilon$, for any $\varepsilon > 0$.*

The proof of the above theorem requires the following lemma, establishing that strategyproof mechanisms are *ordinal*, that is, their decisions are based only on the orderings over the alternative locations induced by the positions of the agents.

Lemma 5. *Let $\mathcal{M}$ be a strategyproof mechanism with finite distortion. Let $\mathbf{x}_d$ and $\mathbf{y}_d$ be two district position profiles for some district d, such that for every agent $i \in \mathcal{N}_d$ and any two alternative locations $\alpha \neq \beta$, $\delta(x_i, \alpha) \neq \delta(x_i, \beta)$ and $\delta(y_i, \alpha) \neq \delta(y_i, \beta)$; Also, if $\delta(x_i, \alpha) < \delta(x_i, \beta)$ then $\delta(y_i, \alpha) < \delta(y_i, \beta)$. The representative of the district chosen by $\mathcal{M}$ must be the same under both $\mathbf{x}_d$ and $\mathbf{y}_d$.*

By Lemma 5, Theorem 5, and the fact DM requires only ordinal information, we obtain the following result.

Corollary 1. *The distortion of any ordinal mechanism is at least $7 - \varepsilon$, for any $\varepsilon > 0$. Moreover, there exists an ordinal mechanism with distortion at most 7.*

5 Mechanisms for the Continuous Setting

We now turn our attention to the continuous setting. Recall that MM chooses the alternative location that minimizes the social cost of the agents, whereas DM chooses the location that is closest to the median agent. In the continuous setting, where the set of alternative locations is $\mathbb{R}$, the location of the median agent is known to minimize the social cost of the agents in a district, and thus the continuous version of DM, which chooses as representative the position of the median agent, is an implementation of MM. So, the continuous version of DM inherits the best properties of MM and the discrete version of DM, leading to the following statement.

Theorem 6. *The continuous version of* DM *is strategyproof and has distortion at most* 3.

The proof of the distortion bound in Theorem 6 also follows from the work of Procaccia and Tennenholtz [29], who considered a setting with agents (or *super-agents*, for clarity) that control multiple locations, and their cost is the total distance between those locations and the facility. They showed that the *median-of-medians* mechanism is 3-approximate. The theorem follows by interpreting the super-agents as district representatives in our case, so that the social cost objectives in the two settings coincide.

We next show a lower bound of almost 3 on the distortion of any strategyproof mechanism, thus showing that the continuous version of DM is actually the best possible among those mechanisms in the continuous setting.

Theorem 7. *In the continuous setting, the distortion of any strategyproof mechanism is at least $3 - \varepsilon$, for any $\varepsilon > 0$.*

We also prove the following unconditional lower bound.

Theorem 8. *In the continuous setting, the distortion of any mechanism is at least* $2 - \varepsilon$*, for any* $\varepsilon > 0$.

Even thought we have been unable to show a matching unconditional upper bound, we believe that this should be possible. To this end, we conjecture that there exists a mechanism with distortion 2 for the continuous setting.

6 Extensions and Open Problems

Asymmetric Districts. Our discussion in the previous sections revolves around the assumption that the districts are symmetric. In general however, the districts might be *asymmetric*, where every district $d \in \mathcal{D}$ might consist of a different number n_d of agents. It is not hard to observe that our mechanisms (MM and DM) can be applied in the asymmetric case as well. In addition, the structure of their worst-case instances defined in Sect. 3.1 is exactly the same; the proof of the lemma *does not* require that $n_d = \lambda$ for every $d \in \mathcal{D}$. Exploiting this, we can show the following result, which generalizes Theorems 2, 3 and 6.

Theorem 9. *Let* $\alpha = \frac{\max_{d \in \mathcal{D}} n_d}{\min_{d \in \mathcal{D}} n_d}$. *The distortion of* MM *is at most* 3α *and the distortion of* DM *is at most* 7α.

Unfortunately, our lower bounds are tailor-made for the symmetric case, and thus it is an interesting open problem to extend them to the case of asymmetric districts. As MM and DM do not take into account the district sizes, it would also be interesting to see whether using this information could lead to mechanisms with improved distortion guarantees (besides the symmetric case).

Proxy Voting. Another ingredient of our distributed setting is that the facility location is chosen from the set of district representatives, thus modeling scenarios in which decisions of independent groups are aggregated into a common outcome. Alternatively, one could assume that the location can be chosen from the set of *all* alternative locations, in which case the district representatives are used as *proxies* in a district-based election (e.g. see [5] and references therein). This captures situations where the alternatives are agents themselves, and the groups select as representatives those alternatives that more closely reflect their collective opinions. Since the set of district representatives is a subset of the alternative locations, it is straightforward to see that our upper bounds also hold for this proxy model. Our lower bounds in the discrete setting extend as well, since there are only two alternative locations in the instances used in the proofs, and each of them is a representative for at least one district. Hence, our mechanisms are best possible for the proxy model in the discrete setting.

Corollary 2. *In the proxy model, the distortion of* MM *is at most* 3 *and the distortion of* DM *is at most* 7. *Furthermore, in the discrete setting,* MM *and* DM *are the best possible among general and strategyproof mechanisms.*

In the continuous setting, our lower bounds do not immediately carry over, and it is an intriguing question to identify the exact bound for general and strategyproof mechanisms.

Other Directions. In terms of extending and generalizing our model, there is ample ground for future work. As is typical in the facility location literature, one could consider objectives different than the social cost, such as the *maximum cost* or the *sum of squares*. Again, the goal would be to show bounds on the distortion, and also design good strategyproof mechanisms. Other possible extensions could include multiple facilities, more general metric spaces, different cost functions, or studying the many different variants of the facility location problem in the distributed setting.

References

1. Abramowitz, B., Anshelevich, E.: Utilitarians without utilities: maximizing social welfare for graph problems using only ordinal preferences. In: Proceedings of the 32nd AAAI Conference on Artificial Intelligence (AAAI), pp. 894–901 (2018)
2. Amanatidis, G., Birmpas, G., Filos-Ratsikas, A., Voudouris, A.A.: Peeking behind the ordinal curtain: improving distortion via cardinal queries. Artif. Intell. **296**, 103488 (2021)
3. Anshelevich, E., Bhardwaj, O., Elkind, E., Postl, J., Skowron, P.: Approximating optimal social choice under metric preferences. Artif. Intell. **264**, 27–51 (2018)
4. Anshelevich, E., Filos-Ratsikas, A., Shah, N., Voudouris, A.A.: Distortion in social choice problems: the first 15 years and beyond. CoRR abs/2103.00911 (2021)
5. Anshelevich, E., Fitzsimmons, Z., Vaish, R., Xia, L.: Representative proxy voting. In: Proceedings of the 35th AAAI Conference on Artificial Intelligence (AAAI), pp. 5086–5093 (2021)
6. Anshelevich, E., Postl, J.: Randomized social choice functions under metric preferences. J. Artif. Intell. Res. **58**, 797–827 (2017)
7. Anshelevich, E., Zhu, W.: Ordinal approximation for social choice, matching, and facility location problems given candidate positions. In: Proceedings of the 14th International Conference on Web and Internet Economics (WINE), pp. 3–20 (2018)
8. Babaioff, M., Feldman, M., Tennenholtz, M.: Mechanism design with strategic mediators. ACM Trans. Econ. Comput. (TEAC) **4**(2), 1–48 (2016)
9. Black, D.: The Theory of Committees and Elections. Kluwer Academic Publishers (1957)
10. Boutilier, C., Caragiannis, I., Haber, S., Lu, T., Procaccia, A.D., Sheffet, O.: Optimal social choice functions: a utilitarian view. Artif. Intell. **227**, 190–213 (2015)
11. Caragiannis, I., Nath, S., Procaccia, A.D., Shah, N.: Subset selection via implicit utilitarian voting. J. Artif. Intell. Res. **58**, 123–152 (2017)
12. Chan, H., Filos-Ratsikas, A., Li, B., Li, M., Wang, C.: Mechanism design for facility location problems: a survey. CoRR abs/2106.03457 (2021)
13. Cheng, Y., Han, Q., Yu, W., Zhang, G.: Obnoxious facility game with a bounded service range. In: Proceedings of the 10th International Conference on Theory and Applications of Models of Computation (TAMC), pp. 272–281 (2013)
14. Cheng, Y., Yu, W., Zhang, G.: Mechanisms for obnoxious facility game on a path. In: Proceedings of the 5th International Conference on Combinatorial Optimization and Applications (COCOA), pp. 262–271 (2011)

15. Deligkas, A., Filos-Ratsikas, A., Voudouris, A.A.: Heterogeneous facility location with limited resources. CoRR abs/2105.02712 (2021)
16. Duan, L., Li, B., Li, M., Xu, X.: Heterogeneous two-facility location games with minimum distance requirement. In: Proceedings of the 18th International Conference on Autonomous Agents and Multiagent Systems (AAMAS), pp. 1461–1469 (2019)
17. Feigenbaum, I., Sethuraman, J., Ye, C.: Approximately optimal mechanisms for strategyproof facility location: minimizing L_p norm of costs. Math. Oper. Res. **42**(2), 434–447 (2017)
18. Feldman, M., Fiat, A., Golomb, I.: On voting and facility location. In: Proceedings of the 2016 ACM Conference on Economics and Computation (EC), pp. 269–286 (2016)
19. Feldman, M., Wilf, Y.: Strategyproof facility location and the least squares objective. In: Proceedings of the 14th ACM Conference on Electronic Commerce (EC), pp. 873–890 (2013)
20. Filos-Ratsikas, A., Micha, E., Voudouris, A.A.: The distortion of distributed voting. Artif. Intell. **286**, 103343 (2020)
21. Fong, C.K.K., Li, M., Lu, P., Todo, T., Yokoo, M.: Facility location games with fractional preferences. In: Proceedings of the 32nd AAAI Conference on Artificial Intelligence (AAAI), pp. 1039–1046 (2018)
22. Fotakis, D., Tzamos, C.: Winner-imposing strategyproof mechanisms for multiple facility location games. Theoret. Comput. Sci. **472**, 90–103 (2013)
23. Lu, P., Sun, X., Wang, Y., Zhu, Z.A.: Asymptotically optimal strategy-proof mechanisms for two-facility games. In: Proceedings of the 11th ACM Conference on Electronic Commerce (EC), pp. 315–324 (2010)
24. Lu, P., Wang, Y., Zhou, Y.: Tighter bounds for facility games. In: Proceedings of the 5th International Workshop on Internet and Network Economics (WINE), pp. 137–148 (2009)
25. Mandal, D., Procaccia, A.D., Shah, N., Woodruff, D.P.: Efficient and thrifty voting by any means necessary. In: Proceedings of the 33rd Conference on Neural Information Processing Systems (NeurIPS), pp. 7178–7189 (2019)
26. Moulin, H.: On strategy-proofness and single peakedness. Public Choice **35**(4), 437–455 (1980)
27. Munagala, K., Wang, K.: Improved metric distortion for deterministic social choice rules. In: Proceedings of the 2019 ACM Conference on Economics and Computation (EC), pp. 245–262 (2019)
28. Procaccia, A.D., Rosenschein, J.S.: The distortion of cardinal preferences in voting. In: Proceedings of the 10th International Workshop on Cooperative Information Agents (CIA), pp. 317–331 (2006)
29. Procaccia, A.D., Tennenholtz, M.: Approximate mechanism design without money. ACM Trans. Econ. Comput. **1**(4), 18:1–18:26 (2013)
30. Schummer, J., Vohra, R.V.: Strategy-proof location on a network. J. Econ. Theory **104**(2), 405–428 (2002)
31. Serafino, P., Ventre, C.: Truthful mechanisms without money for non-utilitarian heterogeneous facility location. In: Proceedings of the 29th AAAI Conference on Artificial Intelligence (AAAI), pp. 1029–1035 (2015)
32. Serafino, P., Ventre, C.: Heterogeneous facility location without money. Theoret. Comput. Sci. **636**, 27–46 (2016)

Prior-Free Clock Auctions for Bidders with Interdependent Values

Vasilis Gkatzelis, Rishi Patel(✉), Emmanouil Pountourakis, and Daniel Schoepflin

Drexel University, Philadelphia, USA
{gkatz,riship,manolis,schoep}@drexel.edu

Abstract. We study the problem of selling a good to a group of bidders with interdependent values in a prior-free setting. Each bidder has a signal that can take one of k different values, and her value for the good is a weakly increasing function of all the bidders' signals. The bidders are partitioned into ℓ expertise-groups, based on how their signal can impact the values for the good, and we prove upper and lower bounds regarding the approximability of social welfare and revenue for a variety of settings, parameterized by k and ℓ. Our lower bounds apply to all ex-post incentive compatible mechanisms and our upper bounds are all within a small constant of the lower bounds. Our main results take the appealing form of ascending clock auctions and provide strong incentives by admitting the desired outcomes as *obvious ex-post equilibria.*

Keywords: Clock auctions · Interdependent values · Obvious ex-post equilibrium

1 Introduction

We study the problem of selling a good to bidders with *interdependent values*, which has received a lot of attention in economics (e.g., see [12, Chapters 6 and 10]), and recently also in computer science (e.g., [3,4,6–9,19,20]). In contrast to the *private values* model, where each bidder knows her value for the good being sold, the interdependent value literature assumes that each bidder has some private signal regarding the value of the good, e.g., through some research or technical expertise, and the actual value of the good to each bidder is a function of all the bidders' signals. For instance, a common motivating example for this problem involves firms competing over the mineral rights of a piece of land [23]: each firm has conducted some tests, trying to estimate the land's capacity in desired minerals, but each of these tests may provide only partial evidence, and the best estimate can be inferred by appropriately aggregating all the test results, e.g., by computing the average across all of these measurements.

The first and last authors were partially supported by NSF grants CCF-2008280 and CCF-1755955. The second author was supported by an REU through CCF-1755955.

I. Caragiannis and K. A. Hansen (Eds.): SAGT 2021, LNCS 12885, pp. 64–78, 2021.
https://doi.org/10.1007/978-3-030-85947-3_5

The main difficulty when designing auctions for bidders with interdependent values arises from the fact that the bidders' signals are not known to the auctioneer, or to the other bidders. Therefore, the auctioneer needs to elicit these signals before deciding who should win the item and what the price should be. But, why would any bidder reveal her true signal to the auctioneer? A sealed-bid auction is said to be *ex-post incentive compatible* if truth-telling, i.e., reporting the true signal to the auctioneer, is an equilibrium for all the bidders. Designing ex-post incentive compatible auctions with non-trivial welfare or revenue guarantees has been a central goal of this line of research.

Prior work has considered several different ways in which the bidders' values can depend on the vector of signals. For example, in the *common value* model all the bidders have the same value for the good but, even in this special case, the design of ex-post incentive compatible auctions is a non-trivial problem. This problem becomes even harder when the bidders' values can differ. To enable the design of efficient incentive compatible mechanisms, prior work has introduced useful restrictions on the structure of these valuation functions, such as *submodularity over signals* (SOS) [1,7], or constraints across pairs of valuation functions, such as the *single-crossing* property [16,17].

In this paper, we consider a variety of settings with interdependent values that are not captured by (approximate) SOS or the single-crossing property. We let k be the number of possible values that a bidder's signal can have, and we partition the bidders into ℓ expertise-groups, depending on the type of information that their signals provide regarding the good being sold. Using these parameters, we prove upper and lower bounds, parameterized by k and ℓ, on the extent to which auctions can approximate the optimal welfare or revenue. All our proposed auctions are ex-post incentive compatible, but our main results also satisfy stronger incentive guarantees: they can be implemented not only as direct-revelation mechanisms (sealed-bid auctions), but also as ascending clock auctions, and they admit the desired outcomes as *obvious ex-post equilibria* [14] which are easy for the bidders to verify, thus leading to more practical solutions.

1.1 Our Results

We begin, in Sect. 3, by considering the interesting case where each bidder's signal regarding the quality of the good can take two possible values, either "low" or "high", and each bidder's value is a weakly increasing function of these signals. If the valuation function of each bidder is symmetric, i.e., every bidder's signal matters the same, then we provide a clock auction that achieves a 5-approximation of the optimal social welfare, and a variation of that auction that guarantees revenue that is a 10-approximation of the optimal social welfare. We then generalize these results to non-symmetric functions, where the bidders are partitioned into ℓ groups based on their expertise, and signals from different groups may have different impact on the values. Our generalization achieves a 5ℓ-approximation for social welfare and a 10ℓ-approximation for revenue.

In Sect. 4, we go beyond the case of binary signals and consider problem instances with k distinct signal value options, $\{0, 1, \ldots, k-1\}$, allowing for the bidders' signals regarding the quality of the good to be more refined. The valuation of each bidder can be an arbitrary weakly increasing function of the average quality estimate of each group. Using a reduction to the binary case, we design a clock auction that achieves a $5\ell(k-1)$-approximation for social welfare and a $10\ell(k-1)$-approximation for revenue. To complement these positive results, we also prove a lower bound of $\ell(k-1)+1$ for the welfare approximation ratio of ex-post incentive compatible auctions.

Our auctions in these two sections achieve signal discovery using random sampling, while minimizing the probability of rejecting the highest value bidder. Unlike prior work, our random sampling process is adaptive, depending on prior signal discovery. Thus, our auction gradually refines our estimate of the item's quality as perceived by the bidders and eventually decides who to allocate to, aiming to achieve high welfare and revenue. Apart from matching the lower bound up to small constants, these auctions crucially also guarantee improved incentives: they admit the desired outcome not just an ex-post equilibrium, but as an *obvious* ex-post equilibrium, making our upper bounds stronger.

Finally, in Sect. 5 we consider the most general setting with any number of signals $k > 2$ and arbitrary quality functions per expert type. We first prove a stronger lower bound of $\ell\binom{k}{2}+1$ for the welfare approximation of ex-post incentive compatible auctions. Then we prove the existence of a universally incentive compatible and individually rational auction that matches this bound.

Due to space constraints, the proofs of some theorems (particularly those which are similar to previous proofs) have been deferred to the full version.

1.2 Related Work

In an interdependent values setting, a bidder's value for a good may depend on how much others value it. This idea is formally captured by the canonical interdependent values model given by Milgrom and Weber [17]. The interdependent values setting has been well-studied in the economics literature for its descriptive ability to capture many real-world scenarios. Noted examples in the literature include the mineral rights [23] and common value (e.g., "wallet game") models [11] discussed above, and the resale model [18] in which the value a bidder has for a good (e.g., a painting) depends on her own value for the good and the amounts others may be willing to pay on its resale.

A common assumption when studying the interdependent values setting in both the computer science and economics literature is that the valuations of the bidders satisfy a *single-crossing* condition. Following the definition of Roughgarden and Talgam-Cohen [20], a set of valuation functions satisfies single-crossing if for all bidders i and j

$$\frac{\partial v_i(s_i, \mathbf{s}_{-i})}{\partial s_i} \geq \frac{\partial v_j(s_i, \mathbf{s}_{-i})}{\partial s_i}.$$

Loosely speaking, single-crossing states that a bidder is more sensitive to her own signal than anyone else is. Using this assumption, many strong results can be obtained for both welfare and revenue. For example, Dasgupta and Maskin [5] demonstrated that the celebrated Vickrey-Clarke-Groves (VCG) mechanism can be adapted and extended into the common value setting to obtain optimal welfare given single-crossing. Ausubel [2] demonstrated that a generalized Vickrey auction can achieve efficiency in a multi-unit setting with single-crossing valuations. For revenue, Li [15] and Roughgarden and Talgam-Cohen [20] gave, independently, auctions extracting near optimal revenue in the interdependent values model for any matroid feasibility constraint when the valuations satisfy single-crossing and the signals are drawn from distributions with a regularity-type condition. Chawla et al. [3] gave an alternative generalization of the VCG auction with reserve prices and random admission which approximates the optimal revenue in any matroid setting without conditions on signal distributions.

On the other hand, it is well-known that without single-crossing, achieving the optimal welfare becomes impossible [5,10]. There have thus been recent efforts to *approximate* the optimal welfare when the single-crossing assumption is relaxed. Eden et al. [6] suggested a notion called "c-single-crossing" wherein each bidder is at most a factor c times less sensitive to changes in her own signal than any other bidder is (exact single-crossing has $c = 1$). They gave a $2c$-approximate randomized mechanism when valuation functions are concave and satisfy c-single-crossing. Eden et al. [7] proposed an alternative notion termed "submodularity over signals" (SOS) which, loosely speaking, stipulates that a valuation function must be less sensitive to increases in any particular signal when the other signals are high. The authors then gave a randomized 4-approximate mechanism for all single-parameter downward-closed settings when valuation functions are SOS; this factor was very recently improved to 2 for the case of binary signals by Amer and Talgam-Cohen [1]. We note that the valuations studied in this paper satisfy neither c-single-crossing nor (approximate) SOS, in general. Our work proposes alternative parameterizations of the valuation functions and it provides another step toward a better understanding of interdependent values beyond the classic, and somewhat restrictive, single-crossing assumption.

In accordance with some recent work in computer science (e.g., see [6,7]), and unlike much of the existing economics literature, we consider a *prior-free* setting where there is no distributional information regarding the signals of the bidders. Thus, our results are in consistent with "Wilson's doctrine" [22], which envisions a mechanism design process that is less reliant on the assumption of common knowledge. Our results are independent of an underlying distribution and do not assume that the auctioneer or the bidders have any information regarding each other's signals.

2 Preliminaries

We consider a setting where a set N of n bidders is competing to receive a good. Each bidder $i \in N$ has a private signal s_i regarding the good being sold, which can take one of k publicly known different values. Her valuation of the good, $v_i(\mathbf{s})$, is a publicly known weakly increasing function of the vector of all the bidders' signals, $\mathbf{s} = (s_1, s_2, \ldots, s_n)$. In many settings of interest it is natural to assume that this is a *symmetric* function over the signals, e.g., when all the bidders have the same access to information, or the same level of expertise. However, we also consider the case when the signal of some bidders may have a different impact than others'. To capture this case we partition the bidders into $\ell > 1$ groups and assume that each group has different types of expertise. In this case, the valuation functions $v_i(\mathbf{s})$ are symmetric with respect to the signals of bidders with the same type of expertise, but arbitrarily non-symmetric across bidders with different types of expertise. Note that this captures arbitrary monotone valuation functions when $\ell = n$, and it also captures several classes of instances where the valuations of different bidders are not (even approximately) single-crossing or SOS. We call a bidder *optimal for some signal vector* $\mathbf{s}$ if i is a highest value bidder for that signal profile, i.e., $i \in \arg\max_{j \in N}\{v_j(\mathbf{s})\}$. We use $h(\mathbf{s})$ to refer to an optimal bidder for signal vector $\mathbf{s}$, breaking ties arbitrarily but consistently if there are multiple optimal bidders for $\mathbf{s}$.

In interdependent value settings, a direct-revelation mechanism receives the bidders' signals as input and outputs a bidder to serve and a vector of prices $\mathbf{p}(\mathbf{s})$ which each bidder is charged. For any bidder i, the utility $u_i(\mathbf{s}) = v_i(\mathbf{s}) - p_i(\mathbf{s})$ if i is served and $u_i(\mathbf{s}) = -p_i(\mathbf{s})$, otherwise. A mechanism is *ex-post individually rational* if $u_i(\mathbf{s}) \geq 0$ for all i, assuming all bidders report their true signals. A mechanism is *ex-post incentive compatible* if the utility that bidder i receives by reporting her true signal is at least as high as the utility she would obtain by reporting any other signal, assuming all the other bidders report their true signals, i.e., $u_i(s_i, \mathbf{s}_{-i}) \geq u_i(s'_i, \mathbf{s}_{-i})$ for all $i, \mathbf{s}_{-i}$. If a mechanism uses randomization, we say that it is *universally ex-post individually rational and ex-post incentive compatible* (universally IC-IR) if it is a distribution over deterministic ex-post individually rational and ex-post incentive compatible mechanisms.

We look to design universally IC-IR randomized mechanisms that aim to serve the bidder with highest realized value given the signal profile. We measure the expected performance of these mechanisms against the optimal solution given full information. Given some instance I, let $\mathcal{A}(I)$ denote the bidder served by auction $\mathcal{A}$. We then say that $\mathcal{A}$ achieves an α-approximation to the optimal welfare for a family of instances $\mathcal{I}$ if

$$\sup_{I \in \mathcal{I}} \frac{\max_{i \in N}\{v_i(\mathbf{s})\}}{\mathbb{E}\left[v_{\mathcal{A}(I)}(\mathbf{s})\right]} \leq \alpha$$

where the expectation is taken over the random coin flips of our mechanism. In terms of revenue, note that for mechanisms that are individually rational (like the ones that we propose in this paper), we know that the revenue of these mechanisms is always a lower bound for their social welfare. We therefore use the optimal social welfare as an upper bound for the optimal revenue and say that $\mathcal{A}$ achieves an α-approximation of revenue for a family of instances $\mathcal{I}$ if

$$\sup_{I \in \mathcal{I}} \frac{\max_{i \in N}\{v_i(\mathbf{s})\}}{\mathbb{E}\left[p_{\mathcal{A}(I)}(\mathbf{s})\right]} \leq \alpha.$$

Our main results in this paper take the form of *clock auctions over signals*. A clock auction over signals is a multi-round dynamic mechanism in which bidders are faced with personalized ascending signal clocks. Throughout the auction, the clocks are non-decreasing and, at any point in the auction, a bidder may choose to permanently exit the auction (thereby losing the good permanently). When a bidder is declared the winner, she is offered a price (greater than or) equal to the value implied by the final clock signals for all bidders. In a clock auction, a bidder exits the auction if and only if her signal clock is greater than her true signal, we refer to this as *consistent bidding*. In particular, we seek to design clock auctions where consistent bidding is an *obvious ex-post equilibrium* (OXP) strategy profile [13]. A strategy profile is an OXP of an auction if for any bidder i, holding all other bidders' strategies fixed (and assuming they are acting truthfully), the best utility i can obtain by deviating from her truthful strategy under any possible type profile of the other bidders consistent with the history (i.e., their clock signals) is worse than the worst utility i can obtain by following her truthful strategy under any possible type profile of the other bidders consistent with the history.

3 Instances with Binary Signal Values

In this section, we consider the natural case where the signal of each bidder regarding the good can take one of two possible values, e.g., "low quality" and "high quality". We first focus on instances where the bidders' valuation functions are symmetric over the signals, and we provide a clock auction which admits an ex-post obvious equilibrium and 5-approximation to the optimal social welfare. We then extend this result to general valuation functions, achieving a 5ℓ-approximation to the optimal social welfare. This auction is then also used as a building block for the results of the next section, which considers a setting with $k > 2$ signal values.

3.1 A Clock Auction for Symmetric Valuation Functions

A central result of this paper is the *signal discovery auction*, which is presented as a sealed-bid auction below (see Mechanism 1), but can also be implemented as a clock auction (see Theorem 3). This auction aims to discover how many bidders have a high signal, while minimizing the probability that the optimal bidder is rejected during the discovery process. Throughout the execution of the auction, the set A includes the bidders that remain active, i.e., the ones that have not been rejected yet. The variables q_{min} and q_{max} provide a lower and an upper bound, respectively, for the number of bidders that have a high signal, based on the signals discovered up to that point. Note that q_{min} is initialized to 0 and q_{max} is initialized to n, corresponding to all bidders having signal 0 or signal 1, respectively. The set R^* contains all the bidders that have been rejected, without first verifying that they are not optimal.

The auction uses randomized sampling in order to initiate this discovery process: it chooses one of the active bidders uniformly at random, it rejects that bidder, and then uses its signal value to narrow down the range $[q_{min}, q_{max}]$. We refer to this as a "costly" signal discovery, because it may lead to the rejection of the highest value bidder. Then, this discovery leads to a sequence of "free" signal discoveries, by using this information to identify active bidders that cannot be optimal, rejecting them, and then using their signal to further narrow down the $[q_{min}, q_{max}]$ range. When no additional free signal discoveries are available, the auction removes any bidder of R^* that is now verified to be non-optimal, and executes another costly signal discovery.

This process continues until there is only one active bidder, at which point this bidder is declared the winner. We say that a signal profile $\mathbf{s}$ is consistent with some $q \in [q_{min}, q_{max}]$ if it contains a number of "high" signals equal to q. If this bidder i is optimal for a signal profile $\mathbf{s}$ consistent with exactly one $q \in [q_{min}, q_{max}]$, then the bidder is awarded the good at price $p = v_i(\mathbf{s})$; if the bidder is optimal for multiple signal profiles consistent with distinct numbers of "high" signal bidders in $[q_{min}, q_{max}]$, she is awarded the good at the price corresponding to a signal profile with the fewest number of "high" signal bidders.

Mechanism 1: Signal discovery auction for binary signal values

1 Let $A \leftarrow N$, $R^* \leftarrow \emptyset$, $q_{min} \leftarrow 0$, and $q_{max} \leftarrow n$
2 **while** $|A| > 1$ **do**
  `// A ''costly'' signal discovery`
3   Select a bidder $i \in A$ uniformly at random
4   Let $A \leftarrow A \backslash \{i\}$ and $R^* \leftarrow R^* \cup \{i\}$
5   **if** $s_i = 0$ **then**
6     $q_{max} \leftarrow q_{max} - 1$
7   **else**
8     $q_{min} \leftarrow q_{min} + 1$
  `// A sequence of ''free'' signal discoveries`
9   **while** $\exists j \in A$ *that is not optimal for any* $\mathbf{s}$ *consistent with some* $q \in [q_{min}, q_{max}]$ **do**
10     $A \leftarrow A \backslash \{j\}$
11   **if** $s_j = 0$ **then**
12       $q_{max} \leftarrow q_{max} - 1$
13   **else**
14       $q_{min} \leftarrow q_{min} + 1$
15   **while** $\exists j \in R^*$ *that is not optimal for any* $\mathbf{s}$ *consistent with some* $q \in [q_{min}, q_{max}]$ **do**
16     $R^* \leftarrow R^* \backslash \{j\}$
17 Let i be the single bidder in A
18 Let s_i' be the smallest signal such that i is optimal for $(s_i', \mathbf{s}_{-i})$
19 **if** $v_i(\mathbf{s}) \geq v_i((s_i', \mathbf{s}_{-i}))$ **then**
20   Award the good to i at price $v_i((s_i', \mathbf{s}_{-i}))$

The following lemma shows that the size of R^* is never more than 2, which allows us to bound the probability that the auction identifies the optimal bidder.

Lemma 1. *Throughout the execution of the signal discovery auction, the size of R^* is never more than 2.*

Proof. We first note that, throughout the auction, the only bidders in $A \cup R^*$ are the potentially optimal bidders (i.e., those which correspond to some possible signal profile) since bidders are removed from $A \cup R^*$ when they are determined to be non-optimal. Initially R^* is empty and at the beginning of each iteration of the outer while-loop, one randomly sampled active bidder i is added to this set, increasing its size by one. The signal of bidder i is then used to update either q_{min} or q_{max}; if $s_i = 0$ the auction can infer that q_{max} is not the true number of high signal bidders, and if $s_i = 1$ the auction can infer that q_{min} is not the true number of high signal bidders. In both of these cases, some possible symmetric signal profile is ruled out, and this may lead to a sequence of "free" signal discoveries, as discussed below.

Whenever a symmetric signal profile $\mathbf{s}$ is ruled out, there are four possibilities regarding the bidder who is optimal for that level, i.e., the bidder $h(\mathbf{s})$:

1. If $h(\mathbf{s})$ is in A and is not optimal for any other $\mathbf{s}'$ consistent with some number q of high signal bidders in the updated interval $[q_{min}, q_{max}]$, then the first inner-while loop of the auction will remove that bidder from A and use its signal to rule out one more quality level.
2. If $h(\mathbf{s})$ is in A and is also optimal for some other $\mathbf{s}'$ consistent with some number q of high signal bidders in the updated interval $[q_{min}, q_{max}]$, then the iteration of the outer while-loop terminates without any additional operations and we proceed to the next iteration.
3. If $h(\mathbf{s})$ is in R^*, and is not optimal for any other $\mathbf{s}'$ consistent with some number q of high signal bidders in the updated interval $[q_{min}, q_{max}]$, then the second inner while-loop removes $h(\mathbf{s})$ from R^* and we proceed to the next iteration of the outer while-loop.
4. If $h(\mathbf{s})$ is in R^*, and is also optimal for some other $\mathbf{s}'$ consistent with some number q of high signal bidders in the updated interval $[q_{min}, q_{max}]$, then the iteration of the outer while-loop terminates without any additional operations and we proceed to the next iteration.

Considering these four possibilities, note that while the first case arises, the execution remains in the first inner while-loop and the size of R^* remains unchanged. When the third case arises, the size of R^* is first reduced by one (because the auction enters the second inner while-loop) and then proceeds to the next iteration of the outer while-loop, which may bring this up to the same size again. As a result, the third case does not increase the size of R^* either.

On the other hand, both cases 2 and 4 may lead to an increase in the size of R^* by 1, since they terminate the current iteration of the outer while-loop and may proceed to the next one, which would add one more bidder to R^*.

However, at the end of each iteration of the outer while-loop, A and R^* contain only bidders that are optimal for some $\mathbf{s}$ consistent with some number of high signal bidders q in $[q_{min}, q_{max}]$ (all the others are removed from A in the first inner while-loop and from R^* in the second inner while-loop). Also, at the end of each iteration of the outer while-loop, we have $q_{max} = q_{min} + |A|$. To verify this fact note that the signal of everyone not in A has already been used to update the interval $[q_{min}, q_{max}]$ and the only signals not used yet are those of the bidders in A. If all the bidders in A have a low signal, then the true $\mathbf{s}$ has q_{min} bidders with high signals. If they all have a high signal (adding $|A|$ bidders with high signal), the true $\mathbf{s}$ has q_{max} bidders with high signals.

Therefore, we know that at the end of each iteration of the outer while-loop, every bidder in A and R^* is optimal for some possible symmetric signal profile with a number of high value bidders in $[q_{min}, q_{max}]$ and there are at most $|A|+1$ such distinct signal profiles. If R^* is empty at that point, this means that there can be at most one bidder in A that is optimal for two distinct signal profiles. If $|R^*| = 1$, then there are $|A| + 1$ optimal bidders and $|A| + 1$ distinct signal profiles, so there is no bidder in A or R^* that is optimal for more than one such profile. This means that in the next iteration of the outer while-loop, cases 2 and 4 listed above cannot arise, and therefore the size of R^* cannot be strictly more than 1 at the end of any iteration of the outer while loop. □

Theorem 1. *The signal discovery auction achieves a 5-approximation of the optimal welfare for instances with binary signals.*

Proof. Let i^* be the optimal bidder and q^* be the true number of high signals. We first observe that a bidder is removed from $A \cup R^*$ only if they are determined to be non-optimal. Thus, we know that $i^* \in A \cup R^*$ throughout the running of the algorithm. By Lemma 1 we know that $|R^*| \leq 2$ throughout the running of the algorithm. There are then at most 5 distinct bidders who can be in $A \cup R^*$ at the end of the algorithm: i^* and the (up to) four other bidders optimal for signal profiles corresponding to $q^* - 2$, $q^* - 1$, $q^* + 1$, or $q^* + 2$ high signal bidders. Provided that these four other bidders enter R^* (or are eliminated) before i^* is added to R^* we then obtain the optimal welfare. We conclude by noting that, since the choices of the bidder to be added to R^* is made uniformly at random, we can envision the order in which bidders are added to R^* as a uniform at random permutation over the bidders fixed at the outset. In a uniform random permutation, i^* follows these four bidders with probability 1/5. □

The signal discovery auction, as presented, achieves no interesting worst-case approximation for revenue when the benchmark is the ex-post optimal welfare. In particular, if there is a single optimal bidder for all the signal profiles corresponding to numbers of high signal bidders in $[q_{min}, q_{max}]$, and the true number of high signal bidders is q_{max}, the mechanism charges the winner i a price of $v_i(\mathbf{s}')$ where $\mathbf{s}'$ is the signal profile obtained by her signal being 0 (corresponding to q_{min} high signal bidders). If the true signal profile $\mathbf{s}''$ corresponds to having q_{max} bidders of high signal, the ex-post optimal welfare is $v_i(\mathbf{s}'')$, which can be arbitrarily higher than $v_i(\mathbf{s}')$. To address this issue, our next result shows that if we slightly modify the pricing rule of the mechanism, then we can achieve revenue which is a 10-approximation of the ex-post optimal welfare (which simultaneously also implies that the welfare we obtain is a 10-approximation).

Theorem 2. *The pricing rule of the signal discovery auction can be adjusted to achieve revenue which is a 10-approximation of the optimal welfare for instances with binary signals.*

Proof. If in line 18 of Mechanism 1 we instead select a $\mathbf{s}$ for which i is optimal consistent with some random $q' \in [q_{min}, q_{max}]$ and $\mathbf{s}$ is the true signal profile, we extract all of the welfare as revenue. Since i is the only bidder with unknown signal value, there are at most two levels for which i is optimal so we select the signal profile with probability 1/2, yielding the 10-approximation. Note that in line 20 we only allocate the item if the price is below the true value of i, so we preserve ex-post IC-IR with this modification. □

We conclude this section by verifying that the outcome of the signal discovery auction can be implemented as an obvious ex-post equilibrium [13].

Theorem 3. *The signal discovery auction can be implemented as an ascending clock auction over the signals wherein consistent bidding is an obvious ex-post equilibrium.*

Proof. Rather than asking bidders to report their signals we may instead equip each bidder with a signal clock. The clocks of all bidders begin at 0 and when bidder i would have her signal discovered by the above mechanism, we instead raise the clock of i to 1. If i rejects the new clock signal level (i.e., permanently exits the auction), she cannot win the item *regardless of her beliefs* about the signals of the remaining bidders.

If the true signal of i is 1, for any profile of signals of the remaining bidders (assuming these signals are true) the worst utility i can obtain by accepting the increased clock signal level is 0 (by losing the item or by winning the item and being charged exactly her welfare). Thus, at any point in the auction, regardless of the history, when i is approached to increase her clock signal level, the best utility i can obtain by not accepting the increased clock signal level (thereby necessarily losing the good) is weakly less than the worst utility i can obtain by accepting the increased clock signal level. On the other hand, if the true signal of i is 0, for any profile of signals of the remaining bidders (assuming these signals are true) if she instead accepts the increased clock signal level she either will continue to lose the auction (thereby obtaining a utility of 0) or win the auction at a quality level higher than the actual underlying quality of the good. Since the threshold signal of i would then be 1, she would necessarily be charged a price weakly higher than her value for the good and she would obtain non-positive utility. Thus, in either case, truthfully responding whether or not the clock signal level is above a bidder's signal is an obvious ex-post equilibrium. □

Corollary 1. *The version of the signal discovery auction which obtains revenue guarantees can also be implemented as an ascending clock auction over the signals wherein consistent bidding is an obvious ex-post equilibrium.*

Proof. The proof follows exactly as above except we raise the clock signal level of the winning bidder to the one corresponding to the randomly selected signal profile (effectively setting a take-it-or-leave-it price at this signal). □

3.2 A Clock Auction for General Valuation Functions

In this section, we demonstrate how our auction for symmetric valuation functions, i.e., the case where $\ell = 1$, above, can be easily extended to handle general valuation functions over binary signals, leading to approximation bounds that depend on the number of expert-groups, ℓ.

The mechanism first uniformly at random selects some $\ell' \in \{1, 2, \dots, \ell\}$, and then assumes that the optimal bidder belongs to expertise type ℓ'. The mechanism rejects all bidders outside expertise-group ℓ' and "learns" their signals. The auction then knows all the signals of bidders not in ℓ' and the problem reduces to also discovering the number of bidders in ℓ' that have a high signal. We can therefore run Mechanism 1 among the bidders in ℓ' to decide the winner among them, and the price offered to her.

Theorem 4. *The above mechanism yields a 5ℓ-approximation of the optimal welfare for instances with binary signals.*

Proof. The probability that the optimal bidder does, indeed, belong to the expertise-group ℓ' is $1/\ell$. If the mechanism guesses the value of ℓ' correctly, then the rejection of all the other bidders comes at no cost, and it reduces the problem to finding the optimal bidder within the group ℓ'. But, since we now know all the signal values of bidders outside the group ℓ', we can use Mechanism 1 to discover the optimal bidder with probability at least $1/5$ (by Theorem 1). Combining these observations, the above mechanism allocates to the optimal bidder with probability at least $1/(5\ell)$. □

Theorem 5. *The pricing rule of the above mechanism can be adjusted to achieve revenue which is a 10ℓ-approximation of the optimal welfare for instances with binary signals.*

4 Shared Quality Functions over k Signal Values

We now move beyond instances with binary signals and consider a class of valuation functions over $k \geq 2$ signal values. Each bidder i's signal can take any value $s_i \in \{0, 1, \dots, k-1\}$ and the average of these signals determines the *quality of the good* $q = \sum_{i \in N} s_i$ (note that the average of the signals can be directly inferred from the sum, so we use the sum for simplicity of notation).[1] This captures a variety of settings where each bidder has some estimate regarding the quality, but the true quality is best approximated by averaging over all the bidders' signals (e.g., see the *wisdom of the crowds* phenomenon [21]). Each bidder i's value for the good is provided by some (arbitrary) weakly increasing function $v_i(q)$, which depends only on q, quantifying how much each bidder values quality.

Apart from these symmetric valuation functions, we also consider non-symmetric ones involving ℓ different classes of experts. The bidders are partitioned into sets $N_1, N_2, \dots, N_\ell$, depending on their expertise, and the quality

[1] Note that the actual k signal values need not be $\{0, 1, \dots, k-1\}$, but we need them to be equidistant for our results to hold.

estimate from each expert group ℓ' is their average signal, i.e., $q_{\ell'} = \sum_{i \in N_{\ell'}} s_i$. In this case, the quality of the good is captured by the *shared quality vector* $\mathbf{q} = (q_1, q_2, \ldots, q_\ell)$, and each bidder's valuation is a function $v_i(\mathbf{q})$. The only restriction on the valuation function is that it is weakly increasing with respect to the underlying signals, but it can otherwise arbitrarily depend on the quality vector. For instance, this allows us to model settings where the signals of each group of experts imply the quality of the good with respect to some dimension, and each bidder can then synthesize this information into a quite complicated valuation function, depending on the aspects that she cares about the most.

In this section, we first provide a lower bound for the approximability of the optimal social welfare by universally ex-post IC-IR auctions, parameterized by k and ℓ. We then provide a way to leverage the ideas from the previous section to achieve essentially matching upper bounds using clock auctions and ensuring incentive guarantees even better than ex-post IC-IR.

4.1 Approximation Lower Bound for Ex-Post IC-IR Auctions

We first prove a lower bound for the welfare approximation that one can achieve for the class of instances of this section involving ℓ types of experts with k signal values each. It is worth noting that the construction for this lower bound is based on a simple class of valuation functions that only depend on the weighted average of the bidder's signals (with each expert group having a different weight coefficient). Also, for the case $k = 2$, i.e., the binary case considered in the previous section, this implies a lower bound of $\ell + 1$.

Theorem 6. *No ex-post incentive compatible auction with ℓ types of experts and shared quality functions can achieve better than an $\ell(k-1)+1$-approximation to the optimal welfare.*

Proof. We consider a particularly simple setting, in which the quality of the good can be summarized as a weighted average of all the bidders' signals (with bidders from different expertise classes given different weights). Note that this is readily captured by the model described above. It follows that when we reduce the signal of i by $d > 0$, the quality of the good changes by dw_i. Note that d can be at most $k-1$ different values. We construct a valuation function as follows. For each $j \in \{0, 1, \ldots, \ell - 1\}$, we define the valuation function of bidder i where $(k-1) \cdot j + 1 \leq i \leq (k-1) \cdot (j+1)$ as follows:

$$v_i(t) = \begin{cases} \Delta_i & \text{if } t \geq S - (i - (k-1) \cdot j) \cdot w_i, \\ 0 & \textit{otherwise.} \end{cases}$$

Finally, for bidder $i' = (k-1) \cdot \ell + 1$ (who has signal 0 in $\mathbf{s}$), $v_{i'}(t) = \Delta_{i'}$ when $t \geq S$ and $v_{i'}(t) = 0$ otherwise. We let $\Delta_1 = 1$ and $\forall i > 1$, $\Delta_i = H\Delta_{i-1}$ (H is arbitrarily large). In other words, at any of these qualities, we must allocate the good to the optimal bidder with probability $1/\alpha$ to obtain an α-approximation to the optimal welfare in the worst case. To obtain a $\ell(k-1)+1-\epsilon$ approximation for $\epsilon > 0$, it then must be that we allocate the good to the optimal bidder at all of these qualities with probability at least $1/(\ell(k-1)+1-\epsilon)$. But then we have that for all $d \in \{1, 2, \ldots, k-1\}$ and $w \in \{1, k, \ldots, k^{\ell-1}\}$ if we allocate the good to the optimal bidder i when the quality is $S - dw$ with probability p, we must continue to allocate the good to i with probability p when the quality is S in order to maintain universal ex-post incentive compatibility (by monotonicity of an allocation rule). Finally, since there are $\ell(k-1)+1$ qualities identified above, each of the distinct optimal bidders at these qualities must be allocated the good with probability at least $1/(\ell(k-1)+1-\epsilon)$ at quality S, a contradiction. □

4.2 A Clock Auction for Instances with Shared Quality Functions

We now provide a way to reduce this problem to the case of binary signals, while losing only a $k-1$ factor in our bounds. As a result, the induced upper bounds closely approximate the lower bound provided above. The majority of this section discusses how Mechanism 2 achieves this reduction for the case where ℓ, and then briefly explain how to generalize our bounds for instances with $\ell > 1$.

Similarly to Mechanism 1 in the binary setting, whose goal is to discover the number of signals that are high, Mechanism 2 aims to discover the value of the sum of the signals. Throughout its execution, the auction maintains an interval $[q_{min}, q_{max}]$ such that the true sum q is guaranteed to be in that interval. It gradually refines this range by discovering bidder signals as in the binary setting. The main difference is that we now need to be more careful in order to ensure that the size of R^* remains low. To achieve this, the auction chooses some $m \in \{0, 1, \ldots, k-2\}$ uniformly at random and assumes that $q \mod (k-1) = m$. It thus randomly reduces the number of values of q that it considers from $n(k-1)+1$ (since the sum can initially range from 0 to $n(k-1)$) to just $n+1$ (which is equal to the length of the $[q_{min}, q_{max}]$ interval in the case of binary signals). Importantly, the values of q that are considered after this sampling are spaced apart by $k-1$, allowing us to upper bound the size of R^*.

Lemma 2. *The set of R^* in Mechanism 2 is never more than 2.*

Theorem 7. *The signal discovery auction achieves a $5(k-1)$ approximation of the optimal welfare for instances with shared quality functions.*

Theorem 8. *The pricing rule of the signal discovery auction can be adjusted to achieve revenue which is a $10(k-1)$ approximation of the optimal welfare for instances with shared quality functions.*

Theorem 9. *Mechanism 2 can be implemented as an ascending clock auction over the signals wherein consistent bidding is an obvious ex-post equilibrium.*

Theorem 10. *Mechanism 2 can be modified to yield a $5(k-1)\ell$-approximation of the optimal welfare and achieve revenue which is a $10(k-1)\ell$-approximation of the optimal welfare for shared quality functions with ℓ expertise types.*

Mechanism 2: Signal discovery auction for k signal values

```
1  Let A ← N, R* ← ∅, q_min ← 0, and q_max ← n(k − 1)
2  Choose some m ∈ {0, 1, ..., k − 2} uniformly at random
3  S ← {q ∈ [q_min, q_max] | q mod (k − 1) = m}
4  while |A| > 1 do
       // A ''costly'' signal discovery
5      Select a bidder i ∈ A uniformly at random
6      Let A ← A\{i} and R* ← R* ∪ {i}
7      q_max ← q_max − (k − 1 − s_i)
8      q_min ← q_min + s_i
       // A sequence of ''free'' signal discoveries
9      while ∃j ∈ A that is not optimal for any q ∈ S ∩ [q_min, q_max] do
10         A ← A\{j}
11         q_max ← q_max + s_j − k + 1
12         q_min ← q_min + s_j
13     while ∃j ∈ R* that is not optimal for any q ∈ S ∩ [q_min, q_max] do
14         R* ← R*\{j}
15 Let i be the single bidder in A
16 Choose the smallest quality level q' ∈ S ∩ [q_min, q_max] for which i is optimal
17 if v_i(q(s)) ≥ v_i(q') then
18     Award the good to i at price v_i(q')
```

5 General Valuation Functions and Signal Values

We now turn to the more general case where the quality of the good is any weakly increasing function of the signals that treats bidders with the same expertise type symmetrically. We provide an approximation lower bound for any allocation function that is *monotone*: a necessary condition of ex-post IC. We conclude our results by proving that there exists a universally IC-IR auction that matches the approximation ratio lower bound. We can adjust the mechanism to achieve revenue that is $k \cdot (\ell\binom{k}{2} + 1)$ approximation of the welfare.

Theorem 11. *No ex-post incentive compatible auction can get more than a $\ell\binom{k}{2}+1$-approximation to the optimal welfare. Also, no universally ex-post IC-IR auction can obtain revenue more than a $\ell\binom{k}{2}+1$ fraction of the revenue.*

Theorem 12. *There exists a universally ex-post IC-IR auction that achieves a $\ell\binom{k}{2}+1$-approximation to the optimal welfare.*

Theorem 13. *There exists a universally ex-post IC-IR auction that obtains revenue that is a $k \cdot (\ell\binom{k}{2}+1)$ approximation to the social welfare.*

References

1. Amer, A., Talgam-Cohen., I.: Auctions with interdependence and SOS: improved approximation. In: Proceedings of the 14th International Symposium on Algorithmic Game Theory (2021, to appear)

2. Ausubel, L.M., et al.: A generalized Vickrey auction. Econometrica (1999)
3. Chawla, S., Fu, H., Karlin, A.: Approximate revenue maximization in interdependent value settings. In: Proceedings of the fifteenth ACM Conference on Economics and Computation, pp. 277–294 (2014)
4. Constantin, F., Ito, T., Parkes, D.C.: Online auctions for bidders with interdependent values. In: Proceedings of the 6th International Joint Conference on Autonomous Agents and Multiagent Systems, pp. 1–3 (2007)
5. Dasgupta, P., Maskin, E.: Efficient auctions. Q. J. Econ. **115**(2), 341–388 (2000)
6. Eden, A., Feldman, M., Fiat, A., Goldner, K.: Interdependent values without single-crossing. In: Proceedings of the 2018 ACM Conference on Economics and Computation, p. 369 (2018)
7. Eden, A., Feldman, M., Fiat, A., Goldner, K., Karlin, A.R.: Combinatorial auctions with interdependent valuations: SOS to the rescue. In: Proceedings of the 2019 ACM Conference on Economics and Computation, pp. 19–20 (2019)
8. Eden, A., Feldman, M., Talgam-Cohen, I., Zviran, O.: PoA of simple auctions with interdependent values. In: Thirty-Fifth AAAI Conference on Artificial Intelligence, AAAI 2021, pp. 5321–5329. AAAI Press (2021)
9. Ito, T., Parkes, D.C.: Instantiating the contingent bids model of truthful interdependent value auctions. In: Proceedings of the fifth international joint conference on Autonomous agents and multiagent systems, pp. 1151–1158 (2006)
10. Jehiel, P., Moldovanu, B.: Efficient design with interdependent valuations. Econometrica **69**(5), 1237–1259 (2001)
11. Klemperer, P.: Auctions with almost common values: the "wallet game" and its applications. Eur. Econ. Rev. **42**(3–5), 757–769 (1998)
12. Krishna, V.: Auction Theory. Academic Press, Cambridge (2009)
13. Li, S.: Obvious ex post equilibrium. Am. Econ. Rev. **107**(5), 230–34 (2017)
14. Li, S.: Obviously strategy-proof mechanisms. Am. Econ. Rev. **107**(11), 3257–87 (2017)
15. Li, Y.: Approximation in mechanism design with interdependent values. Games Econom. Behav. **103**, 225–253 (2017)
16. Maskin, E.: Auctions and Privatization, pp. 115–136. J.C.B. Mohr Publisher (1992)
17. Milgrom, P.R., Weber, R.J.: A theory of auctions and competitive bidding. Econometr.: J. Econ. Soc. **50**, 1089–1122 (1982)
18. Myerson, R.B.: Optimal auction design. Math. Oper. Res. **6**(1), 58–73 (1981)
19. Robu, V., Parkes, D.C., Ito, T., Jennings, N.R.: Efficient interdependent value combinatorial auctions with single minded bidders. In: Proceedings of the 23rd International Joint Conference on Artificial Intelligence, IJCAI, p. 339–345 (2013)
20. Roughgarden, T., Talgam-Cohen, I.: Optimal and robust mechanism design with interdependent values. ACM Trans. Econ. Comput. (TEAC) **4**(3), 1–34 (2016)
21. Surowiecki, J.: The Wisdom of Crowds. Anchor (2005)
22. Wilson, R.: Game-theoretic analyses of trading processes. In: Advances in Economic Theory, Fifth World Congress, pp. 33–70 (1987)
23. Wilson, R.B.: Competitive bidding with disparate information. Manage. Sci. **15**(7), 446–448 (1969)

Incentive Compatible Mechanism for Influential Agent Selection

Xiuzhen Zhang, Yao Zhang, and Dengji Zhao(✉)

ShanghaiTech University, Shanghai, China
{zhangxzh1,zhangyao1,zhaodj}@shanghaitech.edu.cn

Abstract. Selecting the most influential agent in a network has huge practical value in applications. However, in many scenarios, the graph structure can only be known from agents' reports on their connections. In a self-interested setting, agents may strategically hide some connections to make themselves seem to be more important. In this paper, we study the incentive compatible (IC) selection mechanism to prevent such manipulations. Specifically, we model the progeny of an agent as her influence power, i.e., the number of nodes in the subgraph rooted at her. We then propose the Geometric Mechanism, which selects an agent with at least 1/2 of the optimal progeny in expectation under the properties of incentive compatibility and fairness. Fairness requires that two roots with the same contribution in two graphs are assigned the same probability. Furthermore, we prove an upper bound of $1/(1 + \ln 2)$ for any incentive compatible and fair selection mechanisms.

Keywords: Incentive compatibility · Mechanism design · Influence approximation

1 Introduction

The motivation for influential agent selection in a network comes from real-world scenarios, where networks are constructed from the following/referral relationships among agents and the most influential agents are selected for various purposes (e.g., information diffusion [10] or opinion aggregation). However, in many cases, the selected agents are rewarded (e.g., coupons or prizes), and the network structures can only be known from their reports on their following relationships. Hence, agents have incentives to strategically misreport their relationships to make themselves selected, which causes a deviation from the optimal results. An effective selection mechanism should be able to prevent such manipulations, i.e., agents cannot increase their chances to be selected by misreporting, which is a key property called incentive compatibility.

There have been many studies about incentive compatible selection mechanisms with different influential measurements for various purposes (e.g., maximizing the in-degrees of the selected agent [1,5,7]). In this paper, we focus on selecting an agent with the largest progeny. For this purpose, the following two

I. Caragiannis and K. A. Hansen (Eds.): SAGT 2021, LNCS 12885, pp. 79–93, 2021.
https://doi.org/10.1007/978-3-030-85947-3_6

papers are the most related studies. Babichenko et al. [3] proposed the Two Path Mechanism based on random walks. Although their mechanism achieves a good approximation ratio of 2/3 between the expected and the optimal influence in trees, it has no guaranteed performance in forests or general directed acyclic graphs (DAGs). Furthermore, Babichenko et al. [4] advanced these results by proposing another two selection mechanisms with an approximation ratio of about 1/3 in forests. In these two papers, the authors assumed that agents can add their out-edges to any other agents in the network. This strong assumption limited the design of incentive compatible mechanisms. Also, in many cases, agents cannot follow someone they do not know. Therefore, we focus on the manipulation of hiding the connections they already have. In practice, it is possible that two agents know each other, but they are not connected. Then they are more than welcome to connect with each other, which is not harmful for the selection. Moreover, there still exists a noticeable gap between the approximation ratios of existing mechanisms and a known upper bound of 4/5 [4] for all incentive compatible selection mechanisms in forests. Therefore, by restricting the manipulations of agents, we want to investigate whether we can do better.

Furthermore, the previous studies mainly explored the forests, while in this paper, we also looked at DAGs. A DAG forms naturally in many applications because there exist sequential orders for agents to join the network. Each agent can only connect to others who joined the network before her, e.g., a reference or referral relationship network. Then, in a DAG, each node represents an agent, and each edge represents the following relationship between two agents.

In this setting, the action of each agent is to report a set of her out-edges, which can only be a subset of her true out-edges. The goal is to design selection mechanisms to incentivize agents to report their true out-edge sets. Besides the incentive compatibility, we also consider another desirable property called fairness. Fairness requires that two agents with the maximum progeny in two graphs share the same probability of being selected if their progeny make no difference in both graphs (the formal definition is given in Sect. 2). Then, we present an incentive compatible selection mechanism with an approximation ratio of 1/2 and prove an upper bound of $1/(1+\ln 2)$ for any incentive compatible and fair selection mechanism.

1.1 Our Contributions

We focus on the incentive compatible selection mechanism in DAGs. It is natural to assign most of the probabilities to select agents with more progeny to achieve a good approximation ratio. Thus, we identify a special set of agents in each graph, called the influential set. Each agent in the set, called an influential node, is a root with the maximum progeny if deleting all her out-edges in the graph. They are actually the agents who have the chances to make themselves the optimal agent with manipulations. On the other hand, we also define a desirable property based on the graph structure, called fairness. It requires that the most influential nodes (the agents with the maximum progeny) in two graphs have the same probability to be selected if the number of nodes in the two graphs, the subgraphs constructed by the two nodes' progeny, and the influential sets are all the same.

Based on these ideas, we propose the Geometric Mechanism, which only assigns positive probabilities to the influential set. Each influential node will be assigned a selection probability related to her ranking in the influential set. We prove that the Geometric Mechanism satisfies the properties of incentive compatibility and fairness and can select an agent with her progeny no less than $1/2$ of the optimal progeny in expectation. The approximation ratio of the previous mechanisms is at most $1/\ln 16$ (≈ 0.36). Under the constraints of incentive compatibility and fairness, we also give an upper bound of $1/(1 + \ln 2)$ for the approximation ratio of any selection mechanism, while the previous known upper bound for any incentive compatible selection mechanism is $4/5$.

1.2 Other Related Work

Without the Constraint of Incentive Compatibility. Focusing on influence maximization, Kleinberg [11] proposed two models for describing agents' diffusion behaviours in networks, i.e., the linear threshold model and the independent cascade model. It is proved to be NP-hard to select an optimal subset of agents in these two models. Following this, there are studies on efficient algorithms to achieve bounded approximation ratios between the selected agents and the optimal ones under these two models [9,12,15,21].

In cases where only one influential agent can be selected, the most related literature also studied methods to rank agents based on their abilities to influence others in a given network, i.e., their centralities in the network. A common way is to characterize their centralities based on the structure of the network. In addition to the classic centrality measurements (e.g., closeness and betweenness [13,19]) or Shapley value based characterizations [17], there are also other ranking methods in real-world applications, such as PageRank [18] where each node is assigned a weight according to its connected edges and nodes.

With the Constraint of Incentive Compatibility. In this setting, incentive compatible selection mechanisms are implemented in two ways: with or without monetary payments. The first kind of mechanism incentivizes agents to truthfully reveal their information by offering them payments based on their reports. For example, Narayanam et al. [16] considered the influence maximization problem where the network structure is known to the planner, and each agent will be assigned a fixed positive payment based on influence probabilities they reported. With monetary incentives, there are also different mechanisms proposed to prevent agents from increasing their utilities by duplicating themselves or colluding together [6,20,22]. To achieve incentive compatible mechanisms without monetary incentives, the main idea of the existing work is to design probabilistic selection mechanisms and ensure that each agent's selection probability is independent of her report [1,2,7]. For example, Alon et al. [1] designed randomized selection mechanisms in the setting of approval voting, where networks are constructed from agents' reports. Our work belongs to this category.

2 The Model

Let $\mathcal{G}^n$ be the set of all possible directed acyclic graphs (DAGs) with n nodes and $\mathcal{G} = \bigcup_{n \in \mathbb{N}^*} \mathcal{G}^n$ be the set of all directed acyclic graphs. Consider a network represented by a graph $G = (N, E) \in \mathcal{G}$, where $N = \{1, 2, \cdots, n\}$ is the node set and E is the edge set. Each node $i \in N$ represents an agent in the network and each edge $(i, j) \in E$ indicates that agent i follows (quotes) agent j. Let P_i be the set of agents who can reach agent i, i.e., for all agent $j \in P_i$, there exists at least one path from j to i in the network. We assume $i \in P_i$. Let $p_i = |P_i|$ be agent i's progeny and $p^* = \max_{i \in N} |P_i|$ be the maximum progeny in the network.

Our objective is to select the agent with the maximum progeny. However, we do not know the underlying network and can only construct the network from the following/referral relationships declared by all agents, i.e., their out-edges. In a game-theoretical setting, agents are self-interested. If we simply choose an agent $i \in N$ with the maximum progeny, agents who directly follow agent i may strategically misreport their out-edges (e.g., not follow agent i) to increase their chances to be selected. Therefore, in this paper, our goal is to design a selection mechanism to assign each agent a proper selection probability, such that no agent can manipulate to increase her chance to be selected and it can provide a good approximation of the expected progeny in the family of DAGs.

For each agent $i \in N$, her type is denoted by her out-edges $\theta_i = \{(i, j) \mid (i, j) \in E, j \in N\}$, which is only known to her. Let $\theta = (\theta_1, \cdots, \theta_n)$ be the type of all agents and θ_{-i} be the type of all agents expect i. Let θ'_i be agent i's report to the mechanism and $\theta' = (\theta'_1, \cdots, \theta'_n)$ be the report profile of all agents. Note that agents do not know the others except for the agents they follow in the network. Then $\theta'_i \subseteq \theta_i$ should hold for all $i \in N$, which satisfies the Nested Range Condition [8] thus guarantees the revelation principles. Thereby, we focus on direct revelation mechanism design here. Let $\Phi(\theta_i)$ be the space of all possible report profiles of agent i with true type θ_i, i.e., $\Phi(\theta_i) = \{\theta'_i \mid \theta'_i \subseteq \theta_i\}$. Let $\Phi(\theta)$ be the set of all possible report profiles of all agents with true type profile θ.

Given n agents, let Θ^n be the set of all possible type profile of n agents. Given $\theta \in \Theta^n$ and a report profile $\theta' \in \Phi(\theta)$, let $G(\theta') = (N, E')$ be the graph constructed from θ', where $N = \{1, 2, \cdots, n\}$ and $E' = \{(i, j) \mid i, j \in N, (i, j) \in \theta'\}$. Denote the progeny of agent i in graph $G(\theta')$ by $p_i(\theta')$ and the maximum progeny in this graph by $p^*(\theta')$. We give a formal definition of a selection mechanism.

Definition 1. *A selection mechanism $\mathcal{M}$ is a family of functions $f : \Theta^n \to [0, 1]^n$ for all $n \in \mathbb{N}^*$. Given a set of agents N and their report profile θ', the mechanism $\mathcal{M}$ will give a selection probability distribution on N. For each agent $i \in N$, denote her selection probability by $x_i(\theta')$. We have $x_i(\theta') \in [0, 1]$ for all $i \in N$ and $\sum_{i \in N} x_i(\theta') \leq 1$.*

Next, we define the property of *incentive compatibility* for a selection mechanism, which incentivizes agents to report their out-edges truthfully.

Definition 2 (Incentive Compatible). *A selection mechanism $\mathcal{M}$ is **incentive compatible (IC)** if for all N, all $i \in N$, all $\theta \in \Theta^n$, all $\theta'_{-i} \in \Phi(\theta_{-i})$ and all $\theta'_i \in \Phi(\theta_i)$, $x_i((\theta_i, \theta'_{-i})) \geq x_i((\theta'_i, \theta'_{-i}))$.*

An incentive compatible selection mechanism guarantees that truthfully reporting her type is a dominant strategy for all agents. An intuitive realization is a uniform mechanism where each agent gets the same selection probability. However, there exists a case where most of the probabilities are assigned to agents with low progeny, thus leading to an unbounded approximation ratio. We desire an incentive compatible selection mechanism to achieve a bounded approximation ratio for all DAGs. We call this property *efficiency* and define the efficiency of a selection mechanism by its approximation ratio.

Definition 3. *Given a set of agents* $N = \{1, 2, \cdots, n\}$*, their true type profile* $\theta \in \Theta^n$*, the performance of an incentive compatible selection mechanism in the graph* $G(\theta)$ *is defined by*

$$R(G(\theta)) = \frac{\sum_{i \in N} x_i(\theta) p_i(\theta)}{p^*(\theta)}.$$

We say an incentive compatible selection mechanism $\mathcal{M}$ *is* ***efficient with an approximation ratio*** r *if for all* N*, all* $\theta \in \Theta^n$*,* $R(G(\theta)) \geq r$*.*

This property guarantees that the worst-case ratio between the expected progeny of the selected agent and the maximum progeny is at least r for all DAGs. Without the constraint of incentive compatibility, an optimal selection mechanism will always choose the agent with the maximum progeny. While in the strategic setting, an agent with enough progeny can misreport to make herself the agent with the maximum progeny. We define such an agent as *an influential node*. In a DAG, there can be multiple influential nodes. Thus we define them as *the influential set*, denoted by $S^{inf.}$. For example, in the graph shown in Fig. 1, when removing agent 3's out-edge, agent 3 will be the root with the maximum progeny, same for agents 1 and 2. The formal definitions are as follows.

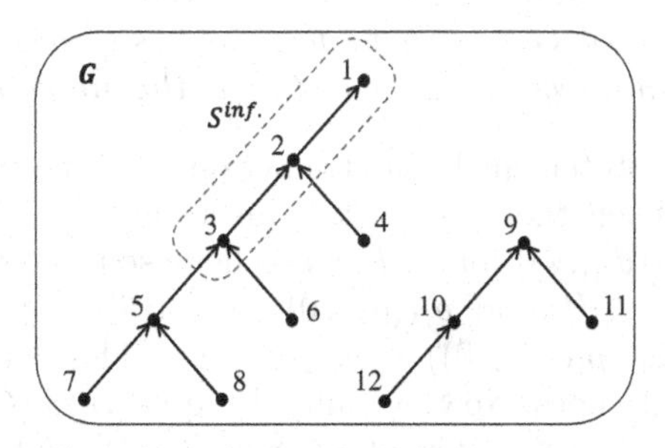

Fig. 1. An example for illustrating the definition of influential nodes: agents 1, 2, 3 are the influential nodes and they form the influential set in the graph G.

Definition 4. *For a set of agents* $N = \{1, 2, \cdots, n\}$*, their true type profile* $\theta \in \Theta^n$ *and their report profile* $\theta' \in \Phi(\theta)$*, an agent* $i \in N$ *is an* ***influential node*** *in the graph* $G(\theta')$ *if* $p_i((\theta'_{-i}, \emptyset)) \succ p_j((\theta'_{-i}, \emptyset))$ *for all* $j \neq i \in N$*, where* $p_i \succ p_j$ *if either* $p_i > p_j$ *or* $p_i = p_j$ *with* $i < j$*.*

Definition 5. *For a set of agents* $N = \{1, 2, \cdots, n\}$*, their true type profile* $\theta \in \Theta^n$ *and their report profile* $\theta' \in \Phi(\theta)$*, the* ***influential set*** *in the graph* $G(\theta')$ *is the set of all influential nodes, denoted by* $S^{inf.}(G(\theta')) = \{s_1, \cdots, s_m\}$*, where* $s_i \succ s_j$ *holds if and only if* $p_i \succ p_j$*,* $s_i \succ s_j$ *holds for all* $m \geq j > i \geq 1$ *and* $m = |S^{inf.}(G(\theta'))|$.

According to the above definitions, we present three observations about the properties of influential nodes.

Observation 1. *Given a set of agents* $N = \{1, 2, \cdots, n\}$*, their true type* $\theta \in \Theta^n$ *and their report profile* $\theta' \in \Phi(\theta)$*, there must exist a path that passes through all agents in* $S^{inf.}(G(\theta'))$ *with an increasing order of their progeny.*

Proof. Let the influential set be $S^{inf.}(G(\theta')) = \{s_1, \cdots, s_m\}$. The statement shows that agent s_j is one of the progeny of agent s_i for all $1 \leq i < j \leq m$, then we can prove it by contradiction.

Assume that there doesn't exist a path passing through all agents in the influential set, then there must be an agent j such that $s_j \notin P_{s_i}$ for all $1 \leq i < j$. Since $s_i, s_j \in S^{inf.}(G(\theta'))$, for all $1 \leq i < j$, we have

$$p_{s_i}((\theta'_{-s_i}, \emptyset)) \succ p_{s_j}((\theta'_{-s_i}, \emptyset)), \tag{1}$$
$$p_{s_j}((\theta'_{-s_j}, \emptyset)) \succ p_{s_i}((\theta'_{-s_j}, \emptyset)). \tag{2}$$

We also have $p_{s_j}((\theta'_{-s_j}, \emptyset)) = p_{s_j}((\theta'_{-s_i}, \emptyset))$ and $p_{s_i}((\theta'_{-s_j}, \emptyset)) = p_{s_i}((\theta'_{-s_i}, \emptyset))$ since $s_j \notin P_{s_i}$ and there is no cycle in the graph. With the lexicographical tie-breaking way, the inequality 1 and 2 cannot hold simultaneously. Therefore, we get a contradiction. □

Observation 2. *Given a set of agents* $N = \{1, 2, \cdots, n\}$*, their true type* $\theta \in \Theta^n$ *and their report profile* $\theta' \in \Phi(\theta)$*, let* $S^{inf.}(G(\theta')) = \{s_1, \cdots, s_m\}$ *be the influential set in the graph* $G(\theta')$*. Then, agent* s_1 *has no out-edges and she is the one with the maximum progeny, i.e., agent* s_1 *is* ***the most influential node****.*

Proof. We prove this statement by contradiction. Assume that agent s_1 has at least one out-edge. Then there must exist an agent $i \in N$ such that $s_1 \in P_i$ and $p_i((\theta'_{-i}, \emptyset)) \succ p_k((\theta'_{-i}, \emptyset))$ for all $k \neq i$, otherwise there must exist an agent $j \in N$ such that $s_1 \notin P_j$ and $p_j((\theta'_{-i}, \emptyset)) \succ p_i((\theta'_{-i}, \emptyset))$, which means that $s_1 \notin S^{inf.}(G(\theta'))$ since $p_i((\theta'_{-i}, \emptyset)) \succ p_{s_1}((\theta'_{-i}, \emptyset))$. Thus, such an i must exist when agent s_1 has out-edges. Now, we must have $i \in S^{inf.}(G(\theta'))$ and $p_i \succ p_{s_1}$, which contradicts with $p_{s_1} \succ p_j$ for all $j \in S^{inf.}(G(\theta'))$ and $j \neq s_1$.

Then we can conclude that agent s_1 has no out-edges. Since $p_{s_1}((\theta'_{-s_1}, \emptyset)) \succ p_k((\theta'_{-s_1}, \emptyset))$ for all $k \neq s_1$, we can get that agent s_1 has the maximum progeny in the graph $G(\theta')$ and she is the most influential node. □

Observation 3. *Given a set of agents* $N = \{1, 2, \cdots, n\}$*, their true type profile* $\theta \in \Theta^n$*, for all agent* $i \in N$*, all* $\theta'_{-i} \in \Phi(\theta_{-i})$*, if agent* i *is not an influential node in the graph* $G((\theta'_{-i}, \theta_i))$*, she cannot make herself an influential node by misreporting.*

Proof. Given other agents' report θ'_{-i}, whether an agent i can be an influential node depends on the relation between $p_i((\theta'_{-i}, \emptyset))$ and $p_j((\theta'_{-i}, \emptyset))$, rather than the out-edges reported by agent i. □

There is one additional desirable property we consider in this paper. Consider two graphs $G, G' \in \mathcal{G}^n$ illustrated in Fig. 2, where they have the same influential set ($S^{inf.}(G) = S^{inf.}(G')$) and s_1 is the most influential node in both graphs. Additionally, the subgraphs constructed by agents in P_{s_1} are the same in both G and G' (The red parts in Fig. 2, represented by $G(s_1) = G'(s_1)$). The only difference between the two graphs lies in the edges that are not in the subgraphs constructed by agents in P_{s_1} (The yellow parts in Fig. 2).

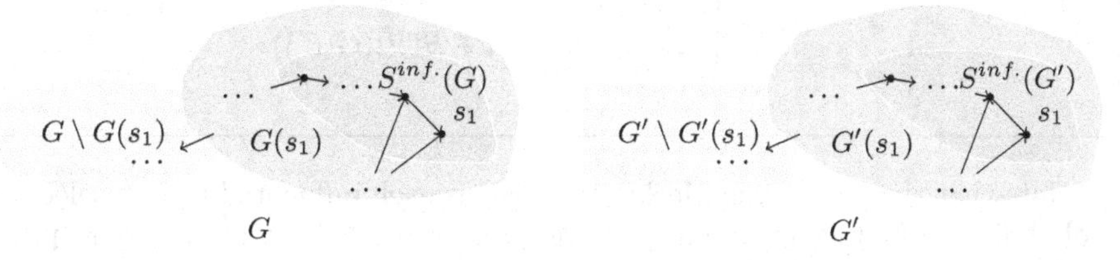

Fig. 2. Example for fairness: in graphs G and G', $S^{inf.}(G) = S^{inf.}(G')$, $G(s_1) = G'(s_1)$; the only difference is in the yellow parts. Fairness requires that $x_{s_1}(G) = x_{s_1}(G')$. (Color figure online)

We can observe that s_1 and all her progeny have the same contributions in the two graphs intuitively. Therefore, it is natural to require that a selection mechanism assigns the same probability to s_1 in the two graphs. We call this property *fairness* and give the formal definition as follows.

Definition 6 (Fairness). *For a graph $G = (N, E) \in \mathcal{G}$, define a subgraph constructed by agent i's progeny as $G(i) = (P_i, E_i)$, where $E_i = \{(j, k) \mid j, k \in P_i, (j, k) \in E\}$ and $i \in N$.*

A selection mechanism $\mathcal{M}$ is ***fair*** *if for all N, for all $G, G' \in \mathcal{G}^n$ where $S^{inf.}(G) = S^{inf.}(G') = \{s_1, \cdots, s_m\}$ and $G(s_1) = G'(s_1)$, then $x_{s_1}(G) = x_{s_1}(G')$.*

3 Geometric Mechanism

In this section, we present the Geometric Mechanism, denoted by $\mathcal{M}_G$. In Observation 3, an agent without enough progeny cannot make herself an influential node by reducing her out-edges. Therefore, to maximize the approximation ratio, we can just assign positive selection probabilities to agents in the influential set. This is the intuition of the Geometric Mechanism.

Geometric Mechanism

1. Given the set of agents $N = \{1, 2, \cdots, n\}$, their true type profile $\theta \in \Theta^n$ and their report profile $\theta' \in \Phi(\theta)$, find the influential set $S^{inf.}$ in the graph $G(\theta')$:
$$S^{inf.}(G(\theta')) = \{s_1, \cdots, s_m\},$$
where $s_i \succ s_{i+1}$ for all $1 \le i \le m-1$.
2. The mechanism gives the selection probability distribution on all agents as the following.
$$x_i = \begin{cases} 1/(2^{m-j+1}), & i = s_j, \\ 0, & i \notin S^{inf.}(G(\theta')). \end{cases}$$

Note that the Geometric Mechanism assigns each influential node a selection probability related to her ranking in the influential set. Besides, an agent' probability is decreasing when her progeny is increasing. This is reasonable because if an influential node j is one of the progeny of another influential node i, the contribution of agent i partially relies on j. To guarantee efficiency and incentive compatibility simultaneously, we assign a higher probability to agent j compared to agent i. We give an example to illustrate how our mechanism works below.

Example 1. Consider the network G shown in Fig. 3. We can observe that only agents 1 and 2 will have the largest progeny in the graphs when they have no out-edges respectively. Thus, the influential set is $S^{inf.}(G) = \{1, 2\}$. Since $p_1 \succ p_2$, then according to the probability assignment defined in the Geometric Mechanism, we choose agent 1 with probability 1/4, choose agent 2 with probability 1/2 and choose no agent with probability 1/4. The expected progeny chosen by the Geometric Mechanism in this graph is

$$\mathbb{E}[p] = \frac{1}{2} \times 6 + \frac{1}{4} \times 8 = 5.$$

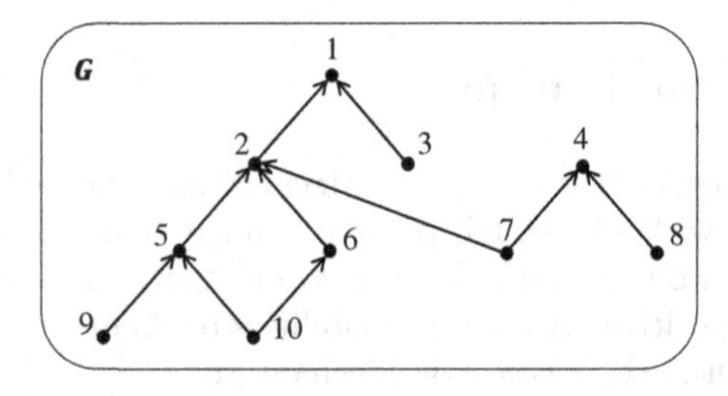

Fig. 3. An example for the geometric mechanism.

On the other hand, the largest progeny is given by agent 1, which is 8, so that the expected ratio of the Geometric Mechanism in this graph is 5/8.

Next, in Theorems 1 and 2, we show that our mechanism satisfies the properties of incentive compatibility and fairness and has an approximation ratio of 1/2 in the family of DAGs.

Theorem 1. *In the family of DAGs, the Geometric Mechanism satisfies incentive compatibility and fairness.*

Proof. In the following, we give the proof for these properties separately.

Incentive Compatibility. Given a set of agents $N = \{1, 2, \cdots, n\}$, their true type $\theta \in \Theta^n$ and their report profile $\theta' \in \Phi(\theta)$, let $G(\theta')$ be the graph constructed by θ', and $S^{inf.}(G(\theta'))$ be the influential set in $G(\theta')$. To prove that the mechanism is incentive compatible, we need to show that $x_i((\theta'_{-i}, \theta_i)) \geq x_i((\theta'_{-i}, \theta'_i))$ holds for all agent $i \in N$.

- According to Observation 3, for agent $i \notin S^{inf.}(G((\theta'_{-i}, \theta_i)))$, she cannot misreport to make herself be an influential node. Thus, her selection probability will always be zero.
- If agent $i \in S^{inf.}(G((\theta'_{-i}, \theta_i)))$, she cannot misreport to make herself be out of the influential set. Suppose $S^{inf.}(G((\theta'_{-i}, \theta_i))) = \{s_1, \cdots, s_m\}$ and $i = s_l$, $1 \leq l \leq m$. Denote the set of influential nodes in her progeny when she truthfully reports by $S_i((\theta'_{-i}, \theta_i)) = \{j \in S^{inf.}(G((\theta'_{-i}, \theta_i))) \mid p_i((\theta'_{-i}, \theta_i)) \succ p_j((\theta'_{-i}, \theta_i))\}$. Then agent i's selection probability in the graph $G((\theta'_{-i}, \theta_i))$ is $x_i((\theta'_{-i}, \theta_i)) = 1/(2^{m-l+1}) = 1/(2^{|S_i((\theta'_{-i}, \theta_i))|+1})$.
  When she misreports her type as $\theta'_i \subset \theta_i$, i.e., deleting a nonempty subset of her real out-edges, $p_j((\theta'_{-j}, \emptyset)) \succ p_k((\theta'_{-j}, \emptyset))$ still holds for all $j \in S_i((\theta'_{-i}, \theta_i))$, all $k \in N$ and $k \neq j$. This can be inferred from Observation 1, agent j is one of the progeny of agent i for all $j \in S_i$. Thus, agent i's report will not change agent j's progeny. Moreover, some other agent $t \in P_i$ may become an influential node in the graph $G((\theta'_{-i}, \theta'_i))$, since $\max_{k \in N, k \neq t} p_k((\theta'_{-t}, \emptyset))$ may be decreased and $p_t((\theta'_{-t}, \emptyset))$ keeps unchanged. Then we can get $S_i((\theta'_{-i}, \theta_i)) \subseteq S_i((\theta'_{-i}, \theta'_i))$, which implies that $x_i((\theta'_{-i}, \theta_i)) = 1/2^{|S_i((\theta'_{-i}, \theta_i))|+1} \geq x_i((\theta'_{-i}, \theta'_i)) = 1/2^{|S_i((\theta'_{-i}, \theta'_i))|+1}$.

Thus, no agent can increase her probability by misreporting her type and the Geometric Mechanism satisfies incentive compatibility.

Fairness. For any two graph $G, G' \in \mathcal{G}^n$, if their influential sets and the subgraphs constructed by the progeny of the most influential node are both the same, i.e., $S^{inf.}(G) = S^{inf.}(G') = \{s_1, \cdots, s_m\}$ and $G(s_1) = G'(s_1)$, according to the definition of Geometric Mechanism, agent s_1 will always get a selection probability of $1/2^m$. Therefore, the Geometric Mechanism satisfies fairness. □

Theorem 2. *In the family of DAGs, the Geometric Mechanism can achieve an approximation ratio of* 1/2.

Proof. Given a graph $G = (N, E) \in \mathcal{G}$ and its influential set $S^{inf.}(G) = \{s_1, \cdots, s_m\}$, the maximum progeny is $p^* = p_{s_1}$. Then the expected ratio should be

$$
\begin{aligned}
R = \frac{\mathbb{E}[p]}{p^*} &= \frac{\sum_{i \in S^{inf.}(G)} x_i p_i}{p^*} \\
&= \frac{\sum_{i=1}^{m} 1/(2^{m-i+1}) \cdot p_{s_i}}{p^*} \\
&= \sum_{i=2}^{m} \frac{1}{2^{m-i+1}} \cdot \frac{p_{s_i}}{p_{s_1}} + \frac{1}{2^m} \cdot \frac{p_{s_1}}{p_{s_1}} \\
&\geq \sum_{j=1}^{m-1} \frac{1}{2^j} \cdot \frac{1}{2} + \frac{1}{2^m} \\
&= \frac{1}{2} - \frac{1}{2^m} + \frac{1}{2^m} = \frac{1}{2}.
\end{aligned}
$$

The inequality holds since $p_{s_i}/p_{s_1} \geq \frac{1}{2}$ holds for all $1 \leq i \leq m-1$. This can be inferred from Observation 1, agent s_i is one of agent s_1's progeny for all $i > 1$. If $p_{s_i}/p_{s_1} < \frac{1}{2}$, then we will have $p_{s_i}((\theta_{-s_i}, \emptyset)) \prec p_{s_1}((\theta_{-s_i}, \emptyset))$, which contradicts with that $s_i \in S^{inf.}(G)$.

The expected ratio holds for any directed acyclic graph, which means that

$$
r_{\mathcal{M}_G} = \min_{G \in \mathcal{G}} R(G) = \frac{1}{2}.
$$

Thus we complete the proof. □

4 Upper Bound and Related Discussions

In this section, we further give an upper bound for any incentive compatible and fair selection mechanisms in Theorem 3. After that, we consider a special class of selection mechanisms, called root mechanisms (detailed in Sect. 4.2), which contains the Geometric Mechanism. Then, we propose two conjectures on whether root mechanisms and fairness will limit the upper bound of the approximation ratio.

4.1 Upper Bound

We prove an upper bound for any IC and fair selection mechanisms as below.

Theorem 3. *For any incentive compatible and fair selection mechanism* $\mathcal{M}$, $r_{\mathcal{M}} \leq \frac{1}{1+\ln 2}$.

Proof. Consider the graph $G = (N, E)$ shown in Fig. 4, the influential set in G is $S^{inf.}(G) = \{2k-1, 2k-2, \cdots, k\}$. When $k \to \infty$, for each agent i, $i \leq k-1$, their contributions can be ignored, it is without loss of generality to assume that they get a probability of zero, i.e., $x_i(G) = 0$. Then, applying a generic incentive compatible and fair mechanism $\mathcal{M}$ in the graph G, assume that $x_i(G) = \beta_{i-k}$ is the selection probability of agent i, $\beta_{i-k} \in [0, 1]$ and $\sum_{i=k}^{2k-1} \beta_{i-k} \leq 1$.

For each agent $i \in N$, set $N_i = P_i(G)$, $N_{-i} = N \setminus N_i$, $E_i = \{(j,k) \mid j,k \in I_i, (j,k) \in E\}$ and $E_{-i} = E \setminus \{E_i \cup \theta_i\}$. Define a set of graphs $\mathcal{G}_i = \{G' = (G(i); G(-i)) \mid G(-i) = (N_{-i}, E'_{-i}), E'_{-i} \subseteq E_{-i}\}$. Then for any graph $G' \in \mathcal{G}_i$, it is generated by deleting agent i's out-edge and a subset of out-edges of agent i's parent nodes, illustrated in Fig. 4. For any $i \geq k$ and any graph $G' \in \mathcal{G}_i$, the influential set in the graph G' should be $S^{inf.}(G') = \{i, i-1, \cdots, k\}$.

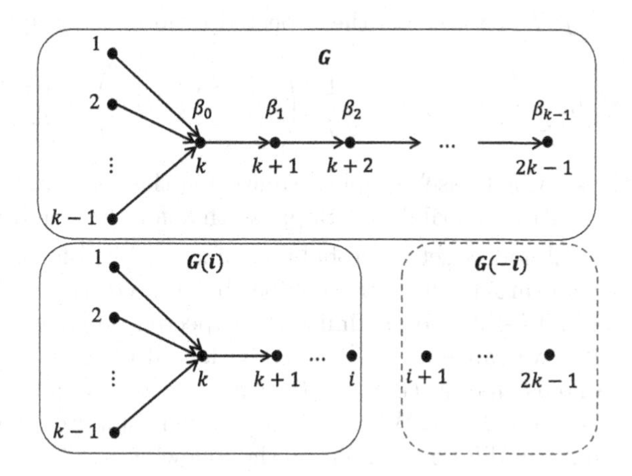

Fig. 4. The upper part is the origin graph G. The bottom part is an example in $\mathcal{G}_i$: for any $i \geq k$, any graph $(G(i); G(-i)) \in \mathcal{G}_i$, the graph $(G(i); G(-i))$ is generated by dividing G into two parts. Then, $G(i)$ is generated by keeping the same as the first part, $G(-i)$ is then generated by deleting some of the edges in the second part. Note that there is no edge between i and $i+1$.

To get the upper bound of the approximation ratio, we consider a kind of "worst-case" graphs where the contributions of agents except influential nodes can be ignored when $k \to \infty$. Since the mechanism $\mathcal{M}$ satisfies the fairness, it holds that $x_i(G') = x_i(G'')$ for any two graphs $G', G'' \in \mathcal{G}_i$. Then for any graph $G' \in \mathcal{G}_k$, agent k is assigned the same probability. Thus, we can find that in the graph set $\mathcal{G}_k$, the "worst-case" graph G_k is a graph where there are only edges between k and i, $1 \leq i \leq k-1$ (shown in Fig. 5).

Since no matter how much the probability the mechanism assigns to other agents, the expected ratio for the graph G_k approaches the probability $x_k(G_k)$ when $k \to \infty$, i.e.,

$$\lim_{k \to \infty} R(G_k) \leq \lim_{k \to \infty} x_k(G_k) + \frac{1}{k} \cdot (1 - x_k(G_k)) = x_k(G_k).$$

The inequality holds since $\sum_{i=1}^{2k-1} x_i(G_k) \leq 1$. Similarly, for any $k < j \leq 2k-2$, the "worst-case" graph G_j in $\mathcal{G}_j$ is the graph where the out-edge of agent i is

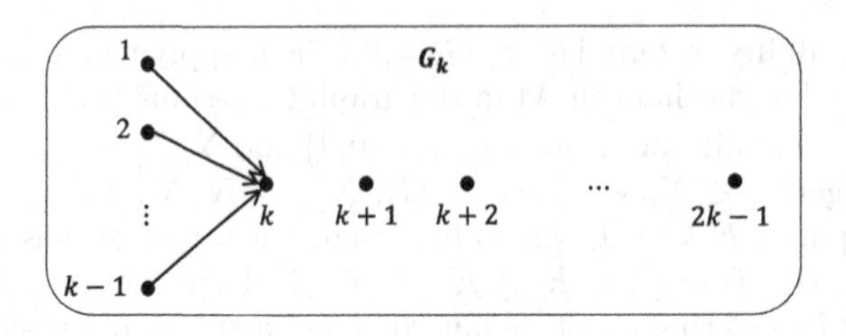

Fig. 5. The "worst-case" graph G_k in the set $\mathcal{G}_k$.

deleted for all $i \geq j$. When $k \to \infty$, the expected ratio in the graph G_j is

$$\lim_{k\to\infty} R(G_j) \leq \lim_{k\to\infty} \sum_{i=k}^{j} x_i(G_j) \cdot \frac{i}{j} + \frac{1}{j} \cdot \left(1 - \sum_{i=k}^{j} x_i(G_j)\right) = \sum_{i=k}^{j} x_i(G_j) \cdot \frac{i}{j}.$$

Therefore, in these "worst-case" graphs, we assume that only influential nodes can be assigned positive probabilities. Suppose that for the graph G_j, $k \leq j \leq 2k-2$, an influential node i gets a probability of $x_i = \beta_{i-k}^{(j-i)}$ for $k \leq i \leq j$.

Since the mechanism $\mathcal{M}$ is incentive compatible, it holds that $x_i(G) \geq x_i(G')$ for all $G' \in \mathcal{G}_i$ and all $i \in N$. To maximize the expected progeny of the selected agent in all graphs, we can set $x_i(G') = x_i(G)$ for all $G' \in \mathcal{G}_i$ and all $i \in N$. Similarly, it also holds that $x_i(G'') \geq x_i(G')$ for any $i \in N$, any $G' \in \mathcal{G}_i$, any $G'' \in \mathcal{G}_j$ and $k \leq i < j \leq 2k-1$. When $k \to \infty$, we can compute the performance of the mechanism $\mathcal{M}$ in different graphs as the following.

$$R(G_j) = \sum_{i=k}^{j} \beta_{i-k}^{(j-i)} \cdot \frac{i}{j}, k \leq j \leq 2k-2,$$

$$R(G) = \sum_{i=k}^{2k-1} \beta_{i-k} \cdot \frac{i}{2k-1},$$

with $\beta_{i-k}^{(j-i)} \geq \beta_{i-k}^{(0)}$, $\beta_{i-k}^{(0)} = \beta_{i-k}$, $k \leq i \leq 2k-1$, $k \leq j \leq 2k-2$. The approximation ratio of the mechanism $\mathcal{M}$ should be at most the minimum of $R(G_j)$ and $R(G)$ for $k \leq j \leq 2k-2$, i.e.,

$$r_{\mathcal{M}} \leq \min\left\{\beta_0^{(0)}, \beta_0^{(1)} \cdot \frac{k}{k+1} + \beta_1^{(0)}, \cdots, \beta_0 \cdot \frac{k}{2k-1} + \beta_1 \cdot \frac{k+1}{2k-1} + \cdots + \beta_{k-1}\right\}.$$

Then we can choose $\beta_{i-k}^{(j-i)}$ to achieve the highest minimum expected ratio. We find that $r_{\mathcal{M}} \leq \frac{1}{1+\ln 2}$ and the equation holds when $k \to \infty$ and

$$\begin{cases} \beta_{i-k}^{(j-i)} = \beta_{i-k}, \\ \beta_0 + \beta_1 + \cdots + \beta_{k-1} = 1, \\ \beta_0^{(0)} = \beta_0^{(1)} \cdot \frac{k}{k+1} + \beta_1^{(0)} = \cdots = \beta_0 \cdot \frac{k}{2k-1} + \beta_1 \cdot \frac{k+1}{2k-1} + \cdots + \beta_{k-1}. \end{cases}$$

□

4.2 Open Questions

Note that the approximation ratio of the Geometric Mechanism is close to the upper bound we prove in Sect. 4.1. However, there is still a gap between them. In this section, we suggest two open questions which narrow down the space for finding the optimal selection mechanism.

Root Mechanism. Recall that our goal in this paper is to maximize the approximation ratio between the expected progeny of the selected agent and the maximum progeny. If requiring incentive compatibility, a selection mechanism cannot simply select the most influential node. However, we can identify a subset of agents who can pretend to be the most influential node. This is the influential set we illustrate in Definition 5, and we show that agents cannot be placed into the influential set by misreporting as illustrated in Observation 3. Utilizing this idea, we see that if we assign positive probabilities only to these agents, then the selected agent has a large progeny, and agents have less chance to manipulate. We call such mechanisms as *root mechanisms.*

Definition 7. *A root mechanism $\mathcal{M}_R$ is a family of functions $f_R : \Theta^n \to [0,1]^n$ for all $n \in \mathbb{N}^*$. Given a set of agents N and their report profile θ', a root mechanism $\mathcal{M}_R$ only assigns positive selection probabilities to agents in the set $S^{inf.}(G(\theta'))$. Let $x_i(\theta')$ be the probability of selecting agent $i \in N$. Then $x_i(\theta') = 0$ for all $i \notin S^{inf.}(G(\theta'))$, $x_i(\theta') \in [0,1]$ for all $i \in N$ and $\sum_{i \in N} x_i(\theta') \leq 1$.*

It is clear that our Geometric Mechanism is a root mechanism, whose approximation ratio is not far from the upper bound of $1/(1+\ln 2)$. We conjecture that an optimal incentive compatible selection mechanism and an optimal incentive compatible root mechanism share the same approximation ratio bound.

Conjecture 1. If an optimal incentive compatible root mechanism $\mathcal{M}_R$ has an approximation ratio of $r^*_{\mathcal{M}_R}$, there does not exist other incentive compatible selection mechanism that can achieve a strictly better approximation ratio.

Proof (Discussion). An optimal incentive compatible selection mechanism will usually try to assign more probabilities to agents with more progeny. Following this way, we assign zero probability to all agents who are not an influential node and find a proper probability distribution for the influential set, rather than giving non-zero probabilities to all agents. Since any agent who is not an influential node cannot make herself in the influential set when other agents' reports are fixed, this method will not cause a failure for incentive compatibility. □

Fairness. Note that the upper bound of $1/(1+\ln 2)$ is for all incentive compatible and fair selection mechanisms. We should also consider whether an incentive compatible selection mechanism can achieve a better approximation ratio without the constraint of fairness. Here, we conjecture that an incentive compatible selection cannot achieve an approximation ratio higher than $1/(1 + \ln 2)$ if the requirement of fairness is relaxed.

Conjecture 2. If an optimal incentive compatible and fair mechanism $\mathcal{M}$ can achieve an approximation ratio of $r^*_{\mathcal{M}}$, there does not exist other incentive compatible mechanism with a strictly higher approximation ratio.

Proof (Discussion). Let $\mathcal{G}_f$ be a set of graphs where for any two graphs $G, G' \in \mathcal{G}_f$, their number of nodes, their influential sets $S^{inf.}(G) = S^{inf.}(G') = \{s_1, \cdots, s_m\}$ and the subgraphs constructed by agent s_1's progeny are same. If an incentive compatible selection mechanism is not fair, there must exist such a set $\mathcal{G}_f$ where the mechanism fails fairness. Then the expected ratios in two graphs in $\mathcal{G}_f$ may be different, and the graph with a lower expected ratio might be improved since these two graphs are almost equivalent. One possible way for proving this conjecture is to design a function that reassigns probabilities for all graphs in $\mathcal{G}_f$ such that x_{s_1} is the same for these graphs without hurting the property of incentive compatibility, and all graphs in $\mathcal{G}_f$ then share the same expected ratio without hurting the efficiency of the selection mechanism. □

5 Conclusion

In this paper, we investigate a selection mechanism for choosing the most influential agent in a network. We use the progeny of an agent to measure her influential level so that there exist some cases where an agent can decrease her out-edges to make her the most influential agent. We target selection mechanisms that can prevent such manipulations and select an agent with her progeny as large as possible. For this purpose, we propose the Geometric Mechanism that achieves at least 1/2 of the optimal progeny. We also show that no mechanism can achieve an expected progeny of the selected agent that is greater than $1/(1+\ln 2)$ of the optimal under the conditions of incentive compatibility and fairness.

There are several interesting aspects that have not been covered in this paper. First of all, there is still a gap between the efficiency of our proposed mechanism and the given upper bound. One of the future work is to find the optimal mechanism if it exists. In this direction, we also leave two open questions for further investigations. Moreover, selecting a set of influential agents rather than a single agent is also important in real-world applications (e.g., ranking or promotion). So another future work is to extend our results to the settings where a set of k ($k > 1$) agents need to be selected.

References

1. Alon, N., Fischer, F., Procaccia, A., Tennenholtz, M.: Sum of us: strategyproof selection from the selectors. In: Proceedings of the 13th Conference on Theoretical Aspects of Rationality and Knowledge, pp. 101–110 (2011)
2. Aziz, H., Lev, O., Mattei, N., Rosenschein, J.S., Walsh, T.: Strategyproof peer selection: mechanisms, analyses, and experiments. In: AAAI, pp. 397–403 (2016)
3. Babichenko, Y., Dean, O., Tennenholtz, M.: Incentive-compatible diffusion. In: Proceedings of the 2018 World Wide Web Conference, pp. 1379–1388 (2018)

4. Babichenko, Y., Dean, O., Tennenholtz, M.: Incentive-compatible selection mechanisms for forests. In: EC 2020: The 21st ACM Conference on Economics and Computation, Virtual Event, Hungary, 13–17 July 2020, pp. 111–131. ACM (2020)
5. Caragiannis, I., Christodoulou, G., Protopapas, N.: Impartial selection with prior information. arXiv preprint arXiv:2102.09002 (2021)
6. Emek, Y., Karidi, R., Tennenholtz, M., Zohar, A.: Mechanisms for multi-level marketing. In: Proceedings of the 12th ACM Conference on Electronic Commerce, pp. 209–218 (2011)
7. Fischer, F., Klimm, M.: Optimal impartial selection. SIAM J. Comput. **44**(5), 1263–1285 (2015)
8. Green, J.R., Laffont, J.J.: Partially verifiable information and mechanism design. Rev. Econ. Stud. **53**(3), 447–456 (1986)
9. Huang, K., et al.: Efficient approximation algorithms for adaptive influence maximization. arXiv preprint arXiv:2004.06469 (2020)
10. Kimura, M., Saito, K., Nakano, R.: Extracting influential nodes for information diffusion on a social network. In: AAAI, vol. 7, pp. 1371–1376 (2007)
11. Kleinberg, J.: Cascading behavior in networks: algorithmic and economic issues. Algorithmic Game Theory **24**, 613–632 (2007)
12. Ko, Y.Y., Cho, K.J., Kim, S.W.: Efficient and effective influence maximization in social networks: a hybrid-approach. Inf. Sci. **465**, 144–161 (2018)
13. Kundu, S., Murthy, C.A., Pal, S.K.: A new centrality measure for influence maximization in social networks. In: Kuznetsov, S.O., Mandal, D.P., Kundu, M.K., Pal, S.K. (eds.) PReMI 2011. LNCS, vol. 6744, pp. 242–247. Springer, Heidelberg (2011). https://doi.org/10.1007/978-3-642-21786-9_40
14. Mohammadinejad, A., Farahbakhsh, R., Crespi, N.: Consensus opinion model in online social networks based on influential users. IEEE Access **7**, 28436–28451 (2019)
15. Morone, F., Makse, H.A.: Influence maximization in complex networks through optimal percolation. Nature **524**(7563), 65–68 (2015)
16. Narahari, Y., Mohite, M.: Incentive compatible influence maximization in social networks. In: International Conference on Autonomous Agents and Multiagent Systems (AAMAS) (2011)
17. Narayanam, R., Narahari, Y.: Determining the top-k nodes in social networks using the shapley value. In: AAMAS (3), pp. 1509–1512 (2008)
18. Page, L., Brin, S., Motwani, R., Winograd, T.: The pagerank citation ranking: bringing order to the web. Technical report, Stanford InfoLab (1999)
19. Pal, S.K., Kundu, S., Murthy, C.: Centrality measures, upper bound, and influence maximization in large scale directed social networks. Fund. Inform. **130**(3), 317–342 (2014)
20. Shen, W., Feng, Y., Lopes, C.V.: Multi-winner contests for strategic diffusion in social networks. In: Proceedings of the AAAI Conference on Artificial Intelligence, vol. 33, pp. 6154–6162 (2019)
21. Zhang, J.X., Chen, D.B., Dong, Q., Zhao, Z.D.: Identifying a set of influential spreaders in complex networks. Sci. Rep. **6**, 27823 (2016)
22. Zhang, Y., Zhang, X., Zhao, D.: Sybil-proof answer querying mechanism. In: Proceedings of the Twenty-Ninth International Joint Conference on Artificial Intelligence, IJCAI 2020, pp. 422–428. Ijcai.org (2020)

Computational Aspects of Games

On Tightness of the Tsaknakis-Spirakis Algorithm for Approximate Nash Equilibrium

Zhaohua Chen[1], Xiaotie Deng[1](✉), Wenhan Huang[2], Hanyu Li[1], and Yuhao Li[1]

[1] Center on Frontiers of Computing Studies, Computer Science Department, Peking University, Beijing 100871, China
{chenzhaohua,xiaotie,lhydave,yuhaoli.cs}@pku.edu.cn
[2] Department of Computer Science and Engineering, Shanghai Jiao Tong University, Shanghai 200240, China
rowdark@sjtu.edu.cn

Abstract. Finding the minimum approximate ratio for Nash equilibrium of bi-matrix games has derived a series of studies, started with 3/4, followed by 1/2, 0.38 and 0.36, finally the best approximate ratio of 0.3393 by Tsaknakis and Spirakis (TS algorithm for short). Efforts to improve the results remain not successful in the past 14 years.

This work makes the first progress to show that the bound of 0.3393 is indeed tight for the TS algorithm. Next, we characterize all possible tight game instances for the TS algorithm. It allows us to conduct extensive experiments to study the nature of the TS algorithm and to compare it with other algorithms. We find that this lower bound is not smoothed for the TS algorithm in that any perturbation on the initial point may deviate away from this tight bound approximate solution. Other approximate algorithms such as Fictitious Play and Regret Matching also find better approximate solutions. However, the new distributed algorithm for approximate Nash equilibrium by Czumaj et al. performs consistently at the same bound of 0.3393. This proves our lower bound instances generated against the TS algorithm can serve as a benchmark in design and analysis of approximate Nash equilibrium algorithms.

Keywords: Approximate Nash equilibrium · Stationary point · Descent procedure · Tight instance

1 Introduction

Computing Nash equilibrium is a problem of great importance in a variety of fields, including theoretical computer science, algorithmic game theory and learning theory.

H. Li—Main technical contributor.

I. Caragiannis and K. A. Hansen (Eds.): SAGT 2021, LNCS 12885, pp. 97–111, 2021.
https://doi.org/10.1007/978-3-030-85947-3_7

It has been shown that Nash equilibrium computing lies in the complexity class PPAD introduced by Papadimitriou [17]. Its approximate solution has been shown to be PPAD-complete for 3NASH by Daskalakis, Goldberg and Papadimitriou [7], and for 2NASH by Chen, Deng and Teng [3], indicating its computational intractability in general. This leads to a great many efforts to find an ϵ-approximate Nash equilibrium in polynomial time for small constant $\epsilon > 0$.

Early works by Kontogiannis et al. [13] and Daskalakis et al. [9], introduce simple polynomial-time algorithms to reach an approximation ratio of $\epsilon = 3/4$ and $\epsilon = 1/2$, respectively. Their algorithms are based on searching strategies of small supports. Conitzer [5] also shows that the well-known fictitious play algorithm [2] gives a 1/2-approximate Nash equilibrium within constant rounds, combining Feder et al.'s result [11]. Subsequently, Daskalakis et al. [8] give an algorithm with an approximation ratio of 0.38 by enumerating arbitrarily large supports. The same result is achieved by Czumaj et al. in 2016 [6] with a totally different approach by solving the Nash equilibrium of two zero-sum games and making a further adjustment. Bosse et al. [1] provide another algorithm based on the previous work by Kontogiannis and Spirakis [14] that reaches a 0.36-approximate Nash equilibrium. Concurrently with them, Tsaknakis and Spirakis [18] establish the currently best-known approximation ratio of 0.3393.

The original paper proves that the algorithm gives an upper bound of 0.3393-approximate Nash equilibrium. However, it leaves the problem open that whether 0.3393 is tight for the algorithm. In literature, the experimental performance of the algorithm is far better than 0.3393 [19]. The worst ratio in an empirical trial by Fearnley et al. shows that there is a game on which the TS algorithm gives a 0.3385-approximate Nash equilibrium [10].

In this work, we prove that 0.3393 is indeed the tight bound for the TS algorithm [18] by giving a game instance, subsequently solving the open problem regarding the well-followed the TS algorithm.

Despite the tightness of 0.3393 for the TS algorithm, our extensive experiment shows that it is rather difficult to find a tight instance in practice by brute-force enumerations. The experiment implies that the practical bound is inconsistent with the theoretical bound. This rather large gap is a result of the instability of both the stationary point and the descent procedure searching for a stationary point.[1]

Furthermore, we mathematically characterize all game instances able to attain the tight bound. We do a further experiment to explore for which games the ratio becomes tight. Based on it, we identify a region that the games generated are more likely tight instances.

We use the generated game instances to measure the worst-case performances of the Czumaj et al.'s algorithm [6], the regret-matching algorithm in online learning [12] and the fictitious play algorithm [2]. The experiments suggest that the regret-matching algorithm and the fictitious play algorithm perform well.

[1] We follow [18] to define a stationary point for a strategy pair of the maximum value of two players' deviations: It is one where the directional derivatives in all directions are non-negative. The formal definition is presented in Definition 2.

Surprisingly, the algorithm of Czumaj et al. always reaches an approximation ratio of 0.3393, implying that the tight instance generator for the TS algorithm beats a totally different algorithm.

This paper is organized in the followings. In Sect. 2, we introduce the basic definitions and notations that we use throughout the paper. In Sect. 3, we restate the TS algorithm [18] and propose two other auxiliary methods which help to analyze the original algorithm. With all preparations, we prove the existence of a game instance on which the TS algorithm reaches the tight bound $b \approx 0.3393$ by giving an example in Sect. 4. Further, we characterize all tight game instances and present a generator that outputs tight game instances in Sect. 5. At last, We conduct extensive experiments to reveal the properties of the TS algorithm, and compare it with other approximate Nash equilibrium algorithms in Sect. 6.

2 Definitions and Notations

We focus on finding an approximate Nash equilibrium on general 2-player games, where the row player and the column player have m and n strategies, respectively. Further, we respectively use $R_{m\times n}$ and $C_{m\times n}$ to denote the payoff matrices of row player and column player. We suppose that both R and C are normalized so that all their entries belong to $[0, 1]$. In fact, concerning Nash equilibrium, any game is equivalent to a normalized game, with appropriate shifting and scaling on both payoff matrices.

For two vectors u and v, we say $u \geq v$ if each entry of u is greater than or equal to the corresponding entry of v. Meanwhile, let us denote by e_k a k-dimension vector with all entries equal to 1. We use a probability vector to define either player's behavior, which describes the probability that a player chooses any pure strategy to play. More specifically, row player's strategy and column player's strategy lie in Δ_m and Δ_n respectively, where

$$\Delta_m = \{x \in \mathbb{R}^m : x \geq 0, x^T e_m = 1\},$$
$$\Delta_n = \{y \in \mathbb{R}^n : y \geq 0, y^T e_n = 1\}.$$

For a strategy pair $(x, y) \in \mathbb{R}^m \times \mathbb{R}^n$, we call it an *$\epsilon$-approximate Nash equilibrium*, if for any $x' \in \Delta_m$, $y' \in \Delta_n$, the following inequalities hold:

$$x'^T Ry \leq x^T Ry + \epsilon,$$
$$x^T Cy' \leq x^T Cy + \epsilon.$$

Therefore, a Nash equilibrium is an ϵ-approximate Nash equilibrium with $\epsilon = 0$.

To simplify our further discussion, for any probability vector u, we use

$$\text{supp}(u) = \{i : u_i > 0\},$$

to denote the support of u, and also

$$\text{suppmax}(u) = \{i : \forall j,\ u_i \geq u_j\},$$
$$\text{suppmin}(u) = \{i : \forall j,\ u_i \leq u_j\},$$

to denote the index set of all entries equal to the maximum/minimum entry of vector u.

At last, we use $\max(u)$ to denote the value of the maximal entry of vector u, and $\max_S(u)$ to denote the value of the maximal entry of vector u confined in index set S.

3 Algorithms

In this section, we first restate the TS algorithm [18], and then propose two auxiliary adjusting methods, which help to analyze the bound of the TS algorithm.

The TS algorithm formulates the approximate Nash equilibrium problem into an optimization problem. Specifically, we define the following functions:

$$\begin{aligned} f_R(x,y) &:= \max(Ry) - x^T Ry, \\ f_C(x,y) &:= \max(C^T x) - x^T Cy, \\ f(x,y) &:= \max\{f_R(x,y), f_C(x,y)\}. \end{aligned}$$

The goal is to minimize $f(x,y)$ over $\Delta_m \times \Delta_n$.

The relationship between the above function f and approximate Nash equilibrium is as follows. Given strategy pair $(x,y) \in \Delta_m \times \Delta_n$, $f_R(x,y)$ and $f_C(x,y)$ are the respective deviations of row player and column player. By definition, (x,y) is an ϵ-approximate Nash equilibrium if and only if $f(x,y) \leq \epsilon$. In other words, as long as we obtain a point with f value no greater than ϵ, an ϵ-approximate Nash equilibrium is reached.

The idea of TS algorithm is to find a stationary point of the objective function f by a descent procedure and make a further adjustment on the stationary point.[2] To give the formal definition of stationary points, we need to define the scaled directional derivative of f as follows:

Definition 1. *Given* $(x,y), (x',y') \in \Delta_m \times \Delta_n$, *the* scaled directional derivative *of* (x,y) *in direction* $(x'-x, y'-y)$ *is*

$$Df(x,y,x',y') := \lim_{\theta \to 0+} \frac{1}{\theta}\left(f(x+\theta(x'-x), y+\theta(y'-y)) - f(x,y)\right).$$

$Df_R(x,y,x',y')$ and $Df_C(x,y,x',y')$ are defined similarly with respect to f_R and f_C.

Now we give the definition of stationary points.

Definition 2. $(x,y) \in \Delta_m \times \Delta_n$ *is a* stationary point *if and only if for any* $(x',y') \in \Delta_m \times \Delta_n$,

$$Df(x,y,x',y') \geq 0.$$

[2] We will see in Remark 1 that finding a stationary point is not enough to reach a good approximation ratio; therefore the adjustment step is necessary.

We use a descent procedure to find a stationary point. The descent procedure is presented in the full version [4] of this paper. It has already been proved that the procedure runs in the polynomial-time of precision δ to find a nearly stationary point [19].

Now let $S_C(x) := \text{suppmax}(C^T x)$, $S_R(y) := \text{suppmax}(Ry)$. To better deal with $Df(x, y, x', y')$, we introduce a new function T as follows:

$$\begin{aligned} T(x, y, x', y', \rho, w, z) := &\rho(w^T Ry' - x^T Ry' - x'^T Ry + x^T Ry) \\ &+ (1-\rho)(x'^T Cz - x^T Cy' - x'^T Cy + x^T Cy), \end{aligned}$$

where $\rho \in [0, 1]$, $w \in \Delta_m$, $\text{supp}(w) \subseteq S_R(y)$, $z \in \Delta_n$, $\text{supp}(z) \subseteq S_C(x)$.[3] One can verify that when $f_R(x, y) = f_C(x, y)$ (which is a necessary condition for a stationary point as proved in Proposition 3),

$$Df(x, y, x', y') = \max_{\rho, w, z} T(x, y, x', y', \rho, w, z) - f(x, y).$$

Now let

$$V(x, y) := \min_{x', y'} \max_{\rho, w, z} T(x, y, x', y', \rho, w, z).$$

By Definition 2, (x, y) is a stationary point if and only if $V(x, y) \geq f(x, y)$. Further, notice that

$$V(x, y) \leq \max_{\rho, w, z} T(x, y, x, y, \rho, w, z) = f(x, y),$$

therefore, we have the following proposition.

Proposition 1. *(x, y) is a stationary point if and only if*

$$V(x, y) = f_R(x, y) = f_C(x, y).$$

In the following context, we use (x^*, y^*) to denote a stationary point. By von Neumann's minimax theorem [16], we have

Proposition 2.

$$V(x^*, y^*) = \max_{\rho, w, z} \min_{x', y'} T(x^*, y^*, x', y', \rho, w, z),$$

and there exist ρ^*, w^*, z^* such that

$$V(x^*, y^*) = \min_{x', y'} T(x^*, y^*, x', y', \rho^*, w^*, z^*).$$

[3] Throughout the paper, we suppose that $(x, y), (x', y') \in \Delta_m \times \Delta_n$, and $\rho \in [0, 1]$, $w \in \Delta_m$, $\text{supp}(w) \subseteq S_R(y)$, $z \in \Delta_n$, $\text{supp}(z) \subseteq S_C(x)$. These restrictions are omitted afterward for fluency.

We call the tuple (ρ^*, w^*, z^*) a *dual solution* as it can be calculated by dual linear programming.

Nevertheless, a stationary point may not be satisfying (i.e., with an approximation ratio of no less than 1/2 in the worst case). In this case, we adjust the stationary point to another point lying in the following rectangle:

$$\Lambda := \{(\alpha w^* + (1-\alpha)x^*, \beta z^* + (1-\beta)y^*) : \alpha, \beta \in [0,1]\}.$$

Different adjustments on Λ derive different algorithms to find an approximate Nash equilibrium. We present three of these methods below, of which the first one is the solution by the TS algorithm, and the other two are for the sake of analysis in Sect. 4. For writing brevity, we define the following two subsets of the boundary of Λ:

$$\Gamma_1 := \{(\alpha x^* + (1-\alpha)w^*, y^*) : \alpha \in [0,1]\} \cup \{(x^*, \beta y^* + (1-\beta)z^* : \beta \in [0,1]\},$$
$$\Gamma_2 := \{(\alpha x^* + (1-\alpha)w^*, z^*) : \alpha \in [0,1]\} \cup \{(w^*, \beta y^* + (1-\beta)z^* : \beta \in [0,1]\}.$$

Method in the TS Algorithm [18]**.** The first method is the original adjustment given by [18] (known as the *TS algorithm* in literature). Define the quantities

$$\lambda := \min_{y':\text{supp}(y')\subseteq S_C(x^*)} \{(w^* - x^*)^T R y'\},$$
$$\mu := \min_{x':\text{supp}(x')\subseteq S_R(y^*)} \{x'^T C(z^* - y^*)\}.$$

The adjusted strategy pair is

$$(x_{\text{TS}}, y_{\text{TS}}) := \begin{cases} \left(\frac{1}{1+\lambda-\mu} w^* + \frac{\lambda-\mu}{1+\lambda-\mu} x^*, z^*\right), & \lambda \geq \mu, \\ \left(w^*, \frac{1}{1+\mu-\lambda} z^* + \frac{\mu-\lambda}{1+\mu-\lambda} y^*\right), & \lambda < \mu. \end{cases}$$

Minimum Point on Γ_2. For the second method, define

$$\alpha^* := \operatorname*{argmin}_{\alpha\in[0,1]} f(\alpha w^* + (1-\alpha)x^*, z^*),$$
$$\beta^* := \operatorname*{argmin}_{\beta\in[0,1]} f(w^*, \beta z^* + (1-\beta)y^*).$$

In a geometric view, our goal is to find the minimum point of f on Γ_2.

The strategy pair given by the second method is

$$(x_{\text{MB}}, y_{\text{MB}}) := \begin{cases} (\alpha^* w^* + (1-\alpha^*)x^*, z^*), & f_C(w^*, z^*) \geq f_R(w^*, z^*), \\ (w^*, \beta^* z^* + (1-\beta^*)y^*), & f_C(w^*, z^*) < f_R(w^*, z^*). \end{cases}$$

Intersection Point of Linear Bound of f_R and f_C on Γ_2. As we will see later, $(x_{\text{MB}}, y_{\text{MB}})$ always behaves no worse than $(x_{\text{TS}}, y_{\text{TS}})$ theoretically.

However, it is rather hard to quantitatively analyze the exact approximation ratio of the second method. Therefore, we propose a third adjustment method. Notice that $f_R(x,y)$, $f_C(x,y)$ and $f(x,y)$ are all convex and linear-piecewise functions with either x or y fixed. Therefore, they are bounded by linear functions on the boundary of Λ. Formally, for $0 \le p, q \le 1$, we have

$$f_R(pw^* + (1-p)x^*, z^*) = (f_R(w^*, z^*) - f_R(x^*, z^*))p + f_R(x^*, z^*), \quad (1)$$
$$f_C(pw^* + (1-p)x^*, z^*) \le f_C(w^*, z^*)p; \quad (2)$$
$$f_C(w^*, qz^* + (1-q)y^*) = (f_C(w^*, x^*) - f_C(w^*, y^*))q + f_C(w^*, y^*), \quad (3)$$
$$f_R(w^*, qz^* + (1-q)y^*) \le f_R(w^*, z^*)q. \quad (4)$$

Taking the minimum of terms on the right hand sides of (1) and (2), (3) and (4) respectively, we derive the following quantities[4]

$$p^* := \frac{f_R(x^*, z^*)}{f_R(x^*, z^*) + f_C(w^*, z^*) - f_R(w^*, z^*)},$$
$$q^* := \frac{f_C(w^*, y^*)}{f_C(w^*, y^*) + f_R(w^*, z^*) - f_C(w^*, z^*)}.$$

The adjusted strategy pair is

$$(x_{\text{IL}}, y_{\text{IL}}) := \begin{cases} (p^* w^* + (1-p^*)x^*, z^*), & f_C(w^*, z^*) \ge f_R(w^*, z^*), \\ (w^*, q^* z^* + (1-q^*)y^*), & f_C(w^*, z^*) < f_R(w^*, z^*). \end{cases}$$

We remark that the outcome of all these three methods can be calculated in polynomial time of m and n.

4 A Tight Instance for All Three Methods

We now show the tight bound of the TS algorithm that we present in the previous section, with the help of two auxiliary adjustment methods proposed in Sect. 3. [18] has shown that the TS algorithm gives an approximation ratio of no greater than $b \approx 0.3393$. In this section, we construct a game on which the TS algorithm attains the tight bound $b \approx 0.3393$. In detail, the game is with payoff matrices (5), where $b \approx 0.3393$ is the tight bound, $\lambda_0 \approx 0.582523$ and $\mu_0 \approx 0.812815$ are the quantities to be defined formally in Lemma 6. The game attains the tight bound $b \approx 0.3393$ at stationary point $x^* = y^* = (1,0,0)^T$ with dual solution $\rho^* = \mu_0/(\lambda_0 + \mu_0)$, $w^* = z^* = (0,0,1)^T$. Additionally, the bound stays $b \approx 0.3393$ for this game even when we try to find the minimum point of f on entire Λ.

$$R = \begin{pmatrix} 0.1 & 0 & 0 \\ 0.1+b & 1 & 1 \\ 0.1+b & \lambda_0 & \lambda_0 \end{pmatrix}, \qquad C = \begin{pmatrix} 0.1 & 0.1+b & 0.1+b \\ 0 & 1 & \mu_0 \\ 0 & 1 & \mu_0 \end{pmatrix}. \quad (5)$$

The formal statement of this result is presented in the following Theorem 1.

[4] The denominator of p^* or q^* may be zero. In this case, we simply define p^* or q^* to be 0.

Theorem 1. *There exists a game such that for some stationary point* (x^*, y^*) *with dual solution* (ρ^*, w^*, z^*), $b = f(x^*, y^*) = f(x_{\text{IL}}, y_{\text{IL}}) = f(x_{\text{MB}}, y_{\text{MB}}) \leq f(\alpha w^* + (1-\alpha)x^*, \beta z^* + (1-\beta)y^*)$ *holds for any* $\alpha, \beta \in [0, 1]$.

The proof of Theorem 1 is done by verifying the tight instance (5) above. Nevertheless, some preparations are required to theoretically finish the verification. They also imply the approach that we find the tight instance.

The preparation work consists of three parts. First, we give an equivalent condition of the stationary point in Proposition 3, which makes it easier to construct payoff matrices with a given stationary point and its corresponding dual solution. Second, we will illustrate a panoramic figure of function f_R and f_C on Λ and subsequently reveal the relationship among the three adjusting strategy pairs presented in Sect. 3. Finally, we give some estimations over f and show when these estimations are exactly tight. Below we present all propositions and lemmas we need. We leave all the proofs in the full version of this paper.

The following proposition shows how to construct payoff matrices with given stationary point (x^*, y^*) and dual solution (ρ^*, w^*, z^*).

Proposition 3. *Let*

$$A(\rho, y, z) := -\rho R y + (1-\rho) C (z - y),$$
$$B(\rho, x, w) := \rho R^T (w - x) - (1-\rho) C^T x.$$

Then (x^*, y^*) *is a stationary point if and only if* $f_R(x^*, y^*) = f_C(x^*, y^*)$ *and there exist* ρ^*, w^*, z^* *such that*

$$\text{supp}(x^*) \subset \text{suppmin}(A(\rho^*, y^*, z^*)), \tag{6}$$
$$\text{supp}(y^*) \subset \text{suppmin}(B(\rho^*, x^*, w^*)). \tag{7}$$

Now we define the following quantities:

$$\lambda^* := (w^* - x^*)^T R z^*,$$
$$\mu^* := {w^*}^T C (z^* - y^*).$$

Lemma 1. *If* $\rho^* \in (0, 1)$, *then* $\lambda^*, \mu^* \in [0, 1]$.

For the sake of brevity below, we define

$$F_I(\alpha, \beta) := f_I(\alpha w^* + (1-\alpha)x^*, \beta z^* + (1-\beta)y^*), I \in \{R, C\}, \alpha, \beta \in [0, 1].$$

Then we have the following lemma:

Lemma 2. *The following two statements hold:*

1. *Given* β, $F_C(\alpha, \beta)$ *is an increasing, convex and linear-piecewise function of* α; $F_R(\alpha, \beta)$ *is a decreasing and linear function of* α.
2. *Given* α, $F_R(\alpha, \beta)$ *is an increasing and convex, linear-piecewise function of* β; $F_C(\alpha, \beta)$ *is a decreasing and linear function of* β.

Recall that the second adjustment method yields the strategy pair $(x_{\mathrm{MB}}, y_{\mathrm{MB}})$. We have the following lemma indicating that (x^*, y^*) and $(x_{\mathrm{MB}}, y_{\mathrm{MB}})$ are the minimum points on the boundary of Λ.

Lemma 3. *The following two statements hold:*

1. (x^*, y^*) *is the minimum point of* f *on* $\Gamma_1 = \{(\alpha x^* + (1-\alpha)w^*, y^*) : \alpha \in [0,1]\} \cup \{(x^*, \beta y^* + (1-\beta)z^* : \beta \in [0,1]\}$.
2. $(x_{\mathrm{MB}}, y_{\mathrm{MB}})$ *is the minimum point of* f *on* $\Gamma_2 = \{(\alpha x^* + (1-\alpha)w^*, z^*) : \alpha \in [0,1]\} \cup \{(w^*, \beta y^* + (1-\beta)z^* : \beta \in [0,1]\}$.

Now we are ready to give an analysis on the third adjusting method.

Lemma 4. *The following two statements hold:*

1. $F_C(\alpha, \beta) = f_C(\alpha w^* + (1-\alpha)x^*, \beta z^* + (1-\beta)y^*)$ *is a linear function of* α *if and only if*
$$S_C(x^*) \cap S_C(w^*) \neq \varnothing. \tag{8}$$
2. $F_R(\alpha, \beta) = f_R(\alpha w^* + (1-\alpha)x^*, \beta z^* + (1-\beta)y^*)$ *is a linear function of* β *if and only if*
$$S_R(y^*) \cap S_R(z^*) \neq \varnothing. \tag{9}$$

With all previous results, we can finally give a comparison on the three adjusting methods we present in Sect. 3.

Proposition 4. $f(x_{\mathrm{TS}}, y_{\mathrm{TS}}) \geq f(x_{\mathrm{MB}}, y_{\mathrm{MB}})$ *and* $f(x_{\mathrm{IL}}, y_{\mathrm{IL}}) \geq f(x_{\mathrm{MB}}, y_{\mathrm{MB}})$ *always hold. Meanwhile,* $f(x_{\mathrm{MB}}, y_{\mathrm{MB}}) = f(x_{\mathrm{IL}}, y_{\mathrm{IL}})$ *holds if and only if*

$$\begin{cases} S_C(x^*) \cap S_C(w^*) \neq \varnothing, & \text{if } f_C(w^*, z^*) > f_R(w^*, z^*), \\ S_R(y^*) \cap S_R(z^*) \neq \varnothing, & \text{if } f_C(w^*, z^*) < f_R(w^*, z^*), \\ f_R(w^*, z^*) = f_C(w^*, z^*). \end{cases}$$

There is a final step to prepare for the proof of the tight bound. We present the following estimations and inequalities.

Lemma 5. *The following two estimations hold:*

1. *If* $f_C(w^*, z^*) > f_R(w^*, z^*)$, *then*
$$f(x_{\mathrm{IL}}, y_{\mathrm{IL}}) = \frac{f_R(x^*, z^*)(f_C(w^*, y^*) - \mu^*)}{f_C(w^*, y^*) + \lambda^* - \mu^*} \leq \frac{1 - \mu^*}{1 + \lambda^* - \mu^*}.$$
And symmetrically, when $f_R(w^*, z^*) > f_C(w^*, z^*)$, *we have*
$$f(x_{\mathrm{IL}}, y_{\mathrm{IL}}) = \frac{f_C(w^*, y^*)(f_R(x^*, z^*) - \lambda^*)}{f_R(x^*, z^*) + \mu^* - \lambda^*} \leq \frac{1 - \lambda^*}{1 + \mu^* - \lambda^*}.$$

Furthermore, if (x^*, y^*) *is not a Nash equilibrium, the equality holds if and only if* $f_C(w^*, y^*) = f_R(x^*, z^*) = 1$.

2. $f(x^*, y^*) \leq \min\{\rho^*\lambda^*, (1-\rho^*)\mu^*\} \leq \frac{\lambda^*\mu^*}{\lambda^*+\mu^*}$.

Remark 1. Lemma 5 tells us that the worst f value of a stationary point could attain is 1/2. In fact, $f(x^*, y^*) \leq \lambda^*\mu^*/(\lambda^*+\mu^*) \leq (\lambda^*+\mu^*)/4 \leq 1/2$. We now give the following game to demonstrate this. Consider the payoff matrices:

$$R = \begin{pmatrix} 0.5 & 0 \\ 1 & 1 \end{pmatrix}, \quad C = \begin{pmatrix} 0.5 & 1 \\ 0 & 1 \end{pmatrix}.$$

One can verify by Proposition 3 that $((1,0)^T, (1,0)^T)$ is a stationary point with dual solution $\rho^* = 1/2, w^* = z^* = (0,1)^T$ and $f(x^*, y^*) = 1/2$. Therefore, merely a stationary point itself cannot beat a straightforward algorithm given by [9], which always finds a 1/2-approximate Nash equilibrium.

Lemma 6 ([18]). *Let*

$$b = \max_{s,t\in[0,1]} \min\left\{\frac{st}{s+t}, \frac{1-s}{1+t-s}\right\},$$

Then $b \approx 0.339321$, *which is attained exactly at* $s = \mu_0 \approx 0.582523$ *and* $t = \lambda_0 \approx 0.812815$.

Finally, we prove Theorem 1 by verifying the tight instance (5) with stationary point $x^* = y^* = (1,0,0)^T$ and dual solution $\rho^* = \mu_0/(\lambda_0+\mu_0)$, $w^* = z^* = (0,0,1)^T$.

Proof Sketch. The verification is divided into 4 steps.

Step 1. Verify that (x^*, y^*) is a stationary point by Proposition 3.
Step 2. Verify that $S_C(x^*) \cap S_C(w^*) \neq \varnothing$ and $f_C(w^*, z^*) > f_R(w^*, z^*)$. Then by Proposition 4, $f(x_{\mathrm{IL}}, y_{\mathrm{IL}}) = f(x_{\mathrm{MB}}, y_{\mathrm{MB}})$ and by Lemma 4, $F_C(\alpha, \beta)$ is a linear function of α.
Step 3. Verify that $\lambda^* = \lambda_0$, $\mu^* = \mu_0$, $f_R(x^*, z^*) = f_C(w^*, y^*) = 1$, and $f(x^*, y^*) = b$. Then by Lemma 5 and Lemma 6, $f(x_{\mathrm{IL}}, y_{\mathrm{IL}}) = f(x_{\mathrm{MB}}, y_{\mathrm{MB}}) = b$.
Step 4. Verify that $b \leq f(\alpha w^* + (1-\alpha)x^*, \beta z^* + (1-\beta)y^*)$ for any $\alpha, \beta \in [0,1]$.

The last step needs more elaboration. First, we do a verification similar to Step 2: $S_R(y^*) \cap S_R(z^*) \neq \varnothing$, and thus $F_R(\alpha, \beta)$ is a linear function of β. Second, we define $g(\beta) := \min_\alpha f(\alpha w^* + (1-\alpha)x^*, \beta z^* + (1-\beta)y^*)$ and prove that $g(\beta) \geq b$ for all $\beta \in [0,1]$, which completes the proof. □

From the proof of Theorem 1, we obtain the following useful corollaries.

Corollary 1. *Suppose $f(x^*, y^*) = f(x_{\rm IL}, y_{\rm IL}) = b$. If either of the following two statements holds:*

1. $S_C(x^*) \cap S_C(w^*) \neq \varnothing$ *and* $f_C(w^*, z^*) > f_R(w^*, z^*)$,
2. $S_R(y^*) \cap S_R(z^*) \neq \varnothing$ *and* $f_R(w^*, z^*) > f_C(w^*, z^*)$,

then for any (x, y) on the boundary of Λ, $f(x, y) \geq b$.

Corollary 2. *Suppose $f(x^*, y^*) = f(x_{\rm IL}, y_{\rm IL}) = b$, $S_C(x^*) \cap S_C(w^*) \neq \varnothing$ and $S_R(y^*) \cap S_R(z^*) \neq \varnothing$. Then for any $\alpha, \beta \in [0, 1]$, $f(\alpha w^* + (1 - \alpha)x^*, \beta z^* + (1 - \beta)y^*) \geq b$.*

It is worth noting that the game with payoff matrices (5) has a pure Nash equilibrium with $x = y = (0, 1, 0)^T$, and the stationary point

$$(x^*, y^*) = ((1, 0, 0)^T, (1, 0, 0)^T)$$

is a strictly-dominated strategy pair. However, a Nash equilibrium never supports on dominated strategies! We can also construct bountiful games that are able to attain the tight bound but own distinct characteristics. For instance, we can give a game with no dominated strategies but attains the tight bound. Some examples are listed in the full version of this paper. Such results suggest that stationary points may not be an optimal concept (in theory) for further approximate Nash equilibrium calculation.

5 Generating Tight Instances

In Sect. 4, we proved the existence of the tight game instance, and we can do more than that. Specifically, we can mathematically profile *all* games that are able to attain the tight bound. In this section, we gather properties in the previous sections and post an algorithm that generates games of this kind. Using the generator, we can dig into the previous three approximate Nash equilibrium algorithms and reveal the behavior of these algorithms and even further, the features of stationary points. Algorithm 1 gives the generator of tight instances, in which the inputs are arbitrary $(x^*, y^*), (w^*, z^*) \in \Delta_m \times \Delta_n$. The algorithm outputs games such that (x^*, y^*) is a stationary point and $(\rho^* = \lambda_0/(\lambda_0 + \mu_0), w^*, z^*)$ is a corresponding dual solution, or outputs "NO" if there is not such a game.

The main idea of the algorithm is as follows. Proposition 3 shows an easier-to-verify equivalent condition of the stationary point; and all additional equivalence conditions required by a tight instance are stated in Proposition 4, Lemma 5 and Lemma 6. Therefore, if we enumerate each pair of possible pure strategies in $S_R(z^*)$ and $S_C(w^*)$ respectively, whether there exists a tight instance solution becomes a linear programming problem.

Algorithm 1. Tight Instance Generator

Input $(x^*, y^*), (w^*, z^*) \in \Delta_m \times \Delta_n$.

1: **if** $\text{supp}(x^*) = \{1, 2, \ldots, m\}$ **or** $\text{supp}(y^*) = \{1, 2, \ldots, n\}$ **then**
2: **Output** "NO"
3: **end if**
4: $\rho^* \leftarrow \mu_0/(\lambda_0 + \mu_0)$.

5: // *Enumerate* $k \in S_R(z^*)$ *and* $l \in S_C(w^*)$.
6: **for** $k \in \{1, \ldots, m\} \setminus \text{supp}(x^*), l \in \{1, \ldots, n\} \setminus \text{supp}(y^*)$ **do**
7: Solve a feasible $R = (r_{ij})_{m \times n}, C = (c_{ij})_{m \times n}$ from the following LP with no objective function:
8: // *basic requirements.*
9: $0 \leq r_{ij}, c_{ij} \leq 1$ for $i \in \{1, \ldots, m\}, j \in \{1, \ldots, n\}$,
10: $\text{supp}(w^*) \subset S_R(y^*)$, $\text{supp}(z^*) \subset S_C(x^*)$,
11: $k \in S_R(z^*)$, $l \in S_C(w^*)$,

12: // *ensure* (x^*, y^*) *is a stationary point.*
13: $\text{supp}(x^*) \subset \text{suppmin}(-\rho^* R y^* + (1 - \rho^*)(C z^* - C y^*))$,
14: $\text{supp}(y^*) \subset \text{suppmin}(\rho^* (R^T w^* - R^T x^*) - (1 - \rho^*) R^T x^*)$,

15: // *ensure* $f(x^*, y^*) = b$.
16: $(w^* - x^*)^T R y^* = {x^*}^T C (z^* - y^*) = b$,

17: // *ensure* $f(x_{\text{IL}}, y_{\text{IL}}) = b$.
18: ${x^*}^T R z^* = {w^*}^T C y^* = 0$,
19: $r_{kj} = 1$ for $j \in \text{supp}(z^*)$, $c_{il} = 1$ for $i \in \text{supp}(w^*)$,
20: ${w^*}^T R z^* = \lambda_0$, ${w^*}^T C z^* = \mu_0$,

21: // *ensure* $f(x_{\text{MB}}, y_{\text{MB}}) = f(x_{\text{IL}}, y_{\text{IL}})$.
22: $l \in S_C(x^*)$.

23: **if** LP is feasible **then**
24: **Output** feasible solutions
25: **end if**
26: **end for**

27: **if** LP is infeasible in any round **then**
28: **Output** "No"
29: **end if**

Proposition 5. *Given* $(x^*, y^*), (w^*, z^*) \in \Delta_m \times \Delta_n$, *all the feasible solutions of the LP in Algorithm 1 are all the games* (R, C) *satisfying*

1. (x^*, y^*) *is a stationary point,*
2. *tuple* $(\rho^* = \mu_0/(\lambda_0 + \mu_0), w^*, z^*)$ *is the dual solution,*[5]

[5] One can verify that the value of ρ^* in the dual solution of any tight stationary point has to be $\mu_0/(\lambda_0 + \mu_0)$, by the second part of Lemma 5.

3. *$f_C(w^*, z^*) > f_R(w^*, z^*)$, and*
4. *$f(x, y) \geq b$ for all (x, y) on the boundary of Λ*

if such a game exists, and the output is "NO" if no such game exists.

The proof of the proposition is presented in the full version of this paper.

For the sake of experiments, there are two concerns of the generator we should take into account.

First, sometimes we want to generate games such that the minimum value of f on entire Λ is also $b \approx 0.3393$. By Corollary 2, it suffices to add a constraint $S_R(y^*) \cap S_R(z^*) \neq \varnothing$ to the LP in Algorithm 1. This is not a necessary condition though.

Second, the dual solution of the LP is usually not unique, and we cannot expect which dual solution the LP algorithm yields. [15] gives some methods to guarantee that the dual solution is unique. For practice, we simply make sure that w^* and z^* are pure strategies. The reason is that even if the dual solution is not unique, the simplex algorithm will end up with some optimal dual solution on a vertex, i.e., w^* and z^* are pure strategies.

6 Experimental Analysis

In this section, we further explore the characteristics of the algorithms presented in Sect. 3 with the help of numerical experiments. Such empirical results may provide us with a deep understanding of the behavior of these algorithms, and specifically, the behavior of stationary points and the descent procedure. Furthermore, we are interested in the tight instance generator itself presented in Sect. 5, particularly, on the probability that the generator outputs an instance given random inputs. At last, we will compare the algorithms with other approximate Nash equilibrium algorithms, additionally showing the potentially implicit relationships among these different algorithms.

Readers can refer to the full version of this paper for the details of the experiments. We now list the key results and insights we gain from these experiments.

1. Our studies on the behavior of algorithms we present in Sect. 3 show that even in a tight game instance, it is almost impossible for a uniformly-picked initial strategy pair to fall into the tight stationary point at the termination. Such results imply the inconsistency of tight instances of stationary point algorithms between theory and practice.
2. Next, we evaluate how the descent procedure behaves on strategy pairs slightly perturbed from a tight stationary point. Surprisingly, as it turns out, even with a minor perturbation, the initial strategy pair will eventually escape from the proximate tight stationary point in almost all cases. Such a consequence shows that the descent procedure is indeed unstable to the perturbation of stationary point and further explains why the tight bound is hardly reached even in tight game instances.

3. We then turn to the tight instance generator we described in Sect. 5. Given two arbitrary strategy pairs (x^*, y^*) and (w^*, z^*) in $\Delta_m \times \Delta_n$, we are interested in whether the generator outputs a tight game instance. The result shows that the intersecting proportion of (x^*, y^*) and (w^*, z^*) plays a vital role in whether a tight game instance can be successfully generated from these two pairs.
4. At last, we measure how other algorithms behave on these tight game instances. Surprisingly, for Czumaj et al.'s algorithm [6], for all cases and all trials, the algorithm terminates on the approximation ratio $b \approx 0.3393$. Meanwhile, regret-matching algorithms [12] always find the pure Nash equilibrium of a 2-player game if there exists, which is the case for all generated tight game instances. At last, fictitious play algorithm [2] behaves well on these instances, with a median approximate ratio of approximately 1×10^{-3} to 1.2×10^{-3} for games with different sizes.

Acknowledgement. This work is supported by Science and Technology Innovation 2030 –"The Next Generation of Artificial Intelligence" Major Project No. (2018AAA0100901). We are grateful to Dongge Wang, Xiang Yan and Yurong Chen for their inspiring suggestions and comments. We thank a number of readers for their revision opinions. We are also grateful to anonymous reviewers for their kindness.

References

1. Bosse, H., Byrka, J., Markakis, E.: New algorithms for approximate nash equilibria in bimatrix games. Theor. Comput. Sci. **411**(1), 164–173 (2010). https://doi.org/10.1016/j.tcs.2009.09.023
2. Brown, G.W.: Iterative solution of games by fictitious play. Act. Anal. Prod. Allocat. **13**(1), 374–376 (1951)
3. Chen, X., Deng, X., Teng, S.H.: Settling the complexity of computing two-player nash equilibria. J. ACM **56**(3), 1–57 (2009). https://doi.org/10.1145/1516512.1516516
4. Chen, Z., Deng, X., Huang, W., Li, H., Li, Y.: On tightness of the Tsaknakis-Spirakis algorithm for approximate nash equilibrium. CoRR abs/2107.01471 (2021). https://arxiv.org/abs/2107.01471
5. Conitzer, V.: Approximation guarantees for fictitious play. In: 2009 47th Annual Allerton Conference on Communication, Control, and Computing (Allerton). IEEE, September 2009. https://doi.org/10.1109/allerton.2009.5394918
6. Czumaj, A., Deligkas, A., Fasoulakis, M., Fearnley, J., Jurdziński, M., Savani, R.: Distributed methods for computing approximate equilibria. Algorithmica **81**(3), 1205–1231 (2018). https://doi.org/10.1007/s00453-018-0465-y
7. Daskalakis, C., Goldberg, P.W., Papadimitriou, C.H.: The complexity of computing a nash equilibrium. SIAM J. Comput. **39**(1), 195–259 (2009). https://doi.org/10.1137/070699652
8. Daskalakis, C., Mehta, A., Papadimitriou, C.: Progress in approximate nash equilibria. In: Proceedings of the 8th ACM Conference on Electronic Commerce - EC 2007. ACM Press (2007). https://doi.org/10.1145/1250910.1250962
9. Daskalakis, C., Mehta, A., Papadimitriou, C.: A note on approximate nash equilibria. Theore. Comput. Sci. **410**(17), 1581–1588 (2009). https://doi.org/10.1016/j.tcs.2008.12.031

10. Fearnley, J., Igwe, T.P., Savani, R.: An empirical study of finding approximate equilibria in bimatrix games. In: Bampis, E. (ed.) SEA 2015. LNCS, vol. 9125, pp. 339–351. Springer, Cham (2015). https://doi.org/10.1007/978-3-319-20086-6_26
11. Feder, T., Nazerzadeh, H., Saberi, A.: Approximating nash equilibria using small-support strategies. In: Proceedings of the 8th ACM Conference on Electronic Commerce - EC 2007. ACM Press (2007). https://doi.org/10.1145/1250910.1250961
12. Greenwald, A., Li, Z., Marks, C.: Bounds for regret-matching algorithms. In: International Symposium on Artificial Intelligence and Mathematics, ISAIM 2006, Fort Lauderdale, Florida, USA, 4–6 January 2006 (2006)
13. Kontogiannis, S.C., Panagopoulou, P.N., Spirakis, P.G.: Polynomial algorithms for approximating nash equilibria of bimatrix games. Theor. Comput. Sci. **410**(17), 1599–1606 (2009). https://doi.org/10.1016/j.tcs.2008.12.033
14. Kontogiannis, S.C., Spirakis, P.G.: Efficient algorithms for constant well supported approximate equilibria in bimatrix games. In: Arge, L., Cachin, C., Jurdziński, T., Tarlecki, A. (eds.) ICALP 2007. LNCS, vol. 4596, pp. 595–606. Springer, Heidelberg (2007). https://doi.org/10.1007/978-3-540-73420-8_52
15. Mangasarian, O.: Uniqueness of solution in linear programming. Linear Algebra Appl. **25**, 151–162 (1979). https://doi.org/10.1016/0024-3795(79)90014-4
16. Neumann, J.: Zur theorie der gesellschaftsspiele. Math. Ann. **100**(1), 295–320 (1928). https://doi.org/10.1007/bf01448847
17. Papadimitriou, C.H.: On the complexity of the parity argument and other inefficient proofs of existence. J. Comput. Syst. Sci. **48**(3), 498–532 (1994). https://doi.org/10.1016/S0022-0000(05)80063-7
18. Tsaknakis, H., Spirakis, P.G.: An optimization approach for approximate nash equilibria. Internet Math. **5**(4), 365–382 (2008). https://doi.org/10.1080/15427951.2008.10129172
19. Tsaknakis, H., Spirakis, P.G., Kanoulas, D.: Performance evaluation of a descent algorithm for bi-matrix games. In: Papadimitriou, C., Zhang, S. (eds.) WINE 2008. LNCS, vol. 5385, pp. 222–230. Springer, Heidelberg (2008). https://doi.org/10.1007/978-3-540-92185-1_29

Prophet Inequality with Competing Agents

Tomer Ezra[1(✉)], Michal Feldman[1,2(✉)], and Ron Kupfer[3(✉)]

[1] Tel Aviv University, Tel Aviv, Israel
michal.feldman@cs.tau.ac.il
[2] Microsoft Research, Herzliya, Israel
[3] The Hebrew University of Jerusalem, Jerusalem, Israel
ron.kupfer@mail.huji.ac.il

Abstract. We introduce a model of *competing* agents in a prophet setting, where rewards arrive online, and decisions are made immediately and irrevocably. The rewards are unknown from the outset, but they are drawn from a known probability distribution. In the standard prophet setting, a single agent makes selection decisions in an attempt to maximize her expected reward. The novelty of our model is the introduction of a competition setting, where multiple agents compete over the arriving rewards, and make online selection decisions simultaneously, as rewards arrive. If a given reward is selected by more than a single agent, ties are broken either randomly or by a fixed ranking of the agents. The consideration of competition turns the prophet setting from an online decision making scenario to a multi-agent game.

For both random and ranked tie-breaking rules, we present simple threshold strategies for the agents that give them high guarantees, independent of the strategies taken by others. In particular, for random tie-breaking, every agent can guarantee herself at least $\frac{1}{k+1}$ of the highest reward, and at least $\frac{1}{2k}$ of the optimal social welfare. For ranked tie-breaking, the ith ranked agent can guarantee herself at least a half of the ith highest reward. We complement these results by matching upper bounds, even with respect to equilibrium profiles. For ranked tie-breaking rule, we also show a correspondence between the equilibrium of the k-agent game and the optimal strategy of a single decision maker who can select up to k rewards.

Keywords: Prophet inequality · Multi-agent system · Threshold-strategy

1 Introduction

In the classical prophet inequality problem a decision maker observes a sequence of n non-negative real-valued rewards $v_1, \ldots, v_n$ that are drawn from known independent distributions $F_1, \ldots, F_n$. At time t, the decision maker observes reward v_t, and needs to make an immediate and irrevocable decision whether or

I. Caragiannis and K. A. Hansen (Eds.): SAGT 2021, LNCS 12885, pp. 112–123, 2021.
https://doi.org/10.1007/978-3-030-85947-3_8

not to accept it. If she accepts v_t, the game terminates with value v_t; otherwise, the reward v_t is gone forever and the game continues to the next round. The goal of the decision maker is to maximize the expected value of the accepted reward.

This family of problems captures many real-life scenarios, such as an employer who interviews potential workers overtime, renters looking for a potential house, a person looking for a potential partner for life, and so on. More recently, starting with the work of Hajiaghayi et al. [5], the prophet inequality setting has been studied within the AI community in the context of market and e-commerce scenarios, with applications to pricing schemes for social welfare and revenue maximization. For a survey on a market-based treatment of the prophet inequality problem, see the survey by Lucier [13].

An algorithm ALG has a guarantee α if the expected value of ALG is at least α, where the expectation is taken over the coin flips of the algorithm, and the probability distribution of the input. Krengel and Sucheston [11,12] established the existence of an algorithm that gives a tight guarantee of $\frac{1}{2}\mathbb{E}[\max_i v_i]$. Later, it has been shown that this guarantee can also be obtained by a single-threshold algorithm—an algorithm that specifies some threshold from the outset, and accepts a reward if and only if it exceeds the threshold. Two such thresholds have been presented by Samuel-Cahn [15], and Kleinberg and Weinberg [10]. Single-threshold algorithms are simple and easy to explain and implement.

Competing Agents. Most attention in the literature has been given to scenarios with a single decision maker. Motivated by the economic aspects of the problem, where competition among multiple agents is a crucial factor, we introduce a multi-agent variant of the prophet model, in which multiple agents compete over the rewards.

In our model, a sequence of n non-negative real-valued rewards $v_1, \ldots, v_n$ arrive over time, and a set of k agents make immediate and irrevocable selection decisions. The rewards are unknown from the outset, but every reward v_t is drawn independently from a known distribution F_t. Upon the arrival of reward v_t, its value is revealed to all agents, and every agent decides whether or not to select it.

One issue that arises in this setting is how to resolve ties among agents. That is, who gets the reward if more than one agent selects it. We consider two natural tie-breaking rules; namely, *random* tie breaking (where ties are broken uniformly at random) and *ranked* tie-breaking (where agents are a-priori ranked by some global order, and ties are broken in favor of higher ranked agents). Random tie-breaking fits scenarios with symmetric agents, whereas ranked tie-breaking fits scenarios where some agents are preferred over others, according to some global preference order. For example, it is reasonable to assume that a higher-position/salary job is preferred over lower-position/salary job, or that firms in some industry are globally ordered from most to least desired. Random and ranked tie-breaking rules were considered in [8] and [9], respectively, in secretary settings.

Unlike the classical prophet scenario, which studies the optimization problem of a single decision maker, the setting of competing agents induces a game among multiple agents, were an agent's best strategy depends on the strategies chosen by others. Therefore, we study the equilibria of the induced games. In particular, we study the structure and quality of equilibrium in these settings and devise simple strategies that give agents high guarantees.

When the order of distributions is unknown in advance, calculating the optimal strategy is computationally hard. This motivates the use of simple and efficiently computed strategies that give good guarantees.

1.1 Main Results and Techniques

For both random and ranked tie-breaking rules, we present simple single-threshold strategies for the agents that give them high guarantees. A single-threshold strategy specifies some threshold T, and selects any reward that exceeds T.

For $j = 1, \ldots, n$, let y_j be the jth highest reward.

Under the random tie-breaking rule, we show a series of thresholds that have the following guarantee:

Theorem *(Theorem 1). For every $\ell = 1, \ldots, n$, let $T^\ell = \frac{1}{k+\ell} \sum_{j=1}^{\ell} \mathbb{E}[y_j]$. Then, for every agent, the single threshold strategy T^ℓ (i.e., select v_t iff $v_t \geq T^\ell$) guarantees an expected utility of at least T^ℓ.*

Two special cases of the last theorem are where $\ell = 1$ and $\ell = k$. The case of $\ell = 1$ implies that every agent can guarantee herself (in expectation) at least $\frac{1}{k+1}$ of the highest reward. The case of $\ell = k$ implies that every agent can guarantee herself (in expectation) at least $\frac{1}{2k}$ of the optimal social welfare (i.e., the sum of the highest k rewards), which also implies that the social welfare in equilibrium is at least a half of the optimal social welfare.

The above result is tight, as shown in Proposition 1.

Similarly, for the ranked tie-breaking rule, we show a series of thresholds that have the following guarantee:

Theorem *(Theorem 2). For every $i \leq n$ and $\ell = 0, \ldots, n-i$, let $\hat{T}_i^\ell = \frac{1}{\ell+2} \sum_{j=i}^{i+\ell} \mathbb{E}[y_j]$. Then, for the i-ranked agent, the single threshold strategy $\hat{T}_i^\ell$ (i.e., select v_t iff $v_t \geq \hat{T}_i^\ell$) guarantees an expected utility of at least $\hat{T}_i^\ell$.*

This result implies that for every i, the i-ranked agent can guarantee herself (in expectation) at least a half of the ith highest reward. In Proposition 2 we show that the last result is also tight.

Finally, we show that under the ranked tie-breaking rule, the equilibrium strategies of the (ordered) agents coincide with the decisions of a single decision maker who may select up to k rewards in an online manner and wishes to maximize the sum of selected rewards. Thus, the fact that every agent is aware of her position in the ranking allows them to coordinate around the socially optimal outcome despite the supposed competition between them.

Theorem *(Corollary 4). Under the ranked tie-breaking rule, in every equilibrium of the k-agent game the expected social welfare is at least* $1 - O(\frac{1}{\sqrt{k}})$ *of the optimal welfare.*

A similar phenomenon was observed in a related secretary setting, where the equilibrium strategy profile of a game with several ranked agents, induces an optimal strategy for a single decision maker who is allowed to choose several rewards and wishes to maximize the probability that the highest reward is selected [14].

1.2 Additional Related Literature

The prophet problem and variants thereof has attracted a vast amount of literature in the last decade. For comprehensive surveys, see, e.g., the survey by Hill and Kertz [6] and the survey by Lucier [13] which gives an economic view of the problem.

A related well-known problem in the optimal stopping theory is the *secretary* problem, where the rewards are arbitrary but arrive in a random order. For the secretary problem a tight $1/e$-approximation has been established; for a survey, see, e.g., [4].

Our work is inspired by a series of studies that consider scenarios where multiple agents compete over the rewards in secretary-like settings, where every agent aims to receive the highest reward. Karlin and Lei [9] and Immorlica et al. [8] considered the ranked- and the random tie-breaking rules, respectively, in secretary settings with competition. For the ranked tie-breaking rule, Karlin and Lei [9] show that the equilibrium strategies take the form of *time-threshold strategies*; namely, the agent waits until a specific time t, thereafter competes over any reward that is the highest so far. The values of these time-thresholds are given by a recursive formula. For the random tie-breaking rule, Immorlica et al. [8] characterize the Nash equilibria of the game and show that for several classes of strategies (such as threshold strategies and adaptive strategies), as the number of competing agents grows, the timing in which the earliest reward is chosen decreases. This confirms the argument that early offers in the job market are the result of competition between employers.

Competition among agents in secretary settings has been also studied by Ezra et al. [3], in a slightly different model. Specifically, in their setting, decisions need not be made immediately; rather, any previous reward can be selected as long as it is still available (i.e., has not been taken by a different agent). Thus, the competition is inherent in the model.

Another related work is the dueling framework by Immorlica et al. [7]. One of their scenarios considers a 2-agent secretary setting, where one agent aims to maximize the probability of getting the highest reward (as in the classical secretary problem), and the other agent aims to outperform her opponent. They show an algorithm for the second agent that guarantees her a winning probability of at least 0.51. They also establish an upper bound of 0.82 on this probability.

Other competitive models have been considered in the optimal stopping theory; see [1] for a survey.

The work of Kleinberg and Weinberg [10] regarding matroid prophet problems is also related to our work. They consider a setting where a single decision maker makes online selections under a matroid feasibility constraint, and show an algorithm that achieve 1/2-approximation to the expected optimum for arbitrary matroids. For the special case of uniform matroids, namely selecting up to k rewards, earlier works of Alaei [2] and Hajiaghayi et al. [5] shows a approximation of $1 - O(\frac{1}{\sqrt{k}})$ for the optimal solution. As mentioned above, the same guarantee is obtained in a setting with k ranked competing agents.

1.3 Structure of the Paper

In Sect. 2 we define our model. In Sects. 3 and 4 we present our results with respect to the random tie-breaking rule, and the ranked tie-breaking rule, respectively. We conclude the paper in Sect. 5 with future directions.

2 Model

We consider a prophet inequality variant, where a set of n rewards, $v_1, \ldots, v_n$, are revealed online. While the values $v_1, \ldots, v_n$ are unknown from the outset, v_t is drawn independently from a known probability distribution F_t, for $t \in [n]$, where $[n] = \{1, \ldots, n\}$. In the classical prophet setting, a single decision maker observes the realized reward v_t at time t, and makes an immediate and irrevocable decision whether to take it or not. If she takes it, the game ends. Otherwise, the reward v_t is lost forever, and the game continues with the next reward.

Unlike the classical prophet setting that involves a single decision maker, we consider a setting with k decision makers (hereafter, agents) who compete over the rewards. Upon the revelation of reward v_t, every active agent (i.e., an agent who has not received a reward yet) may select it. If a reward is selected by exactly one agent, then it is assigned to that agent. If the reward v_t is selected by more than one agent, it is assigned to one of these agents either randomly (hereafter, *random tie-breaking*), or according to a predefined ranking (hereafter, *ranked tie-breaking*). Agents who received rewards are no longer active.

A strategy of agent i, denoted by S_i, is a function that for every $t = 1, \ldots, n$, decides whether or not to select v_t, based on t, the realization of v_t, and the set of active agents[1]. A strategy profile is denoted by $S = (S_1, \ldots, S_k)$. We also denote a strategy profile by $S = (S_i, S_{-i})$, where S_{-i} denotes the strategy profile of all agents except agent i.

Every strategy profile S induces a distribution over assignments of rewards to agents. For ranked tie breaking, the distribution is with respect to the realizations

[1] One can easily verify that in our setting, additional information, such as the history of realizations of $v_1, \ldots, v_{t-1}$, and the history of selections and assignments, is irrelevant for future decision making.

of the rewards, and possibly the randomness in the agent strategies. For random tie breaking, the randomness is also with respect to the randomness in the tie-breaking.

The utility of agent i under strategy profile S, $u_i(S)$, is her expected reward under S; every agent acts to maximize her utility.

We say that a strategy S_i guarantees agent i a utility of α if $u_i(S_i, S_{-i}) \geq \alpha$ for every S_{-i}.

Definition 1. *A single threshold strategy T is the strategy that upon the arrival of reward v, v is selected if and only if the agent is still active and $v_t \geq T$.*

We also use the following equilibrium notions:

- Nash Equilibrium (NE): A strategy profile $S = (S_1, \ldots, S_k)$ is a NE if for every agent i and every strategy S_i', it holds that $u_i(S_i', S_{-i}) \leq u_i(S_i, S_{-i})$.
- Subgame perfect equilibrium (SPE): A strategy profile $S = (S_1, \ldots, S_k)$ is an SPE if S is a NE for every subgame of the game. I.e. for every initial history h, S is a NE in the game induced by history h.

SPE is a refinement of NE; namely, every SPE is a NE, but not vice versa.

In the next sections, we let y_j denote the random variable that equals the jth maximal reward among $\{v_1, \ldots, v_n\}$.

3 Random Tie-Breaking

In this section we consider the random tie-breaking rule.

We start by establishing a series of single threshold strategies that guarantee high utilities.

Theorem 1. *For every $\ell = 1, \ldots, n$, let $T^\ell = \frac{1}{k+\ell} \sum_{j=1}^{\ell} \mathbb{E}[y_j]$. Then, for every agent, the single threshold strategy T^ℓ (i.e., select v_t iff $v_t \geq T^\ell$) guarantees an expected utility of at least T^ℓ.*

Proof. Fix an agent i. Let S_{-i} be the strategies of all agents except agent i, and let $S = (T^\ell, S_{-i})$. Let $A_{i,j}^S$ denote the event that agent i is assigned the reward v_j in strategy profile S. I.e., $A_{i,j}^S$ is the event that agent i competed over reward v_j and received it according to the random tie-breaking rule. For simplicity of presentation, we omit S and write $A_{i,j}$. It holds that

$$
\begin{aligned}
u_i(S) &= \mathbb{E}\left[\sum_{j=1}^{n} v_j \cdot \Pr\left(A_{i,j}\right)\right] \\
&= \mathbb{E}\left[\sum_{j=1}^{n} (T^\ell + v_j - T^\ell) \Pr\left(v_j \geq T^\ell, \forall_{r<j} \overline{A_{i,r}}, A_{i,j}\right)\right].
\end{aligned}
$$

Let $p = \sum_{j=1}^{n} \Pr(v_j \geq T^\ell, \forall_{j'<j}\overline{A_{i,j'}}, A_{i,j})$ (i.e., p is the probability that agent i receives some reward in strategy profile $S = (T^\ell, S_{-i})$), and let $Z^+ = \max\{Z, 0\}$. We can now write $u_i(S)$ as follows:

$$\begin{aligned} u_i(S) &= pT^\ell + \mathbb{E}\left[\sum_{j=1}^{n}(v_j - T^\ell)^+ \Pr\left(\forall_{r<j}\overline{A_{i,r}}, A_{i,j}\right)\right] \\ &= p \cdot T^\ell + \mathbb{E}\Big[\sum_{j=1}^{n}(v_j - T^\ell)^+ \cdot \Pr\left(\forall_{r<j}\overline{A_{i,r}}\right) \cdot \Pr\left(A_{i,j} \mid \forall_{r<j}\overline{A_{i,r}}\right)\Big] \\ &\geq p \cdot T^\ell + \mathbb{E}\Big[\sum_{j=1}^{n}(v_j - T^\ell)^+ \cdot (1-p) \cdot \Pr\left(A_{i,j} \mid \forall_{r<j}\overline{A_{i,r}}\right)\Big] \\ &\geq p \cdot T^\ell + \frac{1-p}{k} \cdot \mathbb{E}\left[\sum_{j=1}^{n}(v_j - T^\ell)^+\right]. \end{aligned}$$

The first inequality holds since the probability of not getting any reward until time j is bounded by $1-p$ (i.e., the probability of not getting any reward). The last inequality holds since if $v_j - T^\ell \geq 0$ and agent i is still active, the reward is selected, thus assigned with probability at least $1/k$. Since each term in the summation is non-negative, we get the following:

$$\begin{aligned} u_i(S) &\geq p \cdot T^\ell + \frac{1-p}{k} \cdot \mathbb{E}\left[\sum_{j=1}^{\ell}(y_j - T^\ell)^+\right] \\ &\geq p \cdot T^\ell + \frac{1-p}{k} \cdot \mathbb{E}\left[\sum_{j=1}^{\ell} y_j - \ell \cdot T^\ell\right] \\ &= p \cdot T^\ell + \frac{1-p}{k} \cdot \left((k+\ell) \cdot T^\ell - \ell \cdot T^\ell\right) = T^\ell, \end{aligned}$$

where the last equality follows by the definition of T^ℓ. □

The special cases of $\ell = 1$ and $\ell = k$ give the following corollaries:

Corollary 1. *The single-threshold strategy T^k guarantees an expected utility of at least $\frac{1}{2k}\mathbb{E}[\sum_{i=1}^{k} y_i]$.*

Corollary 2. *The single-threshold strategy T^1 guarantees an expected utility of at least $\frac{1}{k+1}\mathbb{E}[y_1]$.*

We now show that the bound in Theorem 1 is tight.

Proposition 1. *For every $\epsilon > 0$ there exists an instance such that in the unique equilibrium of the game, no agent gets an expected utility of more than $\frac{1}{k+\ell}\sum_{j=1}^{\ell}\mathbb{E}[y_j] + \epsilon$ for any $\ell \leq n$.*

Proof. Given an $\epsilon > 0$, consider the following instance (depicted in Fig. 1):

$$v_t = 1 \text{ for all } t \leq n-1, \quad \text{and } \; v_n = \begin{cases} \frac{k+\epsilon}{\epsilon} & \text{w.p. } \epsilon \\ 0 & \text{w.p. } 1-\epsilon \end{cases}$$

One can easily verify that in the unique equilibrium S, all agents compete over the last reward, for an expected utility of $1 + \frac{\epsilon}{k}$. It holds that for every agent i:

$$u_i(S) = 1 + \frac{\epsilon}{k} \leq 1 + \epsilon = \frac{\mathbb{E}[\sum_{j=1}^{\ell} y_j]}{k+\ell} + \epsilon.$$

This example also shows that there are instances in which the social welfare in equilibrium is at most half the optimal welfare allocation. □

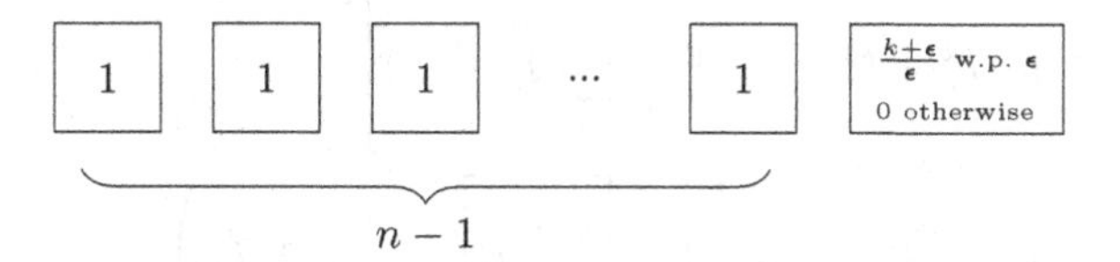

Fig. 1. An example where the expected reward is no more than $\frac{1}{k+\ell}\sum_{j=1}^{\ell} \mathbb{E}[y_j] + \epsilon$

4 Ranked Tie-Breaking

In this section we consider the ranked tie-breaking rule, and present a series of single threshold strategies with their guarantees. We then show an interesting connection to the setting of a single agent that can choose up to k rewards. We start by presenting the single threshold strategies.

Theorem 2. *For every $i \leq n$ and $\ell = 0, \dots, n-i$, let $\hat{T}_i^\ell = \frac{1}{\ell+2}\sum_{j=i}^{i+\ell} \mathbb{E}[y_j]$. The single threshold strategy $\hat{T}_i^\ell$ (i.e., select v_t iff $v_t \geq \hat{T}_i^\ell$) guarantees an expected utility of at least $\hat{T}_i^\ell$ for the i-ranked agent.*

Proof. Fix an agent i. Let S_{-i} be the strategies of all agents except agent i, and let $S = (\hat{T}_i^\ell, S_{-i})$. Let $A_{i,j}^S$ denote the event that agent i is assigned the reward v_j in strategy profile S. I.e., $A_{i,j}^S$ is the event that agent i competed over reward v_j and received it according to the ranked tie-breaking rule. For simplicity of presentation, we omit S and write $A_{i,j}$. We bound the utility of agent i under strategy profile S.

$$\begin{aligned} u_i(S) &= \mathbb{E}\left[\sum_{j=1}^{n} v_j \cdot \Pr(A_{i,j})\right] \\ &= \mathbb{E}\left[\sum_{j=1}^{n} (\hat{T}_i^\ell + v_j - \hat{T}_i^\ell) \Pr\left(v_j \geq \hat{T}_i^\ell, \forall_{r<j} \overline{A_{i,r}}, A_{i,j}\right)\right]. \end{aligned}$$

Let $p = \sum_{j=1}^{n} \Pr(v_j \geq \hat{T}_i^\ell, \forall_{r<j}\overline{A_{i,r}}, A_{i,j})$ (i.e., p is the probability that agent i receives some reward in strategy profile $S = (\hat{T}_i^\ell, S_{-i})$), and let $Z^+ = \max\{Z, 0\}$. We can now write $u_i(S)$ as follows:

$$
\begin{aligned}
u_i(S) &= p \cdot \hat{T}_i^\ell + \mathbb{E}\left[\sum_{j=1}^{n}(v_j - \hat{T}_i^\ell)^+ \Pr\left(\forall_{r<j}\overline{A_{i,r}}, A_{i,j}\right)\right] \\
&\geq p \cdot \hat{T}_i^\ell + \mathbb{E}\Big[\sum_{j=1}^{n}(v_j - \hat{T}_i^\ell)^+ \cdot (1-p) \cdot \Pr\left(A_{i,j} \mid \forall_{r<j}\overline{A_{i,r}}\right)\Big] \\
&\geq p \cdot \hat{T}_i^\ell + (1-p) \cdot \mathbb{E}\left[\sum_{j=i}^{n}(y_j - \hat{T}_i^\ell)^+\right] \qquad (1) \\
&\geq p \cdot \hat{T}_i^\ell + (1-p) \cdot \mathbb{E}\left[\sum_{j=i}^{i+\ell}(y_j - \hat{T}_i^\ell)\right] \\
&= p \cdot \hat{T}_i^\ell + (1-p) \cdot \left(\mathbb{E}\left[\sum_{j=i}^{i+\ell} y_j\right] - (\ell+1)\hat{T}_i^\ell\right) \\
&= p \cdot \hat{T}_i^\ell + (1-p) \cdot \left((\ell+2) \cdot \hat{T}_i^\ell - (\ell+1) \cdot \hat{T}_i^\ell\right) = \hat{T}_i^\ell.
\end{aligned}
$$

Inequality (1) holds since the probability of not getting any reward until time j is bounded by $1-p$ (i.e., the probability of not getting any reward). Inequality (1) holds since there are at most $i-1$ agents that are ranked higher than agent i, therefore there are at most $i-1$ rewards that can be selected but not assigned to agent i. Finally, the last equality holds by the definition of $\hat{T}_i^\ell$. □

The special case of Theorem 2 where $\ell = 0$ gives the following corollary.

Corollary 3. *For every i, the threshold strategy $\hat{T}_i^0$ guarantees an expected utility of $\frac{\mathbb{E}[y_i]}{2}$ for the i-ranked agent.*

We next show that the bound in Theorem 2 is tight.

Proposition 2. *For every $\epsilon > 0$ and every $i \leq n$, there exists an instance such that in the unique equilibrium of the game, the i-ranked agent gets an expected utility of at most $\frac{1}{\ell+2}\sum_{j=i}^{i+\ell}\mathbb{E}[y_j] + \epsilon$ for every $\ell \leq n-i$.*

Proof. Given some $\epsilon > 0$ and $i \leq n$, consider the following instance (depicted in Fig. 2):

$$
v_t = \begin{cases} \infty & \text{for } t < i \\ 1 & \text{for } i \leq t < n \\ \frac{1+\epsilon}{\epsilon} \text{ w.p. } \epsilon, \text{ and } 0 \text{ w.p. } 1-\epsilon & \text{for } t = n \end{cases}
$$

One can easily verify that in the unique equilibrium of the game, agents $1,\ldots,i-1$ will be assigned rewards $v_1,\ldots,v_{i-1}$, and agent i will be assigned the last reward v_n for an expected utility of $1+\epsilon$. It holds that:

$$u_i(S) = 1+\epsilon = \frac{\mathbb{E}[\sum_{j=i}^{i+\ell} y_j]}{2+\ell} + \epsilon.$$

□

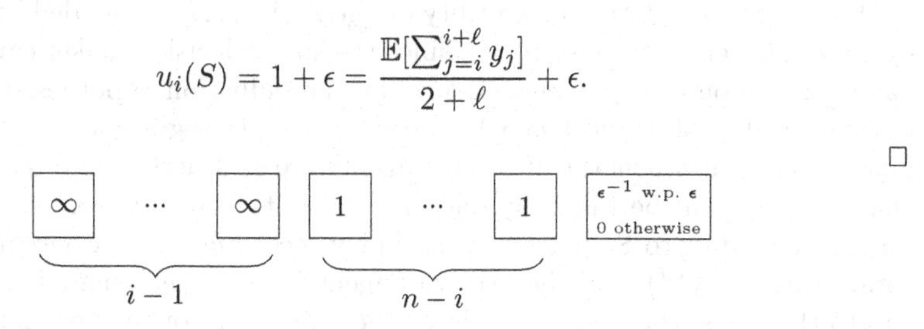

Fig. 2. An example where the expected reward for agent i is no more than $\frac{1}{\ell+2}\sum_{j=i}^{i+\ell}\mathbb{E}[y_j]+\epsilon$

We next show that for any instance, the set of rewards assigned to the k competing agents in equilibrium coincides with the set of rewards that are chosen by the optimal algorithm for a single decision maker who can choose up to k rewards and wishes to maximize their sum. Kleinberg and Weinberg [10] show that the only optimal strategy of such a decision maker, takes the form of nk dynamic thresholds, $\{T_t^i\}_{i,t}$ for all $t \leq n$ and $i \leq k$, so that the agent accepts reward v_t if $v_t \geq T_t^i$, where $k-i$ is the number of rewards already chosen (i.e., i is the number of rewards left to choose)[2]. Moreover, they show that these thresholds are monotone with respect to i.

With the characterization of the strategy of a single decision maker who can choose up to k rewards, we can characterize the unique SPE for the k-agent game[3].

Theorem 3. *Let $\{T_t^i\}_{i\in[k],t\in[n]}$ be the optimal strategy of a single decision maker who may choose up to k rewards and wishes to maximize their sum. The unique SPE of the k-agent game is for agent i to accept v_t iff $v_t \geq T_t^{i'+1}$, where $i' \leq i$ is the rank of agent i among the active agents. This SPE is unique up to cases where $v_t = T_t^{i'}$.*

Proof. Let S^i denote the optimal strategy of the single agent who may choose up to i rewards, as described above. Let S_i be the strategy of agent i as described in the assertion of the theorem. We prove by induction that for every $i \in [k]$, the rewards that are chosen by agents $1,\ldots,i$ correspond to the rewards chosen by a single decision maker, who may choose up to i rewards, and uses strategy S^i. For the case of $i=1$, the claim holds trivially. Assume the claim holds for any

[2] The uniqueness holds for distributions with no mass points. For distributions with mass points, whenever $v_t = T_t^i$, the decision maker is indifferent between selecting and passing.

[3] The SPE is unique up to cases where $T_j^i = v_t$; in these cases the agent is indifferent.

number of agents smaller than i. Since agent i has no influence on the rewards received by agents $1, \ldots, i-1$, we may assume that agents $1, \ldots, i-1$ are playing according to strategies $S_1, \ldots, S_{i-1}$.

For every $i \in [k]$, the total utility of agents $1, \ldots, i$ is bounded by the utility of the single decision maker $u(S^i)$, since the single decision maker can simulate a game with i competing agents. Hence, by the induction hypothesis, agent i can obtain a utility of at most $u(S^i) - u(S^{i-1})$. By playing according to S_i, we are guaranteed that whenever at least j agents are still active, any reward v_t such that $v_t \geq T_t^j$ will be taken by one of the agents. Thus, when every agent i is playing according to S_i, players $1, \ldots, i$ play according to S^i. Consequently, their total utility is $u(S^i)$, and the utility of agent i is then maximal. The uniqueness (up to the cases where $v_j = T_j^{i'}$) is by the uniqueness of the optimal strategy of the single decision maker. □

We note that by Theorem 2 it holds that in the unique SPE described in Theorem 3, every agent i receives at least $\max_{\ell=0}^{n-i} \frac{1}{\ell+2} \sum_{j=i}^{i+\ell} \mathbb{E}[y_j]$.

Using the results of Alaei [2] regarding a single decision maker choosing k rewards, we deduce an approximation of the social welfare in equilibrium:

Corollary 4. *In SPE of the k agent prophet game, the expected social welfare is at least $1 - O(\frac{1}{\sqrt{k}})$ of the optimal welfare.*

5 Discussion and Future Directions

In this work, we study the effect of competition in prophet settings. We show that under both random and ranked tie-breaking rules, agents have simple strategies that grant them high guarantees, ones that are tight even with respect to equilibrium profiles under some distributions.

Under the ranked tie-breaking rule, we show an interesting correspondence between the equilibrium strategies of the k competing agents and the optimal strategy of a single decision maker that can select up to k rewards. It would be interesting to study whether this phenomenon applies more generally, and what are the conditions under which it holds.

Below we list some future directions that we find particularly natural.

- Study competition in additional problems related to optimal stopping theory, such as Pandora's box [16].
- Study competition in prophet (and secretary) settings under additional tie-breaking rules, such as random tie breaking with non-uniform distribution, and tie-breaking rules that allow to split rewards among agents.
- Study competition in scenarios where agents can choose multiple rewards, under some feasibility constraints (such as matroid or downward-closed feasibility constraints).
- Consider prophet settings with the objective of outperforming the other agents, as in [7], or different agents' objectives.
- Consider competition settings with non-immediate decision making, as in [3].

Acknowledgement. The work was partially supported by the European Research Council (ERC) under the European Union's Horizon 2020 research and innovation program (grant agreement No. 866132, 740282), and by the Israel Science Foundation (grant number 317/17).

References

1. Abdelaziz, F.B., Krichen, S.: Optimal stopping problems by two or more decision makers: a survey. Comput. Manage. Sci. **4**(2), 89 (2007)
2. Alaei, S.: Bayesian combinatorial auctions: expanding single buyer mechanisms to many buyers. In: 2011 IEEE 52nd Annual Symposium on Foundations of Computer Science, pp. 512–521. IEEE (2011)
3. Ezra, T., Feldman, M., Kupfer, R.: On a competitive secretary problem with deferred selections (2020)
4. Ferguson, T.S.: Who solved the secretary problem? Stat. Sci. **4**(3), 282–289 (1989)
5. Hajiaghayi, M.T., Kleinberg, R.D., Sandholm, T.: Automated online mechanism design and prophet inequalities. In: Proceedings of the Twenty-Second AAAI Conference on Artificial Intelligence, pp. 58–65 (2007)
6. Hill, T.P., Kertz, R.P.: A survey of prophet inequalities in optimal stopping theory. Contemp. Math. **125**, 191–207 (1992)
7. Immorlica, N., Kalai, A.T., Lucier, B., Moitra, A., Postlewaite, A., Tennenholtz, M.: Dueling algorithms. In: Proceedings of the Forty-third Annual ACM Symposium on Theory of Computing, pp. 215–224. ACM (2011)
8. Immorlica, N., Kleinberg, R., Mahdian, M.: Secretary problems with competing employers. In: Spirakis, P., Mavronicolas, M., Kontogiannis, S. (eds.) WINE 2006. LNCS, vol. 4286, pp. 389–400. Springer, Heidelberg (2006). https://doi.org/10.1007/11944874_35
9. Karlin, A., Lei, E.: On a competitive secretary problem. In: Twenty-Ninth AAAI Conference on Artificial Intelligence (2015)
10. Kleinberg, R., Weinberg, S.M.: Matroid prophet inequalities and applications to multi-dimensional mechanism design. Games Econom. Behav. **113**, 97–115 (2019)
11. Krengel, U., Sucheston, L.: Semiamarts and finite values. Bull. Am. Math. Soc. **83**, 745–747 (1977)
12. Krengel, U., Sucheston, L.: On semiamarts, amarts, and processes with finite value. Adv. Prob. **4**(197–266), 1–5 (1978)
13. Lucier, B.: An economic view of prophet inequalities. ACM SIGecom Exchanges **16**(1), 24–47 (2017)
14. Matsui, T., Ano, K.: Lower bounds for Bruss' odds problem with multiple stoppings. Math. Oper. Res. **41**(2), 700–714 (2016)
15. Samuel-Cahn, E.: Comparison of threshold stop rules and maximum for independent nonnegative random variables. Ann. Probab. **12**(4), 1213–1216 (1984)
16. Weitzman, M.L.: Optimal search for the best alternative. Econometr. **47**(3), 641–654 (1979)

Lower Bounds for the Query Complexity of Equilibria in Lipschitz Games

Paul W. Goldberg and Matthew J. Katzman(✉)

Department of Computer Science, University of Oxford, Oxford, UK
{paul.goldberg,matthew.katzman}@cs.ox.ac.uk

Abstract. Nearly a decade ago, Azrieli and Shmaya introduced the class of λ-Lipschitz games in which every player's payoff function is λ-Lipschitz with respect to the actions of the other players. They showed that such games admit ϵ-approximate pure Nash equilibria for certain settings of ϵ and λ. They left open, however, the question of how hard it is to find such an equilibrium. In this work, we develop a query-efficient reduction from more general games to Lipschitz games. We use this reduction to show a query lower bound for any randomized algorithm finding ϵ-approximate *pure* Nash equilibria of n-player, binary-action, λ-Lipschitz games that is exponential in $\frac{n\lambda}{\epsilon}$. In addition, we introduce "Multi-Lipschitz games," a generalization involving player-specific Lipschitz values, and provide a reduction from finding equilibria of these games to finding equilibria of Lipschitz games, showing that the value of interest is the sum of the individual Lipschitz parameters. Finally, we provide an exponential lower bound on the *deterministic* query complexity of finding ϵ-approximate *correlated* equilibria of n-player, m-action, λ-Lipschitz games for strong values of ϵ, motivating the consideration of explicitly randomized algorithms in the above results. Our proof is arguably simpler than those previously used to show similar results.

Keywords: Query complexity · Lipschitz games · Nash equilibrium

1 Introduction

A Lipschitz game is a multi-player game in which there is an additive limit λ (called the Lipschitz constant of the game) on how much any player's payoffs can change due to a deviation by any other player. Thus, any player's payoff function is λ-Lipschitz continuous as a function of the other players' mixed strategies. Lipschitz games were introduced about ten years ago by Azrieli and Shmaya [1]. A key feature of Lipschitz games is that they are guaranteed to have approximate Nash equilibria *in pure strategies*, where the quality of the approximation depends on the number of players n, the number of actions m, and the Lipschitz constant λ. In particular, [1] showed that this guarantee holds (keeping the number of actions constant) for Lipschitz constants of size $o(1/\sqrt{n \log n})$ (existence of pure approximate equilibria is trivial for Lipschitz constants of size $o(1/n)$

I. Caragiannis and K. A. Hansen (Eds.): SAGT 2021, LNCS 12885, pp. 124–139, 2021.
https://doi.org/10.1007/978-3-030-85947-3_9

since then, players have such low effect on each others' payoffs that they can best-respond independently to get a pure approximate equilibrium). The general idea of the existence proof is to take a mixed Nash equilibrium (guaranteed to exist by Nash's theorem [16]), and prove that there is a positive probability that a pure profile sampled from it will constitute an approximate equilibrium.

As noted in [1] (and elsewhere), solutions in pure-strategy profiles are a more plausible and satisfying model of a game's outcome than solutions in mixed-strategy profiles. On the other hand, the existence guarantee raises the question of how to *compute* an approximate equilibrium. In contrast with potential games, in which pure-strategy equilibria can often be found via best- and better-response dynamics, there is no obvious natural approach in the context of Lipschitz games, despite the existence guarantee. The general algorithmic question (of interest in the present paper) is:

> *Given a Lipschitz game, how hard is it to find a pure-strategy profile that constitutes an approximate equilibrium?*

Recent work [8,10] has identified algorithms achieving additive constant approximation guarantees, but as noted by Babichenko [5], the extent to which we can achieve the pure approximate equilibria that are guaranteed by [1] (or alternatively, potential lower bounds on query or computational complexity) is unknown.

Variants and special cases of this question include classes of Lipschitz games having a concise representation, as opposed to unrestricted Lipschitz games for which an algorithm has query access to the payoff function (as we consider in this paper). In the latter case, the question subdivides into what we can say about the query complexity, and about the computational complexity (for concisely-represented games the query complexity is low, by Theorem 3.3 of [11]). Moreover, if equilibria can be easily computed, does that remain the case if we ask for this to be achievable via some kind of natural-looking decentralized process? Positive results for these questions help us to believe in "approximate pure Nash equilibrium" as a solution concept for Lipschitz games. Alternatively, it is of interest to identify computational obstacles to the search for a Nash equilibrium.

1.1 Prior Work

In this paper we apply various important lower bounds on the query complexity of computing approximate Nash equilibria of *unrestricted* n-player games. In general, lower bounds on the query complexity are known that are exponential in n (which motivates a focus on subclasses of games, such as Lipchitz games, and others). Hart and Mansour [13] showed that the communication (and thus query) complexity of computing an exact Nash equilibrium (pure or mixed) in a game with n players is $2^{\Omega(n)}$. Subsequent results have iteratively strengthened this lower bound. First, Babichenko [3] showed an exponential lower bound on the *randomized* query complexity of computing an ϵ-well supported Nash equilibrium (an approximate equilibrium in which every action in the support of a given player's mixed strategy is an ϵ-best response) for a constant value of ϵ, even when considering δ-distributional query complexity, as defined in Definition 4. Shortly after, Chen, Cheng, and Tang [6] showed a $2^{\Omega(n/\log n)}$ lower bound on the randomized query complexity of computing an ϵ-approximate Nash equilibrium for a constant value of ϵ, which Rubinstein [19] improved to a $2^{\Omega(n)}$ lower bound, even allowing a constant fraction of players to have regret greater than ϵ (taking regret as defined in Definition 1). These intractability results motivate us to consider a restricted class of games (Lipschitz, or large, games) which contain significantly more structure than do general games.

Lipschitz games were initially considered by Azrieli and Shmaya [1], who showed that any λ-Lipschitz game (as defined in Sect. 2.1) with n players and m actions admits an ϵ-approximate *pure* Nash equilibrium for any $\epsilon \geq \lambda\sqrt{8n\log 2mn}$ [1]. In Sect. 3 we provide a lower bound on the query complexity of finding such an equilibrium.

Positive algorithmic results have been found for classes of games that combine the Lipschitz property with others, such as *anonymous* games [7] and *aggregative* games [4]. For anonymous games (in which each player's payoffs depend only on the *number* of other players playing each action, and not *which* players), Daskalakis and Papadimitriou [7] improved upon the upper bound of [1] to guarantee the existence of ϵ-approximate pure Nash equilibria for $\epsilon = \Omega(\lambda)$ (the only dependence on n coming from λ itself). Peretz et al. [17] analyze Lipschitz values that result from δ-perturbing anonymous games, in the sense that every player is assumed to randomize uniformly with probability δ. Goldberg and Turchetta [12] showed that a 3λ-approximate pure Nash equilibrium of a λ-Lipschitz anonymous game can be found querying $O(n\log n)$ individual payoffs.

Goldberg et al. [10] showed a logarithmic upper bound on the randomized query complexity of computing $\frac{1}{8}$-approximate Nash equilibria in binary-action $\frac{1}{n}$-Lipschitz games. They also presented a randomized algorithm finding a $(\frac{3}{4}+\alpha)$-approximate Nash equilibrium when the number of actions is unbounded.

1.2 Our Contributions

The primary contribution of this work is the development and application of a query-efficient version of a reduction technique used in [1,2] in which an algorithm finds an equilibrium in one game by reducing it to a population game with a smaller Lipschitz parameter. As the former is a known hard problem, we prove hardness for the latter.

In Sect. 2 we introduce notation and relevant concepts, and describe the query model assumed for our results. Section 3 contains our main contributions. In particular, Theorem 3 utilizes a query-efficient reduction to a population

[1] General Lipschitz games cannot be written down concisely, so we assume black-box access to the payoff function of a Lipschitz game. This emphasizes the importance of considering *query complexity* in this context. Note that a pure approximate equilibrium can still be checked using mn queries.

game with a small Lipschitz parameter while preserving the equilibrium. Hence, selecting the parameters appropriately, the hardness of finding well-supported equilibria in general games proven in [3] translates to finding approximate pure equilibria in Lipschitz games. Whilst several papers have discussed both this problem and this technique, none has put forward this observation.

In Sect. 3.2 we introduce "Multi-Lipschitz" games, a generalization of Lipschitz games that allows player-specific Lipschitz values (the amount of influence the player has on others). We show that certain results of Lipschitz games extend to these, and the measure of interest is the sum of individual Lipschitz values (in a standard Lipschitz game, they are all equal). Theorem 4 provides a query-efficient reduction from finding equilibria in Multi-Lipschitz games to finding equilibria in Lipschitz games. In particular, if there is a query-efficient approximation algorithm for the latter, there is one for the former as well.

Finally, Sect. 3.3 provides a simpler proof of the result of [14] showing exponential query lower-bounds on finding correlated equilibria with approximation constants better than $\frac{1}{2}$. Theorem 7 provides a more general result for games with more than 2 actions, and Corollary 4 extends this idea futher to apply to Lipschitz games. While [14] relies on a reduction from the `ApproximateSink` problem, we explicitly describe a class of games with vastly different equilibria between which no algorithm making a subexponential number of queries can distinguish. To any weak deterministic algorithm, these games look like pairs of players playing Matching Pennies against each other - however the equilibria are far from those of the Matching Pennies game.

For the sake of brevity, some technical details are omitted from this work, and can be found in the full version [9].

2 Preliminaries

Throughout, we use the following notation.

- Boldface capital letters denote matrices, and boldface lowercase letters denote vectors.
- The symbol $\mathbf{a}$ is used to denote a pure action profile, and $\mathbf{p}$ is used when the strategy profiles may be mixed. Furthermore, $\mathbf{X}$ is used to denote *correlated* strategies.
- $[n]$ and $[m]$ denote the sets $\{1, \ldots, n\}$ of players and $\{1, \ldots, m\}$ of actions, respectively. Furthermore, $i \in [n]$ will always refer to a player, and $j \in [m]$ will always refer to an action.
- Whenever a query returns an approximate answer, the payoff vector $\tilde{\mathbf{u}}$ will be used to represent the approximation and $\mathbf{u}$ will represent the true value.

2.1 The Game Model

We introduce standard concepts of strategy profiles, payoffs, regret, and equilibria for pure, mixed, and correlated strategies.

Types of strategy profile; notation:

- A *pure* action profile $\mathbf{a} = (a_1, \ldots, a_n) \in [m]^n$ is an assignment of one action to each player. We use $\mathbf{a}_{-i} = (a_1, \ldots, a_{i-1}, a_{i+1}, \ldots, a_n) \in [m]^{n-1}$ to denote the set of actions played by players in $[n] \setminus \{i\}$.
- A (possibly *mixed*) strategy profile $\mathbf{p} = (p_1, \ldots, p_n) \in (\Delta[m])^n$ (where $\Delta(S)$ is the probability simplex over S) is a collection of n independent probability distributions, each taken over the action set of a player, where p_{ij} is the probability with which player i plays action j. The set of distributions for players in $[n] \setminus \{i\}$ is denoted $\mathbf{p}_{-i} = (p_1, \ldots, p_{i-1}, p_{i+1}, \ldots, p_n)$. When $\mathbf{p}$ contains just 0-1 values, $\mathbf{p}$ is equivalent to some action profile $\mathbf{a} \in [m]^n$. Furthermore, when considering binary-action games with action set $\{1, 2\}$, we instead describe strategy profiles by $\mathbf{p} = (p_1, \ldots, p_n)$, where p_i is the probability that player i plays action 1.
- A *correlated* strategy profile $\mathbf{X} \in \Delta([m]^n)$ is a single joint probability distribution taken over the space of all pure action profiles $\mathbf{a}$.

Notation for payoffs: Given player i, action j, and pure action profile $\mathbf{a}$,

- $u_i(j, \mathbf{a}_{-i})$ is the payoff that player i obtains for playing action j when all other players play the actions given in $\mathbf{a}_{-i}$.
- $u_i(\mathbf{a}) = u_i(a_i, \mathbf{a}_{-i})$ is the payoff that player i obtains when all players play the actions given in $\mathbf{a}$.
- Similarly for mixed-strategy profiles:
  $u_i(j, \mathbf{p}_{-i}) = \mathbb{E}_{\mathbf{a}_{-i} \sim \mathbf{p}_{-i}}[u_i(j, \mathbf{a}_{-i})]$ and $u_i(\mathbf{p}) = \mathbb{E}_{\mathbf{a} \sim \mathbf{p}}[u_i(\mathbf{a})]$.
- For a given player $i \in [n]$, consider a deviation function $\phi : [m] \to [m]$. Then, similarly, $u_i^{(\phi)}(\mathbf{X}) = \mathbb{E}_{\mathbf{a} \sim \mathbf{X}}[u_i(\phi(a_i), \mathbf{a}_{-i})]$ and $u_i(\mathbf{X}) = \mathbb{E}_{\mathbf{a} \sim \mathbf{X}}[u_i(\mathbf{a})]$. Furthermore, given an event E, $u_i(\mathbf{X} \mid E) = \mathbb{E}_{\mathbf{a} \sim \mathbf{X}}[u_i(\mathbf{a}) \mid E]$.

Definition 1 (Regret).

- *Given a player i and a strategy profile* $\mathbf{p}$, *define the regret*

$$reg_i(\mathbf{p}) = \max_{j \in [m]} u_i(j, \mathbf{p}_{-i}) - u_i(\mathbf{p})$$

  to be the difference between the payoffs of player i's best response to $\mathbf{p}_{-i}$ *and i's strategy p_i.*
- *Given a player i and a correlated strategy profile* $\mathbf{X}$, *define*

$$reg_i^{(\phi)}(\mathbf{X}) = u_i^{(\phi)}(\mathbf{X}) - u_i(\mathbf{X}), \qquad reg_i(\mathbf{X}) = \max_{\phi:[m] \to [m]} reg_i^{(\phi)}(\mathbf{X}),$$

  the regret $reg_i(\mathbf{X})$ being the difference between the payoffs of player i's best deviation from $\mathbf{X}$, *and* $\mathbf{X}$.

Definition 2 (Equilibria).

- *An* ϵ-approximate Nash equilibrium *(ϵ-ANE) is a strategy profile* $\mathbf{p}^*$ *such that, for every player $i \in [n]$, $reg_i(\mathbf{p}^*) \leq \epsilon$.*

- *An* ϵ-well supported Nash equilibrium *(ϵ-WSNE) is an ϵ-ANE* $\mathbf{p}^*$ *for which every action* j *in the support of* p_i^* *is an ϵ-best response to* $\mathbf{p}_{-i}^*$.
- *An* ϵ-approximate pure Nash equilibrium *(ϵ-PNE) is a pure action profile* $\mathbf{a}$ *such that, for every player* $i \in [n]$, $reg_i(\mathbf{a}) \leq \epsilon$.
- *An* ϵ-approximate correlated equilibrium *(ϵ-ACE) is a correlated strategy profile* $\mathbf{X}^*$ *such that, for every player* $i \in [n]$, $reg_i(\mathbf{X}^*) \leq \epsilon$.

Note that any ϵ-PNE is an ϵ-WSNE, and any ϵ-ANE constitutes an ϵ-ACE. Consequently, any algorithmic lower bounds on correlated equilibria also apply to Nash equilibria.

Finally, to end this section, we introduce the class of games that is the focus of this work.

Definition 3 (Lipschitz Games). *For any value $\lambda \in (0, 1]$, a* λ-Lipschitz game *is a game in which a change in strategy of any given player can affect the payoffs of any other player by at most an additive λ, or for every player i and pair of action profiles $\mathbf{a}, \mathbf{a}'$, $|u_i(\mathbf{a}) - u_i(\mathbf{a}')| \leq \lambda ||\mathbf{a}_{-i} - \mathbf{a}'_{-i}||_1$. Here we consider games in which all payoffs are in the range $[0, 1]$ (in particular, note that any general game is, by definition, 1-Lipschitz).*

The set of n-player, m-action, λ-Lipschitz games will be denoted $\mathcal{G}(n, m, \lambda)$.

2.2 The Query Model

This section introduces the model of queries we consider.

Definition 4 (Queries).

- *A* profile query *of a pure action profile* $\mathbf{a}$ *of a game* G*, denoted* $\mathcal{Q}^G(\mathbf{a})$*, returns a vector* $\mathbf{u}$ *of payoffs* $u_i(\mathbf{a})$ *for each player* $i \in [n]$.
- *A* δ-distribution query *of a strategy profile* $\mathbf{p}$ *of a game* G*, denoted* $\mathcal{Q}_\delta^G(\mathbf{p})$*, returns a vector* $\tilde{\mathbf{u}}$ *of* n *values such that* $||\tilde{\mathbf{u}} - \mathbf{u}||_\infty \leq \delta$*, where* $\mathbf{u}$ *is the players' expected utilities from* $\mathbf{p}$*. We also define a* (δ, γ)-distribution query *to be a δ-distribution query of a strategy profile* $\mathbf{p}$ *in which every action* j *in the support of* p_i *is allocated probability at least* γ *for every player* $i \in [n]$.
- *The* (profile) query complexity *of an algorithm* A *on input game* G *is the number of calls* A *makes to* $\mathcal{Q}^G$*. The* δ-distribution query complexity *of* A *is the number of calls* A *makes to* $\mathcal{Q}_\delta^G$.

Babichenko [3] points out that it is uninteresting to consider 0-distribution queries, as any game in which every payoff is a multiple of $\frac{1}{M}$ for some $M \in \mathbb{N}$ can be completely learned by a single 0-distribution query. On the other hand, additive approximations to the expected payoffs can be computed via sampling from $\mathbf{p}$. Indeed, for general binary-action games we have from [11]:

Theorem 1 ([11]). *Take $G \in \mathcal{G}(n, 2, 1), \eta > 0$. Any (δ, γ)-distribution query of G can be simulated with probability at least $1 - \eta$ by*

$$\max\left\{\frac{1}{\gamma\delta^2}\log\left(\frac{8n}{\eta}\right), \frac{8}{\gamma}\log\left(\frac{4n}{\eta}\right)\right\}$$

profile queries.

Corollary 1. *Take $G \in \mathcal{G}(n,2,1), \eta > 0$. Any (δ,γ)-distribution query of G can be simulated with probability at least $1-\eta$ by*

$$\frac{8}{\gamma^2\delta^2}\log^2\left(\frac{8n}{\eta}\right)$$

profile queries. Furthermore, any algorithm making q (δ,γ)-distribution queries of G can be simulated with probability at least $1-\eta$ by

$$\frac{8q}{\gamma^2\delta^2}\log^2\left(\frac{8nq}{\eta}\right) = poly\left(n, \frac{1}{\gamma}, \frac{1}{\delta}, \log\frac{1}{\eta}\right)\cdot q\log q$$

profile queries.

Proof. The first claim is a weaker but simpler version of the upper bound of Theorem 1. The second claim follows from the first by a union bound. □

2.3 The Induced Population Game

Finally, this section introduces a reduction utilized by [1] in an alternative proof of Nash's Theorem, and by [2] to upper bound the support size of ϵ-ANEs.

Definition 5. *Given a game G with payoff function $\mathbf{u}$, we define the* population game *induced by G, $G' = g_G(L)$ with payoff function $\mathbf{u}'$ as follows. Every player i is replaced by a population of L players (v^i_ℓ for $\ell \in [L]$), each playing G against the aggregate behavior of the other $n-1$ populations. More precisely, $u'_{v^i_\ell}(\mathbf{p}') = u_i\left(p'_{v^i_\ell}, \mathbf{p}_{-i}\right)$ where $p_{i'} = \frac{1}{L}\sum_{\ell=1}^{L} p'_{v^{i'}_\ell}$ for all $i' \neq i$.*

Population games date back even to Nash's thesis [15], in which he uses them to justify the consideration of mixed equilibria. To date, the reduction to the induced population game has been focused on proofs of existence. We show that the reduction can be made query-efficient: an equilibrium of $g_G(L)$ induces an equilibrium on G *which can be found with few additional queries.* This technique is the foundation for the main results of this work.

Lemma 1. *Given an n-player, m-action game G and a population game $G' = g_G(L)$ induced by G, if an ϵ-PNE of G' can be found by an algorithm making q (δ,γ)-distribution queries of G', then an ϵ-WSNE of G can be found by an algorithm making $n \cdot m \cdot q$ $(\delta,\gamma/L)$-distribution queries of G.*

The proof can be found in [9].

3 Results

In this section, we present our three main results:

- In Sect. 3.1, Theorem 3 shows a lower bound exponential in $\frac{n\lambda}{\epsilon}$ on the randomized query complexity of finding ϵ-approximate *pure* Nash equilibria of games in $\mathcal{G}(n,2,\lambda)$.

- In Sect. 3.2, we generalize the concept of Lipschitz games. Theorem 4 provides a reduction from finding approximate equilibria in our new class of "Multi-Lipschitz" games to finding approximate equilibria of Lipschitz games.
- In Sect. 3.3, Theorem 7 and Proposition 1 provide a complete dichotomy of the query complexity of deterministic algorithms finding ϵ-approximate correlated equilibria of n-player, m-action games. Corollary 4 scales the lower bound to apply to Lipschitz games, and motivates the consideration of explicitly randomized algorithms for the above results.

These results also use the following simple lemma (which holds for all types of queries and equilibria mentioned in Sect. 2).

Lemma 2. *For any constants $\lambda' < \lambda \leq 1, \epsilon > 0$, there is a query-free reduction from finding ϵ-approximate equilibria of games in $\mathcal{G}(n, m, \lambda)$ to finding $\frac{\lambda'}{\lambda}\epsilon$-approximate equilibria of games in $\mathcal{G}(n, m, \lambda')$.*

In other words, query complexity upper bounds hold as λ and ϵ are scaled up together, and query complexity lower bounds hold as they are scaled down. The proof is very simple - the reduction multiplies every payoff by $\frac{\lambda'}{\lambda}$ (making no additional queries) and outputs the result. Note that the lemma does not hold for $\lambda' > \lambda$, as the reduction could introduce payoffs that are larger than 1.

3.1 Hardness of Approximate Pure Equilibria

In this section we will rely heavily on the following result of Babichenko.

Theorem 2 ([3]). *There is a constant $\epsilon_0 > 0$ such that, for any $\beta = 2^{-o(n)}$, the randomized δ-distribution query complexity of finding an ϵ_0-WSNE of n-player binary-action games with probability at least β is $\delta^2 2^{\Omega(n)}$.*

For the remainder of this work, the symbol ϵ_0 refers to this specific constant. A simple application of Lemma 2 yields

Corollary 2. *There is a constant $\epsilon_0 > 0$ such that, for any $\beta = 2^{-o(n)}$, the randomized δ-distribution query complexity of finding an $\epsilon_0\lambda$-WSNE of games in $\mathcal{G}(n, 2, \lambda)$ with probability at least β is $\delta^2 2^{\Omega(n)}$.*

We are now ready to state our main result – an exponential lower bound on the randomized query complexity of finding ϵ-PNEs of λ-Lipschitz games.

Theorem 3 (Main Result). *There exists some constant ϵ_0 such that, for any $n \in \mathbb{N}, \epsilon < \epsilon_0, \lambda \leq \frac{\epsilon}{\sqrt{8n \log 4n}}$, while every game in $\mathcal{G}(n, 2, \lambda)$ has an ϵ-PNE, any randomized algorithm finding such equilibria with probability at least $\beta = 1/poly(n)$ must make $\lambda^2 2^{\Omega(n\lambda/\epsilon)}$ profile queries.*

The proof follows by contradiction. Assume such an algorithm A exists making $\lambda^2 2^{o(n\lambda/\epsilon)}$ profile queries, convert it to an algorithm B making $\lambda^2 2^{o(n\lambda/\epsilon)}$ δ-distribution queries, then use Lemma 1 to derive an algorithm C finding $\epsilon_0\lambda$-WSNE in λ-Lipschitz games contradicting the lower bound of Corollary 2.

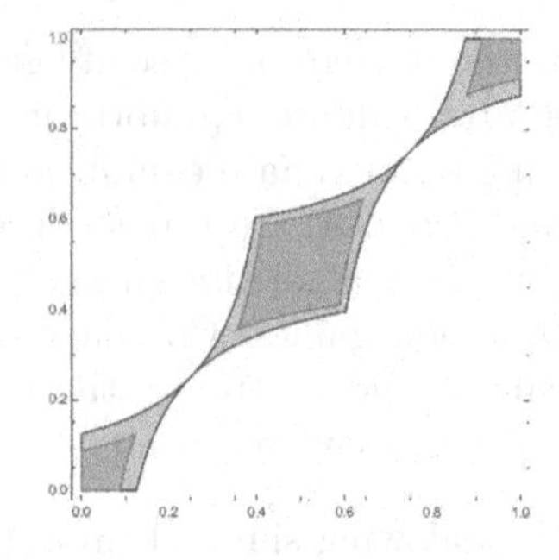

Fig. 1. Taking G' to be the Coordination Game for fixed values of ϵ and δ, the blue region shows the set of ϵ-approximate equilibria of G' (the acceptable outputs of algorithm B) while the orange region shows the set of all $\frac{\epsilon}{2}$-approximate equilibria of any possible game G'' in which each payoff may be perturbed by at most δ (the possible outputs of algorithm B).

Proof. Assume that some such algorithm A exists finding ϵ-PNEs of games in $\mathcal{G}(n, 2, \lambda)$ making at most $\lambda^2 2^{o(n\lambda/\epsilon)}$ profile queries. Consider any $\epsilon < \epsilon_0, \lambda' < \frac{\epsilon}{\sqrt{8n \log 4n}}$, and define $\lambda = \frac{\epsilon}{\epsilon_0}, L = \frac{\lambda}{\lambda'}, N = Ln$. We derive an algorithm C (with an intermediate algorithm B) that contradicts Corollary 2.

A Note that A finds $\frac{\epsilon}{2}$-PNEs of games in $\mathcal{G}\left(N, 2, \frac{3\lambda'}{2}\right)$ with probability at least β making at most $\lambda'^2 2^{o(N\lambda'/\epsilon)}$ profile queries (β can be amplified to constant).

B Let $\delta = \frac{\epsilon_0 \lambda'}{4}$. For any game $G' \in \mathcal{G}(N, 2, \lambda')$, consider an algorithm making δ-distribution queries of *pure action profiles* of G' (introducing the uncertainty without querying mixed strategies).

Claim. There is a game $G'' \in \mathcal{G}\left(N, 2, \frac{3\lambda'}{2}\right)$ that is consistent with all δ-distribution queries (i.e. $\mathbf{u}''(\mathbf{a}) = \tilde{\mathbf{u}}'(\mathbf{a})$ for all queried $\mathbf{a}$) in which no payoff differs from G' by more than an additive δ. Futhermore, any $\frac{\epsilon}{2}$-PNE of G'' is an ϵ-PNE of G'. Figure 1 visually depicts this observation.

The above claim is proven in [9]. Define the algorithm B that takes input G' and proceeds as though it is algorithm A (but makes δ-distribution queries instead). By the claim above, after at most $\lambda'^2 2^{o(N\lambda'/\epsilon)}$ queries, it has found an $\frac{\epsilon}{2}$-PNE of some $G'' \in \mathcal{G}\left(N, 2, \frac{3\lambda'}{2}\right)$ that it believes it has learned, which is also an ϵ-PNE of G'.

C Consider any game $G \in \mathcal{G}(n, 2, \lambda)$, and let $G' = g_G(L)$ be the population game induced by G. There is an algorithm C described by Lemma 1 that takes input G and simulates B on G' (making $2n \cdot \lambda'^2 2^{o(N\lambda'/\epsilon)} = \delta^2 2^{o(n\lambda/\epsilon)}$ δ-distribution queries) and correctly outputs an ϵ-WSNE (i.e. an $\epsilon_0\lambda$-WSNE) of G with probability constant probability (so certainly $2^{-o(n)}$).

The existence of algorithm C directly contradicts the result of Corollary 2, proving that algorithm A cannot exist. □

Remark 1. Note that, if we instead start the proof with the assumption of such an algorithm B, we can also show a $\delta^2 2^{o(n\lambda/\epsilon)}$ lower bound for the δ-distribution query complexity of finding ϵ-PNEs of λ-Lipschitz games.

3.2 Multi-Lipschitz Games

In this section, we consider a generalization of Lipschitz games in which each player $i \in [n]$ has a "player-specific" Lipschitz value λ_i in the sense that, if player i changes actions, the payoffs of all other players are changed by at most λ_i.

Definition 6. *A Λ-Multi-Lipschitz game G is an n-player, m-action game G in which each player $i \in [n]$ is associated with a constant $\lambda_i \leq 1$ such that $\sum_{i'=1}^{n} \lambda_{i'} = \Lambda$ and, for any player $i' \neq i$ and action profiles $\mathbf{a}^{(1)}, \mathbf{a}^{(2)}$ with $\mathbf{a}_{-i}^{(1)} = \mathbf{a}_{-i}^{(2)}$, $\left|u_{i'}\left(\mathbf{a}^{(1)}\right) - u_{i'}\left(\mathbf{a}^{(2)}\right)\right| \leq \lambda_i$. The class of such games is denoted $\mathcal{G}_\Lambda(n, m)$, and for simplicity it is assumed that $\lambda_1 \leq \ldots \leq \lambda_n$.*

The consideration of this generalized type of game allows real-world situations to be more accurately modeled. Geopolitical circumstances, for example, naturally take the form of Multi-Lipschitz games, since individual countries have different limits on how much their actions can affect the rest of the world. Financial markets present another instance of such games; they not only consist of individual traders who have little impact on each other, but also include a number of institutions that might each have a much greater impact on the market as a whole. This consideration is further motivated by the recent GameStop frenzy; the institutions still wield immense power, but so do the aggregate actions of millions of individuals [18].

Notice that a λ-Lipschitz game is a Λ-Multi-Lipschitz game, for $\Lambda = n\lambda$. Any algorithm that finds ϵ-ANEs of Λ-Multi-Lipschitz games is also applicable to Λ/n-Lipschitz games. Theorem 4 shows a kind of converse for query complexity, reducing from finding ϵ-ANE of Λ-Multi-Lipschitz games to finding ϵ-ANE of λ-Lipschitz games, for λ a constant multiple of Λ/n.

Theorem 4. *There is a reduction from computing ϵ-ANEs of games in $\mathcal{G}_\Lambda(n, 2)$ with probability at least $1 - \eta$ to computing $\frac{\epsilon}{2}$-ANEs of games in $\mathcal{G}\left(2n, 2, \frac{3\Lambda}{2n}\right)$ with probability at least $1 - \frac{\eta}{2}$ introducing at most a multiplicative poly$(n, \frac{1}{\epsilon}, \log\frac{1}{\eta})$ query blowup.*

As we now consider ϵ-ANEs, existence is no longer a question: such equilibria are *always* guaranteed to exist by Nash's Theorem [16]. This proof will also utilize a more general population game $G' = g_G(L_1, \ldots, L_n)$ in which player i is replaced by a population of size L_i (where the L_i may differ from each other), and the queries in Lemma 1 become $(\delta, \min_{i \in [n]}\{\gamma/L_i\})$-distribution queries (this will now be relevant, as we need to apply Corollary 1). Otherwise, the proof follows along the same lines as that of Theorem 3.

Proof. Consider some $\epsilon > 0$ and a game $G \in \mathcal{G}_\Lambda(n,2)$ (WLOG take $\lambda_1 \leq \ldots \leq \lambda_n$). First, if $\Lambda < \frac{\epsilon}{n}$, finding an ϵ-ANE is trivial (each player can play their best-response to the uniform mixed strategy, found in $2n$ queries). So assume $\Lambda \geq \frac{\epsilon}{n}$. Define $L_i = \max\{\frac{n\lambda_i}{\Lambda}, 1\}$ and, taking $i' = \max_{i \in [n]}\{i : L_i = 1\}$, note that

$$\sum_{i=1}^{n} L_i = \sum_{i=1}^{i'} 1 + \sum_{i=i'+1}^{n} \frac{n\lambda_i}{\Lambda} = \sum_{i=1}^{i'} 1 + \frac{n}{\Lambda} \sum_{i=i'+1}^{n} \lambda_i \leq \sum_{i=1}^{i'} 1 + \frac{n}{\Lambda}\Lambda \leq 2n.$$

Thus the population game $G' = g_G(L_1, \ldots, L_n) \in \mathcal{G}\left(2n, 2, \frac{\Lambda}{n}\right)$.

A Consider an algorithm A that finds $\frac{\epsilon}{2}$-ANEs of games in $\mathcal{G}\left(2n, 2, \frac{3\Lambda}{2n}\right)$, with probability at least $1 - \frac{\eta}{2}$ making q profile queries.
B Taking $\delta = \frac{\epsilon^2}{4n^2} < \frac{\epsilon\Lambda}{4n}$, the algorithm B from the proof of Theorem 3 that simulates A but makes $(\delta, 1)$-distribution queries finds an ϵ-ANE of G' (The claim in Theorem 3 also holds for these parameters with this choice of δ).
C By Lemma 1, there is an algorithm C on input $G \in \mathcal{G}_\Lambda(n,2)$ that simulates B (replacing each $(\delta, 1)$-distribution query of G' with $2n$ $(\delta, \frac{1}{n})$-distribution queries of G since $\frac{1}{L_n} \geq \frac{1}{n}$) finding an ϵ-ANE with probability at least $1 - \eta$.

Applying Corollary 1 (using $\delta = \frac{\epsilon^2}{4n^2}, \gamma = \frac{1}{n}$) to create a profile-query algorithm from C completes the proof. □

As an example application of Theorem 4, an algorithm of [10] finds $\left(\frac{1}{8} + \alpha\right)$-approximate Nash equilibria of games in $\mathcal{G}\left(n, 2, \frac{1}{n}\right)$; Theorem 5 states that result in detail, and Corollary 3 extends it to Multi-Lipschitz games.

Theorem 5 **([10]).** *Given constants $\alpha, \eta > 0$, there is a randomized algorithm that, with probability at least $1 - \eta$, finds $\left(\frac{1}{8} + \alpha\right)$-approximate Nash equilibria of games in $\mathcal{G}\left(n, 2, \frac{1}{n}\right)$ making $O\left(\frac{1}{\alpha^4}\log\left(\frac{n}{\alpha\eta}\right)\right)$ profile queries.*

We now have some ability to apply this to Multi-Lipschitz games; if $1 \leq \Lambda < 4$ we can improve upon the trivial $\frac{1}{2}$-approximate equilibrium of Proposition 1.

Corollary 3. *For $\alpha, \eta > 0, \Lambda \geq 1, \epsilon \geq \frac{\Lambda}{8} + \alpha$, there is an algorithm finding ϵ-ANEs of games in $\mathcal{G}_\Lambda(n,2)$ with probability at least $1 - \eta$ making at most poly($n, \frac{1}{\alpha}, \log\frac{1}{\eta}$) profile queries.*

Remark 2. This is actually a slight improvement over just combining Theorems 4 and 5, since the choice of δ can be made slightly smaller to shrink α as necessary.

3.3 A Deterministic Lower Bound

We complete this work by generalizing the following result of Hart and Nisan.

Theorem 6 **([14]).** *For any $\epsilon < \frac{1}{2}$, the deterministic profile query complexity of finding ϵ-ACEs of n-player games is $2^{\Omega(n)}$.*

Player 2

Player 1	1	2
1	1, 0	0, 1
2	0, 1	1, 0

(a) The payoff matrix of $G_{1,2}$, the Matching Pennies game.

Player 2

Player 1	1	2	3
1	1, 0	0, 1	0, 1
2	0, 1	1, 0	0, 1
3	0, 1	0, 1	1, 0

(b) The payoff matrix of $G_{1,3}$, the generalized Matching Pennies game.

Fig. 2. The payoff matrices of $G_{1,2}$ and $G_{1,3}$.

While the proof of Theorem 6 utilizes a reduction from `ApproximateSink`, we employ a more streamlined approach, presenting an explicit family of "hard" games that allows us to uncover the optimal value of ϵ as a function of the number of actions:

Theorem 7. *Given some $m \in \mathbb{N}$, for any $\epsilon < \frac{m-1}{m}$, the deterministic profile query complexity of finding ϵ-ACEs of n-player, m-action games is $2^{\Omega(n)}$.*

Furthermore, this value of ϵ cannot be improved:

Proposition 1. *Given some $n, m \in \mathbb{N}$, for any $\epsilon \geq \frac{m-1}{m}$, an ϵ-ANE of an n-player, m-action game can be found making no profile queries.*

The upper bound of Proposition 1 can be met if every player plays the uniform mixed strategy over their actions. Finally, we can apply Lemma 2 to scale Theorem 7 and obtain our intended result:

Corollary 4. *Given some $m \in \mathbb{N}, \lambda \in (0, 1]$, for any $\epsilon < \frac{m-1}{m}\lambda$, the deterministic profile query complexity of finding ϵ-ACEs of n-player, m-action, λ-Lipschitz games is $2^{\Omega(n)}$.*

In order to prove these results, we introduce a family of games $\{G_{k,m}\}$. For any $k, m \in \mathbb{N}$, $G_{k,m}$ is a $2k$-player, m-action generalization of k Matching Pennies games in which every odd player i wants to match the even player $i+1$ and every even player $i+1$ wants to mismatch with the odd player i.

Definition 7. *Define $G_{1,2}$ to be the generalized Matching Pennies game, as described in Fig. 2(a). Define the generalization $G_{k,m}$ to be the $2k$-player m-action game such that, for any $i \in [k]$, player $2i-1$ has a payoff 1 for matching player $2i$ and 0 otherwise (and vice versa for player $2i$) ignoring all other players.*

The critical property of the generalized Matching Pennies game is that we can bound the probability that any given action profile is played in any ϵ-ACE of $G_{k,m}$. If too much probability is jointly placed on matching actions, player 2 will have high regret. Conversely, if too much probability is jointly placed on mismatched actions, player 1 will have high regret.

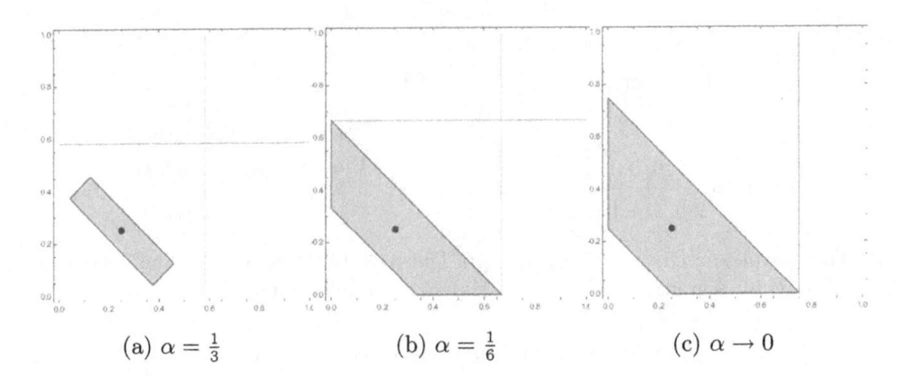

(a) $\alpha = \frac{1}{3}$ (b) $\alpha = \frac{1}{6}$ (c) $\alpha \to 0$

Fig. 3. The region of possible values for $(\Pr_{\mathbf{a}\sim\mathbf{X}^*}(a_1 = 1, a_2 = 1), \Pr_{\mathbf{a}\sim\mathbf{X}^*}(a_1 = 2, a_2 = 2)$ in any $(\frac{1}{2} - \alpha)$-approximate correlated equilibrium of $G_{k,2}$. The only exact correlated equilibrium is shown by the red point, and the corresponding values of ρ are displayed as the orange lines.

Lemma 3. *For any $k, m \in \mathbb{N}, \alpha > 0$, take $\epsilon = \frac{m-1}{m} - \alpha$. In any ϵ-ACE $\mathbf{X}^*$ of $G_{k,m}$, every action profile $\mathbf{a}' \in [m]^n$ satisfies $\Pr_{\mathbf{a}\sim\mathbf{X}^*}(\mathbf{a} = \mathbf{a}') < \rho^{\frac{n}{2}}$ where*

$$\rho = \frac{(2-\alpha)m - 1}{2m}.$$

This phenomenon can be seen in Fig. 3.

Proof. Define $n = 2k$ to be the number of players in $G_{k,m}$, and consider some ϵ-ACE $\mathbf{X}^*$. Now WLOG consider players 1 and 2 and assume, for the sake of contradiction, that there exist some actions j_1, j_2 such that $\Pr_{\mathbf{a}\sim\mathbf{X}^*}(a_1 = j_1, a_2 = j_2) > \rho$. We will need to consider the two cases $j_1 = j_2$ and $j_1 \neq j_2$.

Matching Actions. In this case, WLOG assume $j_1 = j_2 = 1$. We show that player 2 can improve her payoff by more than ϵ. Under $\mathbf{X}^*$, with probability $> \rho$, any random realization $\mathbf{a} \sim \mathbf{X}^*$ will yield player 2 a payoff of 0. In other words, $u_2(\mathbf{X}^*) < 1 - \rho$. Furthermore, considering the marginal distribution over player 1's action, we are guaranteed that

$$\sum_{j=2}^{m} \Pr_{\mathbf{a}\sim\mathbf{X}^*}(a_1 = j) < 1 - \rho$$

so there must exist some action (WLOG action 2) for which $\Pr_{\mathbf{a}\sim\mathbf{X}^*}(a_1 = 2) < \frac{1-\rho}{m-1}$. As such, define $\phi(j) = 2$. Then $u_2^{(\phi)}(\mathbf{X}^*) > 1 - \frac{1-\rho}{m-1}$, so

$$\mathrm{reg}_2^{(\phi)}(\mathbf{X}^*) > \underbrace{\left(1 - \frac{1-\rho}{m-1}\right)}_{u_2^{(\phi)}(\mathbf{X}^*)} - \underbrace{(1-\rho)}_{u_2(\mathbf{X}^*)} = \frac{\rho m - 1}{m-1} \geq \frac{m-1}{m} - \alpha$$

for all $m \geq 2$. This contradicts our assumption of an ϵ-ACE.

Mismatched Actions. In this case, WLOG assume $j_1 = 1, j_2 = 2$. The situation is simpler, taking $\phi(j) = 2$, $u_1(\mathbf{X}^*) < 1 - \rho$ and $u_1^{(\phi)}(\mathbf{X}^*) > \rho$, so $\text{reg}_1^{(\phi)}(\mathbf{X}^*) > \frac{m-1}{m} - \alpha$. This too contradicts our assumption of an ϵ-ACE, and thus completes the proof of Lemma 3. □

We can now prove Theorem 7. The general idea is that, should an efficient algorithm exist, because any equilibrium of $G_{k,m}$ must have large support by Lemma 3, there is significant probability assigned to action profiles that are not queried by the algorithm. We show there is a game that the algorithm cannot distinguish from $G_{k,m}$ that shares no approximate equillibria with $G_{k,m}$.

Proof (Theorem 7). Consider any $\alpha > 0$ and let $\epsilon = \frac{m-1}{m} - \alpha$. Taking ρ as in the statement of Lemma 3, assume there exists some deterministic algorithm A that takes an n-player, m-action game G as input and finds an ϵ-ACE of G querying the payoffs of $q < \frac{\alpha}{2}\rho^{-\frac{n}{2}}$ action profiles. Fix some $k \in \mathbb{N}$ and consider input $G_{k,m}$ as defined in Definition 7. Then $\mathbf{X}^* = A\left(G_{k,m}\right)$ is an ϵ-ACE of $G_{k,m}$. Note that, for some j, $\Pr_{\mathbf{a}\sim\mathbf{X}^*}(a_1 = j) \leq \frac{1}{m}$ (WLOG assume $j = 1$).

Now define the perturbation $G'_{k,m}$ of $G_{k,m}$ with payoffs defined to be equal to $G_{k,m}$ for every action profile queried by A, 1 for every remaining action profile in which player 1 plays action 1 (chosen because it is assigned low probability by $\mathbf{X}^*$ by assumption), and 0 otherwise. Note that, by definition, A cannot distinguish between $G_{k,m}$ and $G'_{k,m}$, so $A(G'_{k,m}) = \mathbf{X}^*$.

Taking the function $\phi(j) = 1$, the quantity we need to bound is $\text{reg}_i'^{(\phi)}(\mathbf{X}^*) \geq u_i'^{(\phi)}(\mathbf{X}^*) - u_i'(\mathbf{X}^*)$. We must bound the components of this expression as follows:

Claim. $u_1'^{(\phi)}(\mathbf{X}^*) > (1 - q\rho^{\frac{n}{2}})$ and $u_1'(\mathbf{X}^*) < (\frac{1}{m} + q\rho^{\frac{n}{2}})$.

Proof Using the claim (proven in [9]) and once again recalling the assumption that $q < \frac{\alpha}{2}\rho^{-\frac{n}{2}}$, we see

$$\text{reg}_1'^{(\phi)}(\mathbf{X}^*) > \left(1 - \frac{1}{m} - 2q\rho^{\frac{n}{2}}\right) = \frac{m-1}{m} - \alpha = \epsilon.$$

So $\mathbf{X}^*$ cannot actually be an ϵ-ACE of $G_{k,m}$. This completes the proof of Theorem 7. □

4 Further Directions

An important additional question is the query complexity of finding ϵ-PNEs of n-player, λ-Lipschitz games in which $\epsilon = \Omega(n\lambda)$. Theorem 3 says nothing in this parameter range, yet Theorem 5 provide a logarithmic upper bound in this regime. The tightness of this bound is of continuing interest. Furthermore, the query- and computationally-efficient reduction discussed in Lemma 1 provides a hopeful avenue for further results bounding the query, and computational, complexities of finding equilibria in many other classes of games.

Acknowledgements. We thank Francisco Marmalejo Cossío and Rahul Santhanam for their support in the development of this work, and the reviewers of an earlier version for helpful comments. Matthew Katzman was supported by an Oxford-DeepMind Studentship for this work.

References

1. Azrieli, Y., Shmaya, E.: Lipschitz games. Math. Oper. Res. **38**(2), 350–357 (2013). https://doi.org/10.1287/moor.1120.0557
2. Babichenko, Y.: Small support equilibria in large games. CoRR abs/1305.2432 (2013). https://arxiv.org/abs/1305.2432
3. Babichenko, Y.: Query complexity of approximate Nash equilibria. J. ACM **63**(4), 1–24 (2016). https://doi.org/10.1145/2908734
4. Babichenko, Y.: Fast convergence of best-reply dynamics in aggregative games. Math. Oper. Res. **43**(1), 333–346 (2018). https://doi.org/10.1287/moor.2017.0868
5. Babichenko, Y.: Informational bounds on equilibria (a survey). SIGecom Exch. **17**(2), 25–45 (2019). https://doi.org/10.1145/3381329.3381333
6. Chen, X., Cheng, Y., Tang, B.: Well-supported vs. approximate Nash equilibria: query complexity of large games. In: Papadimitriou, C.H. (ed.) 8th Innovations in Theoretical Computer Science Conference, ITCS. LIPIcs, vol. 67, pp. 1–9 (2017). https://doi.org/10.4230/LIPIcs.ITCS.2017.57
7. Daskalakis, C., Papadimitriou, C.H.: Approximate Nash equilibria in anonymous games. J. Econ. Theory **156**, 207–245 (2015). https://doi.org/10.1016/j.jet.2014.02.002
8. Deligkas, A., Fearnley, J., Spirakis, P.G.: Lipschitz continuity and approximate equilibria. Algorithmica **82**(10), 2927–2954 (2020). https://doi.org/10.1007/s00453-020-00709-3
9. Goldberg, P.W., Katzman, M.J.: Lower bounds for the query complexity of equilibria in Lipschitz games (2021)
10. Goldberg, P.W., Marmolejo Cossío, F.J., Wu, Z.S.: Logarithmic query complexity for approximate Nash computation in large games. Theory Comput. Syst. **63**(1), 26–53 (2019). https://doi.org/10.1007/s00224-018-9851-8
11. Goldberg, P.W., Roth, A.: Bounds for the query complexity of approximate equilibria. ACM Trans. Econ. Comput. **4**(4), 1–25 (2016)
12. Goldberg, P.W., Turchetta, S.: Query complexity of approximate equilibria in anonymous games. J. Comput. Syst. Sci. **90**, 80–98 (2017). https://doi.org/10.1016/j.jcss.2017.07.002
13. Hart, S., Mansour, Y.: How long to equilibrium? the communication complexity of uncoupled equilibrium procedures. Games Econ. Behav. **69**(1), 107–126 (2010)
14. Hart, S., Nisan, N.: The query complexity of correlated equilibria. Games Econ. Behav. **108**, 401–410 (2018). https://doi.org/10.1016/j.geb.2016.11.003
15. Nash, J.: Non-Cooperative Games. Ph.D. thesis, Princeton University (May 1950)
16. Nash, J.: Non-cooperative games. Annals of mathematics, pp. 286–295 (1951)
17. Peretz, R., Schreiber, A., Schulte-Geers, E.: The Lipschitz constant of perturbed anonymous games. CoRR abs/2004.14741 (2020). https://arxiv.org/abs/2004.14741

18. Phillips, M., Lorenz, T.: 'dumb money' is on gamestop, and it's beating wall street at its own game (Feb 2021). https://www.nytimes.com/2021/01/27/business/gamestop-wall-street-bets.html. Accessed 5 Feb 2021
19. Rubinstein, A.: Settling the complexity of computing approximate two-player Nash equilibria. In: Dinur, I. (ed.) IEEE 57th Annual Symposium on Foundations of Computer Science, FOCS, pp. 258–265. IEEE Computer Society (2016). https://doi.org/10.1109/FOCS.2016.35

Gerrymandering on Graphs: Computational Complexity and Parameterized Algorithms

Sushmita Gupta[1], Pallavi Jain[2], Fahad Panolan[3], Sanjukta Roy[4(✉)], and Saket Saurabh[1]

[1] The Institute of Mathematical Sciences, HBNI, Chennai, India
{sushmitagupta,saket}@imsc.res.in
[2] Indian Institute of Technology Jodhpur, Jodhpur, India
pallavi@iitj.ac.in
[3] Indian Institute of Technology Hyderabad, Hyderabad, India
fahad@cse.iith.ac.in
[4] TU Wien, Vienna, Austria
sanjukta.roy@tuwien.ac.at

Abstract. This paper studies *gerrymandering on graphs* from a computational viewpoint (introduced by Cohen-Zemach et al. [AAMAS 2018] and continued by Ito et al. [AAMAS 2019]). Our contributions are two-fold: conceptual and computational. We propose a generalization of the model studied by Ito et al., where the input consists of a graph on n vertices representing the set of voters, a set of m candidates $\mathcal{C}$, a weight function $w_v : \mathcal{C} \rightarrow \mathbb{Z}^+$ for each voter $v \in V(G)$ representing the preference of the voter over the candidates, a distinguished candidate $p \in \mathcal{C}$, and a positive integer k. The objective is to decide if it is possible to partition the vertex set into k *districts* (i.e., pairwise disjoint connected sets) such that the candidate p *wins* more districts than any other candidate. There are several natural parameters associated with the problem: the number of districts (k), the number of voters (n), and the number of candidates (m). The problem is known to be NP-complete even if $k = 2$, $m = 2$, and G is either a complete bipartite graph (in fact $K_{2,n}$, i.e., partitions of size 2 and n) or a complete graph. Moreover, recently we and Bentert et al. [WG 2021], independently, showed that the problem is NP-hard for paths. This means that the search for FPT algorithms needs to focus either on the parameter n, or subclasses of forest (as the problem is NP-complete on $K_{2,n}$, a family of graphs that can be transformed into a forest by deleting *one* vertex). Circumventing these intractability results we successfully obtain the following algorithmic results.

Sushmita Gupta supported by SERB-Starting Research Grant (SRG/2019/001870). Fahad Panolan supported by Seed grant, IIT Hyderabad (SG/IITH/F224/2020-21/SG-79). Sanjukta Roy supported by the WWTF research grant (VRG18-012). Saket Saurabh supported by European Research Council (ERC) under the European Union's Horizon 2020 research and innovation programme (grant no. 819416), and Swarnajayanti Fellowship grant DST/SJF/MSA-01/2017-18.

I. Caragiannis and K. A. Hansen (Eds.): SAGT 2021, LNCS 12885, pp. 140–155, 2021.
https://doi.org/10.1007/978-3-030-85947-3_10

- A $2^n(n+m)^{\mathcal{O}(1)}$ time algorithm on general graphs.
- FPT algorithm with respect to k (an algorithm with running time $2^{\mathcal{O}(k)}n^{\mathcal{O}(1)}$) on paths in both deterministic and randomized settings, even for arbitrary weight functions. Whether the problem is FPT parameterized by k on trees remains an interesting open problem.

Our algorithmic results use sophisticated technical tools such as representative set family and Fast Fourier Transform based polynomial multiplication, and their (possibly first) application to problems arising in social choice theory and/or algorithmic game theory is likely of independent interest to the community.

Keywords: Gerrymandering · Parameterized complexity · Representative set

1 Introduction

"Elections have consequences" a now-famous adage ascribed to Barack Obama, the former President of U.S.A, brings to sharp focus the high stakes of an electoral contest. Political elections, or decision making in a large organization, are often conducted in a hierarchical fashion. Thus, in order to win the final prize it is enough to manipulate at district/division level, obtain enough votes and have the effect propagate upwards to win finally. Needless to say the ramifications of winning and losing are extensive and possibly long-term; consequently, incentives for *manipulation* are rife.

The objective of this article is to study a manipulation or control mechanism, whereby the manipulators are allowed to create the voting "districts". A well-thought strategic division of the voting population may well result in a favored candidate's victory who may not win under normal circumstances. In a more extreme case, this may result in several favored candidates winning multiple seats, as is the case with election to the US House of Representatives, where candidates from various parties compete at the district level to be the elected representative of that district in Congress. This topic has received a lot of attention in recent years under the name of *gerrymandering.* A New York Times article "How computers turned gerrymandering into science" [16] discusses how Republicans were able to successfully win 65% of the available seats in the state assembly of Wisconsin even though the state has about an equal number of Republican and Democrat voters. The possibility for gerrymandering and its consequences have long been known to exist and have been discussed for many decades in the domain of political science, as discussed by Erikson [17] and Issacharoff [23]. Its practical feasibility and long-ranging implications have become a topic of furious public, policy, and legal debate only somewhat recently [33], driven largely by the ubiquity of computer modelling in all aspects of the election process. Thus, it appears that via the vehicle of gerrymandering the political battle lines have been drawn to (re)draw the district lines.

While gerrymandering has been studied in political sciences for long, it is only rather recently that the problem has attracted attention from the perspective of algorithm design and complexity theory. Lewenberg et al. [26] and Eiben et al. [15] study gerrymandering in a geographical setting in which voters must vote in the closest polling stations and thus problem is about strategic placement of polling stations rather than drawing district lines. Cohen-Zemach et al. [8] modeled gerrymandering using graphs, where vertices represent voters and edges represent some connection (be it familial, professional, or some other kind), and studied the computational complexity of the problem. Ito et al. [24] further extended this study to various classes of graphs, such as paths, trees, complete bipartite graphs, and complete graphs.

In both the papers the following hierarchical voting process is considered: A given set of voters is partitioned into several groups, and each of the groups holds an independent election. From each group, one candidate is elected as a nominee (using the plurality rule). Then, among the elected nominees, the winner is determined by a final voting rule (again by plurality). The formal definition of the problem, termed GERRYMANDERING (GM), considered in [24] is as follows. The input consists of an undirected graph G, a set of candidates $\mathcal{C}$, an approval function $a : V(G) \rightarrow \mathcal{C}$ where $a(v)$ represents the candidate approved by v, a weight function $w: V(G) \rightarrow \mathbb{Z}^+$, a distinguished candidate p, and a positive integer k. We say a candidate q wins a subset $V' \subseteq V(G)$ if $q \in \arg\max_{q' \in \mathcal{C}} \left\{ \sum_{v \in V',\, a(v)=q'} w(v) \right\}$, i.e., the sum of the weights of voters in the subset V' who approve q is not less than that of any other candidate. The objective is to decide whether there exists a partition of $V(G)$ into k non-empty parts $V_1 \uplus \ldots \uplus V_k$ (called *districts*) such that (i) the induced subgraph $G[V_i]$ is connected for each $i \in \{1, \ldots, k\}$, and (ii) the number of districts *won only by* p is more than the districts won by any other candidate alone or with others.

In this paper we continue the line of investigation done in [8,24]. Our contribution is two fold, conceptual and the other is computational. Towards the former, we offer a realistic generalization of GM, named WEIGHTED GERRYMANDERING (W-GM). Towards the latter, we present *fixed parameter tractable* (FPT) algorithms with respect to natural parameters associated with the gerrymandering problem.

Our Model. A natural generalization of GM in real-life is that of a vertex representing a locality or an electoral booth as opposed to an individual citizen. In that situation, however, it is natural that more than one candidate receives votes in a voting booth, and the number of such votes may vary arbitrarily. We can model the number of votes each candidate gets in the voting booth corresponding to booth v by a weight function $w_v : \mathcal{C} \rightarrow \mathbb{Z}^+$, i.e., the value $w_v(c)$ for any candidate $c \in \mathcal{C}$ represents the number of votes obtained by candidate c in booth v. This model is perhaps best exemplified by a nonpartisan "blanket primary" election (such as in California) where all candidates for the same elected post regardless of political parties, compete on the same ballot against each other all at once. In a two-tier system, multiple winners (possibly more than two) are declared and they contest the general election. The idea that one can have multiple candidates earning votes from the same locality and possibly emerging as winners is captured by GM. In [8,24], the vertex v "prefers" only one candidate, and in this sense our model (W-GM) generalizes theirs (GM).

Formally stated, the input to W-GM consists of an undirected graph G, a set of candidates $\mathcal{C}$, a weight function for each vertex $v \in V(G)$, $w_v : \mathcal{C} \to \mathbb{Z}^+$, a distinguished candidate p, and a positive integer k. A candidate q is said to win a subset $V' \subseteq V(G)$ if $q \in \arg\max_{q' \in \mathcal{C}} \left\{\sum_{v \in V'} w_v(q')\right\}$. The objective is to decide whether there exists a partition of the vertex set $V(G)$ into k districts such that (i) $G[V_i]$ is connected for each $i \in [k]$, and (ii) the number of districts *won only by* p is more than the number of districts won by any other candidate alone or with others. GM can be formally shown to be a special case of W-GM since we can transform an instance $\mathcal{I} = (G, \mathcal{C}, a, w, p, k)$ of GM to an instance $\mathcal{J} = (G, \mathcal{C}, \{w_v\}_{v \in V(G)}, p, k)$ of W-GM as follows. For each $v \in V(G)$, let $w_v : \mathcal{C} \to \mathbb{Z}^+$ such that for any $q \in \mathcal{C}$, if $a(v) = q$, then $w_v(q) = w(v)$ and $w_v(q) = 0$, otherwise.

Our Results and Methods. The main open problem mentioned in Ito et al. [24] is the complexity status of GM on paths when the number of candidates is not fixed (for the fixed number of candidates, it is solvable in polynomial time). This question was recently resolved by Bentert et al. [1], and has also been proved independently by us, which is presented in our extended version [22] and omitted from here because of lack of space. Thus, in this article we will focus on designing efficient algorithms. We must remark that Bentert et al. [1] also show that the problem is weakly NP-hard for trees with three or more candidates.

We study the problem from the viewpoint of parameterized complexity. The goal of parameterized complexity is to find ways of solving NP-hard problems more efficiently than brute force: here the aim is to restrict the combinatorial explosion in the running time to a parameter that is expected to be much smaller than the input size. Formally, a *parameterization* of a problem is assigning an integer ℓ to each input instance and we say that a parameterized problem is *fixed-parameter tractable* (FPT) if there is an algorithm that solves the problem in time $f(\ell) \cdot |I|^{\mathcal{O}(1)}$, where $|I|$ is the size of the input and f is an arbitrary computable function depending on the parameter ℓ only. There is a long list of NP-hard problems that are FPT under various parameterizations. For more background, the reader is referred to the monographs [9,14,29].

Our Choice of Parameters. There are several natural parameters associated with the gerrymandering problem: the number of districts the vertex set needs to be partitioned (k), the number of voters (n), and the number of candidates (m). Ito et al. [24] proved that GM is NP-complete even if $k = 2$, $m = 2$, and G is either a complete bipartite graph (in fact $K_{2,n}$) or a complete graph. Thus, we cannot hope for an algorithm for W-GM that runs in $f(k,m) \cdot n^{\mathcal{O}(1)}$ time, i.e., an FPT algorithm with respect to the parameter $k + m$, even on planar graphs. In fact, we cannot hope to have an algorithm with running time $(n + m)^{f(k,m)}$, where f is a function depending only on k and m, as that would imply P=NP. This means that our search for FPT algorithms needs to either focus on the parameter n, or subclasses of planar graphs (as the problem is NP-complete on $K_{2,n}$, which is planar). Furthermore, note that $K_{2,n}$ could be transformed into a forest by deleting a vertex, and thus we cannot even hope to have an algorithm with running time $(n+m)^{f(k,m)}$, where f is a function depending only on k and m, on a family of graphs that can be made acyclic, in fact a star, by *deleting at most one vertex.* This essentially implies that if we wish to design an FPT algorithm for W-GM with respect to the parameter k, or m, or $k+m$, we must restrict input graphs to forests. Circumventing these intractable results, we successfully obtain several algorithmic results. We give deterministic and randomised FPT algorithms for W-GM on paths with respect to k. Since W-GM generalizes GM, the algorithmic results hold for GM as well.

Theorem 1. *There is an algorithm that given an instance of* W-GM *on arbitrary graphs and a tie-breaking rule* η*, solves the instance in time* $2^n(n+m)^{\mathcal{O}(1)}$.

Intuition Behind the Proof of Theorem 1. Suppose that we are given a Yes-instance of the problem. Of the k possibilities, we first "guess" in a solution the number of districts that are won by the distinguished candidate p. Let this number be denoted by $k^\star$. Next, for every candidate $c \in \mathcal{C}$, we consider the family $\mathcal{F}_c$, the set of districts of $V(G)$ in which c wins in each of them. These families are pairwise disjoint because each district has a unique winner. Our goal is to find $k^\star$ disjoint sets from the family $\mathcal{F}_p$ and at most $k^\star-1$ disjoint sets from any other family so that in total we obtain k pairwise disjoint districts that partition $V(G)$. The exhaustive algorithm to find the districts from these families would take time $\mathcal{O}^\star(2^{nmk^\star})$. We reduce our problem to polynomial multiplication involving polynomial-many multiplicands, each with degree at most $\mathcal{O}(2^n)$.

Why Use Polynomial Algebra? Every district S is a subset of $V(G)$. Let $\chi(S)$ denotes the characteristic vector corresponding to S. We view $\chi(S)$ as an n-digit binary number, in particular, if $u_i \in S$, then i^{th} bit of $\chi(S)$ is 1, otherwise 0. A crucial observation guiding our algorithm is that two sets S_1 and S_2 are disjoint if and only if the number of 1 in $\chi(S_1)+\chi(S_2)$ (binary sum/modulo 2) is equal to $|S_1|+|S_2|$. So, for each set $\mathcal{F}_c$, we make a polynomial $P_c(y)$, where for each set $S \in \mathcal{F}_c$, there is a monomial $y^{\chi(S)}$. Let c_1 and c_2 be two candidates, and for simplicity assume that each set in $\mathcal{F}_{c_1}$ has size exactly s and each set in $\mathcal{F}_{c_2}$ has size exactly t (we do not have such assumption in the formal description of the algorithm). Let $P^\star(y)$ be the polynomial obtained by multiplying $P_{c_1}(y)$ and $P_{c_2}(y)$; and let y^z be a monomial of $P^\star(y)$. Then, the z has exactly $s+t$ ones if and only if "the sets which corresponds to z are disjoint". Thus, the polynomial method allows us to capture disjointness and hence, by multiplying appropriate subparts of polynomial described above, we obtain our result. Furthermore, note that $\chi(S) \in \{0,1\}^n$, throughout the process, correspond to some set in $V(G)$, and hence the decimal representation of the maximum degree of the considered polynomials is upper bounded by 2^n. Hence, the algorithm itself is about applying an $\mathcal{O}(d \log d)$ algorithm to multiply two polynomials of degree d; here $d \leq 2^n$. Thus, we obtain Theorem 1.

Theorem 2. *There is a deterministic algorithm that given an instance of* W-GM *on paths and a tie-breaking rule* η *solves in time* $2.619^k(n+m)^{\mathcal{O}(1)}$.

Theorem 3. *There is a randomized algorithm that given an instance of* W-GM *on paths and a tie-breaking rule* η, *solves the instance in time* $2^k(n+m)^{\mathcal{O}(1)}$ *with no false positives and false negatives with probability at most* $1/3$.

Intuition Behind the Proofs of Theorem 2 and 3. Since, the problem is on paths, it boils down to selecting $k-1$ appropriate vertices which divide the path into k subpaths that form the desired districts. This in turn implies that each district can be identified by the leftmost vertex and the rightmost vertex appearing in the district (based on the way vertices appear on the path). Hence, there can be at most $\mathcal{O}(n^2)$ districts in the path graph. Furthermore, since we are on a path, we observe that if we know a district (identified by its leftmost and the rightmost vertices on the path), then we also know the rightmost (and leftmost) vertex of the district adjacent to its left (resp. right). These observations naturally lead us to consider the following graph H: we have a vertex for each possible district and put an edge from a district to another district, if these two districts appear consecutively on the path graph. Thus, we are looking for a path of length k in H such that (a) it covers all the vertices of the input path (this automatically implies that each vertex appears in exactly one district); and (b) the distinguished candidate wins most number of districts. This equivalence allows us to use the rich algorithmic toolkit developed for designing $2^{\mathcal{O}(k)}n^{\mathcal{O}(1)}$ time algorithm for finding a k-length path in a given graph [5,28,32].

The above tractability result for paths cannot be extended to graphs with pathwidth 2, or graphs with feedback vertex set (a subset of vertices whose deletion transforms the graph into a forest) of size 1, because GM is NP-complete on $K_{2,n}$ when $k=2$ and $|\mathcal{C}|=2$ (see [24]). Note that the pathwidth of graph $K_{2,n}$ is 2 and it has feedback vertex set size 1. For trees, it is easy to obtain a $\mathcal{O}(\binom{n}{k-1})$ time algorithm by "guessing" the $k-1$ edges whose deletion yields the k districts that constitute the solution. However, a $f(k)n^{\mathcal{O}(1)}$ algorithm for trees so far eludes us. Thus, whether the problem is FPT parameterized by k on trees remains an interesting open problem.

Unique Winner vs Multiple Winner: The definition of GM [24] or its generalization W-GM put forward by us does not preclude the possibility of multiple winners in a district. The time complexity stated in Theorems 1 and 2 is achieved when only one winner emerges from each district, a condition that is attainable using a tie-breaking rule. Notably, the algorithms in Theorems 2 and 3 can be modified to handle the case when multiple winners emerge in some district(s) [22].

Additionally, using our parameterized algorithms (Theorems 2 and 3), we can improve over Theorem 1 when the input graph is a path. That is, using Theorems 2 and 3, and the fact that there exists an algorithm for paths that runs in time $\mathcal{O}(\binom{n}{k-1})$, we conclude that for W-GM on paths, there exists a deterministic algorithm that runs in $\max_{1\leq k\leq n}\min\{\binom{n}{k}, 2.619^k\}$ time, and a randomized algorithm that runs in $\max_{1\leq k\leq n}\min\{\binom{n}{k}, 2^k\}$ time. Using, standard calculations we can obtain the following result.

Theorem 4. *There is a (randomized) deterministic algorithm that given an instance of* W-GM *on paths and a tie-breaking rule* η, *solves the instance in time* ($1.708^n(n+m)^{\mathcal{O}(1)}$) $1.894^n(n+m)^{\mathcal{O}(1)}$.

It is worth mentioning that our algorithmic results use sophisticated technical tools from parameterized complexity–representative set family and Fast Fourier transform based polynomial multiplication–that have yielded breakthroughs in improving time complexity of many well-known optimization problems. Thus, their (possibly first) application to problems arising in social choice theory and/or algorithmic game theory is likely of independent interest. Due to the constraints on space, proofs marked by ♣ are deferred to the full version [22].

Related Work. In addition to the result discussed earlier Ito et al. [24] also prove that GM is strongly NP-complete when G is a tree of diameter four; thereby, implying that the problem cannot be solved in pseudo-polynomial time unless P = NP. As GM is a special case of W-GM, each of the hardness results for GM carry onto W-GM. They also exhibit several positive results: GM is solvable in polynomial time on stars (i.e., trees of diameter two) and that the problem can be solved in polynomial time on trees when k is a constant. Moreover, when the number of candidates is a constant, then it is solvable in polynomial time on paths and is solvable in pseudo-polynomial time on trees. The running time of the algorithm on paths is $k^{2^{|\mathcal{C}|}}n^{\mathcal{O}(1)}$, where n is the number of vertices in the input graph and $\mathcal{C}$ is the set of the candidates. Bentert et al. [1] proved GM is NP-hard on paths even if all vertices have unit weights and it is weakly NP-hard on trees even if $|\mathcal{C}| = 2$; and that the problem is polynomial time solvable for trees with diameter three. Prior to these Cohen-Zemach et al. [8] studied GM on graphs. In addition to the papers discussed earlier, there are far too many articles to list on this subject. Some of them are [4,6,7,11,20,25,30,31,34]. Parameterized complexity of manipulation has received extensive attention over the last several years, [2,3,12,18,19] are just a few examples.

2 Preliminaries

For our algorithmic results we define a variant of W-GM that we call TARGET WEIGHTED GERRYMANDERING (TW-GM). The input of TW-GM is an instance of W-GM, and a positive integer $k^\star$. The objective is to test whether the vertex set of the input graph can be partitioned into k districts such that the candidate p wins in $k^\star$ districts alone and no other candidate wins in more than $k^\star - 1$ districts. The following simple lemma implies that to design an efficient algorithm for W-GM it is enough to design an efficient algorithm for TW-GM.

Lemma 1. *If there exists an algorithm that given an instance* $(G, \mathcal{C}, \{w_v\}_{v\in V(G)}, p, k, k^\star)$ *of* TW-GM *and a tie-breaking rule* η, *solves the instance* *in $f(z)$ time, then there exists an algorithm that solves the instance $(G, \mathcal{C}, \{w_v\}_{v \in V(G)}, p, k)$ of* W-GM *in $f(z) \cdot k$ time under the tie-breaking rule η.*

Notations and Basic Terminology. In an undirected graph $G = (V, E)$, uv denotes an edge between the vertices u and v which are called the *endpoints* of uv. For a set $X \subseteq V(G)$, $G[X]$ denotes the graph induced on X. We say that set X is connected if $G[X]$ is a connected graph. In a directed graph $G = (V, A)$, we denote an arc (i.e., directed edge) from u to v by $\langle u, v \rangle$, and say that u is an in-neighbor of v and v is an out-neighbor of u. For $x \in V(G)$, $N^-(x) = \{y \in V(G) \colon \langle y, x \rangle \in A(G)\}$. The in-degree (out-degree) of a vertex x in G is the number of in-neighbors (out-neighbors) of x in G. For background on graph theory we refer to [13].

3 FPT Algorithm for General Graphs

We prove Theorem 1 here. Towards that, we use polynomial algebra that carefully keeps track of the number of districts won by each candidate so that nobody wins (if at all possible) more than p. An intuitive idea was presented in the introduction.

Due to Lemma 1 it is sufficient to prove it for TW-GM.

Before we discuss our algorithm, we introduce some notations. The *characteristic vector* of a set $S \subseteq U$, denoted by $\chi(S)$, is an $|U|$-length vector whose i^{th} bit is 1 if $u_i \in S$, otherwise 0. Two binary strings $S_1, S_2 \in \{0,1\}^n$ are said to be disjoint if for each $i \in \{1, \ldots, n\}$, the ith bit of S_1 and S_2 are different. The *Hamming weight* of a binary string S is denoted by $\mathcal{H}(S)$.

Observation 1. *Let S_1 and S_2 be two binary vectors, and let $S = S_1 + S_2$. If $\mathcal{H}(S) = \mathcal{H}(S_1) + \mathcal{H}(S_2)$, then S_1 and S_2 are disjoint binary vectors.*

Proposition 1 [10]. *Let $S = S_1 \cup S_2$, where S_1 and S_2 are two disjoint subsets of the set $V = \{v_1, \ldots, v_n\}$. Then, $\chi(S) = \chi(S_1) + \chi(S_2)$ and $\mathcal{H}(\chi(S)) = \mathcal{H}(\chi(S_1)) + \mathcal{H}(\chi(S_2)) = |S_1| + |S_2|$.*

A monomial x^i, where i is a binary vector, is said to have Hamming weight h, if i has Hamming weight h. The *Hamming projection* of a polynomial $P(x)$ to h, denoted by $\mathcal{H}_h(P(x))$, is the sum of all the monomials of $P(x)$ which have Hamming weight h. We define the representative polynomial of $P(x)$, denoted by $\mathcal{R}(P(x))$, as the sum of all the monomials that have non-zero coefficient in $P(x)$ but have coefficient 1 in $\mathcal{R}(P(x))$, i.e., it only remembers whether the coefficient is non-zero. We say that $P(x)$ *contains the monomial* x^i if its coefficient in $P(x)$ is non-zero. In the zero polynomial, the coefficient of each monomial is 0.

Algorithm. Let $I = (G, \mathcal{C}, \{w_v\}_{v \in V(G)}, p, k, k^\star)$ be an instance of TW-GM. We assume that $k^\star \geq 1$, otherwise $k = 1$ and it is a trivial instance.

For each candidate c_i in $\mathcal{C}$, we construct a family $\mathcal{F}_i$ that contains all possible districts won by c_i. Due to the application of tie-breaking rule, we may assume that every district has a unique winner. Without loss of generality, let $c_1 = p$, the distinguished candidate. Note that we want to find a family $\mathcal{S}$ of k districts, that contains $k^\star$ elements of the family $\mathcal{F}_1$ and at most $k^\star - 1$ elements from each of the other family $\mathcal{F}_i$, where $i > 1$. The union of these districts gives $V(G)$ and any two districts in $\mathcal{S}$ are pairwise disjoint. To find such k districts, we use the method of polynomial multiplication appropriately using the next proposition. Due to Observation 1 and Proposition 1, we know that subsets S_1 and S_2 are disjoint if and only if the Hamming weight of the monomial $y^{\chi(S_1)+\chi(S_2)}$ is $|S_1| + |S_2|$. Here, degree of a polynomial is the decimal representation of its exponent.

Proposition 2 [27]. *There exists an algorithm that multiplies two polynomials of degree d in $\mathcal{O}(d \log d)$ time.*

For every $i \in \{1, \ldots, m\}$, $\ell \in \{1, \ldots, n\}$, if $\mathcal{F}_i$ has a set of size ℓ, then we construct a polynomial $P_i^\ell(y) = \sum_{\substack{Y \in \mathcal{F}_i \\ |Y|=\ell}} y^{\chi(Y)}$. Next, using polynomials $P_1^\ell(y)$, where $\ell \in \{1, \ldots, n\}$, we will create a sequence of polynomials $Q_{1,j}^s$, where $j \in \{1, \ldots, k^\star - 1\}$, $s \in \{j+1, \ldots, n\}$, in the increasing order of j, such that every monomial in the polynomial $Q_{1,j}^s$ has Hamming weight s. For $j = 1$, we construct $Q_{1,1}^s$ by summing all the polynomials obtained by multiplying $P_1^{s'}$ and $P_1^{s''}$, for all possible values of $s', s'' \in \{1, \ldots, n\}$ such that $s' + s'' = s$, and then by taking the representative polynomial of its Hamming projection to s. If $Q_{1,1}^s$ contains a monomial x^t, then there exists a set $S \subseteq V(G)$ of size s such that $t = \chi(S)$ and S is formed by the union of two districts won by c_1. Next, for $j \in \{2, \ldots, k^\star - 1\}$ and $s \in \{j+1, \ldots, n\}$, we create the polynomial $Q_{1,j}^s$ similarly, using $Q_{1,(j-1)}^{s''}$ in place of $P_1^{s''}$. Formally,

$$Q_{1,1}^s = \mathcal{R}\Big(\mathcal{H}_s\Big(\sum_{\substack{1 \le s', s'' \le s \\ s'+s''=s}} P_1^{s'} \times P_1^{s''}\Big)\Big), \; Q_{1,j}^s = \mathcal{R}\Big(\mathcal{H}_s\Big(\sum_{\substack{1 \le s', s'' \le s \\ s'+s''=s}} P_1^{s'} \times Q_{1,(j-1)}^{s''}\Big)\Big).$$

Thus, if $Q_{1,j}^s$ contains a monomial x^t, then there exists a set $S \subseteq V(G)$ of size s such that $t = \chi(S)$ and S is formed by the union of $j+1$ districts won by c_1. In this manner, we can keep track of the number of districts won by c_1. Next, we will take account of the wins of the other candidates.

Towards this we create a family of polynomials $\mathcal{T} = \{T_{k^\star}, \ldots, T_k\}$ such that the polynomial $T_{k^\star+\ell}$, where $\ell \in \{0, \ldots, k - k^\star\}$, encodes the following information: the existence of a monomial x^t in $T_{k^\star+\ell}$ implies that there is a subset $X \subseteq V(G)$ such that $t = \chi(X)$ and X is the union of $k^\star + \ell$ districts in which c_1 wins in $k^\star$ districts and every other candidate wins in at most $k^\star - 1$ districts. Therefore, it follows that if T_k contains the monomial $y^{\chi(V(G))}$ (the all 1-vector) then our algorithm should return "Yes", otherwise it should return "No". We define $T_{k^\star+\ell}$ recursively, with the base case given by $T_{k^\star} = \sum_{s=k^\star}^{n} Q_{1,(k^\star-1)}^s$. If $T_{k^\star} = 0$, then we return "No". We initialize $T_{k^\star+\ell} = 0$, for each $\ell \in \{1, \ldots, k - k^\star\}$. For each $i \in \{2, \ldots, m\}$, we proceed as follows in the increasing order of i.

- For each $j \in \{1, \ldots, \min\{k^\star - 1, k - k^\star\}\}$
  - For each $\ell \in \{j, \ldots, k - k^\star\}$ and $s \in \{k^\star + 1, \ldots, n\}$
    * Compute the polynomial $Q_\ell^s = \sum_{\substack{1 \le s', s'' \le s \\ s' + s'' = s}} P_i^{s'} \times \mathcal{H}_{s''}(T_{k^\star + \ell - 1})$
    * Compute the Hamming projection of Q_ℓ^s to s, that is, $Q_\ell^s = \mathcal{H}_s(Q_\ell^s)$
  - For each $\ell \in \{j, \ldots, k - k^\star\}$
    * Set $T_{k^\star + \ell} = \mathcal{R}(T_{k^\star + \ell} + \sum_{s=k^\star+1}^{n} Q_\ell^s)$

The range of j is dictated by the fact that since c_1 wins $k^\star$ districts, all other candidates combined can only win $k - k^\star$ districts and each individually may only win at most $k^\star - 1$ districts. Thus, overall candidate c_i, for any $i \geq 2$ can win at most $\min\{k^\star - 1, k - k^\star\}$ districts. The range of ℓ is dictated by the fact that (assuming that first $k^\star$ districts are won by c_1) j^{th} district won by c_i is either $(k^\star + j)^{\text{th}}$ district, or $(k^\star + j + 1)^{\text{th}}$ district, ..., or k^{th} district. The range of s is dictated by the fact that the number of vertices in the union of all the districts is at least $k^\star + 1$ as c_1 wins $k^\star$ districts.

Note that Q_ℓ^s is a non-zero polynomial if there exists a subset of vertices of size s that are formed by the union of $k^\star + \ell$ pairwise disjoint districts, $k^\star$ of which are won by c_1 and every other candidate wins at most $k^\star - 1$. Thus, the recursive definition of $T_{k^\star + \ell}$ is self explanatory. Next, we prove the correctness and running time of the algorithm which conclude the proof of Theorem 1.

Correctness. The following lemma proves the completeness of the algorithm.

Lemma 2. *If* $(G, \mathcal{C}, \{w_v : \mathcal{C} \to \mathbb{Z}^+\}_{v \in V(G)}, p, k, k^\star)$ *is a* Yes*-instance of* TW-GM *under a tie-breaking rule, then the above algorithm returns "*Yes*".*

Proof. Suppose that $V_1, \ldots, V_k$ is a solution to $(G, \mathcal{C}, \{w_v : \mathcal{C} \to \mathbb{Z}^+\}_{v \in V(G)}, p, k, k^\star)$. Recall that we assumed that $p = c_1$. Let $\mathcal{V}_i \subseteq \{V_1, \ldots, V_k\}$ be the set of districts won by the candidate c_i. Due to the application of a tie-breaking rule, $\mathcal{V}_i$s are pairwise disjoint. Without loss of generality, let $\mathcal{V}_1 = \{V_1, \ldots, V_{k^\star}\}$. We begin with the following claim that enables us to conclude that polynomial T_k has monomial $y^{\chi(V(G))}$.

Claim 1 (♣). *For each* $i \in \{1, \ldots, m\}$*, polynomial* $T_{\sum |\mathcal{V}_1| + \ldots + |\mathcal{V}_i|}$ *contains the monomial* $y^{\chi(\cup_{Y \in \mathcal{V}_1 \cup \ldots \cup \mathcal{V}_i} Y)}$.

Hence, we can conclude that the polynomial T_k contains the monomial $y^{\chi(V(G))}$. Hence, the algorithm returns Yes. □

In the next lemma, we prove the soundness of the algorithm.

Lemma 3. *If the above algorithm returns "*Yes*" for an instance* $I = (G, \mathcal{C}, \{w_v : \mathcal{C} \to \mathbb{Z}^+\}_{v \in V(G)}, c_1, k, k^\star)$ *for the tie-breaking rule* η*, then* I *is a* Yes*-instance of* TW-GM *under the tie-breaking rule* η.

Proof. We first prove the following claims.

Claim 2 (♣). *If $T_{k^\star}$ has a monomial y^S, then there are $k^\star$ pairwise disjoint districts $Y_1, \ldots, Y_{k^\star}$ such that $\chi(Y_1 \cup \ldots \cup Y_{k^\star}) = S$ and c_1 wins in every district.*

Claim 3 (♣). *For a pair of integer i, j, where $i \in \{2, \ldots, m\}$ and $j \in \{1, \ldots, \min\{k - k^\star, k^\star - 1\}\}$, let $T_{k^\star+1}, \ldots, T_k$ be the family of polynomials constructed in the algorithm at the end of for loops for i and j, in the above algorithm. Let y^S be a monomial in T_t, where $t \in \{k^\star+1, \ldots, k\}$. Then, the following hold:*

- *there are t pairwise disjoint districts $Y_1, \ldots, Y_t$ such that $\chi(Y_1 \cup \ldots \cup Y_t) = S$*
- *c_1 wins in $k^\star$ districts in $\{Y_1, \ldots, Y_t\}$*
- *c_i wins in j districts in $\{Y_1, \ldots, Y_t\}$*
- *for $2 \leq q < i$, c_q wins in at most $k^\star - 1$ districts in $\{Y_1, \ldots, Y_t\}$*
- *for $q > i$, c_q does not win in any district in $\{Y_1, \ldots, Y_t\}$*

The proof of this claim follows by using nested induction on i and j. If the algorithm returns Yes, then we know that there is a monomial $y^{\chi(V(G))}$ in T_k. Therefore, due to Claim 3, there are k districts such that c_1 wins in $k^\star$ districts and all the candidates win in at most $k^\star - 1$ districts. □

Lemma 4 (♣). *The above algorithm runs in $2^n(n+m)^{\mathcal{O}(1)}$ time.*

4 Deterministic Algorithm for Path

In this section, we discuss the proof of Theorem 2, the full details of each proof is in the Appendix. We note that due to Lemma 1, it is sufficient to present a deterministic FPT algorithm parameterized by k for TW-GM when the input is a path. Let $(G, \mathcal{C}, \{w_v : \mathcal{C} \to \mathbb{Z}^+\}_{v \in V(G)}, p, k, k^\star)$ be the input instance where G is the path $(u_1, \ldots, u_n)$. We begin with a simple observation.

Observation 2 *A path G on n vertices has $\mathcal{O}(n^2)$ distinct connected sets.*

Based on the above observation we create an auxiliary directed graph H with parallel arcs on $\binom{n}{2} + n + 2$ vertices, where we have a vertex for each connected set of G. For $\{i, j\} \subseteq [n]$, $i \leq j$, let $P_{i,j}$ denote the subpath of G starting at the ith vertex and ending at the jth vertex. That is $P_{i,j}$ is the subpath $(u_i, \ldots, u_j)$ of G. Formally, we define the auxiliary graph H as follows.

1. For each $\{i, j\} \subseteq \{1, \ldots, n\}$ such that $i \leq j$, create a vertex $v_{i,j}$ corresponding to the subpath $P_{i,j}$. 2. We do the following for each $\{i, j\} \subseteq \{1, \ldots, n\}$. Let c denote the candidate that wins the district $P_{i,j}$, where $i \leq j$. If $c \neq p$, then we do the following. For each $r \in \{j+1, \ldots, n\}$, we add $k^\star - 1$ arcs $\langle v_{i,j}, v_{j+1,r}, 1\rangle, \langle v_{i,j}, v_{j+1,r}, 2\rangle, \ldots, \langle v_{i,j}, v_{j+1,r}, k^\star - 1\rangle$ from vertex $v_{i,j}$ to $v_{j+1,r}$. We label the $k^\star - 1$ arcs from $v_{i,j}$ to $v_{j+1,r}$ with $\langle c, 1\rangle, \langle c, 2\rangle, \ldots, \langle c, k^\star - 1\rangle$. That is, for each $k' \in \{1, \ldots, k^\star - 1\}$, the arc $\langle v_{i,j}, v_{j+1,r}, k'\rangle$ is labeled with $\langle c, k'\rangle$. If $c = p$, then we do the following. For each $r \in \{j+1, \ldots, n\}$, we add an unlabeled

arc from $v_{i,j}$ to $v_{j+1,r}$. 3. Finally, we add two new vertices s and t. Now we add arcs incident to s. Let $i \in \{1, \ldots, n\}$. We add an unlabeled arc from the vertex s to $v_{1,i}$. Next we add arcs incident to t. Let c denote the candidate that wins in $P_{i,n}$. If $c \neq p$, then we add $k^\star - 1$ arcs $\langle v_{i,n}, t, 1\rangle, \langle v_{i,n}, t, 2\rangle, \ldots, \langle v_{i,n}, t, k^\star - 1\rangle$ from $v_{i,n}$ to t and label them with $\langle c, 1\rangle, \langle c, 2\rangle, \ldots, \langle c, k^\star - 1\rangle$, respectively. If $c = p$, then we add an unlabeled arc from $v_{i,n}$ to t.

Lemma 5 (♣). *There is a path on $k+2$ vertices from s to t in H such that the path has $k - k^\star$ labeled arcs with distinct labels and $k^\star + 1$ unlabeled arcs if and only if $V(G)$ can be partitioned into k districts such that p wins in $k^\star$ districts and any other candidate wins in at most $k^\star - 1$ districts.*

Thus, our problem reduces to finding a path on $k + 2$ vertices from s to t in H such that there are $k^\star + 1$ unlabeled arcs, and $k - k^\star$ distinctly labeled arcs.

Theorem 5. *There is an algorithm that given an instance $\mathcal{I}$ of* TW-GM *and a tie-breaking rule, solves the instance $\mathcal{I}$ in time $2.619^{k-k^\star}|\mathcal{I}|^{\mathcal{O}(1)}$.*

Towards proving Theorem 5, we design a dynamic programming algorithm using the concept of *representative family.* We first define representative family.

Let $\mathcal{S}$ be a family of subsets of a universe U; and let $q \in \mathbb{N}$. A subfamily $\widehat{\mathcal{S}} \subseteq \mathcal{S}$ is said to *q-represent* $\mathcal{S}$ if the following holds. For every set B of size q, if there is a set $A \in \mathcal{S}$ such that $A \cap B = \emptyset$, then there is a set $A' \in \widehat{\mathcal{S}}$ such that $A' \cap B = \emptyset$. If $\widehat{\mathcal{S}}$ q-represents $\mathcal{S}$, then we call $\widehat{\mathcal{S}}$ a *q-representative of* $\mathcal{S}$.

Proposition 3. [21] *Let $\mathcal{S} = \{S_1, \ldots, S_t\}$ be a family of sets of size p over a universe of size n and let $0 < x < 1$. For a given $q \in \mathbb{N}$, a q-representative family $\widehat{\mathcal{S}} \subseteq \mathcal{S}$ for $\mathcal{S}$ with at most $x^{-p}(1-x)^{-q} \cdot 2^{o(p+q)}$ sets can be computed in time $\mathcal{O}((1-x)^{-q} \cdot 2^{o(p+q)} \cdot t \cdot \log n)$.*

We introduce the definition of subset convolution on set families which will be used to capture the idea of "extending" a partial solution, a central concept when using representative family. For two families of sets $\mathcal{A}$ and $\mathcal{B}$, we define $\mathcal{A} * \mathcal{B}$ as $\{A \cup B \colon A \in \mathcal{A}, B \in \mathcal{B}, A \cap B = \emptyset\}$.

Proof (Proof sketch of Theorem 5). An instance of TW-GM is given by $\mathcal{I} = (G, \mathcal{C}, \{w_v\}_{v \in V}, p, k, k^\star)$. Additionally, recall the construction of the labeled digraph H with parallel arcs from $\mathcal{I}$. In order to prove Theorem 5, due to Lemma 5, it is enough to decide whether there exists a path on $k + 2$ vertices from s to t in H that satisfies the following properties: **(PI)** there are $k^\star + 1$ unlabeled arcs, and **(PII)** the remaining $k - k^\star$ arcs have distinct labels.

Before presenting our algorithm, we first define some notations. For $i \in \{1, \ldots, k+1\}$ and $r \in \{1, \ldots, k^\star + 1\}$, a path P starting from s on $i+1$ vertices is said to satisfy $\mathscr{P}(i, r)$ if there are r unlabeled arcs (including the arc from s in P), and the remaining $i - r$ arcs have distinct labels. For a subgraph H' of H, we denote the set of labels in the graph H' by $\mathcal{L}(H')$. Recall that each vertex $v \in V(H) \setminus \{s, t\}$ corresponds to a subpath (i.e., a district) of the path

G. Hence, for each $v \in V(H) \setminus \{s,t\}$, we use $\text{win}(v)$ to denote the (unique) candidate that wins[1] the district denoted by v. Equivalently, we say that the candidate $\text{win}(v)$ *wins* the district v in G. For each vertex $v \in V(H)$, and a pair of integers $i \in \{1,\ldots,k+1\}$, $r \in \{1,\ldots,\min\{i,k^\star+1\}\}$, we define a set family $\mathscr{F}[i,r,v] = \{P : P \textit{ is a } s \textit{ to } v \textit{ path in } H \textit{ on } i{+}1 \textit{ vertices satisfying } \mathscr{P}(i,r)\}$.

The following family contains the arc labels on the path in the family $\mathscr{F}[i,r,v]$. $\mathscr{Q}[i,r,v] = \{\mathcal{L}(P)\colon P \in \mathscr{F}[i,r,v]\}$. Note that for each value of $i \in \{1,\ldots,k+1\}$, r defined above and $v \in V(H)$, each set in $\mathscr{Q}[i,r,v]$ is actually a subset of $\mathcal{L}(H)$ of size $i-r$. If there is a path from s to t on $k+2$ vertices with $k-k^\star$ arcs with distinct labels, then $\mathscr{Q}[k+1,k^\star+1,t] \neq \emptyset$ and vice versa. That is, $\mathscr{Q}[k+1,k^\star+1,t] \neq \emptyset$ if and only if $\mathscr{F}[k+1,k^\star+1,t] \neq \emptyset$. Hence, to solve our problem, it is sufficient to check if $\mathscr{Q}[k+1,k^\star+1,t]$ is non-empty. To decide this, we design a dynamic programming algorithm using representative families over $\mathcal{L}(H)$. In this algorithm, for each value of $i \in \{1,\ldots,k+1\}$, $r \in \{1,\ldots,\min\{i,k^\star+1\}\}$, and $v \in V(H)$, we compute a $(k-k^\star-(i-r))$ representative family of $\mathscr{Q}[i,r,v]$, denoted by $\widehat{\mathscr{Q}}[i,r,v]$, using Proposition 3, where $x = \frac{i-r}{2(k-k^\star)-(i-r)}$. Here, the value of x is set with the goal to optimize the running time of our algorithm, as is the case for the algorithm for k-PATH in [21]. Our algorithm outputs "Yes" if and only if $\widehat{\mathscr{Q}}[k+1,k^\star+1,t] \neq \emptyset$.

Algorithm. We now formally describe how we recursively compute the family $\widehat{\mathscr{Q}}[i,r,v]$, for each $i \in \{1,\ldots,k+1\}$, $r \in \{1,\ldots,\min\{i,k^\star+1\}\}$, and $v \in V(H)$.

Base Case: We set $\widehat{\mathscr{Q}}[1,r,v] = \mathscr{Q}[1,r,v]$

$$= \begin{cases} \{\emptyset\} & \text{if } \langle s,v\rangle \text{ is an arc in } H \text{ and } r=1 \\ \emptyset & \text{otherwise} \end{cases} \tag{1}$$

For each $i \in \{1,\ldots,k+1\}$, $r \in \{1,\ldots,k-k^\star\} \cup \{0\}$, and $v \in V(H)$, we set

$$\widehat{\mathscr{Q}}[i,r,v] = \mathscr{Q}[i,r,v] = \emptyset \text{ if } r=0 \text{ or } r>i. \tag{2}$$

We define (2) so that the recursive definition (3) has a simple description.

Recursive Step: For each $i \in \{2,\ldots,k+1\}$, $r \in \{1,\ldots,\min\{i,k^\star+1\}\}$, and $v \in V(H)$, we compute $\widehat{\mathscr{Q}}[i,r,v]$ as follows. We first compute $\mathscr{Q}'[i,r,v]$ from the previously computed families and then we compute a $(k-k^\star-(i-r))$-representative family $\widehat{\mathscr{Q}}[i,r,v]$ of $\mathscr{Q}'[i,r,v]$. The family $\mathscr{Q}'[i,r,v]$ is computed using the representative family as follows: $\mathscr{Q}'[i,r,v] =$

$$\Big(\bigcup_{\substack{w\in N^-(v),\\ \text{win}(w)=p}} \widehat{\mathscr{Q}}[i-1,r-1,w]\Big) \bigcup \Big(\bigcup_{\substack{w\in N^-(v),\\ \text{win}(w)\neq p}} \widehat{\mathscr{Q}}[i-1,r,w] * \{\{\langle \text{win}(w),j\rangle\} : 1\leq j < k^\star\}\Big) \tag{3}$$

Next, we compute a $(k-k^\star-(i-r))$-representative family $\widehat{\mathscr{Q}}[i,r,v]$ of $\mathscr{Q}'[i,r,v]$ using Proposition 3, where $x = \frac{i-r}{2(k-k^\star)-(i-r)}$. Our algorithm works as

[1] We may assume this by applying the tie-breaking rule.

follows: compute $\widehat{\mathcal{Q}}[i,r,v]$ using Eqs. (1)–(3), and Proposition 3. Output "Yes" if and only if $\widehat{\mathcal{Q}}[k+1,k^\star+1,t] \neq \emptyset$.

Correctness Proof. We prove that for every $i \in \{1,\ldots,k+1\}$, $r \in \{1,\ldots,\min\{i,k^\star+1\}\}$, and $v \in V(H)$, $\widehat{\mathcal{Q}}[i,r,v]$ is indeed a $(k-k^\star-(i-r))$ representative family of $\mathcal{Q}[i,r,v]$, and not just that of $\mathcal{Q}'[i,r,v]$. From the definition of 0-representative family of $\mathcal{Q}[k+1,k^\star+1,t]$, we have that $\mathcal{Q}[k+1,k^\star+1,t] \neq \emptyset$ if and only if $\widehat{\mathcal{Q}}[k+1,k^\star+1,t] \neq \emptyset$. Thus, for correctness we prove the following.

Lemma 6 (♣). *For each $i \in \{1,\ldots,k+1\}$, $r \in \{1,\ldots,\min\{i,k^\star+1\}\}$, and $v \in V(H)$, family $\widehat{\mathcal{Q}}[i,r,v]$ is a $(k-k^\star-(i-r))$-representative of $\mathcal{Q}[i,r,v]$.*

We first prove that the following recurrence for $\mathcal{Q}[i,r,v]$ is correct. $\mathcal{Q}[i,r,v] =$

$$\Bigg(\bigcup_{\substack{w\in N^-(v),\\ \mathrm{win}(w)=p}} \mathcal{Q}[i-1,r-1,w]\Bigg)\bigcup\Bigg(\bigcup_{\substack{w\in N^-(v),\\ \mathrm{win}(w)\neq p}} \mathcal{Q}[i-1,r,w]*\{\{\langle \mathrm{win}(w),j\rangle\}:1\leq j<k^\star\}\Bigg) \quad (4)$$

We claim that Eqs. (1), (2), and (4) correctly compute $\mathcal{Q}[i,r,v]$, for each $i \in \{1,\ldots,k+1\}$, $r \in \{1,\ldots,\min\{i,k^\star+1\}\}$, and $v \in V(H)$. This concludes the proof of Lemma 6 by showing subset containment on both sides. □

5 In Conclusion

We have shown that GM on paths is NP-complete, thereby resolving an open question in [24]. This gives parameterized intractability for parameters such as maximum degree of a vertex in the graph. Furthermore, we have presented FPT algorithms for paths when parameterized by the number of districts. We also give an FPT algorithm running in time $2^n(n+m)^{\mathcal{O}(1)}$ on general graphs.

We conclude with a few directions for further research: (i) Does there exist a $\mathcal{O}(c^n)$ algorithm for W-GM when there are possibly multiple winners in a district?; (ii) Is W-GM on paths FPT parameterized by the number of candidates?; (iii) Is W-GM on trees FPT parameterized by the number of districts?

References

1. Bentert, M., Koana, T., Niedermeier, R.: The complexity of gerrymandering over graphs: paths and trees. arXiv preprint arXiv:2102.08905 (2021)
2. Betzler, N., Guo, J., Niedermeier, R.: Parameterized computational complexity of Dodgson and Young elections. Inf. Comput. **208**(2), 165–177 (2010)
3. Betzler, N., Uhlmann, J.: Parameterized complexity of candidate control in elections and related digraph problems. Theor. Comput. Sci. **410**(52), 5425–5442 (2009)
4. Bevern, R.V., Bredereck, R., Chen, J., Froese, V., Niedermeier, R., Woeginger, G.J.: Network-based vertex dissolution. SIDMA **29**(2), 888–914 (2015)
5. Björklund, A., Husfeldt, T., Kaski, P., Koivisto, M.: Narrow sieves for parameterized paths and packings. J. Comput. Syst. Sci. **87**, 119–139 (2017)

6. Brubach, B., Srinivasan, A., Zhao, S.: Meddling metrics: the effects of measuring and constraining partisan gerrymandering on voter incentives. In: Proceedings of EC 2020, pp. 815–833 (2020)
7. Clough, E.: Talking locally and voting globally: Duverger's law and homogeneous discussion networks. Political Res. Q. **3**(60), 531–540 (2007)
8. Cohen-Zemach, A., Lewenberg, Y., Rosenschein, J.S.: Gerrymandering over graphs. In: Proceedings of AAMAS 2018, pp. 274–282 (2018)
9. Cygan, M., et al.: Parameterized Algorithms. Springer, Cham (2015). https://doi.org/10.1007/978-3-319-21275-3
10. Cygan, M., Pilipczuk, M.: Exact and approximate bandwidth. Theor. Comput. Sci. **411**(40–42), 3701–3713 (2010)
11. Dey, P.: Gerrymandering: a briber's perspective. arXiv:1909.01583 (2019)
12. Dey, P., Misra, N., Narahari, Y.: Parameterized dichotomy of choosing committees based on approval votes in the presence of outliers. Theor. Comput. Sci. **783**, 53–70 (2019)
13. Diestel, R.: Graph Theory. Graduate Texts in Mathematics, vol. 173, 4th edn. Springer, Heidelberg (2012)
14. Downey, R.G., Fellows, M.R.: Fundamentals of Parameterized Complexity. Texts in Computer Science, Springer, London (2013). https://doi.org/10.1007/978-1-4471-5559-1
15. Eiben, E., Fomin, F.V., Panolan, F., Simonov, K.: Manipulating districts to win elections: fine-grained complexity. In: Proceedings of AAAI 2020 (2020)
16. Ellenberg, J.: How computers turned gerrymandering into a science. New York Times, October 2017
17. Erikson, R.S.: Malapportionment, gerrymandering, and party fortunes in congressional elections. Am. Political Sci. Rev. **4**(66), 1234–1245 (1972)
18. Faliszewski, P., Hemaspaandra, E., Hemaspaandra, L.A., Rothe, J.: Copeland voting fully resists constructive control. In: Fleischer, R., Xu, J. (eds.) AAIM 2008. LNCS, vol. 5034, pp. 165–176. Springer, Heidelberg (2008). https://doi.org/10.1007/978-3-540-68880-8_17
19. Faliszewski, P., Hemaspaandra, E., Hemaspaandra, L.A., Rothe, J.: Llull and Copeland voting computationally resist bribery and constructive control. JAIR **35**, 275–341 (2009)
20. Fleiner, B., Nagy, B., Tasnádi, A.: Optimal partisan districting on planar geographies. Cent. Eur. J. Oper. Res. **25**(4), 879–888 (2017)
21. Fomin, F.V., Lokshtanov, D., Panolan, F., Saurabh, S.: Efficient computation of representative families with applications in parameterized and exact algorithms. J. ACM **63**(4), 1–60 (2016)
22. Gupta, S., Jain, P., Panolan, F., Roy, S., Saurabh, S.: Gerrymandering on graphs: computational complexity and parameterized algorithms. arXiv preprint arXiv:2102.09889 (2021)
23. Issacharoff, S.: Gerrymandering and political cartels. Harvard Law Rev. **116**, 593–648 (2002)
24. Ito, T., Kamiyama, N., Kobayashi, Y., Okamoto, Y.: Algorithms for gerrymandering over graphs. In: Proceedings of AAMAS 2019 (2019)
25. Xia, L., Zuckerman, M., Procaccia, A.D., Conitzer, V., Rosenschein, J.S.: Complexity of unweighted coalitional manipulation under some common voting rules. In: Proceedings of IJCAI 2019 (2009)
26. Lewenberg, Y., Lev, O., Rosenschein, J.S.: Divide and conquer: using geographic manipulation to win district-based elections. In: Proceedings of AAMAS 2017 (2017)

27. Moenck, R.T.: Practical fast polynomial multiplication. In: Proceedings of SYMSAC 1976, pp. 136–148 (1976)
28. Monien, B.: How to find long paths efficiently. In: North-Holland Mathematics Studies, vol. 109, pp. 239–254. Elsevier (1985)
29. Neidermeier, R.: Invitation to Fixed-Parameter Algorithms. Springer (2006)
30. Puppe, C., Tasnádi, A.: Optimal redistricting under geographical constraints: why "pack and crack" does not work. Econ. Lett. **105**(1), 93–96 (2009)
31. Talmon, N.: Structured proportional representation. Theor. Comput. Sci. **708**, 58–74 (2018)
32. Williams, R.: Finding paths of length k in $O^{\star}(2^k)$ time. Inf. Process. Lett. **109**(6), 315–318 (2009)
33. Wines, M.: What is gerrymandering? And how does it work? New York Times, June 2019
34. Zuckerman, M., Procaccia, A.D., Rosenschein, J.S.: Algorithms for the coalitional manipulation problem. JAIR **173**(2), 392–412 (2009)

Game Theory on the Blockchain: A Model for Games with Smart Contracts

Mathias Hall-Andersen and Nikolaj I. Schwartzbach(✉)

Department of Computer Science, Aarhus University, Aarhus, Denmark
{ma,nis}@cs.au.dk

Abstract. We propose a model for games in which the players have shared access to a blockchain that allows them to deploy smart contracts to act on their behalf. This changes fundamental game-theoretic assumptions about rationality since a contract can commit a player to act irrationally in specific subgames, making credible otherwise non-credible threats. This is further complicated by considering the interaction between multiple contracts which can reason about each other. This changes the nature of the game in a nontrivial way as choosing which contract to play can itself be considered a move in the game. Our model generalizes known notions of equilibria, with a single contract being equivalent to a Stackelberg equilibrium, and two contracts being equivalent to a reverse Stackelberg equilibrium. We prove a number of bounds on the complexity of computing SPE in such games with smart contracts. We show that computing an SPE is PSPACE-hard in the general case. Specifically, in games with k contracts, we show that computing an SPE is Σ_k^{P}-hard for games of imperfect information. We show that computing an SPE remains PSPACE-hard in games of perfect information if we allow for an unbounded number of contracts. We give an algorithm for computing an SPE in two-contract games of perfect information that runs in time $O(m\ell)$ where m is the size of the game tree and ℓ is the number of terminal nodes. Finally, we conjecture the problem to be NP-complete for three contracts.

1 Introduction

This paper is motivated by the games that arise on permissionless blockchains such as Ethereum [22] that offer "smart contract" functionality: in these permissionless systems, parties can deploy smart contracts without prior authorization by buying the "tokens" required to execute the contract. By smart contracts, we mean arbitrary pieces of code written in a Turing-complete language[1] capable of maintaining state (including funds) and interact with other smart contracts by invoking methods on them. Essentially, smart contracts are objects in the Java sense. Parties can also invoke methods on the smart contracts manually. Note that the state of all smart contracts is public and can be inspected by any party

[1] However the running time of the contracts is limited by the execution environment.

I. Caragiannis and K. A. Hansen (Eds.): SAGT 2021, LNCS 12885, pp. 156–170, 2021.
https://doi.org/10.1007/978-3-030-85947-3_11

at any time. This changes fundamental game-theoretic assumptions about rationality: in particular, it might be rational for a player to deploy a contract that commits them to act irrationally in certain situations to make credible otherwise non-credible threats. This gives rise to very complex games in which parties can commit to strategies, that in turn depend upon other players' committed strategies. Reasoning about such equilibria is important when considering games that are meant to be played on a blockchain, since the players - at least in principle - always have the option of deploying such contracts. In the literature, this is known as a Stackelberg equilibrium where a designated leader commits to a strategy before playing the game. In general, because of first-mover advantage, being able to deploy a contract first is never a disadvantage, since a player can choose to deploy the empty contract that commits them to nothing. It is well-known that it is hard to compute the Stackelberg equilibrium in the general case [12], though much less is known about the complexity when there are several of these contracts in play: when there are two contracts, the first contract can depend on the second contract in what is known as a reverse Stackelberg equilibrium [2,9,21]. This is again strictly advantageous for the leader since they can punish the follower for choosing the wrong strategy. In this paper, we present a model that generalizes (reverse) Stackelberg games, that we believe captures these types of games and which may be of wider interest. In practical terms, we believe that our model is of interest when analyzing distributed systems for "game-theoretic security" in settings where the players naturally have the ability to deploy smart contracts. Potential examples include proof-of-stake blockchains themselves and financial applications that build upon these systems.

Contracts	Players	Information	Strategies	Lower bound	Upper bound
0	2	perfect	pure	P-hard [20]	$O(m)$ [16]
0	2	imperfect	mixed	PPAD-complete [7,6]	
1	2	perfect	pure	P-hard [20]	$O(m\ell)$ [4]
1	2	perfect	mixed	NP-complete [13]	
1	2	imperfect	-	NP-complete [13]	
2	2	perfect	pure	P-hard [20]	$O(m\ell)$ [Theorem 3]
3	3	perfect	pure	Conjectured NP-hard	NP [Theorem 3]
k	$2+k$	imperfect	pure	Σ_k^p-hard [Theorem 2]	?
unbounded	-	perfect	pure	PSPACE-hard [Theorem 4]	?

Fig. 1. An overview of some existing bounds on the complexity of computing an SPE in extensive-form games and where our results fit in. Here, m is the size of the tree, and ℓ is the number of terminal nodes.

Our Results. We propose a game-theoretic model for games in which players have shared access to a blockchain that allows the players to deploy smart contracts to act on their behalf in the games. Allowing a player to deploy a smart contract corresponds to that player making a 'cut' in the tree, inducing a new expanded game of exponential size containing as subgames all possible cuts in the game. We show that many settings from the literature on Stackelberg games can be recovered as special cases of our model, with one contract being equivalent to a Stackelberg equilibrium, and two contracts being equivalent to a reverse Stackelberg equilibrium. We prove bounds on the complexity of computing an SPE in these expanded trees. We prove a lower bound, showing that computing an SPE in games of imperfect information with k contracts is Σ_k^{P}-hard by reduction from the true quantified Boolean formula problem. For $k = 1$, it is easy to see that a contract can be verified in linear time, establishing NP-completeness. In general, we conjecture Σ_k^{P}-completeness for games with k contracts, though this turns out to reduce to whether or not contracts can be described in polynomial space. For games of perfect information with an unbounded number of contracts, we also establish PSPACE-hardness from a generalization of 3-COLORING. We show an upper bound for $k = 2$ and perfect information, namely that computing an SPE in a two-contract game of size m with ℓ terminal nodes (and any number of players) can be computed in time $O(m\ell)$. For $k = 3$, the problem is clearly in NP since we can verify a witness using the algorithm for $k = 2$, and we conjecture the problem to be NP-complete. Finally, we discuss various extensions to the model proposed and leave a number of open questions.

2 Games with Smart Contracts

In this section, we give our model of games with smart contracts. We mostly assume familiarity with game theory and refer to [16] for more details. For simplicity of exposition, we only consider a somewhat restricted class of games, namely finite games in extensive form, and consider only pure strategies in these games. In addition, we will assume games are in *generic form*, meaning the utilities of all players are unique. This has the effect that the resulting subgame perfect equilibrium is unique. Equivalently, we use a tie breaking algorithm to decide among the different subgame perfect equilibria, and slightly perturb the utilities of the players to match the subgame perfect equilibrium chosen by the tie breaker.

Formally, an *extensive-form game* G is a finite tree T. We denote by $L \subseteq T$ the set of leaves in T, i.e. nodes with no children, and let m denote the number of nodes in T. Each leaf ℓ is labeled by a vector $u(\ell) \in \mathbb{R}^n$ that denotes the utility $u_i(\ell)$ obtained by party P_i when terminating in the leaf ℓ. In addition, the game consists of a finite set of n players. We consider a fixed partition of the non-leaves into n sets, one for each player. The game is played by starting at the root, letting the player who owns that node choose a child to recurse into, this is called a move. We proceed in this fashion until we reach a leaf and distribute its utility vector to the players. When there is perfect information, a player always knows exactly which subgame they are playing, though more generally we may consider a partition of the non-leafs into *information sets*, where each player is only told the information set to which their node belongs.

When all information sets are singletons we say the game has perfect information. The players are assumed to be *rational*, that is they choose moves to maximize their utility: we say a strategy for each player (a strategy profile) constitutes a *(Nash) equilibrium* if no unilateral deviation by any party results in higher utility for that party. Knowing the other players are rational, for games of perfect information, at each branch a player can anticipate their utility from each of its moves by recursively determining the moves of the other parties. This process is called *backward induction*, and the resulting strategy profile is a *subgame perfect equilibrium*. A strategy profile is an SPE if it is an equilibrium for every subgame of the game. For games of perfect information, computing the SPE takes linear time in the size of the tree and can be shown to be P-complete [20]. Later, we will show a lower bound, namely that adding a contract to the tree moves this computation up (at least) a level in the polynomial hierarchy. Specifically, we show that computing the SPE in k-contract games is Σ_k^{P}-hard in the general case with imperfect information.

2.1 Smart Contract Moves

We now give our definition of smart contracts in the context of finite games. We add a new type of node to our model of games, a *smart contract move*. Intuitively, whenever a player has a smart contract move, they can deploy a contract that acts on their behalf for the rest of the game. The set of all such contracts is countably infinite, but fortunately, we can simplify the problem by considering equivalence classes of contracts which "do the same thing". Essentially, the only information relevant to other players is whether or not a given action is still possible to play: it is only if the contract dictates that a certain action cannot be played, that we can assume a rational player will not play it. In particular, any contract which does not restrict the moves of a player is equivalent to the player not having a contract. Such a restriction is called a *cut*. A cut $c^{(i)}$ for player P_i is defined to be a union of subtrees whose roots are children of P_i-nodes, such that: (1) every node in $T \setminus c^{(i)}$ has a path going to a leaf; a cut is not allowed to destroy the game by removing all moves for a player, and (2) $c^{(i)}$ respects information sets, that is it 'cuts the same' from each node in the same information set.

In other words, deploying a smart contract corresponds to choosing a cut in the game tree. This means that a smart contract node for player P_i in a game T is essentially syntactic sugar for the *expanded tree* that results by applying the set of all cuts $c^{(i)}$ to T and connecting the resulting games with a new node belonging to P_i at the top. Computing the corresponding equilibrium with smart contracts then corresponds to the SPE in this expanded tree. Note that this tree is uniquely determined. See Fig. 2 for an example. We use the square symbol in figures to denote smart contract moves. When a game contains multiple smart contract moves, we expand the smart contract nodes recursively in a depth-first manner using the transformation described above.

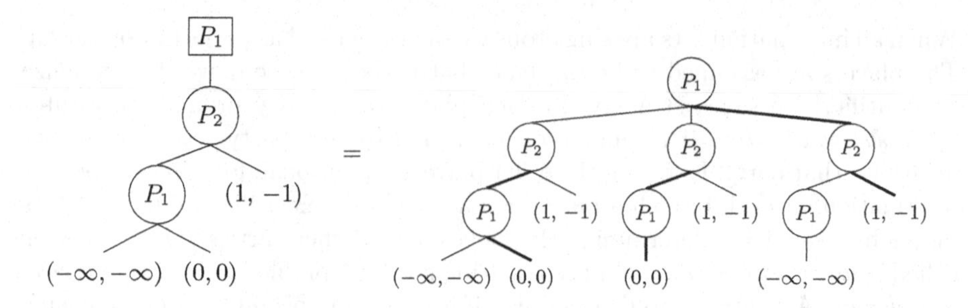

Fig. 2. Expanding a smart contract node for a simple game. The square symbol is a smart contract move for player P_1. We compute all P_1-cuts in the game and connect them with a node belonging to P_1. The first coordinate is the leader payoff, and the second is the follower payoff. The dominating paths are shown in bold. We see that the optimal strategy for P_1 is to commit to choosing $(-\infty, -\infty)$ unless P_2 chooses $(1, -1)$.

2.2 Contracts as Stackelberg Equilibria

As mentioned earlier, the idea to let a party commit to a strategy before playing the game is not a new one: in 1934, von Stackelberg proposed a model for the interaction of two business firms with a designated market leader [18]. The market leader holds a dominant position and is therefore allowed to commit to a strategy first, which is revealed to the follower who subsequently decides a strategy. The resulting equilibrium is called a Stackelberg equilibrium. In this section we show that the Stackelberg equilibrium for a game with leader P_1 and follower P_2 can be recovered as a special case of our model where P_1 has a smart contract. We use the definition of strong Stackelberg equilibria from [5,11]. We note that since the games are assumed to be in generic form, the follower always has a unique response, thus making the requirement that the follower break ties in favor of the leader unnecessary.

Let T be a game tree. A *path* $\mathbf{p} \subseteq T$ is a sequence of nodes such that for each j, $\mathbf{p}_{j+1}$ is a child of $\mathbf{p}_j$. If $\mathbf{p}$ is a path, we denote by $\mathbf{p}^{(i)} \subseteq \mathbf{p}$ the subset of nodes owned by player P_i. Now suppose T has a horizon of h. We let $\mathfrak{p} = (\mathfrak{p}_j)_{j=1}^h \subseteq T$ denote the *dominating path* of the game defined as the path going from the root $\mathfrak{p}_1$ to the terminating leaf $\mathfrak{p}_h$ in the SPE of the game.

Definition 1. *Let $i \in [n]$ be the index of a player, and let $f(s_i)$ be the best response to s_i for players other than P_i. We say $(s_i^*, f(s_i^*))$ is a* Stackelberg equilibrium with leader P_i *if the following properties hold true:*

- Leader optimality. *For every leader strategy s_i, $u_i(s_i^*, f(s_i^*)) \geq u_i(s_i, f(s_i))$.*
- Follower best response. *For every $j \neq i$, and every s_{-i}, $u_j(s_i^*, f(s_i^*)) \geq u_j(s_i^*, s_{-i})$.* ◇

Proposition 1. *The Stackelberg equilibrium with leader P_i is equivalent to P_i having a smart contract move.*

Proof. We show each implication separately:

⇒ SPE in the expanded tree T induces a Stackelberg equilibrium in the corresponding Stackelberg game where P_i commits to all moves in $\mathfrak{p}^{(i)}$. It is not hard to see that the follower best response $f(s_i^*)$ is defined by the SPE of the subgame arising after P_i makes the move $\mathfrak{p}_1$ choosing the contract in T.

⇐ A Stackelberg equilibrium induces a SPE in the expanded tree T with the same utility: let $(s_i^*, f(s_i^*))$ be a Stackelberg equilibrium, observe that s_i^* corresponds to a cut $c^{(i)} \subseteq T$ where P_i cuts away all nodes in T not dictated by s_i^*. By letting the first move $\mathfrak{p}_1$ of P_i correspond to $c^{(i)}$, the best follower response $f(s_i^*)$ is the SPE in the resulting subgame, and hence $u(\mathfrak{p}) = u(s_i^*, f(s_i^*))$. □

Multi-leader/multi-follower Contracts. Several variants of the basic Stackelberg game has been considered in the literature with multiple leaders and/or followers [14,17]. We can model this using smart contracts by forcing some of the contracts to independent of each other: formally, we say a contract is *independent* if it makes the same cut in all subgames corresponding to different contracts. It is not hard to see that multiple leaders can be modelled by adding contracts for each leader, where the contracts are forced to be independent. ◇

Reverse Stackelberg Contracts. The reverse Stackelberg equilibrium is an attempt to generalize the regular Stackelberg equilibrium: here, the leader does not commit to a specific strategy *a priori*, rather they provide the follower with a mapping f from follower actions to best response leader actions, see e.g. [1,19] for a definition in the continuous setting. When the follower plays a strategy s_{-i}, the leader plays $f(s_{-i})$. This is strictly advantageous for the leader since as pointed out in [9], they can punish the follower for choosing the wrong strategy.

In the following, if $\mathbf{p}$ is a path of length ℓ, we denote by $G_s(\mathbf{p})$ the subgame whose root is $\mathbf{p}_\ell$.

Definition 2. *Let i be the index of the leader, and $-i$ the index of the follower. We say $(f(s_{-i}^*), s_{-i}^*)$ is a* reverse Stackelberg equilibrium with leader i *if the following holds for every leader strategy s_i and follower strategy s_{-i}, it holds:*

- Leader best response*:* $u_i(f(s_{-i}^*), s_{-i}^*) \geq u_i(s_i, s_{-i}^*)$.
- Follower optimality*:* $u_{-i}(f(s_{-i}^*), s_{-i}^*) \geq u_{-i}(f(s_{-i}), s_{-i})$. ◇

Proposition 2. *The reverse Stackelberg equilibrium for a two-player game with leader P_i is equivalent to adding two smart contract moves to the game, one for P_i, and another for P_{-i} (in that order).*

Proof. We show each implication separately:

⇒ The SPE in the expanded tree induces a reverse Stackelberg equilibrium: for every possible follower strategy s_{-i}, we define $f(s_{-i})$ as the leader strategy in the SPE in the subgame $G_s(\langle \mathfrak{p}_1, s_{-i} \rangle)$ after the two moves, where we slightly abuse notation to let s_{-i} mean that P_{-i} chooses a cut where their SPE is

s_{-i}. Leader best response follows from the observation that $\mathfrak{p}_1$ corresponds to the optimal set of cuts of P_i moves in response to every possible cut of P_{-i} moves.

$\Leftarrow$ A reverse Stackelberg equilibrium induces an SPE in the expanded tree: let $(f(s^*_{-i}), s^*_{-i})$ be a reverse Stackelberg equilibrium and let f be the strategy of P_i in the reverse Stackelberg game, then P_i has a strategy in the two-contract game with the same utility for both players: namely, P_i's first move is choosing the subgame in which for every second move s_{-i} by P_{-i} they make the cut $f(s_{-i})$. □

3 Computational Complexity

Having defined our model of games with smart contracts, in this section we study the computational complexity of computing equilibria in such games. Note that we can always compute the equilibrium by constructing the expanded tree and performing backward induction in linear time. The problem is that the expanded tree is very large: the expanded tree for a game of size m with a single contract has $2^{O(m)}$ nodes since it contains all possible cuts. For every contract we add, the complexity grows exponentially. This establishes the rather crude upper bound of Σ_k^{EXP} for computing SPE in games with perfect information and k contracts. The question we ask if we can do better than traversing the entire expanded tree.

In terms of feasibility, our results are mostly negative: we show a lower bound that computing an SPE, in general, is infeasible for games with smart contracts. We start by considering the case of imperfect information where information sets allow for a rather straightforward reduction from CircuitSAT to games with one contract, showing NP-completeness for single-contract games of imperfect information. This generalizes naturally to the k true quantified Boolean formula problem (k-TQBF), establishing Σ_k^{P}-hardness for games of imperfect information with k contracts. On the positive side, we consider games of perfect information where we provide an algorithm for games and two contracts that runs in time $O(m\ell)$. However, when we allow for an unbounded number of contracts, we show the problem remains PSPACE-complete by reduction from the generalization of 3-COLORING described in [3]. We conjecture the problem to be NP-complete for three contracts.

3.1 Games with Imperfect Information, NP-Completeness

We start by showing NP-completeness for games of imperfect information by reduction from CircuitSAT. We consider a decision problem version of SPE: namely, whether or not a designated player can obtain a utility greater than the target value.

Reduction. Let C be an instance of CircuitSAT. Note that we can start from any complete basis of Boolean functions, so it suffices to suppose the circuit C consists only of NAND with fanin 2 and fanout 1. We will now construct a game tree for the circuit: we will be using one player to model the assignment of variables, say player 1. The game starts with a contract move for player 1 who can assign values to variables by cutting the bottom of the tree: we construct the game such that player 1 only has moves in the bottom level of the tree. In this way, we ensure that every cut corresponds to assigning truth values to the variables. We adopt the convention that a payoff of 1 for player 1 is *true* ($\top$), while a payoff of 0 for player 1 is *false* ($\bot$). All nodes corresponding to occurrences of the same variable get grouped into the same information set, which enforces the property that all occurrences of the same variable must be assigned the same value (Fig. 3).

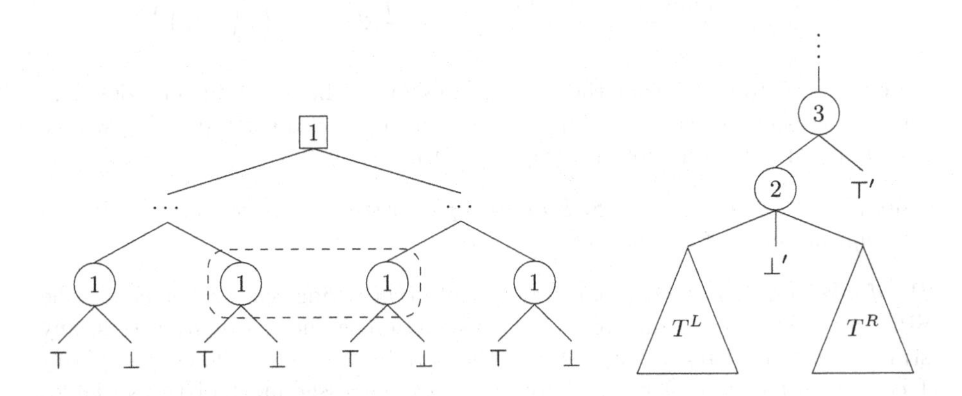

Fig. 3. The basic structure of the reduction. Player 1 has a smart contract that can be used to assign values to the variables. The dashed rectangle denotes an information set and is used when there are multiple occurrences of a variable in the circuit. On the right, we see the NAND-gate gadget connecting the left subgame T^L and the right subgame T^R. We implement the gadget by instantiating the utility vectors such that player 2 chooses $\bot'$ if only if both T^L and T^R propagate a utility vector encoding true.

For the NAND-gate, we proceed using induction: let T^L, T^R be the trees obtained by induction, we now wish to construct a game tree gadget with NAND-gate logic. To do this we require two players which we call player 2 and player 3. Essentially, player 2 does the logic, and player 3 converts the signal to the right format. The game tree will contain multiple different utility vectors encoding true and false, which vary their utilities for players 2 and 3. Each NAND-gate has a left tree and a right tree, each with their own utilities for true and false: $\bot^L, \bot^R; \top^L, \top^R$. The gadget starts with a move for player 3 who can choose to continue the game, or end the game with a true value $\top'$. If they continue the game, player 2 has a choice between false $\bot'$ or playing either T^L or T^R. To make the gadget work like a NAND-gate we need to instantiate the utilities to make backward induction simulate its logic. The idea is to make player 2 prefer both $\bot^L$ and $\bot^R$ to $\bot'$, which they, in turn, prefer to $\top^L$ and $\top^R$. As a result, player 2 propagates $\bot'$ only if both T^L, T^R are true, otherwise, it propagates $\bot^L$ or $\bot^R$. Finally, we must have that player 3 prefers $\top'$ to both $\bot^L$ and $\bot^R$, while they prefer $\bot'$ to $\top'$, $\top^L$ and $\top^R$. This gives rise to a series of inequalities:

$$\bot_2^L > \bot_2' > \top_2^L \qquad \top_3' > \bot_3^L \qquad \bot_3' > \top_3^L \qquad \bot_3' > \top_3'$$
$$\bot_2^R > \bot_2' > \top_2^R \qquad \top_3' > \bot_3^R \qquad \bot_3' > \top_3^R$$

We can instantiate this by defining $\top, \bot$. For the base case corresponding to a leaf, we let $\bot = (0, 1, 0), \top = (1, 0, 0)$. We then define recursively:

$$\top' = \left(1, 0, 1 + \max(\top_3^L, \top_3^R)\right)$$
$$\bot' = \left(0, \frac{\min(\bot_2^L, \bot_2^R) + \max(\top_2^L, \top_2^R)}{2}, 2 + \max\left(\top_3^L, \top_3^R\right)\right)$$

It is not hard to verify that these definitions make the above inequalities hold true. As a result, the gadget will propagate a utility vector corresponding to true if and only if not both subtrees propagate true.

Theorem 1. *Computing an SPE in three-player single-contract games of imperfect information is NP-complete.*

Proof. We consider the decision problem of determining whether or not in the SPE, player 1 has a utility of 1. By construction of the information sets, any strategy is a consistent assignment of the variables. It now follows that player 1 can get a payoff > 0 if and only if there is an assignment of the variables such that the output of the circuit is true. This shows NP-hardness. Now, it is easily seen that this problem is in NP, since a witness is simply a cut that can be verified in linear time in the size of the tree. Completeness now follows using our reduction from CircuitSAT. □

Remark 1. Our reduction also applies to the two-player non-contract case by a reduction from circuit value problem. This can be done in logspace since all the gadgets are local replacements. In doing so, we reestablish the result of [20], showing that computing an SPE on two-player games is P-complete. ◇

3.2 Games with Imperfect Information, PSPACE-Hardness

In this section, we show that computing the SPE in a game with k contract moves is Σ_k^{P}-complete, in the general case with imperfect information. Generalizing the previous result of NP-hardness to k contracts is fairly straightforward. Our claim is that the resulting decision problem is Σ_k^{P}-hard so we obtain a series of hardness results for the polynomial hierarchy. This is similar to the results obtained in [10] where the value problem for a competitive analysis with $k + 1$ players is shown to be hard for Σ_k^{P}.

Formally, we consider the following decision problem with target value V for a game tree T with k contract players: let T' be the expanded tree with contracts for players $P_1, P_2, \dots P_k$ in ascending order. Can player P_1 make a cut in T' such that their payoff is $\geq V$?

To show our claim, we proceed using reduction from the canonical Σ_k^{P}-complete problem k-TQBF, see e.g. [8] for a formal definition.

Theorem 2. *Computing an SPE in* $2+k$ *player games of imperfect information is* Σ_k^{P}*-hard.*

Proof (sketch). We extend our reduction from Theorem 1 naturally to the quantified satisfiability problem. In our previous reduction, the contract player wanted to satisfy the circuit by cutting as to assign values to the variables in the formula. Now, for each quantifier in ψ, we add a new player with a contract, whose moves range over exactly the variables quantified over. The players have contracts in the same order specified by their quantifiers. The idea is that players corresponding to $\forall$ try to sabotage the satisfiability of the circuit, while those corresponding to $\exists$ try to ensure satisfiability. We encode this in the utility vectors by giving $\exists$-players a utility of 1 in $\top$ and 0 utility in $\bot$, while for the $\forall$-players, it is the opposite. It is not hard to see that ψ is true, only if P_1 can make a cut, such that for every cut P_2 makes, there exists a cut for P_3 such that, ..., the utility of P_1 is 1. This establishes our reduction. □

Remark 2. We remark that it is not obvious whether or not the corresponding decision problem is contained within Σ_k^{P}. It is not hard to see we can write a Boolean formula equivalent to the smart contract game in a similar manner as with a single contract. The problem is that it is unclear if the innermost predicate ϕ can be computed in polynomial-time. It is not hard to see that some smart contracts do not have a polynomial description, i.e. we can encode a string $x \in \{0,1\}^*$ of exponential length in the contract. However, there might be an equivalent contract that *does* have a polynomial-time description. By equivalent, we mean one that has the same dominating path. This means that whether or not Σ_k^{P} is also an upper bound essentially boils down to whether or not every contract has an equivalent contract with a polynomial description. ◇

3.3 Games with Perfect Information, Two Contracts, Upper Bound

In this section, we consider two-player games of perfect information and provide a polynomial-time algorithm for computing an SPE in these games. Specifically, for a game tree of size m with ℓ terminal nodes with two contract players (and an arbitrary number of non-contract players), we can compute the equilibrium in time $O(m\ell)$. Our approach is similar to that of [15], in that we compute the inducible region for the first player, defined as the set of leaves they are able to 'induce' by making cuts in the game tree.

Let A, B be two sets. We then define the set of outcomes from A reachable using a threat against player i from outcomes in B as follows:

$$\text{threaten}_i(A, B) = \{x \in A \mid \exists\, y \in B.\, x_i > y_i\}$$

As mentioned, we will compute the *inducible region* for the player with the first contract, defined as the set of outcomes reachable with a contract. Choosing the optimal contract is then reduced to a supremum over this region.

Definition 3. *Let G be a fixed game. We denote by $\mathscr{R}(P_1)$ (resp. $\mathscr{R}(P_1, P_2)$) the* inducible region *of P_1, defined as the set of outcomes reachable by making a cut in G in all nodes owned by P_1. $\mathscr{R}(P_1)$ is a tuple $(\mathbf{u}, c_1)$ where $\mathbf{u} \in \mathbb{R}^n$ is the utility vector, and c_1 is the contract (a cut) of P_i.* ◇

Algorithm. Let G be the game tree in question and let k be a fixed integer. As mentioned, we assume without loss of generality that G is in *generic form*, meaning all non-leafs in G have out-degree exactly two and that all utilities for a given player are distinct such that the ordering of utilities is unique. We denote by P_1, P_2 the players with contracts and assume that P_i has the ith contract. We will compute the inducible regions in G for P_1 (denoted S for *self*), and for (P_1, P_2) (denoted T for *together*) by a single recursive pass of the tree. In the base case with a single leaf with label $\mathbf{u}$ we have $S = T = \{\mathbf{u}\}$. For a non-leaf, we can recurse into left and right child, and join together the results. The procedure is detailed in Algorithm 1.

Algorithm 1: `InducibleRegion`(G)

```
switch G do
    case Leaf(u):
        return ({u}, {u})

    case Node(G^L, G^R, i):
        (S^L, T^L) ← InducibleRegion(G^L)
        (S^R, T^R) ← InducibleRegion(G^R)
        if i = 1 then
            T ← T^L ∪ T^R
            S ← S^L ∪ S^R ∪ threaten_2(T^L ∪ T^R, S^L ∪ S^R)
        else if i = 2 then
            T ← T^L ∪ T^R
            S ← threaten_2(T^L, S^R) ∪ threaten_2(T^R, S^L)
        else
            T ← threaten_i(T^L, T^R) ∪ threaten_i(T^R, T^L)
            S' ← threaten_i(S^L, S^R) ∪ threaten_i(S^R, S^L)
            S ← S' ∪ threaten_2(T, S')
        end
        return (S, T)
end
```

Theorem 3. *An SPE in two-contract games of perfect information can be computed in time $O(m\ell)$.*

Proof. First, the runtime is clearly $O(m\ell)$ since the recursion has $O(m)$ steps where we need to maintain two sets of size at most ℓ. For correctness, we show something stronger: let $\mathscr{R}(P_1)$ be the inducible region for P_1 in the expanded tree and $\mathscr{R}(P_1, P_2)$ be the inducible region of (P_1, P_2). Now, let $(S, T) = \texttt{InducibleRegion}(G)$. Then we show that $S = \mathscr{R}(P_1)$ and $T = \mathscr{R}(P_1, P_2)$. This implies that $\text{argmax}_{u \in S}\, u_1$ is the SPE. The proof is by induction on the height h of the tree. As mentioned, we assume that games are in *generic form*. This base case is trivial so we consider only the inductive step.

Necessity follows using simple constructive arguments: for S and $i = 1$, then for every $(\mathbf{u}, c) \in S^{\ell}$, we can form contract where P_1 chooses left branch and plays c. And symmetrically for S^R. Similarly, for every $(\mathbf{u}, c_1, c_2) \in T^L$ and $(\mathbf{v}, c') \in S^L$ can form contract where P_1 plays c_1 in all subgames where P_2 plays c_2; and plays c' otherwise. Then $\mathbf{u}$ is dominating if and only if $\mathbf{u}_2 > \mathbf{v}_2$. Similar arguments hold for the remaining cases.

For sufficiency, we only show the case of $i = 1$ as the other cases are similar. Assume (for contradiction) that there exists $(\mathbf{u}, c_1) \in \mathscr{R}(P_1) \setminus S$, i.e. there is a P_1-cut c_1 such that $\mathbf{u}$ is dominating. Then,

$$\begin{aligned}(\mathbf{u}, c_1) &\in (T^L \cup T^R) \setminus (S^L \cup S^R \cup \text{threaten}_2(T^L \cup T^R, S^L \cup S^R)) \\ &= \{\mathbf{v} \in (T^L \cup T^R) \setminus (S^L \cup S^R) \mid \forall \mathbf{v}' \in S^L \cup S^R. \mathbf{v}_2 < \mathbf{v}'_2\}\end{aligned}$$

That is, $\mathbf{u}$ must be a utility vector that P_1 and P_2 can only reach in cooperation in a one of the two sub-games, say by P_2 playing c_2. However, for every cut that P_1 makes, the dominating path has utility for P_2 that is $> \mathbf{u}_2$, meaning P_2 strictly benefits by not playing c_2. But this is a contradiction since we assumed $\mathbf{u}$ was dominating. □

3.4 Games with Perfect Information, Unbounded Contracts, PSPACE-Hardness

We now show that computing an SPE remains PSPACE-complete when considering games with an arbitrary number of contract players. We start by showing NP-hardness and generalize to PSPACE-hardness in a similar manner as we did for Theorem 2. The reduction is from 3-COLORING: let (V, E) be an instance of 3-COLORING and assume the colors are $\{R, G, B\}$. The intuition behind the NP-reduction is to designate a coloring player P_{color}, who picks colors for each vertex $u \in V$ by restricting his decision space in a corresponding move using a contract. They are the first player with a contract. This is constructed using a small stump for every edge $e \in E$ with three leaves R_u, G_u, B_u. We also have another player P_{check} whose purpose is to ensure no two adjacent nodes are colored the same. We attach all stumps to a node owned by P_{check} such that P_{check} can choose among the colors chosen by P_{color}. If P_{color} are able to assign colors such that no adjacent nodes share a color, then P_{color} maximizes their utility, however, if no such coloring exists then P_{check} can force a bad outcome for P_{color}. It follows that P_{color} can obtain good utility if and only there is a valid coloring (Fig. 4).

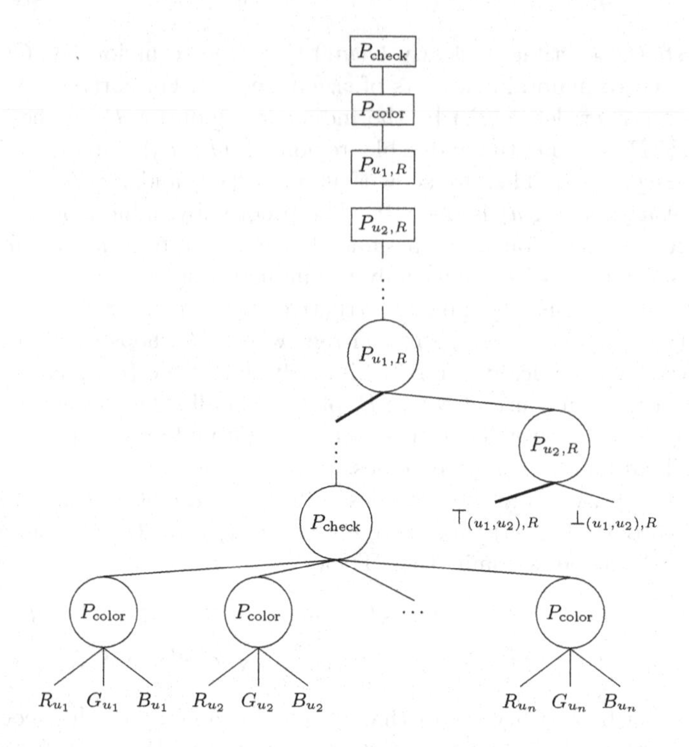

Fig. 4. The structure of the reduction. First, P_{color} is allowed to assign a coloring of all vertices. If there is no 3-COLORING of the graph, there must be some vertex (u_1, u_2) where both vertices are colored the same color c. In this case, P_{check} can force both c_{u_1}, c_{u_2}, which are undesirable to $P_{u_1,c}$, resp. $P_{u_2,c}$: then in every $P_{u_1,c}$-contract where they do not commit to choosing $P_{u_2,c}$, P_{check} cuts as to ensure c_{u_1} and analogously for P_2. It follows that P_{check} can get $\bot$ if and only if the graph is not 3-colored. Then P_{color} can get a different outcome from $\bot$ if and only if they can 3-color the graph.

Reduction. We add six contract players for every edge in the graph. Specifically, for every edge $(u, v) \in E$ and every color $c \in \{R, G, B\}$, we introduce two new contract players $P_{u,c}$ and $P_{v,c}$ who prefer any outcome except c_u (resp. c_v) being colored c. That is, if $c = R$, then the leaf R_u has a poor utility for $P_{u,R}$. We add moves for $P_{u,c}$ and $P_{v,c}$ at the top of the tree, such that if they cooperate, they can get a special utility vector $\bot_{u,v}$ which has a poor utility for P_{color} and great utility for P_{check}, though they themselves prefer any outcome in the tree (except c_u, resp. c_v) to $\bot_{u,v}$. We ensure that P_{check} has a contract directly below P_{color} in the tree. If no coloring exists, then P_{check} can force a bad outcome for both $P_{u,c}, P_{v,c}$ in all contracts where they do not commit to choosing $\bot_{u,v}$. Specifically, P_{check} first threatens $P_{u,c}$ with the outcome c_u, and subsequently threatens $P_{v,c}$ with c_v. Though they prefer any other node in the tree to $\bot_{u,v}$, they still prefer $\bot_{u,v}$ to c_u, c_v, meaning they will comply with the threat. This means P_{color} will receive a poor outcome if the coloring is inconsistent. It follows that P_{color} will only receive a good payoff if they are able to 3-color the graph, see e.g. Sect. 3.4 for an illustration.

Theorem 4. *Computing an SPE in smart contract games of perfect information is PSPACE-hard when we allow for an unbounded number of contract players.*

Proof. Let (V, E) be an instance of 3-COLORING. Our above reduction works immediately for $k = 1$, showing NP-hardness. To show PSPACE-hardness we reduce from a variant of 3-COLORING as described in [3] where players alternately color an edge and use a similar trick as Theorem 2 by introducing new players between P_{color} and P_{check}. □

It remains unclear where the exact cutoff point is, though we conjecture it to be for three contracts: clearly, the decision problem for three-contract games of perfect information is contained in NP as the witness (a cut for the first contract player) can be verified by Algorithm 1.

Conjecture 1. Computing an SPE for three-contract games is NP-complete. ◇

4 Conclusion

In this paper, we proposed a game-theoretic model for games in which players have shared access to a blockchain that allows them to deploy smart contracts. We showed that our model generalizes known notions of equilibria, with a single contract being equivalent to a Stackelberg equilibrium and two contracts equivalent to a reverse Stackelberg equilibrium. We proved a number of bounds on the complexity of computing an SPE in these games with smart contracts, showing, in general, it is infeasible to compute the optimal contract.

References

1. Averboukh, Y.: Inverse stackelberg solutions for games with many followers. Mathematics **6** (2014). https://doi.org/10.3390/math6090151
2. Basar, T., Selbuz, H.: Closed-loop stackelberg strategies with applications in the optimal control of multilevel systems. IEEE Trans. Autom. Control **AC-24**, 166–179 (1979)
3. Bodlaender, H.L.: On the complexity of some coloring games. In: Möhring, R.H. (ed.) WG 1990. LNCS, vol. 484, pp. 30–40. Springer, Heidelberg (1991). https://doi.org/10.1007/3-540-53832-1_29
4. Bošanský, B., Brânzei, S., Hansen, K.A., Lund, T.B., Miltersen, P.B.: Computation of stackelberg equilibria of finite sequential games. ACM Trans. Econ. Comput. **5**(4) (2017). https://doi.org/10.1145/3133242
5. Breton, M., Alj, A., Haurie, A.: Sequential stackelberg equilibria in two-person games. J. Optim. Theory Appl. **59**(1), 71–97 (1988). https://doi.org/10.1007/BF00939867
6. Chen, X., Deng, X.: Settling the complexity of two-player Nash equilibrium. In: 2006 47th Annual IEEE Symposium on Foundations of Computer Science (FOCS 2006), pp. 261–272 (2006). https://doi.org/10.1109/FOCS.2006.69

7. Daskalakis, C., Goldberg, P., Papadimitriou, C.: The complexity of computing a Nash equilibrium. SIAM J. Comput. **39**, 195–259 (2009). https://doi.org/10.1137/070699652
8. Garey, M.R., Johnson, D.S.: Computers and Intractability; A Guide to the Theory of NP-Completeness. W. H. Freeman & Co. (1990)
9. Ho, Y., Olsder, G.: Aspects of the stackelberg problem – incentive, bluff, and hierarchy1. IFAC Proc. Vol. **14**(2), 1359–1363 (1981). 8th IFAC World Congress on Control Science and Technology for the Progress of Society, Kyoto, Japan, 24–28 August 1981
10. Jeroslow, R.G.: The polynomial hierarchy and a simple model for competitive analysis. Math. Program. **32**(2), 146–164 (1985). https://doi.org/10.1007/BF01586088
11. Leitmann, G.: On generalized stackelberg strategies. J. Optim. Theory Appl. **26**(4), 637–643 (1978). https://doi.org/10.1007/BF00933155
12. Letchford, J.: Computational aspects of stackelberg games. Ph.D. thesis, Duke University, Durham, NC, USA (2013)
13. Letchford, J., Conitzer, V.: Computing optimal strategies to commit to in extensive-form games. In: Proceedings of the 11th ACM Conference on Electronic Commerce, EC 2010, pp. 83–92. Association for Computing Machinery, New York (2010). https://doi.org/10.1145/1807342.1807354
14. Liu, B.: Stackelberg-Nash equilibrium for multilevel programming with multiple followers using genetic algorithms. Comput. Math. Appl. **36**(7), 79–89 (1998)
15. Luh, P.B., Chang, S.C., Chang, T.S.: Brief paper: solutions and properties of multi-stage stackelberg games. Automatica 251–256 (1984)
16. Osborne, M.J., Rubinstein, A.: A Course in Game Theory. The MIT Press (1994). Electronic edition
17. Sherali, H.D.: A multiple leader stackelberg model and analysis. Oper. Res. **32**(2), 390–404 (1984)
18. von Stackelberg, H.: Marktform und Gleichgewicht. Verlag von Julius Springer (1934)
19. Stankova, K.: On Stackelberg and Inverse Stackelberg Games & Their Applications in the Optimal Toll Design Problem, the Energy Market Liberalization Problem, and in the Theory of Incentives. Post-Print hal-00391650, HAL, February 2009
20. Szymanik, J.: Backward induction is PTIME-complete. In: Grossi, D., Roy, O., Huang, H. (eds.) LORI 2013. LNCS, vol. 8196, pp. 352–356. Springer, Heidelberg (2013). https://doi.org/10.1007/978-3-642-40948-6_32
21. Tolwinski, B.: Closed-loop stackelberg solution to a multistage linear-quadratic game. J. Optim. Theory Appl. **34**, 484–501 (1981)
22. Wood, G.: Ethereum: a secure decentralised generalised transaction ledger. Ethereum Project Yellow Paper **151**, 1–32 (2014)

On the Complexity of Nucleolus Computation for Bipartite b-Matching Games

Jochen Könemann, Justin Toth, and Felix Zhou(✉)

University of Waterloo, Waterloo, ON N2L 1A2, Canada
{jochen,wjtoth,cfzhou}@uwaterloo.ca

Abstract. We explore the complexity of nucleolus computation in b-matching games on bipartite graphs. We show that computing the nucleolus of a simple b-matching game is $\mathcal{NP}$-hard when $b \equiv 3$ even on bipartite graphs of maximum degree 7. We complement this with partial positive results in the special case where b values are bounded by 2. In particular, we describe an efficient algorithm when a constant number of vertices satisfy $b_v = 2$ as well as an efficient algorithm for computing the non-simple b-matching nucleolus when $b \equiv 2$.

Keywords: Nucleolus · b-Matching · Cooperative game theory · Computational complexity

1 Introduction

Consider a network of companies such that any pair with a pre-existing business relationship can enter into a deal that generates revenue, and at any given time every company has the capacity to fulfill a limited number of deals. This is an example of a scenario that can be modeled as a cooperative b-matching game.

A *cooperative game* is a pair (N, ν) where N is a finite set of *players* and $\nu : 2^N \to \mathbb{R}$ is a *value function* which maps subsets of players, known as *coalitions* to a total value that their cooperation would generate. In the special case of *simple cooperative b-matching games*, we are given an underlying graph $G = (N, E)$, vertex values $b : N \to \mathbb{Z}_+$, and edge weights $w : E \to \mathbb{R}$. The set of *players* in the game corresponds to the vertices N, and $w(uv)$ denotes the value earned when $u, v \in N$ collaborate. For a coalition $S \subseteq N$, $\nu(S)$ corresponds to the maximum weight of a b-matching in $G[S]$ using each edge at most once. More formally, $\nu(S)$ is the optimal value of $w(M)$ where $M \subseteq E[S]$ is subject to $|M \cap \delta(v)| \leq b_v$ for each $v \in S$. On the other hand, in a *non-simple cooperative b-matching game*, $\nu(S)$ is modified to allow M to be a multiset but we still require the underlying set to be a subset of $E[S]$.

We acknowledge the support of the Natural Sciences and Engineering Research Council of Canada (NSERC). Cette recherche a été financée par le Conseil de recherches en sciences naturelles et en génie du Canada (CRSNG).

I. Caragiannis and K. A. Hansen (Eds.): SAGT 2021, LNCS 12885, pp. 171–185, 2021.
https://doi.org/10.1007/978-3-030-85947-3_12

A central question in cooperative game theory is to study how the total revenue generated through the cooperation of the players is shared amongst the players themselves. An *allocation* $x \in \mathbb{R}^N$ is a vector whose entries indicate the value each player should receive. Not all allocations are equally desirable. Cooperative game theory gives us the language to model desirable allocations which capture notions such as fairness and stability.

An allocation $x \in \mathbb{R}^N$ is called *efficient* if its entries sum to $\nu(N)$; i.e., if $\sum_{i \in N} x_i := x(N) = \nu(N)$. Efficiency stipulates that an allocation should distribute the total value generated by the *grand coalition* N. We say x is an *imputation* if it is efficient and satisfies *individual rationality*: $x(i) \geq \nu(\{i\})$ for all i in N. Individual rationality captures the notion that each player should be assigned at least the value they can earn on their own. In a b-matching game, $\nu(\{i\}) = 0$ and individual rationality simplifies to non-negativity.

The natural extension of individual rationality would be coalitional rationality, i.e. stipulating that for any coalition S, $x(S) \geq \nu(S)$. Allocations which satisfy such a property are said to lie in the *core* of the game. Core allocations can be considered highly stable in the sense that no subset of players can earn more value by deviating from the grand coalition.

The core is well-known to be non-empty in *bipartite* b-matching games [10], but may be empty in general matching games. It is in fact known that the core of a matching game instance is non-empty if and only if the *fractional matching linear program* is integral [10]. For example, the core of the matching instance given by an odd-cycle with unit weights is empty.

Since the core may be empty, we need a more robust solution concept. Given an allocation, we let $e(S, x) := x(S) - \nu(S)$ be the *excess* of coalition $S \subseteq N$ with respect to x. Informally, the excess measures the *satisfaction* of coalition S: the higher the excess of S, the more satisfied its players will be. We can rephrase the core as the set of all imputations where all coalitions have non-negative excess.

Instead of requiring all excesses to be non-negative, we can maximize the excess of the worst off coalitions. Consider the following linear program

$$
\begin{aligned}
& \max \epsilon_1 && (P_1) \\
\text{s.t. } & x(N) = N && \\
& x(S) \geq \nu(S) + \epsilon && \forall S \subset N \\
& x(i) \geq \nu(\{i\}) && \forall i \in N
\end{aligned}
$$

and let ϵ^* be its optimal solution. The *least core* is the set of allocations x such that (x, ϵ^*) is optimal for (P_1). The least core is always non-empty.

For b-matching games when the core is non-empty, the least core coincides with the core. When the core is empty, the least core tries to maximize the satisfaction of the coalitions who are worst off in the game. The least core, and by extension the core, both suffer from the fact that they are not in general unique. Furthermore, the least core does nothing to improve the satisfaction of coalitions which are not the worst off. This motivates the definition of the nucleolus, first introduced by Schmeidler [34].

For an allocation x, we write $\theta(x) \in \mathbb{R}^{2^N-2}$ as the vector whose entries are $e(S, x)$ for all $\varnothing \neq S \subsetneq N$ sorted in non-decreasing order. This is a listing of the satisfactions of coalitions from worst off to best off. The *nucleolus* is defined as the allocation which lexicographically maximizes $\theta(x)$ over the imputation set. In a sense, the nucleolus is the most stable allocation. In Schmeidler's paper introducing the nucleolus, the author proved, among other things, that it is unique.

We now have sufficient terminology to state our main result, proven in Sect. 2.

Theorem 1. *The problem of deciding whether an allocation is equal to the nucleolus of an unweighted bipartite 3-matching game is $\mathcal{NP}$-hard, even in graphs of maximum degree 7.*

In the interest of space, we refer the interested reader to the full paper [23] for any omitted proofs.

Kern and Paulusma posed the question of computing the nucleolus for general matching games as an open problem in 2003 [21]. In 2008, Deng and Fang conjectured this problem to be $\mathcal{NP}$-hard [8]. This problem has been reaffirmed to be of interest in 2018 [4]. In 2020, Könemann, Pashkovich, and Toth proved the nucleolus of weighted matching games to be polynomial time computable [24].

On one hand, computing the nucleolus of unweighted b-matching games when $b \geq 3$ is known to be $\mathcal{NP}$-hard for general graphs [3]. However, the gadget graph in their hardness proof has many odd cycles. On the other hand, Bateni et al. provided an efficient algorithm to compute the nucleolus in bipartite b-matching games when one side of the bipartition is restricted to $b_v = 1$ and the other side is unrestricted [2]. Thus it is a natural question whether the nucleolus of bipartite b-matching games is polynomial-time computable. Theorem 1 answers this question in the negative.

The basis of this result is a hardness proof for *core* separation in unweighted bipartite 3-matching games [4]. However, extending this to a hardness proof of nucleolus computation requires significant technical innovation. Towards this end, we introduce a new problem in Sect. 2, a variant of the cubic subgraph problem which is $\mathcal{NP}$-hard. Then, in Sect. 2.1, we reduce the decision variant of nucleolus computation to our new problem, which yields the result.

In Sect. 3, we complement Theorem 1 with efficient algorithms to compute the nucleolus in two relevant cases when $b \leq 2$. Section 3.1 explores the scenario when only a constant number of vertices satisfy $b_v = 2$ and Sect. 3.2 delves into the case when we relax the constraints to allow for non-simple b-matchings.

Theorem 2. *Let G be a bipartite graph with bipartition $N = A\dot{\cup}B$ and $k \geq 0$ a universal constant independent of $|N|$. Let $b \leq 2$ be some node-incidence capacity.*

(i) Suppose $b_v = 2$ for all $v \in A$ but $b_v = 2$ for at most k vertices of B, then the nucleolus of the simple b-matching game on G can be computed in polynomial time.
(ii) If $b \equiv 2$, then the nucleolus of the non-simple b-matching game on G can be computed in polynomial time.

1.1 Related Work

The *assignment game*, introduced by Shapley and Shubik [35], is the special case of simple b-matching games where b is the all ones vector and the underlying graph is bipartite. This was generalized to *matching games* for general graphs by Deng, Ibaraki, and Nagamochi [10]. Solymosi and Raghavan [36] showed how to compute the nucleolus in an unweighted assignment game. Kern and Paulusma [21] later gave a nucleolus computation algorithm for all unweighted matching games. Paulusma [31] extended this result to all node-weighted matching games. An application of assignment games is towards cooperative procurement from the field of supply chain management [11].

The nucleolus is surprisingly ancient, appearing as far back in history as a scheme for bankruptcy division in the Babylonian Talmud [1]. Modern research interest in the nucleolus is not only based on its widespread application [5,28], but also the complexity of computing the nucleolus, which seems to straddle the boundary between $\mathcal{P}$ and $\mathcal{NP}$.

In a similar fashion to how we will define b-matching games, a wide variety of combinatorial optimization games can be defined [10]. Here the underlying structure of the game is based on the optimal solution to some underlying combinatorial optimization problem. One might conjecture that the complexity of nucleolus computation for a combinatorial optimization game lies in the same class as its underlying combinatorial optimization problem. However, this is not in general true. For instance, nucleolus computation is known to be $\mathcal{NP}$-hard for network flow games [9], weighted threshold games [13], and spanning tree games [16,18]. On the other hand, polynomial time algorithms are known for computing the nucleolus in special cases of network flow games [9,33], directed acyclic graph games [37,39], spanning tree games [20,27], b-matching games [2], fractional matching games [17], weighted voting games [14], convex games [17], and dynamic programming games [25].

One possible application of cooperative matching games is to network bargaining [12,40]. In this setting, a population of players are connected through an underlying social network. Each player engages in a profitable partnership with at most one of its neighbours and the profit must be shared between the participating players in some equitable fashion. Cook and Yamagishi [6] proposed a profit-sharing model that generalizes Nash's famous 2-player bargaining solution [30] as well as validates empirical findings from the lab setting.

Both the pre-kernel and least-core are solution concepts which contain the nucleolus. It is well-known that the pre-kernel of a cooperative game may be non-convex and even disconnected [26,38]. Nonetheless, Faigle, Kern, and Kuipers showed how to compute a point in the intersection of the pre-kernel and least-core in polynomial time given a polynomial time oracle to compute the minimum excess coalition for a given allocation [17]. The authors later refined their result to compute a point within the intersection of the core and lexicographic kernel [15], a set which also contains the nucleolus.

The complexity of computing the nucleolus of b-matching games remained open for bipartite graphs, and for b-matching games where $b \leq 2$. In Theorem 1, we show that the former is indeed $\mathcal{NP}$-hard to compute and give an efficient algorithm for a special case of the latter in Sect. 3.

1.2 The Kopelowitz Scheme

A more computational definition of the nucleolus is provided by the Kopelowitz Scheme [26]. Let $\mathrm{OPT}(P)$ denote the set of optimal solutions to the LP (P). Recall the linear program (P_1) and let ϵ_1 be its optimal value. Write $\mathcal{S} := 2^N \setminus \{\varnothing, N\}$ to denote the set of all non-trivial coalitions. Let us write $\mathcal{S}_1 \subseteq \mathcal{S}$ as the set of coalitions satisfying $x(S) = \nu(S) + \epsilon_1$ for every $(x, \epsilon_1) \in \mathrm{OPT}(P_1)$. We say $\mathcal{S}_1$ are the coalitions which are *fixed* in (P_1).

For $\ell \geq 2$, let (P_ℓ) be the linear program

$$\max \epsilon_\ell \qquad (P_\ell) \tag{1}$$

$$x(N) = \nu(N) \tag{2}$$

$$x(S) = \nu(S) + \epsilon_i \qquad \forall i \leq \ell - 1, \forall S \in \mathcal{S}_i \tag{3}$$

$$x(S) \geq \nu(N) + \epsilon \qquad \forall S \in \mathcal{S} \setminus \left(\bigcup_{i=1}^{\ell-1} \mathcal{S}_i \right) \tag{4}$$

Recursively, we set

$$\mathcal{S}_\ell := \{S \in \mathcal{S} : \forall i \leq \ell - 1, S \notin \mathcal{S}_i \wedge \forall (x, \epsilon_\ell) \in \mathrm{OPT}(P_\ell), x(S) = \nu(S) + \epsilon_\ell\}.$$

These are the coalitions which are fixed in (P_ℓ) but not in any (P_i) for $i \leq \ell - 1$. This hierarchy of linear programs terminates when the dimension of the feasible region becomes 0, at which point the unique feasible solution is the nucleolus [7].

Directly solving each (P_ℓ) requires solving a linear program with an exponential number of constraints in terms of the number of players and hence takes exponential time with respect to the input[1]. Moreover, the best general bound on the number of linear programs we must solve until we obtain a unique solution is the naive exponential bound $O(2^{|N|})$. However, we are still able to use the Kopelowitz Scheme as a way to characterize the nucleolus in the proof of Theorem 1.

One way of solving exponentially sized linear programs is to utilize the polynomial time equivalence of optimization and separation [22]. That is, to develop a separation oracle and employ the ellipsoid method. For our positive results, we will take this route.

Indeed, we will develop a polynomial-size formulation of each (P_ℓ) by pruning unnecessary constraints. Not only does this enable us to solve each (P_ℓ) in polynomial time, but we also reduce the number of iterations to a polynomial of

[1] Cooperative games we are interested in have a compact representation roughly on the order of the number of players. For example b-matching games can be specified by a graph, b-values and edge weights rather than explicitly writing out ν.

the input size since at least one inequality constraint is changed to an equality constraint per iteration.

It is of interest to consider a variation of the Kopelowitz Scheme by Maschler [29]. In this variation, the author defines $\mathcal{S}_\ell$ as

$$\mathcal{S}_\ell := \{S \in \mathcal{S} : \forall i \leq \ell - 1, S \notin \mathcal{S}_i \wedge \exists c_S \in \mathbb{R}, \forall (x, \epsilon_\ell) \in \mathrm{OPT}(P_\ell), x(S) = c_S\}.$$

This way, at least 1 equality constraint is added to $(P_{\ell+1})$ which is linearly independent of all equality constraints in (P_ℓ). Hence the feasible region decreases by at least 1 dimension per iteration and there are at most $|N|$ iterations before termination.

2 Hardness

We consider b-matching games for $b \equiv 3$ and uniform weights. The goal of the this section is to prove Theorem 1.

The idea of the proof is inspired the hardness proof of core separation employed in [4] and the hardness proof in [9]. We reduce the problem into a variation of Cubic Subgraph which is $\mathcal{NP}$-hard [32] through a careful analysis of several iterations of the Kopelowitz Scheme. However, it is not clear that our variation of Cubic Subgraph is $\mathcal{NP}$-hard and we significantly extend the techniques from [32] to prove its hardness.

*Problem 1 (*Two from Cubic Subgraph*).* Let G be an arbitrary graph. Decide if G contains a subgraph H with vertices $u \neq v \in V(H)$ satisfying $\deg_H(w) = 2$ if $w = u, v$ and $\deg_H(w) = 3$ otherwise for all $w \in V(H)$. We say that H is a Two from Cubic Subgraph.

Theorem 3. Two from Cubic Subgraph *is $\mathcal{NP}$-hard, even in bipartite graphs of maximum degree 4.*

This theorem is proven in [23] and is crucial to our proof of Theorem 1 in Sect. 2.1.

2.1 The Reduction

Hereinafter, $G = (N, E)$ is a bipartite instance of Two from Cubic Subgraph. We assume that $E \neq \varnothing$ so that $|N| \geq 2$. Take $G^* := (N^*, E^*)$ to be the following bipartite gadget graph depicted in Fig. 1, initially proposed in [4]: For each original vertex $u \in N$, create 5 new vertices v_u, w_u, x_u, y_u, z_u. Then, define $N^* := N \cup \{v_u, w_u, x_u, y_u, z_u : u \in N\}$. To obtain E^* from E, we add edges until $(\{u, v_u, w_u\}, \{x_u, y_u, z_u\})$ is a $K_{3,3}$ subgraph for every $u \in N$.

In Fig. 1, the bigger vertices with bolded edges indicate the original graph and the smaller vertices with thinner edges were added to obtain the gadget graph. The square and circular vertices depict a bipartition of G^*. Observe that the maximum degree of G^* is the maximum degree of G plus 3.

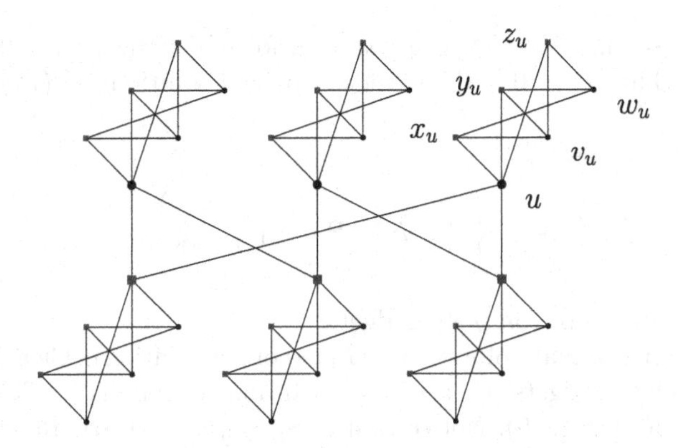

Fig. 1. The gadget graph from [4].

For each $u \in N$, define $T_u := \{v_u, w_u, x_u, y_u, z_u\}$ as well as $V_u := T_u \cup \{u\}$. We say that T_u are the *gadget vertices* of u and V_u is the *complete gadget* of u.

Let $\Gamma = (N^*, \nu)$ be the unweighted 3-matching game on G^*.

In Lemma 2, we show that if no two from cubic subgraph of G exists, then the nucleolus is precisely $x^* \equiv \frac{3}{2}$. Conversely, we prove in Lemma 3 that the existence of a two from cubic subgraph implies that x^* cannot be the nucleolus. The proof of Theorem 1 follows from the above lemmas and the hardness of Two from Cubic Subgraph. Remark that the degree bound follows from the degree bound in Theorem 3 and the fact that our gadget graph increases the maximum degree of the original graph by 3.

Lemma 1. *Let M be a maximum 3-matching in G^*. Let $\mathcal{C}$ be the set of connected components of $G^*[M]$. Then for all core allocations x and every component $C \in \mathcal{C}$, $x(C) = \nu(C)$.*

Proof. Observe that

$$x(N^*) \geq x(M) = \sum_{C \in \mathcal{C}} x(C) \geq \sum_{C \in \mathcal{C}} \nu(C) = \sum_{C \in \mathcal{C}} |M \cap E(C)| = |M| = \nu(N^*)$$

with the first inequality following from the fact that $x \geq 0$ and the second inequality following from the assumption that x is in the core.

Lemma 2. *If G does not contain a two from cubic subgraph, the uniform allocation $x^* \equiv \frac{3}{2}$ is the nucleolus of Γ.*

Proof. We argue using the Kopelowitz Scheme. Put (P_k) as the k-th LP in the Kopelowitz Scheme.

We can check through computation that for all $u \in N$ and $S \subsetneq V_u$, $e(S, x^*) \geq \frac{3}{2}$.

Let ϵ_1 be the optimal objective value to (P_1). We claim that $\epsilon_1 = 0$. By core non-emptiness, we have $\epsilon_1 \geq 0$. Moreover, using Lemma 1, since $E \neq \varnothing$, we can choose $u \in N$ so that $V_u \subsetneq N$ is a coalition for which $e(V_u, x) = 0$ for all core allocations x. Thus $\epsilon_1 = 0$ and the set of optimal solutions to (P_1) is precisely the core.

We now claim that

$$\mathcal{S}_1 = \left\{ S \subseteq N^* : S = \bigcup_{u \in S \cap N} V_u \right\}. \tag{5}$$

These are the unions of complete gadgets.

Let x be an optimal solution to (P_1) (core allocation). Clearly, if S is a union of complete gadgets, then $e(S, x) = 0$ due to Lemma 1. This shows the reverse inclusion in Eq. (5). Notice that $\nu(S) \leq \frac{3}{2}|S| = x^*(S)$ by the definition of a 3-matching, so x^* is an optimal solution to (P_1) and we may assume that $x = x^*$.

We claim that

$$\forall S \notin \mathcal{S}_1, e(S, x^*) \geq \frac{3}{2}. \tag{6}$$

This shows that if S is not a union of complete gadgets, then there is some optimal solution of (P_1) for which S is not fixed in (P_1) and hence $S \notin \mathcal{S}_1$. Thus the inclusion in Eq. (5) would hold.

Equation (6) is true if $S = \{u\}$ for some $u \in N^*$. If $S \subseteq N$ with $|S| \geq 2$, then $\nu(S) \leq \frac{3}{2}|S| - 2$. Otherwise, the edges of a maximum 3-matching in $G[S]$ induce a two from cubic subgraph. Thus $e(S, x) \geq \frac{3}{2}|S| - (\frac{3}{2}|S| - 2) \geq 2$.

It remains to consider the case when there is some $u \in N$ such that $T_u \cap S \neq \varnothing$. The argument here is similar to the reduction employed in [4]: If $S \cap V_u = V_u$, then $e(S \setminus V_u, x^*) \leq e(S, x^*) - 9 + 9 = e(S, x^*)$. We can remove as many of these complete gadgets from S as possible to obtain some coalition S'.

If $S' = \varnothing$, then $S \in \mathcal{S}_1$ by definition. In addition, if $S' \subseteq N$, there is again nothing to prove. Thus there must be some $u' \in S'$ such that $|T_{u'} \cap S'| \geq 1$ and $S' \cap V_{u'} \neq V_{u'}$.

If $|S' \cap T_{u'}| \leq 4$, then

$$e(S' \setminus T_{u'}, x^*) \leq e(S', x^*) - \frac{3}{2}|T_{u'}| + |E^*(S' \cap T_{u'} \cup \{u'\})| \leq e(S', x^*).$$

Finally, if $|S' \cap T_{u'}| = 5$, we are required to have $u' \notin S'$. So

$$e(S' \setminus T_{u'}, x^*) \leq e(S', x^*) - 5 \cdot \frac{3}{2} + 6 \leq e(S', x^*).$$

We may thus again repeatedly remove vertices of $N^* \setminus N$ until we arrive back at the base case of $S' \subseteq N$.

Thus Eq. (6) holds as all other coalitions have strictly greater excess with respect to x^*.

We now argue that $\epsilon_2 = \frac{3}{2}$. Observe $\nu(N^*) = \frac{3}{2}|N^*|$ implies that $\min_{a \in N^*} x(a) \leq \frac{3}{2}$ for any allocation x in the core and thus also for feasible solutions to (P_2) as well as the nucleolus. It follows that $\epsilon_2 \leq \frac{3}{2}$. But Eq. (6) shows that this upper bound is attained by x^*.

For all feasible solutions x to (P_2), $x(a) \geq \frac{3}{2}$ for all $a \in N^*$. But we cannot have some $x(a) > \frac{3}{2}$, or else $x(N^*) > \frac{3}{2}|N^*|$ and x would not be an allocation. Since the singleton coalitions are fixed in (P_2), it must be that $x^* \equiv \frac{3}{2}$ is the nucleolus.

Lemma 3. *If G contains a two from cubic subgraph, then the nucleolus of the gadget graph is not $x^* \equiv \frac{3}{2}$.*

Proof. We will show that $x^* \equiv \frac{3}{2}$ is not an optimal solution to (P_2). Recall that the nucleolus is necessarily an optimal solution to each LP in the Kopelowitz Scheme. This would thus yield the desired result.

Let us introduce a parameter as follows:

$$\Delta := \begin{cases} 0, & G \text{ contains a cubic subgraph} \\ 1, & G \text{ contains a two from cubic subgraph but no cubic subgraphs} \end{cases}$$

Let $N' \subseteq N$ be the vertices in the cubic subgraph or the vertices of the two from cubic subgraph if no cubic subgraph exists. Then

$$e\left(N', x^*\right) = \frac{3}{2}|N'| - \left(\frac{3}{2}|N'| - \Delta\right) = \Delta. \tag{7}$$

In particular, the minimum excess over all coalitions in $\mathcal{S}$ is at most Δ.

For $0 < \delta < \frac{1}{2}$, define $x_\delta(a) := \frac{3}{2} + \delta$ for each $a \in N$ and $x_\delta(a) := \frac{3}{2} - \frac{\delta}{5}$ for $a \in N^* \setminus N$. We can again check by computation that

$$\forall u \in N, \forall S \subsetneq V_u, e(S, x_\delta) \geq \frac{3}{2} - \delta \tag{8}$$

The coalitions with minimum excess among such coalitions is $S = T_u$ for some $u \in N$.

We claim that $\epsilon_1 = 0$ and $\mathcal{S}_1 = \{*\} S \subseteq N^* : S = \bigcup_{u \in S \cap N} V_u$ is again the union of complete gadgets. The fact that $\epsilon_1 = 0$ is clear from our previous lemma. Moreover, it is clear from our prior work that the unions of complete gadgets must be fixed in (P_1). We need only show that $e(S, x_\delta) > 0$ if S is not a union of complete gadgets. This would show that if S is not a union of complete gadgets, then there is some allocation (in particular x_δ) for which S is not fixed in (P_1).

If $S = \{a\}$ for some $a \in N^*$, then $e(S, x_\delta) \geq \frac{3}{2} - \frac{\delta}{5} > 0$.

When $S \subseteq N$. We have

$$e(S, x_\delta) \geq \left(\frac{3}{2} + \delta\right)|S| - \left(\frac{3}{2}|S| - \Delta\right) = \delta|S| + \Delta > 0.$$

Suppose now that there is some $u \in N$ such that $S \cap T_u \neq \varnothing$. Once again, if $S \cap V_u = V_u$, $e(S \setminus V_u, x_\delta) \leq e(S, x_\delta) - 9 + 9 = e(S, x_\delta)$. We can thus remove all complete gadgets from S to obtain another coalition S'. If $S' = \varnothing$, then $S \in \mathcal{S}_1$. Similar to before, if $S' \subseteq N$, we are back at the base case.

Pick some $u' \in S'$ such that $|S' \cap T_{u'}| \geq 1$. Observe that $\nu(S') \leq \nu(S' \setminus T_{u'}) + \nu(S' \cap V_{u'})$. This is because any maximum 3-matching on S' is a disjoint union of 3-matchings on $S' \setminus T_{u'}$ and $S' \cap V_{u'}$.

Suppose $u' \in S'$. We must have $|S' \cap T_{u'}| \leq 4$.

$$\begin{aligned} e(S', x_\delta) &= x(S' \setminus T_{u'}) + x(S' \cap V_{u'}) - x(u') - \nu(S') \\ &\geq x(S' \setminus T_{u'}) - \nu(S' \setminus T_{u'}) + [x(S' \cap V_{u'}) - x(u')] - \nu(S' \cap V_{u'}) \\ &\geq e(S' \setminus T_{u'}, x_\delta) + |S \cap T_{u'}| \left(\frac{3}{2} - \frac{\delta}{5}\right) - |E^*(S' \cap V_{u'})| \\ &\geq e(S' \setminus T_{u'}, x_\delta) - \frac{4}{5}\delta. \end{aligned}$$

Suppose now that $u' \notin S'$. In this case,

$$\begin{aligned} & e(S', x_\delta) \\ &= x(S' \setminus T_{u'}) + x(S' \cap T_{u'}) - \nu(S') \\ &\geq x(S' \setminus T_{u'}) - \nu(S' \setminus T_{u'}) + x(S' \cap T_{u'}) - \nu(S' \cap T_{u'}) \\ &= e(S' \setminus T_{u'}, x_\delta) + e(S' \cap T_{u'}, x_\delta) \\ &\geq e(S' \setminus T_{u'}, x_\delta) + \left(\frac{3}{2} - \delta\right) \qquad \text{by Equation (8)} \end{aligned}$$

By repeatedly removing vertices of $N^* \setminus N$, we see that

$$\begin{aligned} & e(S', x_\delta) \\ &\geq e(S' \cap N, x_\delta) + \sum_{u \in S' \cap N : S' \cap T_u \neq \varnothing} \left[-\frac{4}{5}\delta\right] + \sum_{u \in N \setminus S' : S' \cap T_u \neq \varnothing} \left[\frac{3}{2} - \delta\right] \\ &\geq \delta |S' \cap N| + \Delta - |S' \cap N| \left(\frac{4}{5}\delta\right) + |\{u \in N \setminus S' : S' \cap T_u \neq \varnothing\}| \left(\frac{3}{2} - \delta\right) \\ &= \frac{\delta}{5}|S' \cap N| + \Delta + |\{u \in N \setminus S' : S' \cap T_u \neq \varnothing\}| \left(\frac{3}{2} - \delta\right) \\ &\geq \frac{\delta}{5} + \Delta. \end{aligned}$$

The last inequality follows from the assumption that $S' \neq \varnothing$. In particular, at least one of $S' \cap N$ or $\{u \in N \setminus S' : S' \cap T_u \neq \varnothing\}$ is non-empty. This shows that $\epsilon_1 = 0$ is indeed the optimal solution to (P_1). Moreover, $\mathcal{S}_1$ is again the union of complete gadgets.

As an immediate corollary to the proof, $\epsilon_2 \geq \frac{\delta}{5} + \Delta > \Delta$. Recall there was a coalition $N' \subseteq N$ satisfying Eq. (7). It follows that $x^* \equiv \frac{3}{2}$ is not an optimal solution to (P_2) and therefore cannot be the nucleolus.

3 Positive Results

In the case of $b \leq 2$, we explore several variants of b-matching games for which the nucleolus can be efficiently computed.

First, we will state an important ingredient. Let $\Gamma = (N, \nu)$ be a cooperative game. For $\mathscr{S} \subseteq \mathcal{S}$ and $x \in \mathbb{R}^N$, write $\theta^{\mathscr{S}}(x) \in \mathbb{R}^{\mathscr{S}}$ to denote the restricted vector containing the excess values $e(S, x)$ for $S \in \mathscr{S}$ in non-decreasing order of excess.

Definition 1 (Characterization Set). *Let $\mathscr{S} \subseteq \mathcal{S}$ be a subset of the non-trivial coalitions.*

We say $\mathscr{S}$ is a characterization set for the nucleolus of the cooperative game $\Gamma = (N, \nu)$ if the lexicographic maximizer of $\theta^{\mathscr{S}}(x)$ is a singleton that lexicographically maximizes the unrestricted vector $\theta(x)$.

Intuitively, for $S \in \mathcal{S} \setminus \mathscr{S}$, we can drop the constraint corresponding to S from the Kopelowitz Scheme when computing the nucleolus.

Proposition 1. *Let $\Gamma = (N, \nu)$ be a cooperative game with non-empty core. Suppose $\mathscr{S}$ is a polynomial sized characterization set for the nucleolus of Γ.*

The nucleolus of Γ is polynomial time computable.

Let $\mathscr{S} \subseteq \mathcal{S}$ be a characterization set of the nucleolus of some game Γ. Consider the following tweak of the ℓ-th iteration of Kopelowitz Scheme (P'_ℓ) (with optimal value ϵ'_ℓ) where we only have constraints corresponding to coalitions in the characterization set $\mathscr{S}$ instead of every coalition. The sets $\mathcal{S}_\ell$ are defined in symmetric fashion as the coalitions from $\mathscr{S}$ which are fixed in (P_ℓ) but not at any prior (P_i).

$$\max \epsilon \qquad (P'_\ell) \tag{9}$$

$$x(S) = \nu(S) - \epsilon'_i \qquad \forall 0 \leq i < \ell, \forall S \in \mathcal{S}_i \tag{10}$$

$$x(S) - \nu(S) \geq \epsilon \qquad \forall S \in \mathscr{S} \setminus \bigcup_{i=0}^{\ell-1} \mathcal{S}_i \tag{11}$$

Proof. The tweaked Kopelowitz Scheme computes the lexicographic maximizer of $\theta^{\mathscr{S}}$. Since $\mathscr{S}$ is polynomially sized, each linear program in the scheme can be solved in polynomial time.

We are now ready to state the theorem by found in [19].

Theorem 4 ([19]). *Let $\Gamma = (N, \nu)$ be a cooperative game with non-empty core.*

The non-empty collection $\mathscr{S} \subseteq \mathcal{S}$ is a characterization set for the nucleolus of Γ if for every $S \in \mathcal{S} \setminus \mathscr{S}$ there exists a non-empty subcollection $\mathscr{S}_S$ of $\mathscr{S}$ such that

(i) For all $T \in \mathscr{S}_S$ and core allocations x, $e(T, x) \leq e(S, x)$.
(ii) There are scalars $\lambda_T \in \mathbb{R}$ such that the characteristic vector $\chi_S \in \{0,1\}^N$ of S satisfies $\chi_S = \sum_{T \in \mathscr{S}_S \cup \{N\}} \lambda_T \chi_T$.

Let $\Gamma = (N, \nu)$ be a not necessarily simple weighted b-matching game with non-empty core. Let $\mathscr{S}(\Gamma)$ be the set of all $S \in \mathcal{S}$ such that for all maximum b-matchings M of $G[S]$, $G[S][M]$ is connected.

Corollary 1. *Let $\Gamma = (N, \nu)$ be a not necessarily simple weighted b-matching game with non-empty core. Then $\mathscr{S}(\Gamma)$ is a characterization set for the nucleolus of Γ.*

Proof. Fix $S \in \mathcal{S} \setminus \mathscr{S}$. Suppose M is a maximum b-matching of $G[S]$. Let $T_1, T_2, \ldots, T_k$ be the components of $G[S][M]$ for $k \geq 2$. Suppose x is a core allocation.

Since $x(S) = \sum_{i=1}^{k} x(T_i)$ and $\nu(S) = \sum_{i=1}^{k} \nu(T_i)$, we have $\sum_{i=1}^{k} e(S_i, x) = e(S, x)$. In particular, condition (ii) of Theorem 4 is satisfied. But all excesses are non-negative as x is a core allocation, hence each $e(S_i, x) \leq e(S, x)$ and condition (i) of Theorem 4 is also satisfied.

The result follows by Theorem 4.

Lemma 4. *Let $\Gamma = (N, \nu)$ be a not necessarily simple weighted b-matching game with non-empty core. Suppose $\mathscr{S}(\Gamma)$ is polynomially sized. Then the nucleolus of Γ is polynomial time computable.*

Proof. Apply Proposition 1 and Corollary 1.

3.1 Simple b-Matching Games

We now present a proof for the first claim in Theorem 2.

Proof (Theorem 2(i)). By Lemma 4, it suffices to show that any component of a b-matching in some arbitrary induced subgraph $G[S]$ has at most $2k + 3$ vertices. If we show this, then $\mathscr{S}(\Gamma)$ is polynomially sized since it is contained in the subsets of $V(G)$ of size at most $2k + 3$.

Let C be a component of $G[S][M]$ for some $S \subseteq N$ and maximum b-matching M of $G[S]$. If C is a cycle, then exactly half the vertices of C are from B with $b_v = 2$. It follows that $|C| \leq 2k$. Suppose now that C is some path. By deleting at both endpoints and one more vertex, we may assume that every other vertex in the path are from B with $b_v = 2$. Thus $|C| \leq 2k + 3$ as required.

This result can be modified for the case where at most $O(\log(n + m))$ vertices in total have $b_v = 2$.

3.2 Non-simple 2-Matching Games

In the case where we allow for edges to be included multiple times in a 2-matching, we leverage core non-emptiness and the non-existence of odd cycles to compute the nucleolus in polynomial time.

Lemma 5. *Let G be an arbitrary graph with edge weights $w : E \to \mathbb{R}$. The core of the weighted non-simple 2-matching game on G is non-empty.*

Lemma 6. *For any bipartite graph, there is a maximum weighted non-simple 2-matching consisting only of parallel edges.*

We are now ready to prove the second result in Theorem 2.

Proof (Theorem 2(ii)). By Lemma 4, it suffices to show that if $|S| \geq 3$, then there is a 2-matching in $G[S]$ with multiple components.

But this is precisely what we proved in Lemma 6, concluding the proof.

Unfortunately, Lemma 6 does not hold when the graph is non-bipartite, even when we restrict ourselves to uniform edge weights. Indeed, consider the simple triangle. The maximum non-simple 2-matching has size 3. However, when we restrict ourselves to matchings composed of only parallel edges, the maximum matching we can obtain has cardinality 2.

Similarly, Lemma 6 does not in general hold when there are some vertices v where $b_v = 1$. Consider the path of 3 edges where the endpoints have $b_v = 1$ while the internal vertices have $b_v = 2$. The maximum non-simple 2-matching has size 3. However, if we only allow parallel edges, the maximum matching we can obtain again has cardinality 2.

References

1. Aumann, R.J., Maschler, M.: Game theoretic analysis of a bankruptcy problem from the Talmud. J. Econ. Theory **36**(2), 195–213 (1985)
2. Bateni, M.H., Hajiaghayi, M.T., Immorlica, N., Mahini, H.: The cooperative game theory foundations of network bargaining games. In: Abramsky, S., Gavoille, C., Kirchner, C., Meyer auf der Heide, F., Spirakis, P.G. (eds.) ICALP 2010. LNCS, vol. 6198, pp. 67–78. Springer, Heidelberg (2010). https://doi.org/10.1007/978-3-642-14165-2_7
3. Biró, P., Kern, W., Pálvölgyi, D., Paulusma, D.: Generalized matching games for international kidney exchange. In: Proceedings of the 18th International Conference on Autonomous Agents and MultiAgent Systems, pp. 413–421 (2019)
4. Biró, P., Kern, W., Paulusma, D., Wojuteczky, P.: The stable fixtures problem with payments. Games Econ. Behav. **108**, 245–268 (2018)
5. Brânzei, R., Solymosi, T., Tijs, S.: Strongly essential coalitions and the nucleolus of peer group games. Int. J. Game Theory **33**(3), 447–460 (2005)
6. Cook, K.S., Yamagishi, T.: Power in exchange networks: a power-dependence formulation. Soc. Netw. **14**(3–4), 245–265 (1992)
7. Davis, M., Maschler, M.: The kernel of a cooperative game. Naval Res. Logist. Q. **12**(3), 223–259 (1965)
8. Deng, X., Fang, Q.: Algorithmic cooperative game theory. In: Chinchuluun, A., Pardalos, P.M., Migdalas, A., Pitsoulis, L. (eds.) Pareto Optimality, Game Theory And Equilibria. Springer Optimization and Its Applications, vol. 17, pp. 159–185. Springer, New York (2008). https://doi.org/10.1007/978-0-387-77247-9_7
9. Deng, X., Fang, Q., Sun, X.: Finding nucleolus of flow game. J. Comb. Optim. **18**(1), 64–86 (2009)
10. Deng, X., Ibaraki, T., Nagamochi, H.: Algorithmic aspects of the core of combinatorial optimization games. Math. Oper. Res. **24**(3), 751–766 (1999)
11. Drechsel, J., Kimms, A.: Computing core allocations in cooperative games with an application to cooperative procurement. Int. J. Prod. Econ. **128**(1), 310–321 (2010)

12. Easley, D., Kleinberg, J., et al.: Networks, crowds, and markets: reasoning about a highly connected world. Significance **9**(1), 43–44 (2012)
13. Elkind, E., Goldberg, L.A., Goldberg, P.W., Wooldridge, M.: Computational complexity of weighted threshold games. In: AAAI, pp. 718–723 (2007)
14. Elkind, E., Goldberg, L.A., Goldberg, P.W., Wooldridge, M.: On the computational complexity of weighted voting games. Ann. Math. Artif. Intell. **56**(2), 109–131 (2009)
15. Faigle, U., Kern, W., Kuipers, J.: Computing an element in the lexicographic kernel of a game. Math. Methods Oper. Res. **63**(3), 427–433 (2006)
16. Faigle, U., Kern, W., Kuipers, J.: Note computing the nucleolus of min-cost spanning tree games is NP-hard. Int. J. Game Theory **27**(3), 443–450 (1998)
17. Faigle, U., Kern, W., Kuipers, J.: On the computation of the nucleolus of a cooperative game. Int. J. Game Theory **30**(1), 79–98 (2001)
18. Faigle, U., Kern, W., Paulusma, D.: Note on the computational complexity of least core concepts for min-cost spanning tree games. Math. Methods Oper. Res. **52**(1), 23–38 (2000)
19. Granot, D., Granot, F., Zhu, W.R.: Characterization sets for the nucleolus. Int. J. Game Theory **27**(3), 359–374 (1998)
20. Granot, D., Maschler, M., Owen, G., Zhu, W.R.: The kernel/nucleolus of a standard tree game. Int. J. Game Theory **25**(2), 219–244 (1996)
21. Kern, W., Paulusma, D.: Matching games: the least core and the nucleolus. Math. Oper. Res. **28**(2), 294–308 (2003)
22. Khachiyan, L.G.: A polynomial algorithm in linear programming. Dokl. Akad. Nauk **244**(5), 1093–1096 (1979)
23. Koenemann, J., Toth, J., Zhou, F.: On the complexity of nucleolus computation for bipartite b-matching games. arXiv preprint arXiv:2105.07161 (2021)
24. Könemann, J., Pashkovich, K., Toth, J.: Computing the nucleolus of weighted cooperative matching games in polynomial time. Math. Program. **183**(1), 555–581 (2020)
25. Könemann, J., Toth, J.: A general framework for computing the nucleolus via dynamic programming. In: Harks, T., Klimm, M. (eds.) SAGT 2020. LNCS, vol. 12283, pp. 307–321. Springer, Cham (2020). https://doi.org/10.1007/978-3-030-57980-7_20
26. Kopelowitz, A.: Computation of the kernels of simple games and the nucleolus of n-person games. Technical report, Hebrew University of Jerusalem (Israel), Department of Mathematics (1967)
27. Kuipers, J., Solymosi, T., Aarts, H.: Computing the nucleolus of some combinatorially-structured games. Math. Program. **88**(3), 541–563 (2000)
28. Lemaire, J.: An application of game theory: cost allocation. ASTIN Bull. J. IAA **14**(1), 61–81 (1984)
29. Maschler, M., Peleg, B., Shapley, L.S.: Geometric properties of the kernel, nucleolus, and related solution concepts. Math. Oper. Res. **4**(4), 303–338 (1979)
30. Nash, J.F., Jr.: The bargaining problem. Econom. J. Econom. Soc. **18**, 155–162 (1950)
31. Paulusma, D.: Complexity aspects of cooperative games. Citeseer (2001)
32. Plesník, J.: A note on the complexity of finding regular subgraphs. Discret. Math. **49**(2), 161–167 (1984)
33. Potters, J., Reijnierse, H., Biswas, A.: The nucleolus of balanced simple flow networks. Games Econ. Behav. **54**(1), 205–225 (2006)
34. Schmeidler, D.: The nucleolus of a characteristic function game. SIAM J. Appl. Math. **17**(6), 1163–1170 (1969)

35. Shapley, L.S., Shubik, M.: The assignment game I: the core. Int. J. Game Theory **1**(1), 111–130 (1971)
36. Solymosi, T., Raghavan, T.E.S.: An algorithm for finding the nucleolus of assignment games. Int. J. Game Theory **23**(2), 119–143 (1994)
37. Solymosi, T., Sziklai, B.: Characterization sets for the nucleolus in balanced games. Oper. Res. Lett. **44**(4), 520–524 (2016)
38. Stearns, R.E.: Convergent transfer schemes for n-person games. Trans. Am. Math. Soc. **134**(3), 449–459 (1968)
39. Sziklai, B., Fleiner, T., Solymosi, T.: On the core and nucleolus of directed acyclic graph games. Math. Program. **163**, 243–271 (2016). https://doi.org/10.1007/s10107-016-1062-y
40. Willer, D.: Network Exchange Theory. Greenwood Publishing Group, Westport (1999)

Pure Nash Equilibria in a Generalization of Congestion Games Allowing Resource Failures

Julian Nickerl(✉) and Jacobo Torán

Institute of Theoretical Computer Science, Ulm University, Ulm, Germany
{julian.nickerl,jacobo.toran}@uni-ulm.de
https://www.uni-ulm.de/en/in/theo/

Abstract. We introduce a model for congestion games in which resources can fail with some probability distribution. These games are an extension of classical congestion games, and like these, have exact potential functions that guarantee the existence of pure Nash equilibria (PNE). We prove that the agent's cost functions for these games can be hard to compute by giving an example of a game for which the cost function is hard for Valiant's #**P** class, even in the case when all failure probabilities coincide. We characterize the complexity of computing PNE in congestion games with failures with an extension of the local search class **PLS** that allows queries to a #**P** function, and show examples of games for which the PNE search problem is complete for this class. We also provide a variant of the game with the property that a PNE can be constructed in polynomial time if this also holds in the restricted game without failures.

Keywords: Congestion game · Failure · PLS · #P · Nash equilibrium

1 Introduction

A central question at the verge between game theory and computer science is the study of the complexity for the computation of a Nash equilibrium in a game. This question has motivated many important results in the area and the definitions of several complexity classes (see, e.g., Chap. 2 in [14]) characterizing the complexity of this problem for many classes of games. On the one hand, cases in which such equilibria can be computed efficiently have been identified. More importantly, since reaching an equilibrium can be considered a natural process generated from independent agents' efforts trying to optimize some individual goal, researchers have tried to identify the computational power of such a process. One of the classes for which the study of the complexity of computing Nash equilibria has been particularly fruitful is the class of congestion games introduced by Rosenthal [20]. Congestion games model the behavior of rational and selfish agents that have to share a set of resources to achieve personal goals. The cost of a resource increases with the number of agents using it. Typical

I. Caragiannis and K. A. Hansen (Eds.): SAGT 2021, LNCS 12885, pp. 186–201, 2021.
https://doi.org/10.1007/978-3-030-85947-3_13

settings include, for example, processes competing for computation time or cars participating in a road network.

Rosenthal used what is now known as the *potential function method* to prove the existence guarantee of pure Nash equilibria (PNEs). Essentially, PNEs coincide with minima of a function that always possesses at least one minimum if defined over congestion games. Focusing on the function's behavior, he proposed a simple algorithm based on greedy best-response strategy changes that would always lead to a PNE. However, it was shown that this algorithm might require an exponential number of improvement steps to reach a PNE [4].

The complexity of finding PNEs in congestion games was mainly settled by Fabrikant et al. and Ackermann et al. [1,4], proving completeness for the complexity class *polynomial local search* (**PLS**), previously introduced by Johnson, Papadimitriou, and Yannakakis [7]. **PLS** is a class of total functions characterizing the complexity of finding locally optimal solutions for optimization problems. The problem is to compute any solution that is locally optimal while suboptimal solutions can be locally improved in polynomial time, although it may take an exponential number of improvement steps to reach a local optimum.

We consider a natural extension of congestion games, taking into account that resources in any real-world scenario are subject to failures. For example, roads might be blocked because of accidents, computers links might fail, etc. Motivated by a series of papers by Penn, Polukarov, and Tennenholtz [15–18], we analyze the effect of uncertainty in the resources' availability by assigning a failure probability to each resource. In the mentioned papers, the authors introduced several models incorporating different aspects into congestion games. They manage to show a guarantee of existence of PNEs and even propose polynomial-time algorithms for finding them in restricted scenarios. At the same time, however, they diverged quite far away from the original games, allowing, for example, that agents may add to the congestion of resources even if they do not actively use them. Furthermore, they highly restrict the combinatorial richness of the allowed strategy sets.

We introduce a new model that stays close to the original definition of congestion games. In particular, the congestion of the resources only depends on the number of players using them, like in the original congestion games. In the presence of failures, it is not clear how to define the agents' strategies after some resources fail. Potentially, there can be exponentially many concrete different scenarios in a game. One needs some kind of mechanism to decide which strategy is chosen in each case. We consider two different approaches that allow the agents to define their strategy according to the existing resources. In the first model, we consider a priority list of goals. Once the resource scenario is clear, the first doable strategy in the agent's list is the one chosen. This corresponds to the natural situation of considering a plan A, a plan B (in case A is not possible), etc. In the second model, the agents encode their different goal options succinctly in a Boolean circuit. The input of a circuit is a vector encoding the available resources, and the output is a goal that can be reached with these resources. In both cases, a bound on the sizes of lists or circuits can be considered in order to keep the size of the strategies polynomial in the size of the game description.

We show that the congestion games with resource failures (CGRF) have a potential function and are thus congestion games [12]. This implies that PNEs are guaranteed to exist for both models. Furthermore, we discuss the computational complexity of finding a PNE in these games. We observe that the class **PLS** is not sufficient to describe the complexity anymore, as we prove that calculating the cost of an agent in a strategy combination is #**P**-complete, even in the case in which all resources have the same failure probability. Consequently, we introduce a generalization of **PLS** called $\mathbf{PLS}^{\#\mathbf{P}}$ that has access to a #**P** oracle and prove completeness in this class for the problem of finding a PNE in our model with priority lists, assuming that the lists have a constant length.

Related work: Very recently, Kleinberg et al. [8] have identified several natural complexity classes for search problems beyond **NP**. These classes are included in TFNP^A, the class of total polynomial time relative to an oracle set A in the lower levels of the polynomial time hierarchy. $\mathbf{PLS}^{\#\mathbf{P}}$ is another example of such a complexity class, but with an oracle set in #**P**. Besides the aforementioned models by Penn, Polukarov, and Tennenholtz, several other approaches exist that consider the introduction of uncertainty into congestion games. These approaches can be roughly split into two groups: uncertainty regarding the players, and uncertainty regarding the resources. In the former, there is uncertainty typically in the number, type, or weight of the participating players (see, e.g., [2,3,5, 10,11]). In the latter, the cost functions typically incorporate some randomness [2,6,13,19]). Both approaches motivate, among others, rich fields of study of risk-averse player behavior, and the effect of uncertainty on the existence and quality of equilibria. Our approach is different from these models since in our case both players and cost functions are deterministic.

The remainder of this paper is structured as follows. After some general preliminary definitions in Sect. 2, we define our new model for CGRFs in Sect. 3. We prove the existence of PNEs for the new model in Sect. 4 and discuss the complexity of finding PNEs in Sect. 5. We introduce a generalization of **PLS** and prove completeness in this new class for several problems in Sect. 6.

2 Preliminaries

Let **FP** be the class of polynomial-time computable functions. The complexity class #**P** was introduced by Valiant [24]. It encompasses problems asking the question 'how many' rather than 'is there a solution?'. Formally, it is the class of functions $f : \{0,1\}^* \to \mathbb{N}$ for which there exists a polynomial-time non-deterministic Turing machine M, such that, for each $x \in \{0,1\}^*$, the number of accepting paths of M on input x is exactly $f(x)$. The counting versions of **NP**-complete problems are typically complete for #**P**.

In the problems we consider, part of the input data can consist of rational numbers $\frac{p}{q}$. As usual, we consider that these numbers are given as pairs (p, q) in binary. In some cases we scale up the value of a function f to cf for a suitable number c to ensure that the value of f is a natural number and can be computed by a **#P** function.

Also, in some of the upcoming scenarios, we have to express the expectation $\mathbb{E}(y)$ of a random variable y, as a **#P** function. If the probability for each of the values of y can be computed in polynomial time, we can define a non-deterministic machine whose number of accepting paths is $\mathbb{E}(y)$ (maybe suitably scaled). This machine would non-deterministically guess the possible values of y (thus adding over the possible values the variable can take) and multiply these values by the corresponding probability of y occurring.

$\mathbf{FP}^{\#\mathbf{P}}$ is the class of functions that can be computed in polynomial time, allowing queries to a **#P** function. This class is very powerful and contains all the functions that can be computed within the polynomial-time hierarchy [23]. To show that a function is complete for **#P** we use the concept of metric reducibility between functions [9]. A function f is metric reducible to a function g if there are polynomial time computable functions h_1 and h_2 such that for all x, $f(x) = h_2(g(h_1(x)))$.

#DNF-SAT is the functional problem to compute the number of satisfying assignments for a Boolean formula in disjunctive normal form. It is known to be complete for **#P** under metric reductions.

The class **PLS** [7] encompasses local search problems for which any given solution can be improved within a neighborhood in polynomial time if such an improvement is possible.

Definition 1. *Let L be an optimization problem with a set of instances D_L, and for each $x \in D_L$ a set of solutions $F_L(x)$ of length $p(|x|)$ for some polynomial p. For each solution $s \in F_L(x)$ there is a non negative integer cost $c_L(x, s)$ as well as a subset $N(s, x) \subseteq F_L(x)$ of neighboring solutions. We say that L is in* ***PLS*** *if the following conditions hold:*

1. *The relation $R = \{(x, s) \mid x \in D_L, s \in F_L(x)\}$ is in* ***P****.*
2. *There is a function $\alpha : D_L \to F_L(x)$, $\alpha \in$* ***FP****. (α produces a solution).*
3. *The cost function $c_L(x, s)$ lies within* ***FP****.*
4. *There is a function $\gamma \in$* ***FP*** *that on input $x \in D_L$ and $s \in F_L(x)$ computes a neighboring solution $s' \in N(x, s)$ with better cost than s ($c_L(x, s') < c_L(x, s)$ for a minimization problem and $c_L(x, s') > c_L(x, s)$ for maximization) in case such a solution s' exists. Otherwise it outputs some special symbol.*

The standard way of comparing the complexity of **PLS** problems is the **PLS**-reduction [7].

Definition 2. *A problem A in* ***PLS*** *is reducible to another problem B if there are functions f and g in* ***FP*** *such that*

1. *f maps instances x of A to instances $f(x)$ of B,*
2. *g maps (solution of $f(x)$, x) pairs to solutions of x, and*

3. $\forall x \in A$, if s is a local optimum for $f(x)$ in B, then $g(s,x)$ is a local optimum for x in A.

Note that while each individual improvement step in a **PLS**-problem can be done in polynomial time, finding a local optimum is, in general, assumed to be a difficult problem. It was shown in [7] that the number of necessary improvement steps before convergence may be exponential. Clearly, **FP** $\subseteq$ **PLS** $\subseteq$ **FNP**. It is currently unclear whether **PLS** might be equivalent to **FP** or **FNP**, however, both seem unlikely. In particular, it was proven that if there exists a problem in **PLS** that is **NP**-hard, then **NP** = **coNP** [7].

3 Model

A CGRF is defined by a tuple $\Gamma = (A, E, (p_e)_{e\in E}, (G_i)_{i\in A}, (c_e)_{e\in E}, (w_i)_{i\in A})$ where $A = \{1, \ldots, n\}$ is a set of agents and $E = \{1, \ldots, m\}$ is a set of resources, with each $e \in E$ being associated with a non-failure probability p_e. We say a resource *exists* (with the given probability), or *fails* otherwise. Each agent i has a set of goals $G_i \subseteq \mathcal{P}(E)$ (corresponding to the set of strategies for a classical congestion game). If no goal can be reached (because of failing resources), the agent must pay a cost of failure w_i. We say that a goal $g_j \in G_i$ can be reached or is reachable if none of the resources $e \in g_j$ fail. The strategies of i are not the elements of G_i. Rather, a strategy s_i for agent i describes for each set $I \subseteq E$ of non-failing resources a goal $s_i(I) = g_i^I \in G_i$ reachable with the available resources in I if such a goal exists, or $s_i(I) = \emptyset$ if no such goal can be reached. We call $S = (s_1, ..., s_n)$ a state or strategy vector.

As in classical congestion games, let $n_e(S, I) = |\{i \mid e \in s_i(I)\}|$ be the number of agents using resource e in case the resources in I are non-failing, and let $c_e : \mathbb{N} \to \mathbb{Q}^+$ be the cost function for resource e. The cost u_i of a strategy combination S for an agent i is defined as the expectation of the costs over all subsets of resources

$$u_i(S) = \sum_{I\subseteq E} p(I)u_i(S, I),$$

where $p(I)$ is the probability that the resources available are exactly those in I, $p(I) = \prod_{e\in I} p_e \prod_{e\notin I}(1 - p_e)$, and $u_i(S, I)$ is the cost of agent i in the case that the non-failing set of resources is I,

$$u_i(S, I) = \sum_{e\in s_i(I)} c_e(n_e(S, I)).$$

It is $u_i(S, I) = w_i$ if $s_i(I) = \emptyset$.

Observe that if all resources are guaranteed to exist, many practical choices of strategy functions coincide with the definition of strategies in the classical congestion games. This implies, in particular, that CGRFs generalize classical congestion games in these cases.

4 Existence of Pure Nash Equilibria

We show that these games have an exact potential function, and therefore they always have a PNE. For a combination of strategies S, define the potential function $\varphi(S)$ as

$$\varphi(S) := \sum_{I \subseteq E} p(I) \left[\sum_{e \in E} \left(\sum_{k=1}^{c_e(n_e(S,I))} k \right) + \sum_{\{i \mid s_i(I)=\emptyset\}} w_i \right].$$

Imagine that agent i changes her strategy from s_i to s_i'. For a set of resources I, we consider different cases for the difference $\Delta_i(I)$ in the payoff for agent i and the difference in the potential function $\Delta_\varphi(I)$ on the set of non-failing resources I:

Case 1: $s_i(I), s_i'(I) \neq \emptyset$ are different from $\emptyset$.

$$\Delta_i(I) = \sum_{e \in s_i'(I) \setminus s_i(I)} (c_e(n_e(S,I)+1) - \sum_{e \in s_i(I) \setminus s_i'(I)} c_e(n_e(S,I)) = \Delta_\varphi(I)$$

Case 2: $s_i(I) = \emptyset \neq s_i'(I)$ (the symmetric case is analogous).

$$\Delta_i(I) = \sum_{e \in s_i'(I)} (c_e(n_e(S,I)+1) - w_i = \Delta_\varphi(I)$$

Case 3: $s_i(I) = \emptyset = s_i'(I)$.

$$\Delta_i(I) = w_i - w_i = 0 = \Delta_\varphi(I)$$

As the differences coincide for every set I, minima of φ correspond to PNEs. Since φ can only take finitely many values, the case distinction implies:

Theorem 1. *CGRFs always posses a PNE.*

5 Fixing a Strategy

The introduced model of CGRFs remains very general in the definition of a strategy. In essence, we require a strategy to be any mechanism that, given a set of existing resources, returns a valid goal or the empty set. Since the set $I \subseteq E$ of possible non-failing resources is exponential in $|E|$ the description of such a mechanism can be exponential in size. Note in particular that besides the probabilities the set of existing resources is unknown prior to the choice of strategy. In the remainder of the paper, we elaborate on the properties of two explicit ways to provide a strategy succinctly:

- The agent provides a priority list L_i of goals. For a set I of non-failing resources, the goal selected by i would be the first one in the list that can be accomplished with the resources in I. If no goal can be reached with the resources in I, the agent plays $\emptyset$.

- The agent provides a succinct description of the goals in terms of a Boolean circuit C_i. On the input of a 0–1 vector encoding I, C_i computes a goal g_i^I doable with the available resources or states that no goal can be reached.

To guarantee the succinctness of the strategy, an additional parameter restricting the size can be introduced into the game description. Note that in the previous section, we made no assumptions on the exact mechanism of the strategy. Hence, for both approaches, the existence of a PNE is still guaranteed, and both variants generalize the original congestion games. Because of this generalization finding a PNE in CGRFs is **PLS**-hard.

5.1 Given a Boolean Circuit

The Boolean circuits as described above let us mitigate some of the complexity introduced by the uncertainty in the resources. The only challenge is to keep the circuit of polynomial size in the size of the game description. If an algorithm exists that finds a PNE in these classical congestion games in polynomial time, then the same holds for the CGRFs with circuits.

Theorem 2. *Let Γ be a CGRF with n players and m resources and let $\mathcal{T}$ be a subclass of congestion games for which a polynomial time algorithm $\mathcal{A}$ for finding a PNE exists. Furthermore, let Γ^I be the classical congestion game based on Γ where only the resources in I exist. If it holds for any set $I \subseteq E$ that $\Gamma^I \in \mathcal{T}$, then calculating a circuit C_i for each player such that $C = (C_1, \ldots, C_n)$ is a PNE is possible in polynomial time.*

Proof. For an input size k, any polynomial-time algorithm for a problem can be transformed in polynomial time into a Boolean circuit solving the problem for instances of size k (see e.g. [21]). Let $C^{\mathcal{A}}$ be such a circuit finding PNEs for games in $\mathcal{T}$ of a fixed size. Let furthermore $C_i^{\mathcal{A}}$ be a restriction of $C^{\mathcal{A}}$ that selects the set of resources chosen by player i in the calculated equilibrium. Note that $C_i^{\mathcal{A}}$ can be chosen as a strategy of player i in the circuit model and can be constructed in polynomial time. Now consider the state $(C_1^{\mathcal{A}}, \ldots, C_n^{\mathcal{A}})$. For any set I, the goals chosen by the circuits correspond to a PNE in the classical congestion game Γ^I. In particular, this means that there does not exist a set I such that a player could reduce his cost in Γ^I through a strategy change. Therefore, $(C_1^{\mathcal{A}}, \ldots, C_n^{\mathcal{A}})$ is a PNE.

A class of games for which the above theorem applies is the symmetric network congestion games. A polynomial-time algorithm based on min-cost flow was introduced by Fabrikant, Papadimitriou, and Talwar [4]. Independent of the set of non failing resources, the game remains symmetric (source and sink nodes of the players do not change), hence a PNE can be computed in polynomial time.

Corollary 1. *A PNE in symmetric network CGRF with circuit strategies can be computed in polynomial time.*

Note that while a PNE can be stated in polynomial time, it may not be possible to calculate each player's cost in the equilibrium efficiently, as shown in Theorem 4 for the list model. The hardness result can be applied to the circuit model as well since as shown in the next result, a circuit representing a list of goals can be created in polynomial time. The converse is not necessarily true, i.e., not every circuit can be represented as a list.

Theorem 3. *Any list strategy can be represented as a circuit strategy in polynomial time.*

Proof Sketch. Let $L_i = (g_1, ..., g_k)$ be the given list strategy. Interpreting any g_i as a Boolean variable (it is 1 if goal g_i can be reached and 0 otherwise), consider Boolean formulas of the form $F_j = \neg g_1 \wedge \neg g_2 \wedge \cdots \wedge g_j$. F_j is evaluated to 1 iff goal g_j is chosen by the list. Note that at most, one such formula can be evaluated to 1 at the same time. Each formula can be represented as a Boolean circuit, and from the results of the circuits, the appropriate goal can be returned efficiently.

5.2 Given a List of Goals

We now focus on priority lists of goals as strategies. As mentioned before, these CGRFs generalize the classical congestion games; hence it is unlikely that a polynomial-time algorithm for finding a PNE exists unless **P** = **PLS**. However, finding an equilibrium in the list model may not be significantly harder than finding it in the classical congestion games. In other words, it may be complete for **PLS**. For this to be the case, the cost of a solution (i.e., the value of the potential function) must be computable in polynomial time for any given state. However, we prove that calculating a given player's cost is #**P**-hard.

Theorem 4. *Determining a player's cost in the list model is complete for the metric closure of #**P**.*

Proof. The cost function is not necessarily in #**P** because depending on the game, the cost function might be a rational number, and #**P** functions only range over natural numbers. What the statement of the theorem means is that the cost function, properly scaled, is in #**P**, or more formally, for any CGRF Γ, there is a scaling function $h \in \mathbf{FP}$ such that if $u_i(S, \Gamma)$ is the cost of agent i under strategy combination S in the game, then $u_i(S, \Gamma) \cdot h(\Gamma)$ is a #**P** function. $u_i(S, \Gamma) \cdot h(\Gamma)$ can be computed by a function in #**P** because for a concrete set of available resources, calculating the cost for player i can be done in polynomial time. As indicated in Sect. 2, a non-deterministic machine can guess the available resources, compute the corresponding cost and then multiply this number by the probability of these resources occurring (properly scaled) and produce this number of accepting paths. The sum over all combinations of resources of the accepting paths of the machine is exactly $u_i(S, \Gamma) \cdot h(\Gamma)$.

We prove #**P**-hardness through a reduction from #DNF-SAT. Let $F = \bigvee_{i=1}^{m} C_i$ be a Boolean formula in DNF with $n \leq m$ variables $v_1, \ldots, v_n$ and the

C_i its conjunctions. By $\#SAT(F)$ we represent the number of satisfying assignments of F. We construct the state of a game of an instance of a CGRF based on F as follows:

The set of resources is $E = \{v^i, \overline{v^i} \mid i \in \{1,\ldots,n\}\} \cup \{\mathrm{r}\}$, two for each variable, and an additional auxiliary resource. The variable resources each exist with probability $\frac{1}{2}$, whereas the auxiliary resource always exists ($p_r = 1$). The cost functions are defined as $c_{v^i}(x) = c_{\overline{v^i}}(x) = 0$, and $c_r(x) = \begin{cases} 0, & \text{if } x \leq m \\ M, & \text{otherwise} \end{cases}$, where M is a large number to be fixed later. The cost of failure can be arbitrary, as we will design the strategies such that no agent can fail. There are three different types of agents, each choosing a different list as their strategy. The goals for the players coincide with the subsets of resources in the lists:

1. One agent a that chooses $(\{r\})$.
2. For each $i \in \{1,\ldots,n\}$ there are m agents that choose the list $(\{v_1,\overline{v_1}\},\ldots,\{v_n,\overline{v_n}\},\{v_i\},\{\overline{v_i}\},\{r\})$.
3. For each conjunction C_i, there is one agent choosing $(\{v_1,\overline{v_1}\},\ldots,\{v_n,\overline{v_n}\},C_i,\{r\})$, where by C_i we mean the set of resources corresponding to the literals appearing in C_i.

In total, we consider a game with $(n+1)m+1$ agents. Note, that the largest list consists of $n+3 = \mathcal{O}(\sqrt{n \cdot m})$ elements. We now calculate the expected cost of agent a in this state through a case distinction:

1. There exists at least one index i, such that v_i and $\overline{v_i}$ both exist, and, without loss of generality, assume that i is the lowest index of a resource with that property. All agents but agent a choose the resource set $\{v_i,\overline{v_i}\}$. Agent a is the only one on resource a, hence her cost is 0. This case occurs with probability $1-(\frac{3}{4})^n$
2. Case 1 does not apply, and there exists at least one index i, such that v_i and $\overline{v_i}$ both do not exist. Then there are at least m agents of type 2 that have to use resource r. The total number of agents on resource r is then above m, and the cost of agent a for using the resource is M. This case occurs with probability $(\frac{3}{4})^n - (\frac{1}{2})^n$.
3. For each i, either v_i or $\overline{v_i}$ exist, but not both. The existing resources can then be interpreted as an assignment for the original formula F: If v_i exists, then the respective variable is assigned the value 1; otherwise $\overline{v_i}$ exists, and the variable is assigned the value 0. In this scenario, all agents of type 2 do not choose resource r. Agents of type 3 choose resource r if and only if the conjunction they represent is not satisfied by the assignment. Hence, only if the assignment is not a satisfying one for formula F, all m agents of type 3 choose resource r. Only then does the total number of agents on r increase above m, and the cost of agent a increases from 0 to M. There are 2^n possible assignments, each occurring with equal probability $\frac{1}{4}^n$. The number of assignments for which agent a has non-zero cost is $2^n - \#SAT(F)$.

Going over all cases and choosing $M = 4^n$, the expected cost of agent a is:

$$M \cdot ((\frac{3}{4})^n - (\frac{1}{2})^n) + M \cdot (\frac{1}{4})^n \cdot (2^n - \#SAT(F)) = 3^n - \#SAT(F)$$

Hence, subtracting the expected cost of player a from 3^n gives us the number of satisfying assignments for the original formula F. This implies that any **#P** function is metrically reducible to the cost.

Remark: The probabilities in the proof of Theorem 4 were chosen for ease of calculation. They can be chosen arbitrarily as long as the accumulated probabilities for each case can be efficiently computed. This includes the probability for the auxiliary resource (which is currently set to 1). If the probability of r is not 1, an additional case has to be considered where a pays the cost of failure (which could be set to $2M$). This shows that the hardness of the problem stems not from some property encoded in the probabilities but originates in the high combinatorial complexity introduced by the uncertainty.

6 The Class $\mathbf{PLS}^{\#\mathbf{P}}$

Clearly **PLS** does not suffice to describe the complexity of finding a PNE in CGRFs. However, structurally, PNEs in classical, and congestion games with failures are very similar.

6.1 Allowing Operations in #P

We introduce a generalization of **PLS** capable of dealing with more complex cost functions. For this, we remind the reader of the definition of **PLS** as introduced in Definition 1. We only change the definition in two places:

3. The cost function c_L can be expressed as $c_L(x,s) = \frac{f(x,s)}{g(x)}$ for two functions $f \in \#\mathbf{P}$ and $g \in \mathbf{FP}$. (g is a scaling function depending on x and not on s).
4. There is a function $\gamma \in \mathbf{FP}^{\#\mathbf{P}}$ that on input $x \in D_L$ and $s \in F_L(x)$ computes a neighboring solution $s' \in N(x,s)$ with better cost than s in case such a solution s' exists. Otherwise it outputs some special symbol.

The only difference with the definition of **PLS** is the complexity of the functions c_L and γ, which were polynomial-time computable functions in the original definition. We will see that $\mathbf{PLS}^{\#\mathbf{P}}$ captures the complexity of several natural local search problems by showing that these problems are complete for the class under the original **PLS**-reducibility.

Conceptually, the difference between the classes $\mathbf{FP}^{\#\mathbf{P}}$ and $\mathbf{PLS}^{\#\mathbf{P}}$ is analogous to that between **FP** and **PLS**. We do not expect the classes to coincide since in order to reach a local optimum one might need an exponential number of improvement steps (using the **#P** oracle each time to compute the improvement), which is doable in $\mathbf{PLS}^{\#\mathbf{P}}$ but not clear how to achieve in $\mathbf{FP}^{\#\mathbf{P}}$.

6.2 Some $\mathbf{PLS}^{\#\mathbf{P}}$-complete Problems

In this section, we introduce the problem $\mathbb{E}$-FLIP, a first $\mathbf{PLS}^{\#\mathbf{P}}$-complete problem that will be reduced to all the other $\mathbf{PLS}^{\#\mathbf{P}}$-complete problems considered in this paper. $\mathbb{E}$-FLIP is a version of the canonical **PLS**-complete problem FLIP from [7] in which some of the circuit inputs can be probabilistic bits, and we ask for the expected value of the output.

An instance of $\mathbb{E}$-FLIP consists of a Boolean circuit C with two kinds of input gates, l normal input gates $x_1, \ldots, x_l$, and m further probabilistic input gates $y_1, \ldots, y_m$, plus n output gates $z_1, \ldots, z_l$. Additionally, for each random input gate y_i, a probability $p_i = \text{Prob}[y_i = 1]$ is given. For an input x and a fixed choice for the random bits y, $C(x, y)$ computes Boolean values for the output bits z in the usual way.

Any input vector $x \in \{0,1\}^l$ is a valid solution for $\mathbb{E}$-FLIP. Let $f(x, y) = \sum_{i=1}^{n} z_i 2^{i-1}$, where the z_i are the outputs of C with inputs x and y. For a solution x let $\mathbb{E}(C, x)$ be the expectation of f over all random inputs y, $\mathbb{E}(C, x) := \sum_{y \in \{0,1\}^m} p(y) f(x, y)$. The neighborhood of a solution x is the set of vectors at Hamming distance 1 from x.

Definition 3. *The local search problem $\mathbb{E}$-FLIP for a probabilistic circuit C as described above consists of finding a solution x such that $\mathbb{E}(C, x)$ is a local maximum.*

Some considerations have to be made in order to force the expected value of the circuit to be an integer. We suppose the probabilities p_i to be rational numbers $\frac{q_i}{r_i}$ with the integers q_i and r_i encoded in binary. y_i has the value 1 with probability p_i, and value 0 with probability $1 - p_i$. For any $y = y_1, \ldots, y_m$ we denote by $p(y)$ the probability of the random bits being equal to y, $p(y) = \prod_{\{i : y_i = 1\}} p_i \cdot \prod_{\{i : y_i = 0\}} (1 - p_i)$. Because this number is not necessarily an integer, we multiply this function by a large number in order to force this property. We define the cost function $c(C, x) := \pi \cdot \mathbb{E}(C, x)$ where $\pi = \prod_{i=1}^{m} r_i$. Observe that $c(C, x)$ is an integer and π can be computed in polynomial time in the size of instance C.

Theorem 5. *$\mathbb{E}$-FLIP is $\mathbf{PLS}^{\#\mathbf{P}}$-complete under **PLS**-reductions.*

Proof Sketch. We show first that $\mathbb{E}$-FLIP belongs to the class $\mathbf{PLS}^{\#\mathbf{P}}$. We will see that the cost function is, in fact, a #P function in this case. Once this is established, given a problem instance C and a solution x for it, a better neighboring solution (if it exists) can be obtained by computing the costs on the l neighbors of x and selecting one with the highest cost. This is clearly in $\mathbf{FP}^{\#\mathbf{P}}$. The cost of a solution x is $c(C, x) = \pi \sum_{y \in \{0,1\}^m} p(y) f(x, y)$. We describe a non-deterministic polynomial time machine M that on input (C, x) has exactly $c(C, x)$ accepting paths. Observe that $c(C, x)$ is bounded by $\pi 2^n$, and let s be the smallest integer such that $\pi 2^n \leq 2^s$. s is polynomial in the input size. M computes π and s, in a non-deterministic way, chooses a $y \in \{0,1\}^l$ and a number k, $0 < k \leq 2^s$ and accepts if and only if $k \leq \pi p(y) f(x, y)$. The number of accepting paths of M is then $\sum_y \pi p(y) f(x, y) = c(C, x)$.

The hardness proof for $\mathbb{E}$-FLIP in $\mathbf{PLS}^{\#\mathbf{P}}$ follows the same steps as the proof of FLIP for **PLS** from [7], reducing a generic problem L in $\mathbf{PLS}^{\#\mathbf{P}}$ to $\mathbb{E}$-FLIP over two intermediate problems M and Q so that Q is a version of L with the same neighboring structure as $\mathbb{E}$-FLIP. We refer to the original proof for the details. The only difference here is that in the reduction from Q to $\mathbb{E}$-FLIP, the cost function $c(x, s)$ on an instance x in the original proof has to be transformed into the cost function of an instance of $\mathbb{E}$-FLIP so that local optimality is preserved. By assumption $c(x, s) = \frac{f(x,s)}{g(x)}$ for functions $f \in \#\mathbf{P}$ and $g \in \mathbf{FP}$. Since g does not depend on the solution s, it is enough if the value of the circuit is $f(x, s)$ to preserve local optimality. f can be computed as the number of accepting paths of a non-deterministic polynomial time Turing machine N. We can assume that for some polynomial q, all the computation paths of this machine on an input (x, s) have exactly $q(|x|)$ non-deterministic binary choices. These choices can be encoded as a binary string y of length $q(|x|)$. By standard techniques, for inputs of length $|x|$, N can be transformed into a polynomial sized circuit C with a single output bit b so that on input x, y, it is $b = 1$ if and only if N accepts with input x and non-deterministic choices y. If we consider the input bits for y as probabilistic bits, each one with probability $\frac{1}{2}$, then $\pi = 2^{q(|x|)}$ and the output expectation of the circuit over all $y's$ is exactly $c(x, s)2^{-q(|x|}$ and the expectation of $b\pi$ is exactly $f(x, s)$.

Based on this initial complete problem, we can determine the completeness of other interesting problems. Due to space restrictions, we only give a sketch of the proofs. Finally, we will show the completeness of a natural variant of our CGRFs. We start introducing the problem $\mathbb{E}$-Pos-NAE-3SAT, a version of positive not-all-equal 3-SAT with random variables.

An instance of $\mathbb{E}$-Pos-NAE-3SAT consists of a Boolean formula $F(x, y)$ in 3-CNF, with two sets of variables: $y_1,...,y_m$ are random, with $p(y_i)$ denoting the probability that $y_i = 1$. Again, for a given vector y, we denote the probability of it occurring as $p(y) = \prod_{\{i\,:\,y_i=1\}} p(y_i) \cdot \prod_{\{i\,:\,y_i=0\}} (1 - p(y_i))$. Variables $x_1, ..., x_l$ can be chosen freely. There are n clauses C_i each containing only positive literals. A clause is considered satisfied, if at least one variable in it is assigned 0 and one is assigned 1. If C_i is satisfied, we say that $C_i(x, y) = 1$ (for a fixed y), and $C_i(x, y) = 0$ otherwise. Each clause C_i has an associated weight $w_i \in \mathbb{N}$.

A solution for the problem is any string $x \in \{0, 1\}^n$ and the neighborhood of a solution x is the set of vectors at Hamming distance 1 from x. For a fixed vector y, let $f(x, y) = \sum_{i=1}^{n} C_i(x, y) w_i$. The weight of a solution x is the expected weight over all vectors y: $c(x) = \sum_{y \in \{0,1\}^m} p(y) f(x, y)$. A local optimum is reached, when $c(x)$ cannot be further increased by a single flip of a variable in x.

Definition 4. *The local search problem for a formula $F(x, y)$ as described above, consists of finding a solution x such that $c(x)$ is a local maximum.*

Theorem 6. *$\mathbb{E}$-Pos-NAE-3SAT is $\boldsymbol{PLS}^{\#\boldsymbol{P}}$-complete.*

Proof Sketch. Using the same techniques as before, it is easy to see that $\mathbb{E}$-Pos-NAE-3SAT lies within $\mathbf{PLS}^{\#\mathbf{P}}$. To show hardness, the reduction is analogous to the one from FLIP to Pos-NAE-3SAT found in [22]. Observe, that their *test circuits* now simply compare the expected cost of a flip instead of the exact cost.

From $\mathbb{E}$-Pos-NAE-3SAT, we can reduce to our newly introduced problem:

Theorem 7. *Finding a PNE in CGRFs with list strategies and constant list lengths is* $\boldsymbol{PLS}^{\#\boldsymbol{P}}$*-complete.*

Proof Sketch. We show in section Theorem 4 that calculating the cost of a solution is #**P**-complete, so in particular, it lies in #**P**. It is then easy to see that finding a PNE in CGRFs lies within $\mathbf{PLS}^{\#\mathbf{P}}$. For hardness, we reduce from $\mathbb{E}$-Pos-NAE-3SAT to CGRFs. We adapt the proof that showed the **PLS**-completeness of finding a PNE in classical congestion games by Fabrikant et al. [4]. Consider an instance of $\mathbb{E}$-Pos-NAE-3SAT defined through a formula F as described above Definition 4.

We now create a CGRF out of F. For each clause C_i, we introduce two resources e_i^1 and e_i^0. The cost of both resources is w_i for three players, and 0 otherwise. Both resources exist with probability 1. For each probabilistic variable y_i, we also introduce two resources, namely r_i^1 and r_i^0. It is $p(r_i^1) = p(y_i)$, and $p(r_i^0) = 1$. The cost of r_i^1 is constant 0, while the cost of r_i^0 is always a very large number M. The variables (both probabilistic and standard ones) are identified with one player each. We address them through their variable names in F. We continue to call players from standard variables *standard players* and those from probabilistic variables *probabilistic players*. Each standard player x_i has two goals, $g_{x_i} = \{\{e_j^1 : x_i \in c_j\}, \{e_j^0 : x_i \in c_j\}\}$. One goal represents assigning 1 to the variable, the other 0. The goals of a probabilistic player y_i are similar, however they are enriched with the r_i^b resources: $g_{y_i} = \{\{e_j^1 : y_i \in c_j\} \cup \{r_i^1\}, \{e_j^0 : y_i \in c_j\} \cup \{r_i^0\}\}$. The cost of failure is $2M$.

Observe now that due to the high cost of the r_i^0 resources, it is a dominant strategy for the probabilistic players to play the list $(\{e_j^1 : y_i \in c_j\} \cup \{r_i^1\}, \{e_j^0 : y_i \in c_j\} \cup \{r_i^0\})$. This exactly simulates the random variable: Resource r_i^1 exists with probability $p(y_i)$, thus player y_i plays the according goal with probability $p(y_i)$, assigning the variable the value 1. Otherwise, she assigns it the value 0. The cost of failure is chosen big enough to ensure that a list containing both goals is always a dominant strategy.

Players choosing a resource e_i^0 or e_i^1 suffer non-zero cost from this resource if and only if the clause is not satisfied. This is the case when all variables in the clause are assigned the same value. The cost incurred by that scenario is the weight of the clause. A standard player can reduce her expected cost if and only if the same change in the $\mathbb{E}$-Pos-NAE-3SAT instance increases the weight of the solution. Therefore, maxima in the instance of $\mathbb{E}$-Pos-NAE-3SAT and PNEs of the CGRF coincide.

7 Conclusion

We have introduced a general description of congestion games with uncertainty in the availability of resources. The description leaves the exact specification of a strategy open, however using a potential function, we proved a guarantee of existence of PNEs independent of the specification. For many practical definitions of a strategy, our model generalizes classical congestion games.

We then focused on explicit definitions of a strategy: In one, a strategy is given as a Boolean circuit that returns a set of played resources for any set of existing resources. In the other, the strategy is displayed through a priority list, giving preference to resource sets further up in the list. Drawing similarities with the **PLS**-complete problem in classical congestion games, we discussed the complexity of finding PNEs.

Using the Boolean circuit model, a PNE can be stated in polynomial time if, for any fixed set of existing resources, the resulting classical congestion game can be solved in polynomial time as well. This implies, in particular, that a PNE in symmetric network CGRF can be computed in polynomial time. Peculiarly, while we can state the equilibrium in polynomial time, it is unlikely that the players' cost in the equilibrium can be efficiently calculated.

Using the priority lists, we proved that calculating the cost of a player is #**P**-complete under metric reductions, demonstrating that **PLS** is insufficient to describe the complexity of this type of CGRF. We introduced a generalization of **PLS** called $\mathbf{PLS}^{\#\mathbf{P}}$, which has access to a #**P** oracle when improving a given solution. We proved completeness of this problem for finding a PNE in the list model, given that the length of the list is a constant.

We leave several open ends for further research. While we proved the $\mathbf{PLS}^{\#\mathbf{P}}$-completeness of finding pure Nash equilibria in CGRFs with a list of constant size, the classification of the general case remains unknown. We would be very interested to see useful definitions for strategies that allow the feasible calculation of PNEs, or more feasible special cases for the strategy definitions introduced in this work, including different types of cost and utility functions. Regarding the properties of the newly introduced class, similar to **PLS**, it would be interesting to compare $\mathbf{PLS}^{\#\mathbf{P}}$ to $\mathbf{FP}^{\#\mathbf{P}}$ and $\mathbf{FNP}^{\#\mathbf{P}}$. Furthermore, it may be fruitful to consider approximate variants of equilibria, as the hardness of finding PNEs may be due to a large number of possible strategy changes with little effect on the cost. Besides existence and complexity it would be interesting to analyze the inefficiency of equilibria in the form of the price of anarchy.

References

1. Ackermann, H., Röglin, H., Vöcking, B.: On the impact of combinatorial structure on congestion games. J. ACM **55**(6), 1–22 (2008)
2. Angelidakis, H., Fotakis, D., Lianeas, T.: Stochastic congestion games with risk-averse players. In: Vöcking, B. (ed.) SAGT 2013. LNCS, vol. 8146, pp. 86–97. Springer, Heidelberg (2013). https://doi.org/10.1007/978-3-642-41392-6_8

3. Beier, R., Czumaj, A., Krysta, P., Vöcking, B.: Computing equilibria for congestion games with (im)perfect information. In: Proceedings of the Fifteenth Annual ACM-SIAM Symposium on Discrete Algorithms, pp. 746–755. Citeseer (2004)
4. Fabrikant, A., Papadimitriou, C., Talwar, K.: The complexity of pure nash equilibria. In: Proceedings of the Thirty-sixth Annual ACM Symposium on Theory of Computing, pp. 604–612 (2004)
5. Gairing, M.: Malicious Bayesian congestion games. In: Bampis, E., Skutella, M. (eds.) WAOA 2008. LNCS, vol. 5426, pp. 119–132. Springer, Heidelberg (2009). https://doi.org/10.1007/978-3-540-93980-1_10
6. Georgiou, C., Pavlides, T., Philippou, A.: Selfish routing in the presence of network uncertainty. Parallel Process. Lett. **19**(01), 141–157 (2009)
7. Johnson, D.S., Papadimitriou, C.H., Yannakakis, M.: How easy is local search? J. Comput. Syst. Sci. **37**(1), 79–100 (1988)
8. Kleinberg, R., Korten, O., Mitropolsky, D., Papadimitriou, C.H.: Total functions in the polynomial hierarchy. In: Lee, J.R. (ed.) 12th Innovations in Theoretical Computer Science Conference. LIPIcs, vol. 185, pp. 44:1–44:18. Schloss Dagstuhl - Leibniz-Zentrum für Informatik (2021). https://doi.org/10.4230/LIPIcs.ITCS.2021.44
9. Krentel, M.W.: The complexity of optimization problems. J. Comput. Syst. Sci. **36**(3), 490–509 (1988). https://doi.org/10.1016/0022-0000(88)90039-6
10. Meir, R., Parkes, D.: Congestion games with distance-based strict uncertainty. In: Proceedings of the AAAI Conference on Artificial Intelligence, vol. 29 (2015)
11. Meir, R., Tennenholtz, M., Bachrach, Y., Key, P.: Congestion games with agent failures. In: Proceedings of the AAAI Conference on Artificial Intelligence, vol. 26 (2012)
12. Monderer, D., Shapley, L.S.: Potential games. Games Econ. Behav. **14**(1), 124–143 (1996)
13. Nikolova, E., Stier-Moses, N.E.: Stochastic selfish routing. In: Persiano, G. (ed.) SAGT 2011. LNCS, vol. 6982, pp. 314–325. Springer, Heidelberg (2011). https://doi.org/10.1007/978-3-642-24829-0_28
14. Nisan, N., Roughgarden, T., Tardos, É., Vazirani, V.V. (eds.): Algorithmic Game Theory. Cambridge University Press, Cambridge (2007). https://doi.org/10.1017/CBO9780511800481
15. Penn, M., Polukarov, M., Tennenholtz, M.: Congestion games with failures. In: Proceedings of the 6th ACM Conference on Electronic Commerce, pp. 259–268 (2005)
16. Penn, M., Polukarov, M., Tennenholtz, M.: Congestion games with load-dependent failures: identical resources. Games Econom. Behav. **67**(1), 156–173 (2009)
17. Penn, M., Polukarov, M., Tennenholtz, M.: Random order congestion games. Math. Oper. Res. **34**(3), 706–725 (2009)
18. Penn, M., Polukarov, M., Tennenholtz, M.: Taxed congestion games with failures. Ann. Math. Artif. Intell. **56**(2), 133–151 (2009)
19. Piliouras, G., Nikolova, E., Shamma, J.S.: Risk sensitivity of price of anarchy under uncertainty. ACM Trans. Econ. Comput. **5**(1), 1–27 (2016)
20. Rosenthal, R.W.: A class of games possessing pure-strategy Nash equilibria. Int. J. Game Theory **2**(1), 65–67 (1973)
21. Savage, J.E.: Models of Computation - Exploring the Power of Computing. Addison-Wesley, Boston (1998)
22. Schäffer, A.A., Yannakakis, M.: Simple local search problems that are hard to solve. SIAM J. Comput. **20**(1), 56–87 (1991)

23. Toda, S.: PP is as hard as the polynomial-time hierarchy. SIAM J. Comput. **20**(5), 865–877 (1991). https://doi.org/10.1137/0220053
24. Valiant, L.G.: The complexity of computing the permanent. Theor. Comput. Sci. **8**(2), 189–201 (1979)

Markets and Matchings

On (Coalitional) Exchange-Stable Matching

Jiehua Chen(✉), Adrian Chmurovic, Fabian Jogl, and Manuel Sorge(✉)

TU Wien, Vienna, Austria
jiehua.chen@tuwien.ac.at, manuel.sorge@ac.tuwien.ac.at

Abstract. We study *(coalitional) exchange stability*, which Alcalde [Economic Design, 1995] introduced as an alternative solution concept for matching markets involving property rights, such as assigning persons to two-bed rooms. Here, a matching of a given STABLE MARRIAGE or STABLE ROOMMATES instance is called *coalitional exchange-stable* if it does not admit any *exchange-blocking coalition*, that is, a subset S of agents in which everyone prefers the partner of some other agent in S. The matching is *exchange-stable* if it does not admit any *exchange-blocking pair*, that is, an exchange-blocking coalition of size two.

We investigate the computational and parameterized complexity of the COALITIONAL EXCHANGE-STABLE MARRIAGE (resp. COALITIONAL EXCHANGE ROOMMATES) problem, which is to decide whether a STABLE MARRIAGE (resp. STABLE ROOMMATES) instance admits a coalitional exchange-stable matching. Our findings resolve an open question and confirm the conjecture of Cechlárová and Manlove [Discrete Applied Mathematics, 2005] that COALITIONAL EXCHANGE-STABLE MARRIAGE is NP-hard even for complete preferences without ties. We also study bounded-length preference lists and a local-search variant of deciding whether a given matching can reach an exchange-stable one after at most k *swaps*, where a swap is defined as exchanging the partners of the two agents in an exchange-blocking pair.

1 Introduction

An instance in a matching market consists of a set of agents that each have preferences over other agents with whom they want to be matched with. The goal is to find a matching, i.e., a subset of disjoint pairs of agents, which is *fair*. A classical notion of fairness is *stability* [14], meaning that no two agents can form a *blocking pair*, i.e., they would prefer to be matched with each other rather than with the partner assigned by the matching. This means that a matching is fair if the agents cannot take local action to improve their outcome. If we assign property rights via the matching, however, then the notion of blocking pairs may not be actionable, as Alcalde [3] observed: For example, if the matching

JC was supported by the WWTF research project (VRG18-012). MS was supported by the Alexander von Humboldt Foundation.

I. Caragiannis and K. A. Hansen (Eds.): SAGT 2021, LNCS 12885, pp. 205–220, 2021.
https://doi.org/10.1007/978-3-030-85947-3_14

represents an assignment of persons to two-bed rooms, then two persons in a blocking pair may not be able to deviate from the assignment because they may not find a new room that they could share. Instead, we may consider the matching to be *fair* if no two agents form an *exchange-blocking pair*, i.e., they would prefer to have each other's partner rather than to have the partner given by the matching [3]. In other words, they would like to *exchange* their partners. Note that such an exchange would be straightforward in the room-assignment problem mentioned before. We refer to the work of Alcalde [3], Cechlárová [9], and Cechlárová and Manlove [10] for more discussion and examples of markets involving property rights.

If a matching does not admit an exchange-blocking pair, then the matching is *exchange-stable*. If we also want to exclude the possibility that several agents may collude to favorably exchange partners, then we arrive at *coalitional exchange-stability* [3]. In contrast to classical stability and exchange-stability for perfect matchings (i.e., everyone is matched), it is not hard to observe that coalitional exchange-stability implies *Pareto-optimality*, another fairness concept which asserts that no other matching can make at least one agent better-off without making some other agent worse-off (see also Abraham and Manlove [2]). Cechlárová and Manlove [10] showed that the problem of deciding whether an exchange-stable matching exists is NP-hard, even for the marriage case (where the agents are partitioned into two subsets of equal size such that each agent of either subset has preferences over the agents of the other subset) with complete preferences but without ties. They left open whether the NP-hardness transfers to coalitional exchange-stability, but observed NP-containment.

In this paper, we study the algorithmic complexity of problems revolving around (coalitional) exchange-stability. In particular, we establish a first NP-hardness result for deciding coalitional exchange-stability, confirming a conjecture of Cechlárová and Manlove [10]. The NP-hardness reduction is based on a novel *switch-gadget* wherein each preference list contains at most three agents. Utilizing this, we can carefully complete the preferences so as to prove the desired NP-hardness. We then investigate the impact of the maximum length d of a preference list. We find that NP-hardness for both exchange-stability and coalitional exchange-stability starts already when $d = 3$, while it is fairly easy to see that the problem becomes polynomial-time solvable for $d = 2$. For $d = 3$, we obtain a fixed-parameter algorithm for exchange-stability regarding a parameter which is related to the number of switch-gadgets.

Finally, we look at a problem variant, called PATH TO EXCHANGE-STABLE MARRIAGE (P-ESM), for uncoordinated (or decentralized) matching markets. Starting from an initial matching, in each iteration the two agents in an exchange-blocking pair may swap their partners. An interesting question regarding the behavior of the agents in uncoordinated markets is whether such iterative swap actions can reach a stable state, i.e., exchange-stability, and how hard is it to decide. It is fairly straight-forward to verify that if the number k of swaps is bounded by a constant, then P-ESM is polynomial-time solvable since there are only polynomially many possible sequences of exchanges to be checked. From the parameterized complexity point of view, we obtain an XP algorithm for k, i.e., the exponent in the polynomial running time depends on k. We further show that the dependency of the exponent on k is unlikely to be removed by showing W[1]-hardness with respect to k.

Related Work. Alcalde [3] introduced (coalitional) exchange stability and discussed restricted preference domains where (coalitional) exchange stability is guaranteed to exist. Abizada [1] showed a weaker condition (on the preference domain) to guarantee the existence of exchange stability. Cechlárová and Manlove [10] proved that it is NP-complete to decide whether an exchange-stable matching exists, even for the marriage case with complete preferences without ties. Aziz and Goldwasser [4] introduced several relaxed notions of coalitional exchange-stability and discussed their relations.

The P-ESM problem is inspired by the PATH-TO-STABILITY VIA DIVORCES (PSD) problem, originally introduced by Knuth [16], see also Biró and Norman [5] for more background. Very recently, Chen [11] showed that PSD is NP-hard and W[1]-hard when parameterized by the number of divorces. P-ESM can also be considered as a local search problem and is a special case of the LOCAL SEARCH EXCHANGE-STABLE SEAT ARRANGEMENT (LOCAL-STA) problem, introduced by Bodlaender et al. [6]: Given a a set of agents, each having cardinal preferences (i.e., real values) over the other agents, an undirected graph G with the same number of vertices as agents, and an initial assignment (bijection) of the agents to the vertices in G, is it possible to swap two agents' assignments iteratively so as to reach an exchange-stable assignment? Herein an assignment is called *exchange-stable* if no two agents can each have a higher sum of cardinal preferences over the other's neighboring agents. P-ESM is a restricted variant of LOCAL-STA, where G consists of disjoint edges and the agents have ordinal preferences. Bodlaender et al. [7] showed that LOCAL-STA is W[1]-hard wrt. the number k of swaps. Their reduction relies on the fact that the given graph contains cliques and stars, and the preferences of the agents may contain ties. Our results for P-ESM that LOCAL-STA is W[1]-hard even if the given graph consists of disjoint edges and the preferences do not have ties. Finally, we mention that Irving [15] and McDermid et al. [17] studied the complexity of computing stable matchings in the marriage setting with preference lists, requiring additionally that the matching should be man-exchange stable, i.e., no two men form an exchange-blocking pair, obtaining hardness and tractability results.

Organization. In Sect. 2, we introduce relevant concepts and notation, and define our central problems. In Sect. 3, we investigate the complexity of deciding (coalitional) exchange-stability, both when the preferences are complete and when the preferences length are bounded. In Sect. 4, we provide algorithms for profiles with preference length bounded by three. In Sect. 5, we turn to the local search variant of reaching exchange-stability. Section 6 concludes with open questions. Due to space constraints, results marked by $\star$ are deferred to [12].

2 Basic Definitions and Observations

For each natural number t, we denote the set $\{1, 2, \ldots, t\}$ by $[t]$.

Let $V = \{1, 2, \ldots, 2n\}$ be a set of $2n$ agents. Each agent $i \in V$ has a nonempty subset of agents $V_i \subseteq V$ which he finds *acceptable* as a partner and has a *strict preference list* $\succ_i$ on V_i (i.e., a linear order on V_i). The *length* of preference list $\succ_i$ is defined as the number of acceptable agents of i, i.e., $|V_i|$. Here, $x \succ_i y$ means that i *prefers* x to y.

We assume that the acceptability relation among the agents is *symmetric*, i.e., for each two agents x and y it holds that x is acceptable to y if and only if y is acceptable to x. For two agents x and y, we call x *most acceptable* to y if x is a maximal element in the preference list of y. For notational convenience, we write $X \succ_i Y$ to indicate that for each pair of agents $x \in X$ and $y \in Y$ it holds that $x \succ_i y$.

A *preference profile* $\mathcal{P}$ is a tuple $(V, (\succ_i)_{i \in V})$ consisting of an agent set V and a collection $(\succ_i)_{i \in V}$ of preference lists for all agents $i \in V$. For a graph G, by $V(G)$ and $E(G)$ we refer to its vertex set and edge set, respectively. Given a vertex $v \in V(G)$, by $N_G(v)$ and $d_G(v)$ we refer to the neighborhood and degree of v in G, respectively. To a preference profile $\mathcal{P}$ with agent set V we assign an *acceptability graph* $G(\mathcal{P})$ which has V as its vertex set and two agents are connected by an edge if they find each other acceptable. A preference profile $\mathcal{P}$ may have the following properties: Profile $\mathcal{P}$ is *bipartite*, if the agent set V can be partitioned into two agent sets U and W of size n each, such that each agent from one set has a preference list over a subset of the agents from the other set. Profile $\mathcal{P}$ has *complete* preferences if the underlying acceptability graph $G(\mathcal{P})$ is a complete graph or a complete bipartite graph on two disjoint sets of vertices of equal size; otherwise it has *incomplete* preferences. Profile $\mathcal{P}$ has *bounded length* d if each preference list in $\mathcal{P}$ has length at most d.

(Coalitional) Exchange-stable Matchings. A *matching* M for a given profile $\mathcal{P}$ is a subset of disjoint edges from the underlying acceptability graph $G(\mathcal{P})$. Given a matching M for $\mathcal{P}$, and two agents x and y, if it holds that $\{x, y\} \in M$, then we use $M(x)$ (resp. $M(y)$) to refer to y (resp. x), and we say that x and y are their respective assigned *partners* under M and that they are *matched* to each other; otherwise we say that $\{x, y\}$ is an *unmatched pair* under M. If an agent x is *not* assigned any partner by M, then we say that x is *unmatched by* M and we put $M(x) = x$. We assume that each agent x prefers to be matched than remaining unmatched. To formalize this, we will always say that x prefers all acceptable agents from V_x to himself x. A matching M is *perfect* if every agent is assigned a partner. It is *maximal* if for each unmatched pair $\{x, y\} \in E(G(\mathcal{P})) \setminus M$ it holds that x or y is matched under M. For two agents x, y, we say that x *envies* y *under* M if x prefers the partner of y, i.e., $M(y)$, to his partner $M(x)$. We omit the "under M" if it is clear from the context.

Matching M admits an *exchange-blocking coalition* (in short *ebc*) if there exists a sequence $\rho = (x_0, x_1, \ldots, x_{r-1})$ of r agents, $r \geq 2$, such that each agent x_i envies her successor x_{i+1} in ρ (index $i + 1$ taken modulo r). The *size* of an ebc is defined as the number of agents in the sequence. An *exchange-blocking pair* (in short *ebp*) is an ebc of size two. A matching M of $\mathcal{P}$ is *exchange-stable* (resp. *coalitional exchange-stable*) if it does not admit any ebp (resp. ebc). Note that a coalitional exchange-stable matching is exchange-stable. For an illustration, let us consider the following example.

Example 1. The following bipartite preference profile $\mathcal{P}$ with agent sets $U = \{x, y, z\}$ and $W = \{a, b, c\}$ admits 2 (coalitional) exchange-stable matchings M_1 and M_2 with $M_1 = \{\{x, c\}, \{y, b\}, \{z, a\}\}$ (marked in red boxes) and $M_2 = \{\{x, b\}, \{y, c\}, \{z, a\}\}$ (marked in blue boxes).

$$x: a \succ \boxed{b} \succ \boxed{c},\quad a: y \succ x \succ \boxed{\boxed{z}},$$
$$y: \boxed{b} \succ a \succ \boxed{c},\quad b: \boxed{x} \succ \boxed{y} \succ z,$$
$$z: \boxed{\boxed{a}} \succ c \succ b,\quad c: \boxed{x} \succ \boxed{y} \succ z.$$

Matching M_3 with $M_3 = \{\{x, c\}, \{y, a\}, \{z, b\}\}$ is not exchange-stable and hence not coalitional exchange-stable since for instance (y, z) is an exchange-blocking pair of M_3.

As already observed by Cechlárová and Manlove [10], exchange-stable (or coalitional exchange-stable) matchings may not exist, even for bipartite profiles with complete preferences. Every coalitional exchange-stable matching is maximal ($\star$).

We are interested in the computational complexity of deciding whether a given profile admits a coalitional exchange-stable matching.

Coalitional Exchange-Stable Roommates (CESR)
Input: A preference profile $\mathcal{P}$.
Question: Does $\mathcal{P}$ admit a coalitional exchange-stable matching?

The bipartite restriction of CESR, called Coalitional Exchange-Stable Marriage (CESM), has as input a *bipartite* preference profile. Exchange-Stable Roommates (ESR) and Exchange-Stable Marriage (ESM) are defined analogously.

We are also interested in the case when the preferences have bounded length. In this case, not every coalitional exchange-stable (or exchange-stable) matching is perfect. In keeping with the literature [9,10], we focus on the perfect case.

d-Coalitional Exchange-Stable Roommates (d-CESR)
Input: A preference profile $\mathcal{P}$ with preferences of bounded length d.
Question: Does $\mathcal{P}$ admit a coalitional exchange-stable and *perfect* matching?

We analogously define the bipartite restriction d-Coalitional Exchange-Stable Marriage *(d-CESM)*, and the exchange-stable variants d-Exchange-Stable Roommates (d-ESR) and d-Exchange-Stable Marriage (d-ESM). Note that the above problems are contained in NP [10].

Finally, we investigate a local search variant regarding exchange-stability. To this end, given two matchings M and N of the same profile $\mathcal{P}$, we say that M is *one-swap reachable* from N if there exists an exchange-blocking pair (x, y) of N such that $M = (N \setminus \{\{x, N(x)\}, \{y, N(y)\}\}) \cup \{\{x, y\}, \{N(x), N(y)\}\}$. Accordingly, we say that M is *k-swaps reachable* from N if there exists a sequence $(M_0, M_1, \cdots, M_{k'})$ of k' *matchings* of profile $\mathcal{P}$ such that (a) $k' \leq k$, $M_0 = N$, $M_{k'} = M$, and (b) for each $i \in [k']$, M_i is one-swap reachable from M_{i-1}.

The local search problem variant is defined as follows:

PATH TO EXCHANGE-STABLE MARRIAGE (P-ESM)
Input: A bip. preference profile $\mathcal{P}$, a matching M_0 of $\mathcal{P}$, and an integer k.
Question: Is there an exchange-stable matching M for $\mathcal{P}$ that is k-swap reachable from M_0?

3 Deciding (Coalitional) Exchange-Stability is NP-complete

Cechlárová and Manlove [10] proved NP-completeness for ESM. It is, however, not immediate how to adapt Cechlárová and Manlove's proof to show hardness for coalitional exchange-stability since their constructed exchange-stable matching is not always coalitional exchange-stable. To obtain a hardness reduction for CESM, we first study the case when the preferences have length bounded by three, and show that 3-CESM is NP-hard, even for strict preferences. We reduce from an NP-complete ($\star$) variant of 3SAT, called (2,2)-3SAT: Is there a satisfying truth assignment for a given Boolean formula $\phi(X)$ with variable set X in 3CNF (i.e., a set of clauses each containing at most 3 literals) where no clause contains both the positive and the negated literal of the same variable, and each literal appears *exactly* two times?

A crucial ingredient for our reduction is the following *switch-gadget* which enforces that each exchange-stable matching results in a valid truth assignment. The gadget and its properties are summarized in the following lemma.

Lemma 1 ($\star$). *Let $\mathcal{P}$ be a bipartite preference profile on agent sets U and W. Let $A = \{a^z \mid z \in \{0, 1, \ldots, 6\}\}$ and $B = \{b^z \mid z \in \{0, 1, \ldots, 6\}\}$ be two disjoint sets of agents, and let $Q = \{\alpha, \beta, \gamma, \delta\}$ be four further distinct agents with $A \cup \{\alpha, \gamma\} \subseteq U$ and $B \cup \{\beta, \delta\} \subseteq W$. The preferences of the agents from A and B are as follows; the preferences of the other agents are arbitrary but fixed.*

- a^0: $\boxed{b^1} \succ \boxed{\boxed{\beta}}$,
- a^1: $\boxed{b^0} \succ \boxed{\boxed{b^2}} \succ b^1$,
- a^2: $\boxed{b^3} \succ \boxed{\boxed{b^1}} \succ b^2$,
- a^3: $\boxed{b^2} \succ \boxed{b^3} \succ \boxed{b^4}$,
- a^4: $b^4 \succ \boxed{b^3} \succ \boxed{\boxed{b^5}}$,
- a^5: $\boxed{b^6} \succ \boxed{\boxed{b^4}} \succ b^5$,
- a^6: $\boxed{b^5} \succ \boxed{\boxed{\delta}}$,

- b^0: $\boxed{a^1} \succ \boxed{\boxed{\alpha}}$,
- b^1: $\boxed{a^0} \succ \boxed{\boxed{a^2}} \succ a^1$,
- b^2: $a^2 \succ \boxed{a^3} \succ \boxed{\boxed{a^1}}$,
- b^3: $\boxed{a^4} \succ \boxed{a^3} \succ \boxed{a^2}$,
- b^4: $\boxed{a^3} \succ \boxed{\boxed{a^5}} \succ a^4$,
- b^5: $\boxed{a^6} \succ \boxed{\boxed{a^4}} \succ a^5$,
- b^6: $\boxed{a^5} \succ \boxed{\boxed{\gamma}}$.

Every perfect *matching* M *of* $\mathcal{P}$ *satisfies the following, where*

$$N^1 := \{\{\alpha, b^0\}, \{a^6, \delta\}\} \cup \{\{a^{z-1}, b^z\} \mid z \in [6]\},$$
$$N^2 := \{\{a^0, \beta\}, \{\gamma, b^6\}\} \cup \{\{a^z, b^{z-1}\} \mid z \in [6]\}, \text{ and}$$
$$N^D := \{\{\alpha, b^0\}, \{a^0, \beta\}, \{a^6, \delta\}, \{\gamma, b^6\}, \{a^1, b^2\}, \{a^2, b^1\}, \{a^3, b^3\}, \{a^4, b^5\}, \{a^5, b^4\}\}.$$

(1) If M is exchange-stable, then either (i) $N^1 \subseteq M$, or (ii) $N^2 \subseteq M$, or (iii) $N^D \subseteq M$.
(2) If $N^1 \subseteq M$, then every ebc of M which involves an agent from A (resp. B) also involves α (resp. δ).
(3) If $N^2 \subseteq M$, then every ebc of M which involves an agent from A (resp. B) also involves γ (resp. β).
(4) If $N^D \subseteq M$, then every ebc of M which involves an agent from A (resp. B) also involves an agent from $\{\alpha, \gamma\}$ (resp. $\{\beta, \delta\}$).

Using Lemma 1, we can show NP-hardness for bounded preference length.

Theorem 1. 3-CESM, 3-ESM, 3-CESR, *and* 3-ESR *are NP-complete.*

Proof. As already mentioned [10], by checking for cycles in the envy graph all discussed problems are in NP ($\star$). For the NP-hardness, it suffices to show that 3-CESM and 3-ESM are NP-hard. We use the same reduction from (2,2)-3SAT for both. Let (X, C) be an instance of (2,2)-3SAT where $X = \{x_1, x_2, \cdots, x_{\hat{n}}\}$ is the set of variables and $\phi = \{C_1, C_2, \cdots, C_{\hat{m}}\}$ the set of clauses.

We construct a bipartite preference profile on two disjoint agent sets U and W. The set U (resp. W) will be partitioned into three different agent-groups: the variable-agents, the switch-agents, and the clause-agents. The general idea is to use the variable-agents and the clause-agents to determine a truth assignment and satisfying literals, respectively. Then, we use the switch-agents from Lemma 1 to make sure that the selected truth assignment is consistent with the selected satisfying literals. For each literal $\mathsf{lit}_i \in X \cup \overline{X}$ that appears in two different clauses C_j and C_k with $j < k$, we use $\mathsf{o}_1(\mathsf{lit}_i)$ and $\mathsf{o}_2(\mathsf{lit}_i)$ to refer to the indices j and k; recall that in ϕ each literal appears exactly two times.

The Variable-agents. For each variable $x_i \in X$, introduce 6 *variable-agents* v_i, w_i, x_i, $\overline{x}_i$, y_i, $\overline{y}_i$. Add $v_i, x_i, \overline{x}_i$ to U, and $w_i, y_i, \overline{y}_i$ to W. For each literal $\mathsf{lit}_i \in X \cup \overline{X}$ let $y(\mathsf{lit}_i)$ denote the corresponding Y-variable-agent, that is, $y(x_i) = y_i$ and $y(\overline{x}_i) = \overline{y}_i$. Define $\overline{X} := \{\overline{x}_i \mid i \in [\hat{n}]\}$, and $\overline{Y} := \{\overline{y}_i \mid i \in [\hat{n}]\}$.

The Clause-agents. For each clause $C_j \in C$, introduce two *clause-agents* c_j, d_j. Further, for each literal $\mathsf{lit}_i \in C_j$ with $\mathsf{lit} \in \{x, \overline{x}\}$, introduce two more *clause-agents* e_j^i, f_j^i. Add c_j, f_j^i to U, and d_j, e_j^i to W. For each clause $C_j \in \phi$, define $E_j := \{e_j^i \mid \mathsf{lit}_i \in C_j\}$, and $F_j := \{f_j^i \mid \mathsf{lit}_i \in C_j\}$. Moreover, define $E := \bigcup_{C_j \in \phi} E_j$ and $F := \bigcup_{C_j \in \phi} F_j$.

The Switch-agents. For each clause $C_j \in C$, and each literal $\mathsf{lit}_i \in C_j$ introduce fourteen *switch-agents* $a^z_{i,j}, b^z_{i,j}$, $z \in \{0, 1, \cdots, 6\}$. Define $A_{i,j} = \{a^z_{i,j} \mid z \in \{0, 1, \ldots, 6\}\}$ and $B_{i,j} = \{b^z_{i,j} \mid z \in \{0, 1, \ldots, 6\}\}$. Add $A_{i,j}$ to U and $B_{i,j}$ to W.

In total, we have the following agent sets:

$U := \{v_i \mid i \in [\hat{n}]\} \cup X \cup \overline{X} \cup \{c_j \mid j \in [\hat{m}]\} \cup F \cup \bigcup_{C_j \in \phi \wedge \mathsf{lit}_i \in C_j} A_{i,j}$, and
$W := \{w_i \mid i \in [\hat{n}]\} \cup Y \cup \overline{Y} \cup \{d_j \mid j \in [\hat{m}]\} \cup E \cup \bigcup_{C_j \in \phi \wedge \mathsf{lit}_i \in C_j} B_{i,j}$.

The Preference Lists. The preference lists of the agents are shown in Fig. 1. Herein, the preferences of the switch-agents of each occurrence of the literal correspond to those given in Lemma 1. Note that all preferences are specified except those of $\alpha_{i,j}$ and $\delta_{i,j}$, which we do now. Defining them in an appropriate way will connect the two groups of switch-agents that correspond to the same literal as well as literals to clauses. For each literal $\mathsf{lit}_i \in X \cup \overline{X}$, recall that $\mathsf{o}_1(\mathsf{lit}_i)$ and $\mathsf{o}_2(\mathsf{lit}_i)$ are the indices of the clauses which contain lit_i with $\mathsf{o}_1(\mathsf{lit}_i) < \mathsf{o}_2(\mathsf{lit}_i)$. Let

$$\alpha_{i,\mathsf{o}_1(\mathsf{lit}_i)} := \mathsf{lit}_i,\ \delta_{i,\mathsf{o}_1(\mathsf{lit}_i)} := b^0_{i,\mathsf{o}_2(\mathsf{lit}_i)}, \alpha_{i,\mathsf{o}_2(\mathsf{lit}_i)} := a^6_{i,\mathsf{o}_1(\mathsf{lit}_i)},\ \delta_{i,\mathsf{o}_2(\mathsf{lit}_i)} := y(\mathsf{lit}_i). \quad (1)$$

$$\begin{array}{lll}
\forall i \in [\hat{n}]: & \bullet\ v_i : \boxed{y_i} \succ \boxed{\overline{y}_i}, & \blacksquare\ w_i : \boxed{x_i} \succ \boxed{\overline{x}_i}, \\
& \bullet\ x_i : \boxed{w_i} \succ \boxed{b^0_{i,\mathsf{o}_1(x_i)}}, & \blacksquare\ y_i : \boxed{v_i} \succ \boxed{a^6_{i,\mathsf{o}_2(x_i)}}, \\
& \bullet\ \overline{x}_i : \boxed{w_i} \succ \boxed{b^0_{i,\mathsf{o}_1(\overline{x}_i)}}, & \blacksquare\ \overline{y}_i : \boxed{v_i} \succ \boxed{a^0_{i,\mathsf{o}_2(\overline{x}_i)}}, \\
\forall j \in [\hat{m}]: & \bullet\ c_j : [E_j], & \blacksquare\ d_j : [F_j], \\
\forall i \in [\hat{n}], \forall j \in [\hat{m}] \text{ with } \mathsf{lit}_i \in C_j: & \bullet\ f^i_j : d_j \succ b^6_{i,j} & \blacksquare\ e^i_j : c_j \succ a^0_{i,j} \\
& \bullet\ a^0_{i,j} : \boxed{b^1_{i,j}} \succ \boxed{e^i_j}, & \blacksquare\ b^0_{i,j} : \boxed{a^1_{i,j}} \succ \boxed{\alpha_{i,j}}, \\
& \bullet\ a^1_{i,j} : \boxed{b^0_{i,j}} \succ \boxed{b^2_{i,j}} \succ b^1_{i,j}, & \blacksquare\ b^1_{i,j} : \boxed{a^0_{i,j}} \succ \boxed{a^2_{i,j}} \succ a^1_{i,j}, \\
& \bullet\ a^2_{i,j} : \boxed{b^3_{i,j}} \succ \boxed{b^1_{i,j}} \succ b^2_{i,j}, & \blacksquare\ b^2_{i,j} : a^2_{i,j} \succ \boxed{a^3_{i,j}} \succ \boxed{a^1_{i,j}}, \\
& \bullet\ a^3_{i,j} : \boxed{b^2_{i,j}} \succ \boxed{b^3_{i,j}} \succ \boxed{b^4_{i,j}}, & \blacksquare\ b^3_{i,j} : \boxed{a^4_{i,j}} \succ \boxed{a^3_{i,j}} \succ \boxed{a^2_{i,j}}, \\
& \bullet\ a^4_{i,j} : b^4_{i,j} \succ \boxed{b^3_{i,j}} \succ \boxed{b^5_{i,j}}, & \blacksquare\ b^4_{i,j} : \boxed{a^3_{i,j}} \succ \boxed{a^5_{i,j}} \succ a^4_{i,j}, \\
& \bullet\ a^5_{i,j} : \boxed{b^6_{i,j}} \succ \boxed{b^4_{i,j}} \succ b^5_{i,j}, & \blacksquare\ b^5_{i,j} : \boxed{a^6_{i,j}} \succ \boxed{a^4_{i,j}} \succ a^5_{i,j}, \\
& \bullet\ a^6_{i,j} : \boxed{b^5_{i,j}} \succ \boxed{\delta_{i,j}}, & \blacksquare\ b^6_{i,j} : \boxed{a^5_{i,j}} \succ \boxed{f^i_j}.
\end{array}$$

Fig. 1. The preferences constructed in the proof for Theorem 1. Recall that for each literal $\mathsf{lit}_i \in X \cup \overline{X}$, expressions $\mathsf{o}_1(\mathsf{lit}_i)$ and $\mathsf{o}_2(\mathsf{lit}_i)$ denote the two indices $j < j'$ of the clauses that contain lit_i. For each clause $C_j \in \phi$, the expression $[E_j]$ (resp. $[F_j]$) denotes an arbitrary but fixed order of the agents in E_j (resp. F_j).

This completes the construction of the instance for 3-CESM, which can clearly be done in polynomial-time. Let $\mathcal{P}$ denote the constructed instance with $\mathcal{P} = (U \uplus W, (\succ_x)_{x \in U \cup W})$. It is straight-forward to verify that $\mathcal{P}$ is bipartite and contains no ties and each preference list $\succ_x$ has length bounded by three. Before we give the correctness proof, for each literal $\mathsf{lit}_i \in X \cup \overline{X}$ and each clause C_j with $\mathsf{lit}_i \in C_j$ we define the following three matchings:

$$\begin{aligned}
N^1_{i,j} &:= \{\{\alpha_{i,j}, b^0_{i,j}\}, \{a^6_{i,j}, \delta_{i,j}\}\} \cup \{\{a^{z-1}_{i,j}, b^z_{i,j}\} \mid z \in [6]\},\\
N^2_{i,j} &:= \{\{a^0_{i,j}, e^i_j\}, \{b^6_{i,j}, f^i_j\}\} \cup \{\{a^z_{i,j}, b^{z-1}_{i,j}\} \mid z \in [6]\}, \text{ and}\\
N^D_{i,j} &:= \{\{\alpha_{i,j}, b^0_{i,j}\}, \{a^0_{i,j}, e^i_j\}, \{a^6_{i,j}, \delta_{i,j}\}, \{f^i_j, b^6_{i,j}\},\\
&\quad \{a^1_{i,j}, b^2_{i,j}\}, \{a^2_{i,j}, b^1_{i,j}\}, \{a^3_{i,j}, b^3_{i,j}\}, \{a^4_{i,j}, b^5_{i,j}\}, \{a^5_{i,j}, b^4_{i,j}\}\}.
\end{aligned} \tag{2}$$

Now we show the correctness, i.e., ϕ admits a satisfying assignment if and only if $\mathcal{P}$ admits a perfect and coalitional exchange-stable (resp. exchange-stable) matching. For the "only if" direction, assume that $\sigma\colon X \to \{\mathsf{true}, \mathsf{false}\}$ is a satisfying assignment for ϕ. Then, we define a perfect matching M as follows.

- For each variable $x_i \in X$, let $M(\overline{x}_i) := w_i$ and $M(v_i) := \overline{y}_i$ if $\sigma(x_i) = \mathsf{true}$; otherwise, let $M(x_i) := w_i$ and $M(v_i) := y_i$.
- For each clause $C_j \in \phi$, fix an arbitrary literal whose truth value satisfies C_j and denote the index of this literal as $\mathsf{s}(j)$. Then, let $M(c_j) := e^{\mathsf{s}(j)}_j$ and $M(f^{\mathsf{s}(j)}_j) := d_j$.
- Further, for each literal $\mathsf{lit}_i \in X \cup \overline{X}$ and each clause C_j with $\mathsf{lit}_i \in C_j$, do:
  (a) If $\mathsf{s}(j) = i$, then add to M all pairs from $N^1_{i,j}$.
  (b) If $\mathsf{s}(j) \neq i$ and lit_i is set true under σ (i.e., $\sigma(x_i) = \mathsf{true}$ iff. $\mathsf{lit}_i = x_i$), then add to M all pairs from $N^D_{i,j}$.
  (c) If $\mathsf{s}(j) \neq i$ and lit_i is set to false under σ (i.e., $\sigma(x_i) = \mathsf{true}$ iff. $\mathsf{lit}_i = \overline{x}_i$), then add to M all pairs from $N^2_{i,j}$.

One can verify that M is perfect. Hence, it remains to show that M is coalitional exchange-stable. Note that this would also imply that M is exchange-stable.

Suppose, for the sake of contradiction, that M admits an ebc ρ. First, observe that for each variable-agent $z \in X \cup \overline{X} \cup Y \cup \overline{Y}$ it holds that $M(z)$ either is matched with his most-preferred partner (i.e., either v_i or w_i) or only envies someone who is matched with his most-preferred partner. Hence, no agent from $X \cup \overline{X} \cup Y \cup \overline{Y}$ is involved in ρ. Analogously, no agent from $E \cup F$ is involved in ρ. Next, we claim the following.

Claim 1 (⋆). *For each literal* $\mathsf{lit}_i \in X \cup \overline{X}$ *and each clause* C_j *with* $\mathsf{lit}_i \in C_j$*, it holds that neither* $\alpha_{i,j}$ *nor* $\delta_{i,j}$ *is involved in* ρ.

Using the above observations and claim, we continue with the proof. We successively prove that no agent is involved in ρ, starting with the agents in U.

- If v_i is involved in ρ for some $i \in [\hat{n}]$, then he only envies someone who is matched with y_i. By the preferences of y_i, this means that $M(y_i) = a^6_{i,\mathsf{o}_2(x_i)}$

and v_i envies $a^6_{i,o_2(x_i)}$. Hence, $a^6_{i,o_2(x_i)}$ is also involved in ρ. Moreover, since $M(a^6_{i,o_2(x_i)}) = y_i$, we have $N^1_{i,o_2(x_i)} \subseteq M$ or $N^D_{i,o_2(x_i)} \subseteq M$. By Lemma 1(2) and Lemma 1(4) (setting $\alpha = \alpha_{i,o_2(x_i)}$, $\beta = e^i_{o_2(x_i)}$, $\gamma = f^i_{o_2(x_i)}$, and $\delta = \delta_{i,o_2(x_i)}$), ρ involves an agent from $\{\alpha_{i,o_2(x_i)}, f^i_{o_2(x_i)}\}$. Since no agent from F is involved in ρ, it follows that ρ involves $\alpha_{i,o_2(x_i)}$, a contradiction to Claim 1.
- Analogously, if $c_j \in \rho$ for some $j \in [\hat{m}]$, then this means that E_j contains two agents e^i_j and e^t_j such that $M(c_j) = e^t_j$ but c_j prefers e^i_j to e^t_j, and $M(e^i_j) \in \rho$. Since M is perfect and c_j is not available, it follows that $M(e^i_j) = a^0_{i,j}$, implying that $a^0_{i,j} \in \rho$. Moreover, by the definition of M we have that $N^2_{i,j} \subseteq M$ or $N^D_{i,j} \subseteq M$. By Lemmas 1(3)–(4) (setting $\alpha = \alpha_{i,j}$, $\beta = e^i_j$, $\gamma = f^i_j$, and $\delta = \delta_{i,j}$), ρ involves an agent from $\{\alpha_{i,j}, f^i_j\}$, a contradiction since no agent from F_j is involved in ρ and by Claim 1 $\alpha_{i,j}$ is not in ρ.
- Analogously, we can obtain a contradiction if w_i with $i \in [\hat{n}]$ is in ρ: By the definition of M, if $w_i \in \rho$, then $M(x_i) = b^0_{i,o_1(x_i)}$ and w_i envies $b^0_{i,o_1(x_i)}$. Hence, $b^0_{i,o_1(x_i)}$ is also involved in ρ. Moreover, since $M(b^0_{i,o_1(x_i)}) = x_i$, it follows that $N^1_{i,o_1(x_i)} \subseteq M$ or $N^D_{i,o_1(x_i)} \subseteq M$. By Lemmas 1(2) and (4) (setting $\alpha = \alpha_{i,o_1(x_i)}$, $\beta = e^i_{o_1(x_i)}$, $\gamma = f^i_{o_1(x_i)}$, and $\delta = \delta_{i,o_1(x_i)}$), ρ involves an agent from $\{e^i_{o_1(x_i)}, \delta_{i,o_1(x_i)}\}$. Since no agent from E is involved in ρ, it follows that ρ involves $\delta_{i,o_1(x_i)}$, a contradiction to Claim 1.
- Again, analogously, if $d_j \in \rho$ for some $j \in [\hat{m}]$, then we obtain that $\delta_{i,j}$ is involved in ρ, which is a contradiction to Claim 1.
- Finally, if ρ involves an agent from $A_{i,j}$ (resp. $B_{i,j}$), then by Lemma 1(2) and (4) (setting $\alpha = \alpha_{i,j}$, $\beta = e^i_j$, $\gamma = f^i_j$, and $\delta = \delta_{i,j}$), it follows that ρ involves an agent from $\{\alpha_{i,j}, f^i_j\}$ (resp. $\{\beta_{i,j}, e^i_j\}$), a contradiction to our observation and to Claim 1.

Summarizing, M is coalitional exchange-stable and exchange-stable.

For the "if" direction, assume that M is a perfect and exchange-stable matching for $\mathcal{P}$. We show that there is a satisfying assignment for ϕ. Note that this then also implies that, if M is perfect and coalitional exchange-stable, then there is a satisfying assignment for ϕ.

We claim that the selection of the partner of w_i defines a satisfying truth assignment for ϕ. More specifically, define a truth assignment $\sigma\colon X \to \{\mathsf{true}, \mathsf{false}\}$ with $\sigma(x_i) = \mathsf{true}$ if $M(w_i) = \overline{x}_i$, and $\sigma(x_i) = \mathsf{false}$ otherwise. We claim that σ satisfies ϕ. To this end, consider an arbitrary clause C_j and the corresponding clause-agent. Since M is perfect, it follows that $M(c_j) = e^i_j$ for some $\mathsf{lit}_i \in C_j$. Since e^i_j is not available, it also follows that $M(a^0_{i,j}) = b^1_{i,j}$. By Lemma 1(1) (setting $\alpha = \alpha_{i,j}$, $\beta = e^i_j$, $\gamma = f^i_j$, and $\delta = \delta_{i,j}$), it follows that $N^1_{i,j} \subseteq M$. In particular, $M(\alpha_{i,j}) = b^0_{i,j}$ so that $\alpha_{i,j}$ is not available to other agents anymore.

Now, if we can show that $\mathsf{lit}_i = \alpha_{i,o_1(\mathsf{lit}_i)}$ is matched to $b^0_{i,o_1(\mathsf{lit}_i)}$, then since M is perfect, we have $M(w_i) = \overline{x}_i$ if $\mathsf{lit}_i = x_i$, and $M(w_i) = \overline{x}_i$ otherwise By definition, we have $\sigma(x_i) = \mathsf{true}$ if $\mathsf{lit}_i = x_i$ and $\sigma(x_i) = \mathsf{false}$ otherwise. Thus,

C_j is satisfied under σ, implying that σ is a satisfying assignment. It remains to show that lit_i is matched to $b^0_{i,\mathsf{o}_1(\mathsf{lit}_i)}$. We distinguish between two cases;

- If $j = \mathsf{o}_1(\mathsf{lit}_i)$, then $\mathsf{lit}_i = \alpha_{i,\mathsf{o}_1(\mathsf{lit}_i)}$ is matched to $b^0_{i,\mathsf{o}_1(\mathsf{lit}_i)}$, as required.
- If $j = \mathsf{o}_2(\mathsf{lit}_i)$, then by definition, it holds that $\alpha_{i,j} = a^6_{i,\mathsf{o}_1(\mathsf{lit}_i)}$ and $\delta_{i,\mathsf{o}_1(\mathsf{lit}_i)} = b^0_{i,j}$. In other words, $M(a^6_{i,\mathsf{o}_1(\mathsf{lit}_i)}) = \delta_{i,\mathsf{o}_1(\mathsf{lit}_1)}$. By Lemma 1(1) (setting $\alpha = \alpha_{i,\mathsf{o}_1(\mathsf{lit}_i)}$, $\beta = e^i_{\mathsf{o}_1(\mathsf{lit}_i)}$, $\gamma = f^i_{\mathsf{o}_1(\mathsf{lit}_i)}$, and $\delta = \delta_{i,\mathsf{o}_1(\mathsf{lit}_i)}$), it follows that $N^1_{i,j} \subseteq M$ or $N^D_{i,j} \subseteq M$. In both cases, it follows that $\alpha_{i,\mathsf{o}_1(i)}$ is matched to $b^0_{i,\mathsf{o}_1(i)}$. □

Next, we show how to complete the preferences of the agents constructed in the proof of Theorem 1 to show hardness for complete and strict preferences.

Theorem 2. CESM *and* CESR *are NP-complete even for complete and strict preferences.*

Proof. We only show NP-hardness for CESM as the hardness for CESR will follow immediately by using the same approach as [10, Lemma 3.1]. To show hardness for CESM, we adapt the proof of Theorem 1. In that proof, given (2,2)-3SAT instance (X, ϕ) with $X = \{x_1, x_2, \cdots, x_{\hat{n}}\}$ and $\phi = \{C_1, C_2, \cdots, C_{\hat{m}}\}$, we constructed two disjoint agent sets U and W with $U := \{v_i \mid i \in [\hat{n}]\} \cup X \cup \overline{X} \cup \{c_j \mid j \in [\hat{m}]\} \cup F \cup \bigcup_{C_j \in \phi \wedge \mathsf{lit}_i \in C_j} A_{i,j}$ and $W := \{w_i \mid i \in [\hat{n}]\} \cup Y \cup \overline{Y} \cup \{d_j \mid j \in [\hat{m}]\} \cup E \cup \bigcup_{C_j \in \phi \wedge \mathsf{lit}_i \in C_j} B_{i,j}$. For each agent $z \in U \cup W$ let L_z denote the preference list of z constructed in the proof. The basic idea is to extend the preference list L_z by appending to it the remaining agents appropriately.

We introduce some more notations. Let $\rhd_U$ and $\rhd_W$ denote two arbitrary but fixed linear orders of the agents in U and W, respectively. Now, for each subset of agents $S \subseteq U$ (resp. $S \subseteq W$), let $[S]_\rhd$ denote the fixed order of the agents in S induced by $\rhd_U$ (resp. $\rhd_W$), and let $S \setminus \mathsf{L}_z$ denote the subset $\{t \in S \mid t \notin \mathsf{L}_z\}$, where $z \in W$ (resp. $z \in U$). Finally, for each agent $z \in U$ (resp. $z \in W$), let R_z denote the subset of agents which do not appear in L_z or in $Y \cup \overline{Y} \cup E$ (resp. $X \cup \overline{X} \cup F$). That is, $\mathsf{R}_z := \big(W \setminus (Y \cup \overline{Y} \cup F)\big) \setminus \mathsf{L}_z$ (resp. $\mathsf{R}_z := \big(U \setminus (X \cup \overline{Y} \cup F)\big) \setminus \mathsf{L}_z$).
Now, we define the preferences of the agents as follows.

$$\forall z \in U,\ z\colon \mathsf{L}_z \succ [Y \cup \overline{Y} \cup E \setminus \mathsf{L}_z]_\rhd \succ [\mathsf{R}_z]_\rhd, \quad \text{and}$$
$$\forall z \in W,\ z\colon \mathsf{L}_z \succ [X \cup \overline{X} \cup F \setminus \mathsf{L}_z]_\rhd \succ [\mathsf{R}_z]_\rhd.$$

Let $\mathcal{P}'$ denote the newly constructed preference profile. Clearly, the constructed preferences are complete and strict. Before we show the correctness, we claim the following for each coalitional exchange-stable matching of $\mathcal{P}'$.

Claim 2 ($\star$). *If M is a coalitional exchange-stable matching for $\mathcal{P}'$, then*

(i) for each agent $z \in U \cup W$ it holds that $M(z) \notin \mathsf{R}_z$, and
(ii) for each agent $z \in U \cup W \setminus (X \cup \overline{X} \cup F \cup Y \cup \overline{Y} \cup E)$ it holds that $M(z) \in \mathsf{L}_z$.

Now we are ready to show the correctness, i.e., ϕ admits a satisfying assignment if and only if $\mathcal{P}'$ admits a coalitional exchange-stable matching.

For the "only if" direction, assume that ϕ admits a satisfying assignment, say $\sigma\colon X \to \{\mathsf{true}, \mathsf{false}\}$. We claim that the coalitional exchange-stable matching M for $\mathcal{P}$ that we defined in the "only if" direction of the proof for Theorem 1 is a coalitional exchange-stable matching for $\mathcal{P}'$. Clearly, M is a perfect matching for $\mathcal{P}'$ since $G(\mathcal{P}')$ is a supergraph of $G(\mathcal{P})$. Since each agent $z \in U \cup W$ has $M(z) \in \mathsf{L}_z$, for every two agents $z, z' \in U$ (resp. W), it holds that z envies z' only if $M(z') \in \mathsf{L}_z$. In other words, if M would admit an ebc $\rho = (z_0, z_1, \cdots, z_{r-1})$ $(r \geq 2)$ for $\mathcal{P}'$, then for each $i \in \{0, 1, \ldots, r-1\}$ it must hold that $M(z_i) \in \mathsf{L}_{z-1}$ ($z-1$ taken modulo r). But then, ρ is also an ebc for $\mathcal{P}$, a contradiction to our "only if" part of the proof for Theorem 1.

For the "if" direction, let M be a coalitional exchange-stable matching for $\mathcal{P}'$. Note that in the "if" part of the proof of Theorem 1 we heavily utilize the properties given in Lemma 1(1). Now, to construct a satisfying assignment for ϕ from M, we will prove that the lemma also holds for profile $\mathcal{P}'$. To this end, for each literal $\mathsf{lit}_i \in X \cup \overline{X}$ and each clause C_j with $\mathsf{lit}_i \in C_j$, recall the three matchings $N^1_{i,j}$, $N^2_{i,j}$, $N^D_{i,j}$ and the agents $\alpha_{i,j}$ and $\delta_{i,j}$ that we have defined in Eqs. (2) and (1) .

Claim 3 ($\star$). *Matching M satisfies for each literal $\mathsf{lit}_i \in X \cup \overline{X}$ and each clause $C_j \in \phi$ with $\mathsf{lit}_i \in C_j$, either (i) $N^1_{i,j} \subseteq M$, or (ii) $N^2_{i,j} \subseteq M$, or (iii) $N^D_{i,j} \subseteq M$.*

Now we show that the function $\sigma\colon X \to \{\mathsf{true}, \mathsf{false}\}$ with $\sigma(x_i) = \mathsf{true}$ if $M(w_i) = \overline{x}_i$, and $\sigma(x_i) = \mathsf{false}$ otherwise is a satisfying truth assignment for ϕ. Clearly, ϕ is a valid truth assignment since by Claim 2(ii) every variable agent w_i is matched to either x_i or $\overline{x}_i$. We claim that σ satisfies ϕ. Consider an arbitrary clause C_j and the corresponding clause-agent c_j. By Claim 2(ii), we know that $M(c_j) = e^i_j$ for some $\mathsf{lit}_i \in C_j$. Since e^i_j is not available, by Claim 2(ii), it also follows that $M(a^0_{i,j}) = b^1_{i,j}$. By Claim 3, it follows that $N^1_{i,j} \subseteq M$. In particular, $M(\alpha_{i,j}) = b^0_{i,j}$ so that $\alpha_{i,j}$ is not available to other agents anymore.

We aim to show that $\alpha_{i,\mathsf{o}_1(\mathsf{lit}_i)}$ is matched to $b^0_{i,\mathsf{o}_1(\mathsf{lit}_i)}$ by M, which implies that lit_i is not available to w_i since $\alpha_{i,\mathsf{o}_1(\mathsf{lit}_i)} = \mathsf{lit}_i$ by the definition of $\alpha_{i,\mathsf{o}_1(\mathsf{lit}_i)}$. We distinguish two cases: If $j = \mathsf{o}_1(\mathsf{lit}_i)$, then by the definition of $\alpha_{i,j}$, it follows that $\alpha_{i,\mathsf{o}_1(\mathsf{lit}_i)}$ is matched to $b^0_{i,\mathsf{o}_1(\mathsf{lit}_i)}$. If $j = \mathsf{o}_2(\mathsf{lit}_i)$, then by the definition of $\alpha_{i,j}$, we have $\alpha_{i,j} = a^6_{i,\mathsf{o}_1(\mathsf{lit}_i)}$ and by the definition of $\delta_{i,\mathsf{o}_1(\mathsf{lit}_i)}$ we have $\delta_{i,\mathsf{o}_1(\mathsf{lit}_i)} = b^0_{i,\mathsf{o}_2(\mathsf{lit}_i)} = b^0_{i,j}$. In particular, since $M(\alpha_{i,j}) = b^0_{i,j}$ we have $M(a^6_{i,\mathsf{o}_1(\mathsf{lit}_i)}) = \delta_{i,\mathsf{o}_1(\mathsf{lit}_1)}$. By Claim 3, it follows that $N^1_{i,\mathsf{o}_1(\mathsf{lit}_i)} \subseteq M$ or $N^D_{i,\mathsf{o}_1(\mathsf{lit}_i)} \subseteq M$. In both cases, it follows that $\alpha_{i,\mathsf{o}_1(\mathsf{lit}_i)}$ is matched to $b^0_{i,\mathsf{o}_1(\mathsf{lit}_i)}$. We have just shown that lit_i is *not* available to w_i. Hence, by Claim 2(ii), $M(w_i) = \overline{x}_i$ if $\mathsf{lit}_i = x_i$, and $M(w_i) = \overline{x}_i$ otherwise. By definition, we have that $\sigma(x_i) = \mathsf{true}$ if $\mathsf{lit}_i = x_i$ and $\sigma(x_i) = \mathsf{false}$ otherwise. Thus, C_j is satisfied under σ, implying that σ is a satisfying assignment. □

4 Algorithms for Bounded Preferences Length

When bounding the preference length by two it is not hard to show that (coalitional) exchange-stability can be decided in linear time.

Theorem 3 (⋆). 2-ESM, 2-ESR, 2-CESM, *and* 2-CESR *can be solved in linear time.*

Fixed-parameter Algorithm for 3-ESR. We now turn to preference length at most three . In Theorem 1 we have seen that even this case remains NP-hard, even for bipartite preference profiles. Moreover, the proof suggests that a main obstacle that one has to deal with when solving 3-ESM (and hence 3-ESR) are the switch gadgets. Here we essentially show that they are indeed the *only* obstacles, that is, if there are few of them present in the input, then we can solve the problem efficiently. We capture the essence of the switch gadgets with the following structure that we call hourglasses.

Definition 1. *Let $\mathcal{P}$ be a preference profile and $V_H \subseteq V$ a subset of $2h$ agents with $V_H = \{u_i, w_i \mid 0 \leq i \leq h-1\}$. We call the subgraph $G(\mathcal{P})[V_H]$ induced by V_H an* hourglass *of height h if it satisfies the following:*

- *For each $i \in \{0, h-1\}$ the degrees of u_i and w_i are both at least two in $G(\mathcal{P})[V_H]$;*
- *For each $i \in [h-2]$, the degrees of u_i and w_i are exactly three in $G(\mathcal{P})[V_H]$;*
- *For each $i \in \{0, 1, \ldots, h-1\}$ we have $\{u_i, w_i\} \in E(G(\mathcal{P})[V_H])$;*
- *For each $i \in \{0, 1, \ldots, h-2\}$ we have $\{u_i, w_{i+1}\}, \{u_{i+1}, w_i\} \in E(G(\mathcal{P})[V_H])$.*

We refer to the agents u_i and w_i from V_H as layer-i *agents. We call an hourglass H* maximal *if no larger agent subset $V' \supsetneq V(H)$ exists that induces an hourglass.*

Given an hourglass H in $G(\mathcal{P})$, we call a matching M for $\mathcal{P}$ perfect for H *if for each agent $v \in V(H)$ we have $M(v) \in V(H) \setminus \{v\}$. Further, M is* exchange-stable for H *if no two agents from $V(H)$ can form an exchange-blocking pair.*

Notice that the smallest hourglass has height two and is a cycle with four vertices. We are ready to show the following fixed-parameter tractability result.

Theorem 4 (⋆). *An instance of* 3-ESR *with $2n$ agents and ℓ maximal hourglasses can be solved in $O(6^{\ell} \cdot n\sqrt{n})$ time.*

The main ideas are as follows. The first observation is that a matching for a maximal hourglass can interact with the rest of the graph in only six different ways: The only agents in an hourglass H of height h that may have neighbors outside H are the layer-0 and layer-$(h-1)$ agents; let us call them *connecting agents* of H. A matching M may match these agents either to agents inside or outside H. Requiring M to be perfect means that an even number of the connecting agents has to be matched inside H. This then gives a bound of at most six different possibilities of the matching M with respect to whether the connecting agents are matched inside or outside H. Let us call this the *signature* of M with respect to H. Hence, we may try all 6^ℓ possible combinations of signatures for all hourglasses and check whether one of them leads to a solution (i.e., an exchange-stable matching).

The second crucial observation is that each exchange-blocking pair of a perfect matching yields a four-cycle and hence, is contained in some maximal hourglass. Thus, the task of checking whether a combination of signatures leads to a solution decomposes into (a) checking whether each maximal hourglass H allows for an exchange-stable matching adhering to the signature we have chosen for H and (b) checking whether the remaining acceptability graph after deleting all agents that are in hourglasses or matched by the chosen signatures admits a perfect matching.

Task (b) can clearly be done in $O(n \cdot \sqrt{n})$ time by performing any maximum-cardinality matching algorithm (note that the graph $G(\mathcal{P})$ has $O(n)$ edges). We then prove that task (a) for all six signatures can be reduced to checking whether a given hourglass admits a perfect and exchange-stable matching. This, in turn, we show to be linear-time solvable by giving a dynamic program that fills a table, maintaining some limited but crucial facts about the structure of partial matchings for the hourglass.

5 Paths to Exchange-Stability

We now study the parameterized complexity of P-ESM with respect to the number of swaps. Observe that it is straightforward to decide an instance of P-ESM with $2n$ agents in $O((2n)^{2k+2})$ time by trying k times all of the $O(n^2)$ possibilities for the next swap and then checking whether the resulting matching is exchange-stable. The next theorem shows that the dependency of the exponent on k in the running time cannot be removed unless FPT = W[1].

Theorem 5 ($\star$). Path to Exchange-Stable Marriage *is W[1]-hard with respect to the number k of swaps.*

Proof (Sketch). We provide a parameterized reduction from the W[1]-complete Independent Set problem, parameterized by the size of the independent set [13]: Therein, given a graph H and an integer h, we want to decide whether G admits an h-vertex *independent* set, i.e., a subset of h pairwise nonadjacent vertices.

Let $I = (H, h)$ be an instance of Independent Set with vertex set $V(H) = \{v_1, v_2, \ldots, v_n\}$ and edge set $E(H)$. We construct an instance $I' = (\mathcal{P}, M_0, 2h)$ of P-ESM where $\mathcal{P}$ has two disjoint agent sets U and W, each of size $2n + h$. Both U and W consist of h *selector-agents* and $2n$ *vertex-agents* with preferences which encode the adjacency of the vertices in $V(H)$. More precisely, for each $j \in [h]$, we create two selector-agents, called s_j and t_j, and add them to U and W, respectively. For each $i \in [n]$, we create four vertex-agents, called x_i, u_i, y_i, w_i, add x_i and u_i to U, and add y_i and w_i to W. Altogether, we have $U = \{s_j \mid j \in [h]\} \cup \{u_i, x_i \mid i \in [n]\}$ and $W = \{t_j \mid j \in [h]\} \cup \{w_i, y_i \mid i \in [n]\}$.

Now we define the preferences of the agents from $U \cup W$. For notational convenience, we define two subsets of agents which shall encode the neighborhood of a vertex: For each vertex $v_i \in V(H)$, define $Y(v_i) := \{y_z \mid \{v_i, v_z\} \in E(H)\}$ and $U(v_i) := \{u_z \mid \{v_i, v_z\} \in E(H)\}$.

$$\begin{aligned}
&\forall j \in [h]\colon\ s_j : w_1 \succ \cdots \succ w_n \succ t_j, \quad && t_j : u_1 \succ \cdots \succ u_n \succ x_1 \succ \cdots \succ x_n \succ s_j,\\
&\forall i \in [n]\colon\ x_i : t_1 \succ \cdots \succ t_h \succ y_i, && y_i : u_i \succ x_i \succ [U(v_i)],\\
&\forall i \in [n]\colon\ u_i : w_i \succ [Y(v_i)] \succ y_i \succ t_1 \succ \cdots \succ t_h, && w_i : s_1 \succ \cdots \succ s_h \succ u_i\ .
\end{aligned}$$

Herein, $[Y(v_i)]$ (resp. $[U(v_i)]$) denotes the unique preference list where the agents in $Y(v_i)$ (resp. $U(v_i)$) are ordered ascendingly according to their indices. Observe that the acceptability graph $G(\mathcal{P})$ includes the following edges:

- For all $i \in [h]$ and $j \in [n]$, the edges $\{s_i, t_i\}$, $\{s_i, w_j\}$, $\{t_i, x_j\}$, $\{t_i, u_j\}$, $\{w_j, u_j\}$, $\{y_j, x_j\}$, $\{y_j, u_j\}$ are in $E(G(\mathcal{P}))$.
- For all edges $\{v_i, v_{i'}\} \in E(G)$, the edges $\{u_i, y_{i'}\}$ and $\{u_{i'}, y_i\}$ are in $E(G(\mathcal{P}))$.

We define an initial matching M_0 on $G(\mathcal{P})$ as $M_0 = \{\{s_j, t_j\} \mid j \in [h]\} \cup \{\{w_i, u_i\}, \{y_i, x_i\} \mid i \in [n]\}$. This completes the construction of I', which can clearly be done in polynomial time. It is straight-forward to check that that $\mathcal{P}$ is bipartite and the construction can be done in linear time. The correctness proof is given in the full version [12]. □

6 Conclusion

Regarding preference restrictions [8], it would be interesting to know whether deciding (coalitional) exchange-stability for complete preferences would be become tractable for restricted preferences domains, such as single-peakedness or single-crossingness. Further, the NP-containment of the problem of checking whether a given matching may reach an exchange-stable matching is open.

References

1. Abizada, A.: Exchange-stability in roommate problems. Rev. Econ. Des. **23**, 3–12 (2019). https://doi.org/10.1007/s10058-018-0217-0
2. Abraham, D.J., Manlove, D.F.: Pareto optimality in the roommates problem. Technical report, University of Glasgow, Department of Computing Science (2004). tR-2004-182
3. Alcalde, J.: Exchange-proofness or divorce-proofness? Stability in one-sided matching markets. Econ. Des. **1**, 275–287 (1995)
4. Aziz, H., Goldwaser, A.: Coalitional exchange stable matchings in marriage and roommate market. In: Proceedings of the 16th International Conference on Autonomous Agents and Multiagent Systems (AAMAS 2017), pp. 1475–1477 (2017). extended Abstract
5. Biró, P., Norman, G.: Analysis of stochastic matching markets. Int. J. Game Theory **42**(4), 1021–1040 (2012). https://doi.org/10.1007/s00182-012-0352-8

6. Bodlaender, H.L., Hanaka, T., Jaffke, L., Ono, H., Otachi, Y., van der Zanden, T.C.: Hedonic seat arrangement problems. In: Proceedings of the 19th International Conference on Autonomous Agents and Multiagent Systems (AAMAS 2020), pp. 1777–1779 (2020). extended Abstract
7. Bodlaender, H.L., Hanaka, T., Jaffke, L., Ono, H., Otachi, Y., van der Zanden, T.C.: Hedonic seat arrangement problems. Technical report, arXiv:2002.10898 (cs.GT) (2020)
8. Bredereck, R., Chen, J., Finnendahl, U.P., Niedermeier, R.: Stable roommate with narcissistic, single-peaked, and single-crossing preferences. Auton. Agent. Multi-Agent Syst. **34**(53), 1–29 (2020)
9. Cechlárová, K.: On the complexity of exchange-stable roommates. Discret. Appl. Math. **116**(3), 279–287 (2002)
10. Cechlárová, K., Manlove, D.F.: The exchange-stable marriage problem. Discret. Appl. Math. **152**(1–3), 109–122 (2005)
11. Chen, J.: Reaching stable marriage via divorces is hard. Technical report, arXiv:1906.12274(cs.GT) (2020)
12. Chen, J., Chmurovic, A., Jogl, F., Sorge, M.: On (coalitional) exchange-stable matching. Technical report, arXiv:2105.05725(cs.GT) (2021)
13. Cygan, M., et al.: Lower bounds for kernelization. In: Parameterized Algorithms, pp. 523–555. Springer, Cham (2015). https://doi.org/10.1007/978-3-319-21275-3_15
14. Gale, D., Shapley, L.S.: College admissions and the stability of marriage. Am. Math. Mon. **120**(5), 386–391 (1962)
15. Irving, R.W.: Stable matching problems with exchange restrictions. J. Comb. Optim. **16**(4), 344–360 (2008)
16. Knuth, D.: Mariages Stables. Les Presses de L'Université de Montréal (1976)
17. McDermid, E., Cheng, C.T., Suzuki, I.: Hardness results on the man-exchange stable marriage problem with short preference lists. Inf. Process. Lett. **101**(1), 13–19 (2007)

Optimal Revenue Guarantees for Pricing in Large Markets

José Correa[1], Dana Pizarro[2], and Victor Verdugo[3(✉)]

[1] Department of Industrial Engineering, Universidad de Chile, Santiago, Chile
correa@uchile.cl

[2] Toulouse School of Economics, Université Toulouse 1 Capitole, Toulouse, France
dana.pizarro@tse-fr.eu

[3] Institute of Engineering Sciences, Universidad de O'Higgins, Rancagua, Chile
victor.verdugo@uoh.cl

Abstract. Posted price mechanisms (PPM) constitute one of the predominant practices to price goods in online marketplaces and their revenue guarantees have been a central object of study in the last decade. We consider a basic setting where the buyers' valuations are independent and identically distributed and there is a single unit on sale. It is well-known that this setting is equivalent to the so-called i.i.d. prophet inequality, for which optimal guarantees are known and evaluate to 0.745 in general (equivalent to a PPM with dynamic prices) and $1 - 1/e \approx 0.632$ in the fixed threshold case (equivalent to a fixed price PPM). In this paper we consider an additional assumption, namely, that the underlying market is very large. This is modeled by first fixing a valuation distribution F and then making the number of buyers grow large, rather than considering the worst distribution for each possible market size. In this setting Kennedy and Kertz [Ann. Probab. 1991] breaks the 0.745 fraction achievable in general with a dynamic threshold policy. We prove that this large market benefit continue to hold when using fixed price PPMs, and show that the guarantee of 0.632 actually improves to 0.712. We then move to study the case of selling k identical units and we prove that the revenue gap of the fixed price PPM approaches $1 - 1/\sqrt{2k\pi}$. As this bound is achievable without the large market assumption, we obtain the somewhat surprising result that the large market advantage vanishes as k grows.

Keywords: Prophet inequalities · Pricing · Large markets

1 Introduction

Understanding the worst case revenue obtained by simple pricing mechanisms is a fundamental question in Economics and Computation [2,3,10,16,18]. In this context probably the most basic setting corresponds to selling a single item to n buyers with valuations given by independent and identically distributed random variables. Here the simplest possible mechanism is that of setting a fixed price

I. Caragiannis and K. A. Hansen (Eds.): SAGT 2021, LNCS 12885, pp. 221–235, 2021.
https://doi.org/10.1007/978-3-030-85947-3_15

(a.k.a. anonymous price) for the item and the benchmark, to which we want to compare to, is the revenue obtained by Myerson's optimal mechanism [25]. Through the well established connection between posted pricing mechanisms and prophet inequalities [5,7,15], evaluating this revenue gap is equivalent to determining the best possible *single threshold* prophet inequality in the i.i.d. case. Thus, a result of Ehsani et al. [9] establishes that the performance of a fixed threshold policy when facing i.i.d. samples is at least a fraction $1 - 1/e$ of that of the optimal mechanism, and the bound is best possible.[1] [2] In this paper, we explore this basic question under an additional *large markets* assumption that is relevant to most modern online marketplaces.

In our study we take the viewpoint of prophet inequalities rather than that of pricing mechanisms, mostly because this has become the standard in the literature. Let us thus briefly recall some of the basics. For a fixed positive integer n, let $X_1, \dots, X_n$ be a non-negative, independent random variables and $\mathcal{S}_n$ their set of stopping rules. A classic result of Krengel and Sucheston, and Garling [22,23] asserts that $\mathbb{E}(\max\{X_1, \dots, X_n\}) \leq 2 \sup\{\mathbb{E}(X_s) : s \in \mathcal{S}_n\}$ and that two is the best possible bound. The study of this type of inequalities, known as prophet inequalities, was initiated by Gilbert and Mosteller [13] and attracted a lot of attention in the eighties [17,20,21,27,28]. In particular, Samuel-Cahn [28] noted that rather than looking at the set of all stopping rules one can obtain the same result by using a single threshold stopping rule in which the decision to stop depends on whether the value of the currently observed random variable is above a certain threshold. A natural restriction of this setting, which we consider here, is the case in which the random variables are identically distributed. This problem was studied by Hill and Kertz [17] who provided the family of worst possible instances from which Kertz [20] proved that no stopping rule can extract more than a fraction of roughly 0.745 of the expectation of the maximum. Later, Correa et al. [6] proved that in fact this value is tight. We note, however, that the optimal stopping rule in this i.i.d. case cannot be achieved by a fixed threshold policy. Indeed, the best such policy has an approximation guarantee of $1 - 1/e \approx 0.632$ [9].

In the last two decades, prophet inequalities gained particular attention due to its close connection with online mechanisms. The connection involves mapping the random variables in the prophet inequality setting to the virtual valuations in the pricing setting and the expectation of the maximum value in the prophet inequality setting to revenue of the optimal mechanism in the pricing setting. This relation was firstly studied by Hajiaghayi et al. [15], who showed that prophet inequalities can be interpreted as posted price mechanisms for online

[1] Here the mild technical condition that the distribution is continuous is needed. Otherwise the mechanism would need some randomization.

[2] Ehsani et al. [9] actually prove a more general prophet inequality, namely, that the bound of $1 - 1/e$ holds even if the distributions are nonidentical. However, this more general result does not translate into a fixed price policy (if the distributions are not identical, neither are the virtual values and then this single threshold will be mapped to different prices for different distributions).

selection problems. Later, Chawla et al. [5] proved that any prophet inequality can be turned into a posted price mechanism with the same approximation guarantee. The reverse direction was proven by Correa et al. [7] and thus the guarantees for optimal stopping problems are in fact equivalent to the problem of designing posted price mechanisms. Furthermore, in the i.i.d. setting, fixed threshold stopping rules become equivalent to fixed price policies.

In this work we study single threshold prophet inequalities in a large market regime, where the random variables arriving over time are i.i.d. according to a known and fixed distribution. The essential difference with the classic setting is that rather than considering the worst distribution for each possible market size n, we first fix the distribution and then take n grow to infinity. Our main question is thus to understand to what extent one can obtain improved single threshold prophet inequalities (or fixed price policies) when the market is very large. Interestingly, this setting, though with general stopping rules, was considered three decades ago by Kennedy and Kertz [19]. They prove that the optimal stopping rule recovers at least a 0.776 fraction of the expectation of the maximum, establishing that there is a sensible advantage when compared to the 0.745 bound of classic i.i.d. setting [17,20]. Kennedy and Kertz realize that the limit problem may be ill behaved and thus impose an *extreme value condition.*[3] This condition is, essentially, the equivalent of a central limit theorem for the maximum of an i.i.d. sample, and it is the cornerstone of the extreme value theory.

Then, a natural question that arises is whether the result obtained by Kennedy and Kertz [19] for the optimal stopping rule also holds for the much simpler single threshold policies. We answer this question on the positive proving that the large market assumption allows to obtain a guarantee of 0.712 significantly improving the bound of $1 - 1/e$ [9]. We further consider the case of selecting k items (or selling k items) with a fixed threshold policy and prove that this large market advantage vanishes as k grows.

1.1 Our Results

For every positive integer n, consider an i.i.d. sample $X_1, X_2, \dots, X_n$ with X_j distributed according to F for every $j \in \{1, \dots, n\}$, where F is a distribution over the non-negative reals. Given a value T, consider the simple algorithm given by stopping the first time that a random variable exceeds T. Then, for each distribution F, we are interested in understanding the limit ratio between the reward of this simple stopping rule which is simply given by the probability of having an X_i above T, $1-F^n(T)$ times the expected value of this X_i conditioned on it being larger than T, and the expectation of the maximum X_i, denoted as M_n. Our quantity of interest is thus:

[3] This is a classic condition in extreme value theory and it is satisfied by essentially any distribution that may have a practical use. The characterization of this condition is known as the Fisher-Tippett-Gnedenko Theorem.

$$\mathrm{apx}(F) = \liminf_{n\to\infty} \sup_{T\in\mathbb{R}_+} \frac{1-F^n(T)}{\mathbb{E}(M_n)} \left(T + \frac{1}{1-F(T)} \int_T^\infty (1-F(s))\mathrm{d}s \right). \quad (1)$$

Our first main result shows that 0.712 is a tight lower bound for $\mathrm{apx}(F)$ when the distribution satisfies the extreme value condition. This value is substantially better than the known bound of $1 - 1/e$ by Ehsani et al. [9] and thus represents a significant advantage for the large markets setting. We remark that we are mainly interested in the case of distributions F with unbounded support, since one can show that $\mathrm{apx}(F) = 1$ when F is of bounded support.

A natural and practically relevant extension of the single selection prophet inequality is to consider the setting in which we can select up to k different samples (or sell k items). We call this problem k-selection problem and we study whether the large market advantage continues to be significant beyond the single selection case. To this end, we provide a lower bound for the approximation factor achievable by the best single threshold policy, again under the extreme value condition. More specifically, for each value of k, the approximation factor is bounded by a (computationally) simple optimization problem. In particular, the bound presented when $k = 1$ follows by obtaining the exact solution of the optimization problem. The performance obtained by our characterization yields prophet inequalities that represent an advantage for the k-selection problem. However, we also show that this advantage vanishes as $k \to \infty$. Indeed, we prove that for each integer k, the approximation factor is more than $1 - 1/\sqrt{2k\pi}$, but there exists F such that this lower bound is asymptotically tight in k. This tightness, together with the recent result of Duetting et al. [8] establishing that the approximation ratio of the k-selection problem (without the large market assumption) is almost exactly $1-1/\sqrt{2k\pi}$,[4] implies that the large market advantage vanishes as $k \to \infty$. For an illustration, Fig. 1 depicts the bound obtained by our optimization problem and compares it with $1-1/\sqrt{2k\pi}$. We finally note that as a direct corollary, when F satisfies the extreme value condition and for large markets, we can derive the worst case ratio between the optimal single threshold prophet inequality obtained by our characterization theorem and the value obtained by the optimal dynamic policy of Kennedy and Kertz, the *adaptivity gap*. This value is, roughly, at most 1.105.

As already mentioned, our main result for the multiple selection problem translates into a fixed price policy when the buyers' valuations are identically and independently distributed, say according to F.[5] Of course, this works as long as the distribution of the virtual values of F, call it G, satisfies the extreme value condition. This motivates the following question: When F satisfies the extreme value condition, can we guarantee that the distribution of the virtual valuation G also does? And, if this is the case, does G and F fall in the same extreme value family? We answer these questions in the positive under some mild assumptions.

[4] Slightly weaker bounds are also known for the case in which the random variables are just independent but not necessarily identical [1,4].

[5] Recall that single threshold policies map to fixed price mechanisms.

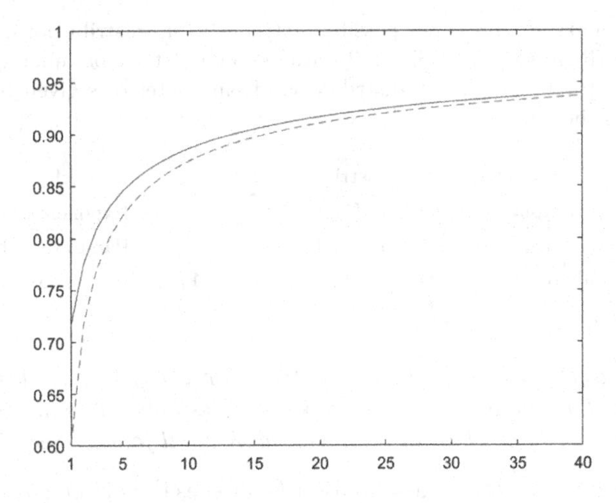

Fig. 1. Our optimal revenue guarantee over k (continuous line) vs. the bound of $1 - 1/\sqrt{2k\pi}$ (dashed line).

2 Preliminaries

We recall that F is a distribution if it is a right-continuous and non-decreasing function, with limit equal to zero in $-\infty$ and equal to one in $+\infty$. We consider F to be absolutely continuous in $\mathbb{R}$, and we denote its density by f or F', depending on the context. In general, F is not invertible, but we work with its generalized inverse, given by $F^{-1}(y) = \inf\{t \in \mathbb{R} : F(t) \geq y\}$. We denote by $\omega_0(F) = \inf\{t \in \mathbb{R} : F(t) > 0\}$ and $\omega_1(F) = \sup\{t \in \mathbb{R} : F(t) < 1\}$, and we call the interval $(\omega_0(F), \omega_1(F))$ the support of F. Given a sequence $\{X_j\}_{j\in\mathbb{N}}$ of i.i.d. random variables with distribution F, we denote by $M_n = \max_{j\in\{1,\ldots,n\}} X_j$.

One of the main goals in the extreme value theory is to understand the limit behavior of the sequence $\{M_n\}_{n\in\mathbb{N}}$. As the central limit theorem characterizes the convergence in distribution of the average of random variables to a normal distribution, a similar result can be obtained for the sequence of maxima $\{M_n\}_{n\in\mathbb{N}}$, but this time there are three possible limit situations. One of the possible limits is the Gumbel distribution $\Lambda(t) = \exp(-e^{-t})$; we call these distributions the Gumbel family. Given $\alpha > 0$, the second possible limit is the Fréchet distribution of parameter α, defined by $\Phi_\alpha(t) = \exp(-t^{-\alpha})$ if $t \geq 0$, and zero otherwise; we call these distributions the Fréchet family. Finally, given $\alpha > 0$, the third possibility is the reversed Weibull distribution of parameter α, defined by $\Psi_\alpha(t) = \exp(-(-t)^\alpha)$ if $t \leq 0$, and one otherwise; we call these distributions the reversed Weibull family. We now state formally the extreme values theorem, result due independently to Gnedenko [14] and Fisher & Tippett [11].

Theorem 1 (see [26]). *Let F be a distribution for which there exists a positive real sequence $\{a_n\}_{n\in\mathbb{N}}$ and other sequence $\{b_n\}_{n\in\mathbb{N}}$ such that $(M_n - b_n)/a_n$ converges in distribution to a random variable with distribution H, namely,*

Table 1. Summary of the three possible extreme value distributions. The Fréchet family and the Reversed Weibull family are associated to a parameter $\alpha \in (0,\infty)$. Recall that for $\alpha > 0$, the Pareto distribution of parameter α is given by $1 - t^{-\alpha}$ for $t \geq 1$ and zero otherwise.

Extreme type	Parameter	Limit distribution	Example
Gumbel	None	$\exp(-e^{-t})$	Exponential distribution
Fréchet	$\alpha \in (0,\infty)$	$\exp(-t^{-\alpha}) \cdot \mathbf{1}_{[0,\infty)}$	Pareto distribution
Reversed Weibull	$\alpha \in (0,\infty)$	$\exp(-(-t)^{\alpha}) \cdot \mathbf{1}_{(-\infty,0)} + \mathbf{1}_{[0,\infty)}$	Uniform distribution

$\mathbb{P}(M_n - b_n \leq a_n t) = F^n(a_n t + b_n) \to H(t)$ for every $t \in \mathbb{R}$ when $n \to \infty$. Then we have that one of the following possibilities hold: H is the Gumbel, H is in the Fréchet family or H is in the reversed Weibull family (see Table 1).

In the following, we say that a distribution F satisfies the *extreme value condition* if there exist sequences $\{a_n\}_{n\in\mathbb{N}}$, that we call the *scaling sequence*, and $\{b_n\}_{n\in\mathbb{N}}$, that we call the *shifting sequence*, satisfying the condition of Theorem 1.[6] It can be shown that for every distribution F with extreme type in the reversed Weibull family we have $\omega_1(F) < \infty$ [26, Proposition 1.13, p. 59]. When F has extreme type Fréchet, we have $\omega_1(F) = \infty$ [26, Proposition 1.11, p. 54]. For the distributions with extreme type Gumbel the picture is not so clear since $\omega_1(F)$ is neither finite nor unbounded in general. In our analysis we need a tool from the extreme value theory related to the order statistics of a sample according to F. We denote the order statistics of a sample of size n by $M_n = M_n^1 \geq M_n^2 \geq \cdots \geq M_n^n$.

Theorem 2 (see [24]). *Let F be a distribution satisfying the extreme value condition with the scaling and shifting sequences $\{a_n\}_{n\in\mathbb{N}}$ and $\{b_n\}_{n\in\mathbb{N}}$ such that $\mathbb{P}(M_n - b_n \leq a_n t) \to H(t)$ for every $t \in \mathbb{R}$ when $n \to \infty$. Then, for each $j \in \{1,2,\ldots,n\}$ and every $t \in \mathbb{R}$ we have*

$$\lim_{n\to\infty} \mathbb{P}\left(M_n^j - b_n \leq a_n t\right) = H(t) \sum_{s=0}^{j-1} \frac{(-\log H(t))^s}{s!}.$$

A distribution V is in the *Von Mises family* if there exist $z_0 \in \mathbb{R}$, a constant $\theta > 0$ and a function $\mu : (\omega_0(V), \infty) \to \mathbb{R}_+$ absolutely continuous with $\lim_{u\to\infty} \mu'(u) = 0$, such that for every $t \in (z_0, \infty)$ we have

$$1 - V(t) = \theta \exp\Big(-\int_{z_0}^{t} \frac{1}{\mu(s)} \mathrm{d}s\Big). \tag{2}$$

We call such μ an *auxiliary function* of V. We summarize next some technical results related to the Von Mises family of distributions that we use in our analysis.

[6] Examples of continuous distributions not satisfying this extreme value condition include distributions with odd behavior such as $F(x) = \exp(-x - \sin(x))$.

Lemma 1 (see [26]). *Let V be in the Von Mises family with auxiliary function μ and such that $\omega_1(V) = \infty$. Then, V has extreme type Gumbel, and the shifting and scaling sequences may be chosen respectively as $b_n = V^{-1}(1 - 1/n)$ and $a_n = \mu(b_n)$ for every n. Furthermore, we have $\lim_{t\to\infty} \mu(t)/t = 0$ and $\lim_{t\to\infty}(t+x\mu(t)) = \infty$ for every $x \in \mathbb{R}$.*

For example, the exponential distribution of parameter λ is in the Von Mises family, with auxiliary constant function $1/\lambda$, $\theta = 1$ and $z_0 = 0$. Furthermore, for every positive integer n we have $b_n = F^{-1}(1 - 1/n) = (\log n)/\lambda$ and $a_n = \mu(b_n) = 1/\lambda$. We need a few results from the extreme value theory. In particular, a relevant property states that every distribution with extreme type Gumbel can be represented by a distribution in the Von Mises family in the following precise sense.

Lemma 2 (see [26]). *Let F be a distribution satisfying the extreme value condition with $\omega_1(F) = \infty$. Then, F has extreme type Gumbel if and only if there exists V in the Von Mises family and a positive function $\eta : (\omega_0(F), \infty) \to \mathbb{R}_+$ with $\lim_{t\to\infty} \eta(t) = \eta^\star > 0$ such that $1 - F(t) = \eta(t)(1 - V(t))$ for every $t \in (\omega_0(F), \infty)$.*

Then, whenever F has extreme Gumbel there exists a pair (V, η) satisfying the condition guaranteed in Lemma 2, and in this case we say that (V, η) is a Von Mises representation of the distribution F.

3 Prophet Inequalities in Large Markets Through Extreme Value Theory

We say that a stopping rule for the k-selection problem with an i.i.d. sample $X_1, X_2, \ldots, X_n$ is a *single threshold policy* if there exists a threshold value T such that we select the first $\min\{k, |Q|\}$ samples attaining a value larger than T, where Q is the subset of samples attaining a value larger than T. Consider the random variable $\mathcal{R}^n_{k,T}$ equal to the summation of the first $\min\{k, |Q|\}$ samples attaining a value larger than T. In particular, this value is completely determined by the sample size n, the distribution F and the threshold T. We are interested in understanding the value

$$\text{apx}_k(F) = \liminf_{n\to\infty} \sup_{T\in\mathbb{R}_+} \frac{\mathbb{E}(\mathcal{R}^n_{k,T})}{\sum_{j=1}^k \mathbb{E}(M^j_n)},$$

where $M^1_n \geq M^2_n \geq \cdots \geq M^n_n$ are the order statistics of a sample of size n according to F. We observe that when $k = 1$ the value $\text{apx}_k(F)$ corresponds to the value $\text{apx}(F)$ in (1). Now we present formally our main results for prophet inequalities in the k-selection problem.

Theorem 3. *Let F be a distribution over the non-negative reals that satisfies the extreme value condition. Then, the following holds.*

(a) *When F has an extreme type Fréchet of parameter α, we have that $apx_k(F) \geq \varphi_k(\alpha)$, where $\varphi_k : (1, \infty) \to \mathbb{R}_+$ is given by*

$$\varphi_k(\alpha) = \frac{\Gamma(k)}{\Gamma(k+1-1/\alpha)} \max_{x \in (0,\infty)} x \exp(-x^{-\alpha}) \sum_{j=1}^{k} \sum_{s=j}^{\infty} \frac{x^{-s\alpha}}{s!}. \tag{3}$$

In particular, we have $apx_k(F) \geq 1 - 1/\sqrt{2\pi k}$ for every distribution F with extreme type in the Fréchet family.

(b) *When F has extreme type in the Gumbel or reversed Weibull families, we have that $apx_k(F) = 1$ for every positive integer k.*

Theorem 4. *Let F be the Pareto distribution with parameter $\alpha = 2$. Then, for every $\varepsilon > 0$ there exists a positive integer k_ε such that for every $k \geq k_\varepsilon$ it holds that $apx_k(F) \leq 1 - (1-\varepsilon)/\sqrt{2\pi k}$.*

Observe that by Theorem 3 we have that for each integer k the approximation factor is more than $1 - 1/\sqrt{2k\pi}$ under the large market assumption. Moreover, by Theorem 4 this lower bound is in fact asymptotically tight in k for the distributions with extreme type Fréchet of parameter $\alpha = 2$. This tightness, together with the recent result of Duetting et al. [8] establishing that the approximation ratio of the k-selection problem without the large market assumption is almost $1 - 1/\sqrt{2k\pi}$, allows us to obtain the surprising result that the large market advantage vanishes as $k \to \infty$.

Despite the tightness result established in Theorem 4, for small values of k this bound is in fact substantially better. Consider a distribution F with extreme type Fréchet of parameter $\alpha \in (1, \infty)$. By Theorem 3 (a), when $k = 1$ it holds that

$$\varphi_1(\alpha) = \frac{1}{\Gamma(2-1/\alpha)} \sup_{x \in (0,\infty)} x\Big(1 - \exp(-x^{-\alpha})\Big),$$

for every $\alpha \in (1, \infty)$. The optimum for the above optimization problem as a function of α is attained at the smallest real non-negative solution $U^*(\alpha)$ of the first order condition $U^\alpha + \alpha = U^\alpha \exp(U^{-\alpha})$, which is given by

$$U^*(\alpha) = \left(-\frac{1}{\alpha}\left(\alpha W_{-1}\left(-\frac{1}{\alpha}e^{-1/\alpha}\right) + 1\right)\right)^{-1/\alpha},$$

where W_{-1} is the negative branch of the Lambert function. Therefore, we have

$$\varphi_1(\alpha) = \frac{\alpha}{\Gamma(2-1/\alpha)} \cdot \frac{U^*(\alpha)}{U^*(\alpha)^\alpha + \alpha}.$$

The minimum value is at least 0.712 and it is attained at $\alpha^* \approx 1.656$. Note that when α approaches to zero or ∞, the function φ_1 goes to one and thus the unique minimizer is given by $\alpha^* \approx 1.656$.

We highlight here that, even though Theorem 3 implies that $\text{apx}_1(F)$ is at least $\varphi_1(\alpha^*) \approx 0.712$ when F has extreme type Fréchet, this bound is in fact reached by the Pareto distribution with parameter α^* and therefore this bound is tight.

Given our closed expression for the function φ_1, we can compare it with the closed expression obtained Kennedy and Kertz for the revenue guarantees of the optimal dynamic policy [19]. Given a distribution F, for every positive integer n let $v_n = \sup\{\mathbb{E}(X_\tau) : \tau \in \mathcal{T}_n\}$ and consider the stopping time given by $\tau_n = \min\{k \in \{1, \ldots, n\} : X_k > v_{n-k}\}$. In particular, $v_n = \mathbb{E}(X_{\tau_n})$ for every positive integer n. The following summarizes the result of Kennedy and Kertz [19] for the optimal dynamic policy: When F is a distribution in the Fréchet family, there exists $\nu : (1, \infty) \to (0, 1)$ such that $\lim_{n\to\infty} v_n/\mathbb{E}(M_n) = \nu(\alpha)$ when F has an extreme type Fréchet of parameter α. Furthermore, $\lim_{\alpha\to\infty} \nu(\alpha) = \lim_{\alpha\to 1} \nu(\alpha) = 1$ and $\nu(\alpha) \geq 0.776$ for every $\alpha \in (1, \infty)$. The function ν is given by

$$\nu(\alpha) = \frac{1}{\Gamma(2 - 1/\alpha)} \left(1 - \frac{1}{\alpha}\right)^{1-\frac{1}{\alpha}},$$

and we have $\varphi_1(\alpha) \leq \nu(\alpha)$ for every $\alpha \in (1, \infty)$. Kennedy and Kertz show that the asymptotic approximation obtained by their multi-threshold policy when the distribution has an extreme type in the Gumbel and reversed Weibull family is equal to one. Our Theorem 3 (b) shows that for both such families we can attain this value by using just single threshold policies. The *adaptivity gap* is equal to the ratio between the optimal prophet inequality obtained by a single threshold policy and the value obtained by the multi-threshold policy of Kennedy and Kertz. As a corollary of our result for $k = 1$, we obtain an upper bound on the adaptivity gap for the case of distributions with extreme value. For the family of distributions over the non-negative reals and satisfying the extreme value condition we have that the adaptivity gap is at most $\max_{\alpha\in(1,\infty)} \nu(\alpha)/\varphi_1(\alpha) \approx 1.105$ and is attained at $\alpha \approx 1.493$.

4 Analysis of the k-Selection Prophet Inequalities

In this section we prove Theorem 3. Throughout the section we introduce some necessary technical results, whose proof can be found in the full version paper. The following proposition gives an equivalent expression for the value $\text{apx}_k(F)$, which is useful in our analysis.

Proposition 1. *Let F be a distribution, let T be a real value and let $X_1, \ldots, X_n$ be an i.i.d. sample according to F. Then, for every positive integer k we have $\mathbb{E}(\mathcal{R}^n_{k,T}) = \mathbb{E}(X_1|X_1 > T) \sum_{j=1}^{k} \mathbb{P}(M^j_n > T)$.*

Using Proposition 1 we have that $\text{apx}_k(F)$ is therefore given by

$$\text{apx}_k(F) = \liminf_{n\to\infty} \sup_{T\in\mathbb{R}_+} \mathbb{E}(X|X > T) \frac{\sum_{j=1}^{k} \mathbb{P}(M^j_n > T)}{\sum_{j=1}^{k} \mathbb{E}(M^j_n)}, \tag{4}$$

where X is a random variable distributed according to F.

4.1 Proof of Theorem 3 (a): The Fréchet Family

In what follows we restrict to the case in which the distribution F has extreme type in the Fréchet family. We remark that if $\alpha \in (0, 1]$ the expected value of a random variable with distribution Fréchet Φ_α is not finite. Therefore, we further restrict to the Fréchet family where $\alpha \in (1, \infty)$. To prove Theorem 3 (a) we require a technical lemma, where we exploit the structure given by the existence of an extreme value and we show how to characterize the approximation factor of a distribution in the Fréchet family for large values of n. Before stating this lemma, let us introduce a few results about the Fréchet family that will be required.

We say that a positive measurable function $\ell : (0, \infty) \to \mathbb{R}$ is *slowly varying* if for every $u > 0$ we have $\lim_{t\to\infty} \ell(ut)/\ell(t) = 1$. For example, the function $\ell(t) = \log(t)$ is slowly varying, since $\ell(ut)/\ell(t) = \log(u)/\log(t) + 1 \to 1$ when $t \to \infty$. On the other hand, the function $\ell(t) = t^\gamma$ is not slowly varying, since for every $u > 0$ we have $\ell(ut)/\ell(t) = u^\gamma$. The following lemma shows the existence of a strong connection between the distributions with extreme type in Fréchet family and slowly varying functions. Recall that for $\alpha > 0$, the Pareto distribution of parameter α is given by $P_\alpha(t) = 1 - t^{-\alpha}$ for $t \geq 1$ and zero otherwise.

Lemma 3 ([26]). *Let F be a distribution with extreme type in the Fréchet family. Then, for every positive integer n, we have $a_n = F^{-1}(1 - 1/n)$ and $b_n = 0$ are scaling and shifting sequences for F. Furthermore, there exists a slowly varying function ℓ_F such that $1 - F(t) = t^{-\alpha}\ell_F(t)$, for every $t \in \mathbb{R}_+$. In particular, we have $1 - F(t) = (1 - P_\alpha(t)) \cdot \ell_F(t)$ for every $t \in \mathbb{R}_+$.*

Observe that this lemma says that if F has extreme type Fréchet of parameter α, then it essentially corresponds to a perturbation of a Pareto distribution with parameter α by some slowly varying function. Let $\{a_n\}_{n\in\mathbb{N}}$ be a scaling sequence for the distribution F in the Fréchet family. Thanks to Lemma 3, we have the shifting sequence in this case is zero. We are now ready to state the main technical lemma.

Lemma 4. *Let F be a distribution with extreme type Fréchet of parameter α and let $\{a_n\}_{n\in\mathbb{N}}$ be an appropriate scaling sequence. Consider a positive sequence $\{T_n\}_{n\in\mathbb{N}}$ with $T_n \to \infty$ and for which there exists $U \in \mathbb{R}_+$ such that $T_n/a_n \to U$. Then, we have*

$$\lim_{n\to\infty} \mathbb{E}(X|X > T_n) \frac{\sum_{j=1}^k \mathbb{P}(M_n^j > T_n)}{\sum_{j=1}^k \mathbb{E}(M_n^j)} = \frac{\Gamma(k)}{\Gamma(k + 1 - 1/\alpha)} U \exp(-U^{-\alpha}) \sum_{j=1}^{k} \sum_{s=j}^{\infty} \frac{U^{-s\alpha}}{s!}.$$

We use this lemma to prove Theorem 3 (a).

Proof (Proof of Theorem 3 (a)). Let F be a distribution with extreme type Fréchet of parameter α. We first prove that for each positive integer k it holds that $\text{apx}_k(F) \geq \varphi_k(\alpha)$. To this end, for each positive integer n and positive real number U, let T_n be the threshold given by $T_n = a_n \cdot U$, where $\{a_n\}_{n\in\mathbb{N}}$ is the scaling sequence for the distribution F given by Lemma 3. Then,

$$\operatorname{apx}_k(F) \geq \liminf_{n\to\infty} \mathbb{E}(X|X > T_n) \frac{\sum_{j=1}^k \mathbb{P}(M_n^j > T_n)}{\sum_{j=1}^k \mathbb{E}(M_n^j)}. \tag{5}$$

Note that $\liminf_{n\to\infty} T_n = \infty$ (and thus $T_n \to \infty$), since $U \in \mathbb{R}_+$ and $a_n \to \infty$. Furthermore, $\lim_{n\to\infty} T_n/a_n = U$ and then applying Lemma 4 together with inequality (5) we obtain that

$$\operatorname{apx}_k(F) \geq \frac{\Gamma(k)}{\Gamma(k+1-1/\alpha)} U \exp(-U^{-\alpha}) \sum_{j=1}^{k} \sum_{s=j}^{\infty} \frac{U^{-s\alpha}}{s!}.$$

Given that the inequality above holds for every positive real number U, we have

$$\operatorname{apx}_k(F) \geq \frac{\Gamma(k)}{\Gamma(k+1-1/\alpha)} \max_{U\in\mathbb{R}_+} U \exp(-U^{-\alpha}) \sum_{j=1}^{k} \sum_{s=j}^{\infty} \frac{U^{-s\alpha}}{s!} = \varphi_k(\alpha).$$

In the rest of the proof we show that, for each positive real number k and each $\alpha \in (1,\infty)$, $\varphi_k(\alpha)$ is lower bounded by $1 - 1/\sqrt{2k\pi}$. To this end, we just need to evaluate the objective function of our optimization problem in a well chosen value. One of the Gautschi inequalities for the Gamma function states that for every $s \in (0,1)$ and every $x \geq 1$ we have $\Gamma(x+1) > x^{1-s} \cdot \Gamma(x+s)$ [12]. Then, setting $x = k$ and $s = 1 - 1/\alpha$ yields $\Gamma(k+1) > k^{1/\alpha}\Gamma(k+1-1/\alpha)$. Since $\Gamma(k) = \Gamma(k+1)/k$, we therefore obtain $k^{1-1/\alpha} > \Gamma(k+1-1/\alpha)/\Gamma(k)$. On the other hand, note that for each $U \in (0,\infty)$ we have

$$\begin{aligned} U \exp(-U^{-\alpha}) \sum_{j=1}^{k} \sum_{s=j}^{\infty} \frac{U^{-s\alpha}}{s!} &= U \exp(-U^{-\alpha}) \left(\sum_{s=1}^{k} \frac{s \cdot U^{-s\alpha}}{s!} + k \sum_{s=k+1}^{\infty} \frac{U^{-s\alpha}}{s!} \right) \\ &= U \exp(-U^{-\alpha}) \left(U^{-\alpha} \sum_{s=0}^{k-1} \frac{U^{-s\alpha}}{s!} + k \sum_{s=k+1}^{\infty} \frac{U^{-s\alpha}}{s!} \right). \end{aligned}$$

In particular, by taking $U_{k,\alpha} = k^{-1/\alpha}$ we get that

$$\begin{aligned} \varphi_k(\alpha) \cdot \frac{\Gamma(k+1-1/\alpha)}{\Gamma(k)} &\geq U_{k,\alpha} \cdot k \exp(-U_{k,\alpha}^{-\alpha}) \left(\sum_{s=0}^{k-1} \frac{U_{k,\alpha}^{-s\alpha}}{s!} + \sum_{s=k+1}^{\infty} \frac{U_{k,\alpha}^{-s\alpha}}{s!} \right) \\ &= U_{k,\alpha} \cdot k \exp(-U_{k,\alpha}^{-\alpha}) \left(\exp(U_{k,\alpha}^{-\alpha}) - \frac{U_{k,\alpha}^{-\alpha k}}{k!} \right) \\ &= k^{1-1/\alpha} \left(1 - \frac{e^{-k}k^k}{k!} \right) \geq \frac{\Gamma(k+1-1/\alpha)}{\Gamma(k)} \left(1 - \frac{1}{\sqrt{2\pi k}} \right), \end{aligned}$$

where the first inequality follows since the value of $\varphi_k(\alpha)$ involves the maximum over $(0,\infty)$, the first equality from the Taylor series for the exponential function and the last inequality is obtained by applying Stirling's approximation inequality. This concludes the proof of the theorem. □

4.2 Proof of Theorem 3 (b): Gumbel and Reversed Weibull Family

In what follows we consider a distribution F with extreme type Gumbel or in the reversed Weibull family. We consider both cases separately. Recall that if F has extreme type in the reversed Weibull family then it holds that $\omega_1(F) < \infty$, that is, F has bounded support.

We start by showing that when $\omega_1(F) < \infty$ we have $\text{apx}_k(F) = 1$ for every positive integer k. In particular, the approximation result follows directly from this in the case of a distribution F with extreme type in the reversed Weibull family. When the support of F is upper bounded by $\omega_1(F) < \infty$, we have $\mathbb{E}(M_n^j) \leq \omega_1(F)$ for every $j \in \{1, \ldots, k\}$. For every $\varepsilon > 0$ consider $T_\varepsilon = (1 - \varepsilon) \cdot \omega_1(F)$. Then, by the expression in (4) we have that $\text{apx}_k(F)$ can be lower bounded as $\text{apx}_k(F) \geq (1-\varepsilon) \cdot \omega_1(F) \cdot \liminf_{n\to\infty} \sum_{j=1}^{k} \mathbb{P}(M_n^j > T_\varepsilon)/(k \cdot \omega_1(F)) = 1 - \varepsilon$, and we conclude that $\text{apx}_k(F) = 1$.

In what follows we restrict attention to the distributions F with extreme type Gumbel where $\omega_1(F) = \infty$. Key to our analysis are the result presented in the Preliminaries Sect. 2 about Von Mises representations for distributions in the Gumbel family. We need some lemmas about the structure of a distribution in the Gumbel family before proving the theorem.

Lemma 5. *Let F be a distribution with extreme type in the Gumbel family such that $\omega_1(F) = \infty$ and let (V, η) be a Von Mises representation of F such that $\lim_{t\to\infty} \eta(t) = \eta^\star$. Let $\{a_n\}_{n\in\mathbb{N}}$ and $\{b_n\}_{n\in\mathbb{N}}$ be scaling and shifting sequences, respectively, for V. For every positive integer n consider $b_n^\eta = b_n + a_n \log \eta^\star$. Then, the following holds:*

(a) $\{a_n\}_{n\in\mathbb{N}}$ and $\{b_n^\eta\}_{n\in\mathbb{N}}$ are scaling and shifting sequences, respectively, for F.
(b) For every $U \in \mathbb{R}$ we have $\lim_{n\to\infty}(a_n U + b_n^\eta) = \infty$.
(c) For every $U \in \mathbb{R}$ and every positive integer k we have that $\lim_{n\to\infty}(a_n U + b_n^\eta)/\sum_{j=1}^{k} \mathbb{E}(M_n^j) = 1/k$, where $M_n^1, \ldots, M_n^n$ are the order statistics for F.

Lemma 6. *Let F be a distribution with extreme type in the Gumbel family and let $\{\Theta_n\}_{n\in\mathbb{N}}$ be a sequence of real values such that $\Theta_n \to \infty$. Then, we have $\lim_{n\to\infty} \frac{1}{\Theta_n} \mathbb{E}(X|X > \Theta_n) = 1$, where X is distributed according to F.*

We are now ready to prove Theorem 3 (b) for the Gumbel family.

Proof (Proof of Theorem 3 (b) *for the Gumbel family).* Let F be a distribution with extreme type in the Gumbel family and such that $\omega_1(F) = \infty$. Consider a Von Mises pair (V, η) that represents F and such that $\lim_{t\to\infty} \eta(t) = \eta^\star > 0$, guaranteed to exist by Lemma 2. Let $\{a_n\}_{n\in\mathbb{N}}$ and $\{b_n\}_{n\in\mathbb{N}}$ be scaling and shifting sequences, respectively, for V. For every positive integer n consider $b_n^\eta = b_n + a_n \log \eta^\star$. We can lower bound the value of $\text{apx}_k(F)$ by

$$\sup_{U\in\mathbb{R}} \liminf_{n\to\infty} \frac{\mathbb{E}\left(X|X > a_n U + b_n^\eta\right)}{a_n U + b_n^\eta} \cdot \frac{a_n U + b_n^\eta}{\sum_{j=1}^{k} \mathbb{E}(M_n^j)} \cdot \sum_{j=1}^{k} \mathbb{P}(M_n^j > a_n U + b_n^\eta).$$

By Lemma 5 (b), we have $a_n U + b_n^\eta \to \infty$ for every U when $n \to \infty$, and therefore from Lemma 6 we obtain

$$\lim_{n\to\infty} \frac{\mathbb{E}\left(X | X > a_n U + b_n^\eta\right)}{a_n U + b_n^\eta} = 1,$$

for every U. Furthermore, Lemma 5 (c) implies that for every U and every positive integer k it holds $(a_n U + b_n^\eta)/\sum_{j=1}^k \mathbb{E}(M_n^j) \to 1/k$. We conclude that for every U

$$\lim_{n\to\infty} \frac{\mathbb{E}\left(X | X > a_n U + b_n^\eta\right)}{a_n U + b_n^\eta} \cdot \frac{a_n U + b_n^\eta}{\sum_{j=1}^k \mathbb{E}(M_n^j)} = \frac{1}{k}.$$

By Lemma 5 (a), $\{a_n\}_{n\in\mathbb{N}}$ and $\{b_n^\eta\}_{n\in\mathbb{N}}$ are scaling and shifting sequences, respectively, for F. Therefore, by Theorem 2 we have

$$\begin{aligned}
\lim_{n\to\infty} \sum_{j=1}^k \mathbb{P}(M_n^j > a_n U + b_n^\eta) &= \lim_{n\to\infty} \sum_{j=1}^k \mathbb{P}\left(\frac{M_n^j - b_n^\eta}{a_n} > U\right) \\
&= \sum_{j=1}^k \left(1 - \exp\Big(-e^{-U}\Big) \sum_{s=0}^{j-1} \frac{e^{-sU}}{s!}\right) \\
&= k - \exp\Big(-e^{-U}\Big) \sum_{j=1}^k \sum_{s=0}^{j-1} \frac{e^{-sU}}{s!}.
\end{aligned}$$

Note that the last term is non-negative for every U. Furthermore, we get that

$$\lim_{U\to\infty} \exp\Big(-e^{-U}\Big) \sum_{j=1}^k \sum_{s=0}^{j-1} \frac{e^{-sU}}{s!} = \inf_{U\in\mathbb{R}} \exp\Big(-e^{-U}\Big) \sum_{j=1}^k \sum_{s=0}^{j-1} \frac{e^{-sU}}{s!} = 0$$

since $\sum_{s=0}^\infty e^{-sU}/s! = \exp(-e^{-U})$. We conclude that

$$\sup_{U\in\mathbb{R}} \lim_{n\to\infty} \frac{\mathbb{E}\left(X | X > a_n U + b_n^\eta\right)}{a_n U + b_n^\eta} \cdot \frac{a_n U + b_n^\eta}{\sum_{j=1}^k \mathbb{E}(M_n^j)} \cdot \sum_{j=1}^k \mathbb{P}(M_n^j > a_n U + b_n^\eta) = \frac{1}{k} \cdot k = 1,$$

and therefore $\text{apx}_k(F) = 1$. That concludes the proof for the Gumbel family. □

5 Extreme Types and Virtual Valuations

The *virtual valuation* associated to a distribution G is given by $\phi_G(t) = t - (1 - G(t))/g(t)$, where g is the density function of G. When v is distributed according to G, we denote by G_ϕ the distribution of $\phi_G(v)$ and by G_ϕ^+ the distribution of $\phi_G^+(v) = \max\{0, \phi_G(v)\}$. Using Theorem 3 we can apply the existing reductions in the literature [5,7,15] to translate our optimal guarantees for single threshold prophet inequalities to optimal fixed price mechanisms as long as G_ϕ^+ satisfies the extreme value condition. If G_ϕ^+ has extreme value Fréchet, the revenue gap of the fixed price PPM for the k-selection problem is bounded by a limit of the maximization problem (3) and, for every k, this revenue gap is more than $1 - 1/\sqrt{2k\pi}$ and asymptotically tight in k. When $k = 1$ we further have that the revenue gap is roughly 0.712. When G_ϕ^+ is in the Gumbel or reversed Weibull families, we have that with fixed prices a PPM is able to recover the same revenue of that of the optimal mechanism for the k-selection problem, for every positive integer k.

In what follows, we say that a pair (V, η) *smoothly represents* a distribution G if it satisfies the conditions in (2) where V is in the Von Mises family and $\lim_{t \to \omega_1(F)} \eta'(t) = 0$. We say that a distribution G with extreme type Fréchet of parameter α satisfies the *asymptotic regularity condition* if $\lim_{t \to \infty}(1 - G(t))/(tg(t)) = 1/\alpha$, where g is the density of the distribution G. This holds, for example, every time that g is non-decreasing [26, Proposition 1.15]. In our next result we show that if a distribution G with extreme type satisfies any of these two conditions, the distribution G_ϕ^+ has an extreme type as well, and furthermore, it belongs to the same family.

Theorem 5. *Let G be a distribution satisfying the extreme value condition. Then, the following holds:*

(a) *When G has extreme type in the Fréchet family and if it satisfies the asymptotic regularity condition, then G_ϕ^+ has extreme type in the Fréchet family as well.*
(b) *When G has extreme type Gumbel and if it can be smoothly represented, then G_ϕ^+ has extreme type Gumbel as well.*

References

1. Alaei, S.: Bayesian combinatorial auctions: expanding single buyer mechanisms to many buyers. SIAM J. Comput. **43**(2), 930–972 (2014)
2. Alaei, S., Fu, H., Haghpanah, N., Hartline, J.: The simple economics of approximately optimal auctions. In: 2013 IEEE 54th Annual Symposium on Foundations of Computer Science, pp. 628–637. IEEE (2013)
3. Alaei, S., Hartline, J., Niazadeh, R., Pountourakis, E., Yuan, Y.: Optimal auctions vs. anonymous pricing. Games Econ. Behav. **118**, 494–510 (2019)
4. Chawla, S., Devanur, N., Lykouris, T.: Static pricing for multi-unit prophet inequalities. arXiv preprint arXiv:2007.07990 (2020)
5. Chawla, S., Hartline, J.D., Malec, D.L., Sivan, B.: Multi-parameter mechanism design and sequential posted pricing. In: Proceedings of the 42th ACM Symposium on Theory of Computing, STOC 2010 (2010)
6. Correa, J., Foncea, P., Hoeksma, R., Oosterwijk, T., Vredeveld, T.: Posted price mechanisms for a random stream of customers. In: Proceedings of the ACM Conference on Economics and Computation, EC 2017 (2017)
7. Correa, J., Foncea, P., Pizarro, D., Verdugo, V.: From pricing to prophets, and back! Oper. Res. Lett. **47**(1), 25–29 (2019)
8. Dütting, P., Fischer, F., Klimm, M.: Revenue gaps for static and dynamic posted pricing of homogeneous goods. arXiv preprint arXiv:1607.07105 (2016)

9. Ehsani, S., Hajiaghayi, M., Kesselheim, T., Singla, S.: Prophet secretary for combinatorial auctions and matroids. In: Proceedings of the Twenty-Ninth Annual ACM-SIAM Symposium on Discrete Algorithms, pp. 700–714. SIAM (2018)
10. Feng, Y., Hartline, J.D., Li, Y.: Optimal auctions vs. anonymous pricing: beyond linear utility. In: Proceedings of the 2019 ACM Conference on Economics and Computation, pp. 885–886 (2019)
11. Fisher, R.A., Tippett, L.H.C.: Limiting forms of the frequency distribution of the largest or smallest member of a sample. In: Mathematical Proceedings of the Cambridge Philosophical Society, vol. 24, pp. 180–190. Cambridge University Press (1928)
12. Gautschi, W.: Some elementary inequalities relating to the gamma and incomplete gamma function. J. Math. Phys. **38**(1), 77–81 (1959)
13. Gilbert, J.P., Mosteller, F.: Recognizing the maximum of a sequence. J. Am. Stat. Assoc. **61**, 35–76 (1966)
14. Gnedenko, B.: Sur la distribution limite du terme maximum d'une serie aleatoire. Ann. Math. 423–453 (1943)
15. Hajiaghayi, M.T., Kleinberg, R., Sandholm, T.: Automated online mechanism design and prophet inequalities. AAAI **7**, 58–65 (2007)
16. Hartline, J.D., Lucier, B.: Non-optimal mechanism design. Am. Econ. Rev. **105**(10), 3102–24 (2015)
17. Hill, T.P., Kertz, R.P.: Comparisons of stop rule and supremum expectations of i.i.d. random variables. Ann. Probab. **10**(2), 336–345 (1982)
18. Jin, Y., Lu, P., Tang, Z.G., Xiao, T.: Tight revenue gaps among simple mechanisms. SIAM J. Comput. **49**(5), 927–958 (2020)
19. Kennedy, D.P., Kertz, R.P.: The asymptotic behavior of the reward sequence in the optimal stopping of i.i.d. random variables. Ann. Probab. **19**, 329–341 (1991)
20. Kertz, R.P.: Stop rule and supremum expectations of i.i.d. random variables: a complete comparison by conjugate duality. J. Multivar. Anal. **19**, 88–112 (1986)
21. Kleinberg, R., Weinberg, S.M.: Matroid prophet inequalities. In: Proceedings of the 44th ACM Symposium on Theory of Computing, STOC 2012 (2012)
22. Krengel, U., Sucheston, L.: Semiamarts and finite values. Bull. Amer. Math. Soc. **83**, 745–747 (1977)
23. Krengel, U., Sucheston, L.: On semiamarts, amarts, and processes with finite value. Adv. Probab. **4**, 197–266 (1978)
24. Leadbetter, M.R., Lindgren, G., Rootzén, H.: Extremes and Related Properties of Random Sequences and Processes. Springer, New York (2012). https://doi.org/10.1007/978-1-4612-5449-2
25. Myerson, R.B.: Optimal auction design. Math. Oper. Res. **6**(1), 58–73 (1981)
26. Resnick, S.I.: Extreme Values, Regular Variation and Point Processes. Springer, New York (2013). https://doi.org/10.1007/978-0-387-75953-1
27. Saint-Mont, U.: A simple derivation of a complicated prophet region. J. Multivar. Anal. **80**, 67–72 (2002)
28. Samuel-Cahn, E.: Comparisons of threshold stop rule and maximum for independent nonnegative random variables. Ann. Probab. **12**(4), 1213–1216 (1983)

Approximate Competitive Equilibrium with Generic Budget

Amin Ghiasi[1] and Masoud Seddighin[2(✉)]

[1] University of Maryland, College Park, USA
[2] School of Computer Science, Institute for Research in Fundamental Sciences (IPM), P. O. Box 19395-5746, Tehran, Iran

Abstract. We study the existence of approximate competitive equilibrium in the Fisher market with generic budgets. We show that for any number of buyers and any number of goods, when the preferences are identical and budgets are generic, a 2 approximation of competitive equilibrium (2-CE) always exists. By 2-CE we mean that every buyer receives a bundle with a value at least half of the value of her most desirable bundle that fits within her budget, and the market clears. We also present a polynomial time algorithm to obtain a 2-CE.

1 Introduction

Competitive equilibrium is a central concept from the general equilibrium theory for allocating resources among agents with different preferences. Consider a simple Fisher market [14]: a seller with m goods and n buyers each of whom holds a certain budget. The seller desires money and buyers only desire good. For such a market, an allocation of the goods to the buyers along with a price for each good constitutes a competitive equilibrium, if supply meets demand and each buyer believes that her share is the best she could obtain under her budget.

Competitive equilibrium (CE) is well known to be a remarkable solution to the efficient and fair allocation problem, and "the description of perfect justice" [2]. By the first welfare theorem, a CE allocation is Pareto efficient, and when buyers have equal budgets, competitive equilibrium implies envy-freeness[1]. For unequal budgets, CE can be interpreted as a generalized fairness criterion for the agents with different entitlements which maps to many real-life scenarios such as dividing cabinet ministries among political parties, distributing the inherited wealth among heirs, and allocating university courses to the students.

Perhaps the most remarkable breakthrough in general equilibrium theory is establishing mild conditions under which a competitive equilibrium tends to exist in different markets of divisible goods [4,29,33]. These existential proofs are sometimes accompanied by constructive algorithms [24].

In contrast to divisible goods, once there are indivisible goods in the market, CE might fail to exist even in very simple cases. For example, when we have one

[1] An allocation is envy-free if each agent prefers her share to other agent's share.

I. Caragiannis and K. A. Hansen (Eds.): SAGT 2021, LNCS 12885, pp. 236–250, 2021.
https://doi.org/10.1007/978-3-030-85947-3_16

Table 1. A summary of the results for CE in Fisher markets. Symbol ✓ means that a CE exists for the corresponding setting while ✘ refers to the non-existence of CE. A preference $\prec$ is leveled, if for every two bundles S, T with $|S| < |T|$, $S \prec T$ holds.

n	m	Preferences	Budgets	Result
2	≥ 1	-	Almost equal	✓ [7]
2	≥ 1	Identical	Generic	✓ [7]
≥ 1	≤ 3	General Ordinal	Generic	✓ [5]
≤ 4	2	General Ordinal	Generic	✓ [5]
2	Any	Leveled Ordinal	Generic	✓ [5]
5	≥ 2	-	Generic	✘ [5]
3	4	-	Generic	✓ [32]
≥ 4	4	-	Generic	✘ [32]

item and two buyers with identical preferences and equal budgets, no CE exists: based on the price of the item, demand is either 0 or 2, and supply is always 1. However, this example is only a knife-edge phenomenon in the sense that requires exact equal budgets; even a very slight difference in the budgets yields the existence of CE. Indeed, CE exists in almost all the income-space, except a subset of measure zero which includes equal incomes. This motivates studying markets with budgets that are almost (but not exactly) equal or unequal (generic). These forms of budget constraints are recently considered in several studies [5,6,16,32]. A summary of the results of these papers is outlined in Table 1.

As is clear from Table 1, when the number of items and buyers are not too small (e.g. ≥ 4), no allocation algorithm can guarantee CE even with generic budgets assumption. In addition, the only positive result for markets with more than four items is when there are two buyers with identical valuations. In light of these negative results, we wish to mitigate this barrier by introducing the approximate version of competitive equilibrium in Fisher market. We give the exact definition of approximate-CE in Sect. 2. Roughly speaking, by approximate-CE we mean that each buyer gets a share which is approximately the best she can obtain within her budget, and the market clears.

Consideration of approximately fair allocations has been a fruitful approach in allocation problems, especially in recent years [1,9,16,31]. These approximations are with respect to various fairness objectives, including envy-freeness, proportionality, maximin-share, maximin-fairness, etc. In particular, in a work more related to ours, Budish [16] circumvents the non-guaranteed existence of CEEI due to indivisibilities by weakening the equilibrium concept and introducing (α, β)-approximate competitive equilibrium from equal incomes in a sense that the market approximately clears with error α and budgets are not exactly equal, but within a ratio β of each other. He proved the existence of an (α, β)-approximate CE for $\alpha = \sqrt{m/2}$ and some small $\beta > 0$. His proof is nonconstructive as it relies on a fixed-point argument. Indeed, it is shown that finding a solution with the same approximation ratio as [16] is PPAD-complete [30].

Our main result is a method to obtain a 2-approximate competitive equilibrium (2-CE) for Fisher markets with any number of indivisible goods and any number of buyers with generic budgets, when preferences are identical. We also show how to find such a solution in polynomial time. This result establishes a clear separation between unconstrained and generic budgets, because even for the case of two agents with equal budgets and identical preferences, neither CE nor any approximation of CE exists.

Even though quite restrictive, the assumption that the goods have the same value to all the agents applies to many economic situations, in particular, when buyers can further sell their goods to the other parties in the market [11]. Identical preferences assumption is observed for various allocation objectives, including envy-freeness, Pareto efficiency, and equity [12,15,23], competitive equilibrium [5,13], Nash welfare [10], and other notions [25,28]. Exploring the case of symmetric preferences is posed by Babaioff et al. [6,7] as a promising step toward achieving more positive results for Fisher markets with generic budgets.

In Sect. 1.1 we overview the techniques we used in our methods.

1.1 Our Results and Techniques

Our focus in this paper is Fisher markets with indivisible goods and generic budgets, that is, arbitrary budgets (possibly far from equal) to which tiny random perturbations are added. We show that for a Fisher market with any number of buyers and any number of indivisible goods when preferences are identical and budgets are generic, a 2 approximation of competitive equilibrium (2-CE) always exists. By 2-CE we mean that every buyer receives a bundle with value at least half of the value of her most desirable bundle which fits within her budget, and the market completely clears, meaning that no good is left behind.

Theorem 1. *Given any one-sided Fisher market with generic budgets and an additive valuation V for all the buyers, there exists a pricing* p *and an allocation S such that* (p, S) *constitutes a* 2-CE.

To prove Theorem 1, we propose a basic pricing rule, namely *linear pricing.* Roughly speaking, a pricing is β-linear, if for each good offers a price which is β times its value. We then choose maximum β, such that an allocation exists for β-linear pricing rule which clears the market and respects the budget constraints (but not necessarily is a 2-CE). Let β_m be this value. We propose a greedy allocation algorithm and show that running this algorithm for β_m-linear pricing, either allocates the entire set of goods[2], or enables us to reduce the problem into a smaller one by removing a subset of the buyers and their allocated goods. For the former case, we use another algorithm to convert the obtained greedy allocation into a 2-CE. For latter, we use induction to prove the existence of 2-CE. Note that, the final pricing of the goods is not necessarily β_m-linear. However, our method guarantees that during the algorithm, the price of each good never decreases. Thus, at the end the price of a good with value v is at least $\beta_m \cdot v$.

[2] Note that our greedy algorithm might leave some of the goods un-allocated.

Next, we investigate the computational aspect of our method. In Lemma 10, we show that finding the exact value of β_m is not polynomial-time tractable, unless P=NP. To circumvent this hurdle, we introduce another linear pricing, namely β_g-linear pricing, which refers to the maximum β such that our greedy algorithm clears the market. We show how to use this allocation to obtain the same approximation guarantee.

Notice that finding the exact value of β_g is also not trivial. In Sect. 4, we provide a process to learn the value of β_g. Roughly, we start by an estimation of β_g and make our estimation more and more accurate by iteratively simulating the greedy allocation algorithm. At the end of this process, we obtain a value β_ℓ which is close to β_g (but not necessarily equal) and an allocation exactly similar to what we obtained if we ran the greedy algorithm with β_g-linear pricing. Finally, by a simple observation about the remaining budget of the agents, we find the exact value of β_g. This implies Theorem 2.

Theorem 2. *Given any one-sided Fisher market with unequal budgets and an additive valuation V for all the agents, we can find a* 2-CE *in polynomial time.*

1.2 Related Work

Fisher market [14] is among the most attractive and well-studied models within mathematical economics which enjoys desirable existential and computational characteristics: for convex utility functions a competitive equilibrium is guaranteed to exist [4]; and for many utility classes such as Constant Elasticity of Substitution functions (e.g., Linear and Leontief), competitive equilibrium can be computed efficiently [3,20,27].

To circumvent the non-guaranteed existence of equilibrium in markets with indivisibilities, several studies consider the relaxed versions of equilibrium notions [19,21,22]. Apart from the work of Budish [16] mentioned in the introduction, eliminating the market-clearing property yields another concept called envy-free pricing which has attracted attention in the past decade [8,26].

Competitive equilibrium can be considered as a fairness criterion. For indivisible goods, CEEI implies envy-freeness, proportionality and maximin-share [16]. For divisible goods, CEEI happens to coincide with Nash Social Welfare (NSW) maximizing allocation[3]. Several recent works provide approximations for NSW objective by rounding modifications of EG program [17,18].

2 Preliminaries and Overview of the Ideas

In this section, we provide definitions and basic observations that are used in Sect. 3. In addition, we expose the main ideas and techniques that help us prove our main results. This section is divided into four parts: first, we define Fisher market and (approximate) competitive equilibrium, next we dedicate two subsections to demonstrate the related concepts of pricing and allocation. Finally, we define cuts and satisfying cuts.

[3] allocation that maximizes the utility product of the agents.

2.1 Fisher Market and Competitive Equilibrium

A **Fisher market** consists of a set of n buyers and a set $\mathcal{M}$ of m goods. Each buyer has a valuation over the goods. Our assumption is that the valuations of the buyers are identical, that is, every bundle S of goods has the same value to all the buyers. We denote by $V(S)$, the value of bundle S to each buyer. In addition, we suppose that V is additive, i.e., for two disjoint bundles S and T we have $V(S \cup T) = V(S) + V(T)$. Furthermore, each buyer i has a budget b_i. In this paper, we suppose that the budgets are not equal and, without loss of generality, $b_1 < b_2 < \ldots < b_n$.

Any solution to a Fisher market is a pair (S, p), where $S = \langle S_1, S_2, \ldots, S_n \rangle$ is an allocation of $\mathcal{M}$ to the buyers (S_i is the bundle allocated to buyer i) and p is a pricing rule which attributes a price to each good. We denote by $\mathsf{p}(T)$ the price of bundle T which by additivity assumption we have $\mathsf{p}(T) = \sum_{q \in T} \mathsf{p}(\{q\})$.

Definition 1. *A competitive equilibrium* (CE) *is a pair* (S, p) *with these properties:*

- ***Market Clearance:*** *all the items are allocated, i.e.,* $\bigcup_i S_i = \mathcal{M}$.
- ***Budget Feasibility:*** *for every buyer* i*, price of the bundle allocated to* i *is within her budget, i.e.,* $\mathsf{p}(S_i) \leq b_i$.
- ***Satisfaction:*** *For every buyer* i*,* S_i *is her preferred bundle among all sets whose price is within her budget, i.e., for every subset* T *with* $\mathsf{p}(T) \leq b_i$ *we have* $V(S_i) \geq V(T)$.

Definition 2. *A pair* (S, p) *is a* 2-CE*, if the following properties hold:*

- ***Market Clearance:*** S *allocates all the goods.*
- ***Budget Feasibility:*** *each buyer receives a bundle with price at most her budget.*
- ***Approximate Satisfaction:*** *For every buyer* i *and bundle* T *where* $\mathsf{p}(T) \leq b_i$*, we have* $V(S_i) \geq V(T)/2$.

For an allocation S we say buyer i is satisfied with bundle S_i, if $\mathsf{p}(S_i) \leq b_i$ and $V(S_i) \geq V(\mathcal{B}_i)/2$, where $\mathcal{B}_i$ is the best bundle of $\mathcal{M}$ for buyer i under pricing p. We also say buyer i is completely satisfied, if she is satisfied and $\mathsf{p}(S_i) = b_i$.

2.2 Pricing: Feasible, Linear, and Maximum-Linear

Definition 3. *A pricing* p *is feasible, if there exists an allocation* S *such that* (S, p) *clears the market and each buyer gets a share with price within her budget.*

Definition 4. *A pricing* p *is* β*-linear, if for every item* x*,* $\mathsf{p}(\{x\}) = \beta V(\{x\})$. *For brevity, for two pricings* p *and* p' *which are respectively* β*-linear and* β'*-linear, we say* $\mathsf{p} > \mathsf{p}'$*, if and only if* $\beta > \beta'$.

Lemma 1 indicates that for a β-linear pricing p, buyer i is satisfied, if she spends a factor $1/2$ of her budget. Thus, when the pricing is β-linear and we want to find an 2-CE, it suffices to allocate each buyer i a bundle with price at least a factor $1/2$ of her budget.

Lemma 1. *For a β-linear pricing p, buyer i is satisfied with bundle S_i, if $\mathsf{p}(S_i) \geq b_i/2$.*[4]

Definition 5. *Define β_m as the maximum possible value such that β_m-linear pricing is feasible. Furthermore, denote by p_m the pricing corresponding to β_m-linear pricing.*

By definition, for any $\beta > \beta_m$, β-linear pricing is not feasible. Another important property of β_m is stated in Observation 1.

Observation 1. *For pricing p_m, every allocation that clears the market admits at least one completely satisfied agent.*

2.3 Allocation: Admissible, Sorted, and Right-Sided Sorted

Definition 6. *Given a pricing p, an allocation S is admissible, if it allocates all the items and the price of the share allocated to each agent is within her budget.*

Notice that, when an allocation S is admissible, its corresponding pricing is feasible. Indeed, the admissibility of an allocation relies on the prices attributed to the goods. Throughout the paper, when the price vector is clear from the context, we only say "S is admissible", without mentioning the pricing vector.

We now define sorted and right-sided sorted allocations and show how to convert an allocation to a sorted and right-sided sorted one.

Definition 7. *For a pricing p, an admissible allocation S is sorted, if $\mathsf{p}(S_1) \leq \mathsf{p}(S_2) \leq \ldots \leq \mathsf{p}(S_n)$.*

Lemma 2. *Given a pricing p and an admissible allocation S, there exists an admissible and sorted allocation S' with the same pricing.*

Definition 8. *For a pricing p, a sorted allocation S is a right-sided sorted allocation, if for every i such that $S_i \neq \emptyset$, buyer $i+1$ satisfies with her share.*

Lemma 3. *Given a sorted allocation S with pricing vector p, we can convert S to a right-sided sorted allocation.*

Algorithm 1 shows the method by which we convert a sorted allocation to a right-sided sorted allocation in the proof [6] of Lemma 3.

2.4 Cuts and Satisfying Cuts

We now introduce cuts and satisfying cuts which play a key role in our method. But before defining these concepts, in Observation 2 we first show that for any buyer i which is satisfied with a bundle of goods, increasing the prices of the other goods does not affect her satisfaction.

[4] For the missing proofs, we refer to the complete version of the paper.

Algorithm 1: Right Sided Sorter algorithm

Data: S : the allocations, b : budgets, p : the pricing vector
Result: Right sided sorted allocation
while $\exists i$ *such that* $S_i \neq \emptyset$ *and* $\mathsf{p}(S_{i+1}) < b_{i+1}/2$ **do**
    $S_{i+1} = S_{i+1} \cup S_i$;
    $S_i = \emptyset$;
    Sort(S);
end

Observation 2. *Suppose buyer i is satisfied with bundle S_i. If we increase the prices of the goods in $\mathcal{M} \setminus S_i$ or reallocate them, buyer i remains satisfied.*

The reason that Observation 2 holds is that after increasing the prices, whatever buyer i can buy, she could also buy before. Intuitively Observation 2 indicates that for an allocation S, we might be able to increase the number of satisfied buyers by increasing the prices of some items.

For an allocation S, denote by S_{i^+}, the set of items in $S_{i+1} \cup S_{i+2} \cup \dots S_n$. In addition, let $S_{i^-} = \mathcal{M} \setminus S_{i^+}$.

Definition 9 (Cut). *For a pricing p and an admissible allocation S, we say there is a cut on buyer i, if $\min_{x \in S_{i^+}} \mathsf{p}(\{x\}) > b_i$. We also call buyers $\{1, 2, \dots, i\}$, the left-side of the cut and the rest of the buyers, the right-side of the cut.*

By definition, if there exists a cut C for some pair (S, p), the left-side buyers of C cannot own any of the goods in S_{i^+} since the price of each good in S_{i^+} is higher than their budget.

Definition 10 (Satisfying cut). *Cut C is a* satisfying cut, *if all the buyers on right-side of C are satisfied.*

As mentioned, satisfying cuts play a key role in our method since they reduce the problem into a smaller sub-problem. Consider pricing p and an admissible allocation S, and assume that C is a satisfying cut. By definition, the price of each good in S_{i^+} is too high for the buyers on the left side. Furthermore, the right-side buyers are currently satisfied with their share. This allows us to put the right-side buyers and their allocated goods aside and solve the problem recursively for the left-side buyers and goods in S_{i^-}. However, note that if we decrease the prices of the goods in S_{i^-}, a right-side buyer may become unsatisfied, because her preferred bundle may change. Fortunately, our method has the property that the prices obtained for the goods in S_{i^-} by recursively solving the problem for left-hand side buyers are at least as their initial prices in the satisfying cut. Therefore, we can use satisfying cuts to reduce our instances.

The main body of Sect. 3 is devoted to proving Lemma 4.

Lemma 4. *For any instance of the problem, there exists an allocation S, such that the pair (S, p_m) either is a 2-CE, or admits a satisfying cut.*

Based on Lemma 4, we operate as follows: we find p_m and allocation S satisfying the condition of Lemma 4. If (S, p_m) constitutes a 2-CE, we are done. Otherwise, we satisfy the agents in the right-side of the produced satisfying cut and remove these agents and their corresponding goods. Next, we repeat the same process for the remaining goods and agents. Observation 3 assures that during the further steps, the price of no good decreases which means that the agents satisfied in the previous steps remain satisfied.

Observation 3. *Let S be an admissible allocation for p_m. Furthermore, for some $1 \leq i < n$, let p'_m be the maximum linear pricing for the Fisher market containing goods in S_{i^-} and buyers $1, 2, \ldots, i$. Then we have $\mathsf{p}'_m \geq \mathsf{p}_m$.*

Lemma 4 with Observation 3 implies Theorem 1.

3 Existence of Approximate CE

In this section, we prove the existence of a 2-CE for every instance of Fisher market with generic budgets, when preferences are identical. As said before, to show this, we prove Lemma 4 which states that there is an allocation S such that (S, p_m) either is a 2-CE or has a satisfying cut. For this purpose, we introduce the greedy allocation.

Definition 11. *Greedy allocation for a pricing p, denoted by $\mathcal{G}(\mathsf{p})$, is the allocation obtained from the following n-step greedy algorithm: in the i'th step, ask buyer i to iteratively pick the most valuable remaining good which fits into her remaining budget. When no good could be selected by i, we head to the next step.*

See Algorithm 2 for a pseudo-code of this algorithm. When the price is clear from the context, we use $\mathcal{G}_i$ to refer to the bundle of buyer i in the greedy allocation. Note that for a pricing vector p, $\mathcal{G}(\mathsf{p})$ does not necessarily allocate all the goods, even for a β-linear pricing with $\beta \leq \beta_m$. However, these allocations are attractive because of the property we show in Lemma 5: in a greedy allocation, for each buyer, either she is satisfied or there is a cut on that buyer[5].

Lemma 5. *For any price p, in $\mathcal{G}(\mathsf{p})$ each buyer i is either satisfied or there is a cut on buyer i.*

Now, we show that if $\mathcal{G}(\mathsf{p}_m)$ clears the market, then one can refine $\mathcal{G}(\mathsf{p}_m)$ to obtain an allocation satisfying the condition of Lemma 4. But before proving this, in Lemma 6 we prove another property for $\mathcal{G}(\mathsf{p}_m)$.

Lemma 6. *If $\mathcal{G}(\mathsf{p}_m)$ is admissible, at least one good is allocated to buyer n.*

We remark that the property proved in Lemma 6 essentially relies on generic budgets. To see how Lemma 6 fails for equal budgets, consider n buyers with budget 10 and $n+1$ identical goods with value 5. For this case, the price of each good in p_m is 5 and the last $\lfloor (n-1)/2 \rfloor$ buyers receive no good in $\mathcal{G}(\mathsf{p}_m)$.

[5] Here we need to extend Definition 9 for $i = n$: there is a cut on buyer n if she is not satisfied.

Algorithm 2: Greedy Allocation algorithm

```
Data: b: set of budgets , M: set of goods , p: pricing
Result: Allocation G
for i : 1 → n do
    G_i = ∅ ;
    F = {x ∈ M | p({x}) ≤ b_i − p(G_i)};
    while F ≠ ∅ do
        y ← arg max_{x∈F} p({x});
        G_i ← G_i ∪ {y} ;
        M ← M \ {y};
        F = {x ∈ M | p(x) ≤ b_i − p(G_i)};
    end
end
```

Lemma 7. *If $\mathcal{G}(\mathsf{p}_m)$ is admissible, we can convert it to an allocation which is either a* 2-CE *or admits a satisfying cut.*

Proof. First, note that if $\mathcal{G}(\mathsf{p}_m)$ admits a satisfying cut then $\mathcal{G}(\mathsf{p}_m)$ itself is the desired allocation. Therefore, without loss of generality, we assume that $\mathcal{G}(\mathsf{p}_m)$ admits no satisfying cut. By Lemma 5, in $\mathcal{G}(\mathsf{p}_m)$ each buyer is either satisfied or the allocation admits a cut on that buyer. In particular, the rightmost unsatisfied buyer (i.e., the buyer with the largest index, say j which is not satisfied with $\mathcal{G}_j$) admits a satisfying cut, unless $j = n$. Therefore, if $\mathcal{G}(\mathsf{p}_m)$ admits no satisfying cut, buyer n is not satisfied with $\mathcal{G}_n$. Note that by Lemma 6, we know this buyer has at least one good.

Now, assume that there are k completely satisfied buyers $i_1 < i_2 < \ldots < i_k$. By Observation 1, we know there is at least one, i.e., $k \geq 1$.

Proposition 1. *For any buyer $j > i_k$ such that j is not satisfied with $\mathcal{G}_j$, we have $\mathsf{p}(\mathcal{G}_{i_k} \cup \mathcal{G}_j) \geq b_j$.*

We now claim that no unsatisfied buyer exists between buyers i_k and n. In other words, n is the only unsatisfied buyer after i_k.

Proposition 2. *For every buyer $i_k < j < n$ we have $b_j/2 \leq \mathsf{p}(\mathcal{G}_j) < b_j$.*

Since buyer n is not satisfied, by Proposition 1 we have

$$\mathsf{p}(\mathcal{G}_{i_k}) > b_n/2. \tag{1}$$

To refine the allocation, we exchange the bundles of buyers n and i_k, i.e., we allocate $\mathcal{G}_n$ to buyer i_k and allocate $\mathcal{G}_{i_k}$ to buyer n. Since before this exchange Inequality (1) holds, buyer n becomes satisfied but not completely satisfied since $b_{i_k} < b_n$. Therefore, if buyer i_k satisfies with her new bundle, we have obtained the desired allocation, since buyers $i_k \ldots n$ are satisfied with their bundles and the bundles allocated to the buyers $1 \ldots i_k - 1$ are the same as the greedy allocation. Recall that by Lemma 5, either all the buyers $1, \ldots, i_k - 1$ are also satisfied, or there is a satisfying cut on buyer ℓ where $\ell = \arg\max_j$ such that buyer j is not satisfied. For the case that buyer i_k is not satisfied after the exchange, we use induction on k (number of completely-satisfied buyers) to prove the statement of Lemma 7. If $k = 1$, after this exchange buyer i_k must remain completely-satisfied, otherwise none of the buyers in the refined allocation is completely-satisfied and therefore we can increase the value of β_m. This contradicts our assumption that i_k is not satisfied after the exchange.

Now, assume $k > 1$, and consider the sub instance containing buyers $[1..i_k]$ and goods $\bigcup_{1 \le l \le i_k} \mathcal{G}_l$. Note that if we ran the greedy algorithm on these sets of buyers and goods with the same pricing, the resulted allocation would be the same as their current bundles. Furthermore, for this sub-instance, the value of β_m is the same as the original instance; otherwise, we can increase β_m in the original instance, because none of the buyers after i_k are completely-satisfied. For this sub-instance, we have $k-1$ completely-satisfied buyers (i.e., $i_1, i_2, \dots, i_{k-1}$) and an un-satisfied buyer i_k. By the induction hypothesis, we can convert the greedy allocation of this sub-instance into an allocation which is either a 2-CE or admits a satisfying cut for buyers $[1..i_k]$. Since all the buyers after i_k are satisfied with their bundles, this allocation combined with the bundles allocated to the buyers after i_k yields an allocation which fulfills the requirements of Lemma 4. □

Algorithm 3: Refine

Data: $\mathcal{G}$: an admissible greedy allocation, b : set of budgets, p : pricing.
Result: An allocation satisfying everyone or having a satisfying cut
initialization;
$i = n$;
Let i_k be the largest index such that $\mathsf{p}(\mathcal{G}_{i_k}) = b_{i_k}$;
Swap $\mathcal{G}_{i_k}$ and $\mathcal{G}_n$;
if $\mathsf{p}(\mathcal{G}_{i_k}) > b_{i_k}/2$ **then**
  return $\mathcal{G}$;
else
  return Refine($\{\mathcal{G}_1, \mathcal{G}_2, \dots, \mathcal{G}_{i_k}\}, [b_1..b_{i_k}], \mathsf{p}) \oplus \langle \mathcal{G}_{i_k+1}, \mathcal{G}_{i_k+2}, \dots, \mathcal{G}_{i_n} \rangle$
end

Algorithm 3 represents the method by which we convert an admissible greedy allocation into an allocation desired by Lemma 4. Thus, the only remaining case is when the greedy allocation does not clear the market. To address this case, we introduce the *Best Fit First* (BFF) algorithm. Roughly, BFF takes p_m and an admissible allocation S for p_m as input and refines S to be as much similar to $\mathcal{G}(\mathsf{p}_m)$ as possible. In what follows, we formally describe BFF.

3.1 BFF Algorithm

BFF algorithm takes p_m and an admissible allocation S as input and updates S according to the following procedure: in the i'th step, let S_i be the current goods allocated to buyer i in S and let $\mathcal{G}_i$ be the goods that would be allocated to buyer i if we ran the greedy allocation for p_m. BFF iteratively selects the good with the maximum value in $\mathcal{G}_i \setminus S_i$. If $\mathcal{G}_i \setminus S_i = \emptyset$, we head to step $i+1$; otherwise, let x be the selected good. We carefully update S, so that these properties hold:

- After the update, x belongs to S_i.
- The update keeps $S_1, S_2, \dots, S_{i-1}$ and the goods in $S_i \cap \mathcal{G}_i$ intact.
- The allocation remains admissible.

In the rest of this section, we show how to perform BFF. We show that we can either perform such an update, or instantly return an allocation satisfying the condition of Lemma 4. We start by Lemma 8 which states that in the i'th step of BFF, for every $j < i$, bundle S_{ij} is exactly the same as $\mathcal{G}_j$.[6]

Lemma 8. *If* BFF *heads to the i'th step, for every $j < i$ we have $\mathcal{G}_j = S_j$.*

Now, suppose that the algorithm is at step i, and we want to transfer a good $x \in \mathcal{G}_i \setminus S_i$ to S_i. We show that either such a transformation is possible, or we can instantly return an allocation satisfying the conditions of Lemma 4.

Lemma 9. *Assume that at the i'th step of* BFF, *we want to transfer a good $x \in \mathcal{G}_i \setminus S_i$ to S_i. Either such a transformation is possible, or we can instantly return an allocation satisfying the requirements of Lemma 4.*

Proof. Suppose that for some x, such a transformation is not possible. Let $R = \mathcal{G}_i \cap S_i$, and consider the following sub-instance:

- buyers $\{i, i+1, \dots, n\}$ with budgets $\{b_i - \mathsf{p}(R), b_{i+1}, b_{i+2}, \dots, b_n\}$.
- $S' = \langle S'_i, S'_{i+1}, S'_{i+2}, \dots, S'_n \rangle$ where $S'_i = S_i \setminus R$ and for every $j > i$, $S'_i = S_i$.

By Lemma 2, we can convert S' to a right-sided sorted allocation. After this conversion, suppose that good x is in some bundle S'_j. Based on the goods in S'_i, one of the following cases occur:

- S'_i is empty: in this case, we can simply update S to allocation: $\langle S_1, S_2, \dots, S_{i-1}, R \cup \{x\}, S'_{i+1}, \dots, S'_j \setminus \{x\}, \dots, S'_n \rangle$. One can easily verify that all 3 properties hold for this allocation which contradicts our assumption that transferring x to the bundle of buyer i is not possible.
- S'_i is not empty and buyer i satisfies with $R \cup S'_i$: by definition of right-sided sorted allocation, since S'_i is not empty, buyers $i+1, i+2, \dots, n$ are satisfied by their bundles in S'. Therefore, if buyer i also satisfies with $R \cup S'_i$, since the first $i-1$ bundles of S are the same as the bundles returned by greedy algorithm, by Lemma 2, in allocation $S'' = \langle S_1, S_2, \dots, S_{i-1}, R \cup S'_i, S'_{i+1}, \dots, S'_n \rangle$, either all the buyers in $\{1, 2, \dots, i-1\}$ are also satisfied which means S'' is a 2-CE, or S'' admits a satisfying cut.

[6] Note that Lemma 8 holds regardless of the method by which we update the bundles.

- S'_i is not empty and buyer i does not satisfy with $R \cup S'_i$: since buyer i does not satisfy with $R \cup S'_i$, we have $\mathsf{p}(R \cup S'_i) < b_i/2$. In addition, we have $\mathsf{p}(\{x\}) \leq \mathsf{p}(S'_i)$, otherwise we can exchange x and S'_i and update allocation S to allocation $\langle S_1, S_2, \ldots, S_{i-1}, R \cup \{x\}, S'_{i+1}, \ldots, S'_{j-1}, S'_j \cup S'_i \setminus \{x\}, S'_{j+1}, \ldots, S'_n \rangle$. Note that this allocation respects the budget constraints since we know $\mathsf{p}(R \cup \{x\}) \leq \mathsf{p}(\mathcal{G}_i) < b_i$ and furthermore $\mathsf{p}(S'_j \cup S'_i \setminus \{x\}) \leq \mathsf{p}(S'_j) \leq b_j$. It is easy to check that this allocation satisfies all 3 properties which contradicts our assumption that transferring x to the bundle of buyer i is not possible. Therefore, we have $\mathsf{p}(\{x\}) \leq \mathsf{p}(S'_i)$. Since buyer i does not satisfy with bundle $R \cup S'_i$, we have

$$\begin{aligned}\mathsf{p}(R \cup S'_i \cup \{x\}) &= \mathsf{p}(R) + \mathsf{p}(S'_i) + \mathsf{p}(\{x\}) \\ &\leq \mathsf{p}(R) + 2\mathsf{p}(S'_i) \\ &\leq 2\mathsf{p}(S'_i \cup R) \leq b_i.\end{aligned}$$

Thus, allocation

$$\langle S_1, S_2, \ldots, S_{i-1}, R \cup S'_i \cup \{x\}, S'_{i+1}, \ldots, S'_{j-1}, S'_j \setminus \{x\}, S'_{j+1}, \ldots, S'_n \rangle$$

satisfies all the desired properties which again contradicts the assumption that moving x to the bundle of buyer i is not possible. □

In conclusion, we can either perform the exchange operation or instantly return an allocation which satisfies the condition of Lemma 4. Note that BFF starts with an admissible allocation and keeps the allocation admissible during the algorithm. Therefore, since $\mathcal{G}(\mathsf{p}_m)$ is not admissible, at some point of the algorithm, we are unable to perform the exchange operation. By Lemma 9, in such case we can return a proper allocation. This completes the proof of Lemma 4 and as a consequence, Theorem 1 holds.

Theorem 1. *Given any one-sided Fisher market with generic budgets and an additive valuation V for all the buyers, there exists a pricing* p *and an allocation* S *such that* (p, S) *constitutes a* 2-CE.

4 Polytime Algorithm

Recall that an essential part of the proof in the previous section was finding the value of β_m and an admissible allocation for pricing p_m. However, as we show in Lemma 10, finding p_m is NP-hard.

Lemma 10. *Finding the exact value of β_m is NP-hard.*

In this section, we show that we can bypass this hardness using another pricing rule. As discussed earlier, for a feasible linear pricing rule p, there is no guarantee that $\mathcal{G}(\mathsf{p})$ allocates all the goods, even if $\mathsf{p} < \mathsf{p}_m$. We define the maximum greedy pricing as a maximum linear pricing p, such that $\mathcal{G}(\mathsf{p})$ is admissible.

Definition 12. *Maximum greedy pricing, denoted by* p_g *is defined as the linear pricing with the maximum* β*, such that* $\mathcal{G}(\mathsf{p}_g)$ *is admissible.*

Trivially, we have $\mathsf{p}_g \leq \mathsf{p}_m$. In the rest of this section, we provide a method to learn the value of p_g.

Observation 4. *If* $\mathcal{G}(\mathsf{p})$ *is admissible, then for any price* $\mathsf{p}' < \mathsf{p}$, $\mathcal{G}(\mathsf{p}')$ *is also admissible. In addition, if* $\mathcal{G}(\mathsf{p})$ *is not admissible, then for any price* $\mathsf{p}' > \mathsf{p}$, $\mathcal{G}(\mathsf{p}')$ *is not admissible.*

Lemma 11. *With* β*-linear pricing* p*, we can either find a solution, or tell if* $\beta < \beta_g$ *or* $\beta > \beta_g$.

Note that Lemma 11 immediately suggests a simple binary search to find the value of β_g. However, the running time of binary search depends on the value of budgets and goods, which is not desirable. Here, we show how to learn β_g by somewhat simulating the greedy algorithm. To illustrate our method, suppose that the value of β_g is unknown to us, and we want to guess β_g based on the operations made in the greedy algorithm. In the beginning we only know that β_g is a value in $(0, +\infty)$. Now, suppose that an oracle tells us which good is the first good that buyer 1 selects in the greedy algorithm and assume that the value of this good is v. Since any allocation must respect the budget constraints, we can immediately conclude that the value of β_g is upper bounded by b_1/v since buyer 1 was able to select an good with value v, and lower bounded b_1/v' where

$$v' = \min_{x \in \mathcal{M}, V(\{x\}) > v} V(\{x\}). \tag{2}$$

This lower bound stems from the fact that in the greedy algorithm, buyer 1 must choose the most valuable good whose price fits into her budget, and we know her choice was the good with value v, which indicates that the price of the more valuable goods are higher than her budget. Therefore, this information from the oracle limits to possible values of β_g to the interval $(b_1/v', b_1/v]$.

Note that, there is also another possibility: buyer 1 passes her turn since none of the goods fits into her budget. If oracle tells that buyer 1 passes without selecting any good, we can conclude that the value of β_g is lower bounded by b_1/v where v is the value of the least valuable good. This limits β_g to $(b_1/v, +\infty)$.

Based on the above discussion, we can discover the first step of the greedy algorithm by considering all $m+1$ possibilities: selecting each one of the goods as the first choice of buyer 1 or passing to the second buyer. Each one of these possibilities determines an interval to limit the value of β_g and therefore we have $m+1$ intervals $[0, \beta_1], (\beta_1, \beta_2], (\beta_2, \beta_3], \ldots, (\beta_m, +\infty)$. Among these intervals we select the correct interval using Lemma 11 and consider allocating its corresponding good to buyer 1 (or passing to buyer 2 if the last interval was correct) as the first event that happens in the greedy algorithm and limit β_g to some interval $(\beta_{i-1}, \beta_i]$. With a similar argument, we can trace the goods allocated at each step of the greedy algorithm one by one, while increasing the accuracy of our estimation of β_g. After finding the greedy allocation, we can use

Observation 1 to determine the exact value of p_g, since at least one of the buyers in this allocation is fully satisfied in $\mathcal{G}(\mathsf{p}_g)$. This concludes Lemma 12.

Lemma 12. *We can find the value of* p_g *in polynomial time.*

Finally, we show that using p_g instead of p_m leads us finding a 2-CE.

Theorem 2. *Given any one-sided Fisher market with unequal budgets and an additive valuation* V *for all the agents, we can find a* 2-CE *in polynomial time.*

References

1. Amanatidis, G., Markakis, E., Nikzad, A., Saberi, A.: Approximation algorithms for computing maximin share allocations. ACM Trans. Algorithms (TALG) **13**(4), 52 (2017)
2. Arnsperger, C.: Envy-freeness and distributive justice. J. Econ. Surv. **8**(2), 155–186 (1994)
3. Arrow, K.J., Chenery, H.B., Minhas, B.S., Solow, R.M.: Capital-labor substitution and economic efficiency. Rev. Econ. Stat. **43**, 225–250 (1961)
4. Arrow, K.J., Debreu, G.: Existence of an equilibrium for a competitive economy. Econ. J. Econ. Soc. **22**, 265–290 (1954)
5. Babaioff, M., Nisan, N., Talgam-Cohen, I.: Competitive equilibria with indivisible goods and generic budgets. arXiv preprint arXiv:1703.08150 (2017)
6. Babaioff, M., Nisan, N., Talgam-Cohen, I.: Competitive equilibrium with generic budgets: beyond additive. arXiv preprint arXiv:1911.09992 (2019)
7. Babaioff, M., Nisan, N., Talgam-Cohen, I.: Fair allocation through competitive equilibrium from generic incomes. In: FAT, p. 180 (2019)
8. Balcan, M.-F., Blum, A., Mansour, Y.: Item pricing for revenue maximization. In: Proceedings of the 9th ACM Conference on Electronic Commerce, pp. 50–59 (2008)
9. Barman, S., Krishnamurthy, S.K.: Approximation algorithms for maximin fair division. In: Proceedings of the 2017 ACM Conference on Economics and Computation, pp. 647–664. ACM (2017)
10. Barman, S., Krishnamurthy, S.K., Vaish, R.: Greedy algorithms for maximizing Nash social welfare. In: Proceedings of the 17th International Conference on Autonomous Agents and MultiAgent Systems, pp. 7–13. International Foundation for Autonomous Agents and Multiagent Systems (2018)
11. Bergemann, D., Brooks, B.A., Morris, S.: Selling to intermediaries: optimal auction design in a common value model (2017)
12. Bouveret, S., Lang, J.: Efficiency and envy-freeness in fair division of indivisible goods: logical representation and complexity. J. Artif. Intell. Res. **32**, 525–564 (2008)
13. Bouveret, S., Lemaître, M.: Characterizing conflicts in fair division of indivisible goods using a scale of criteria. Auton. Agents Multi-agent Syst. **30**(2), 259–290 (2016)
14. Brainard, W.C., Scarf, H.E., et al.: How to Compute Equilibrium Prices in 1891. Cowles Foundation for Research in Economics (2000)
15. Brams, S.J., Fishburn, P.C.: Fair division of indivisible items between two people with identical preferences: envy-freeness, pareto-optimality, and equity. Soc. Choice Welf. **17**(2), 247–267 (2000)

16. Budish, E.: The combinatorial assignment problem: approximate competitive equilibrium from equal incomes. J. Polit. Econ. **119**(6), 1061–1103 (2011)
17. Cole, R.: Convex program duality, fisher markets, and Nash social welfare. In: Proceedings of the 2017 ACM Conference on Economics and Computation, pp. 459–460 (2017)
18. Cole, R., Gkatzelis, V.: Approximating the Nash social welfare with indivisible items. In: Proceedings of the Forty-Seventh Annual ACM Symposium on Theory of Computing, pp. 371–380 (2015)
19. Cole, R., Rastogi, A.: Indivisible markets with good approximate equilibrium prices. In: Electronic Colloquium on Computational Complexity (ECCC), vol. 14, p. 017. Citeseer (2007)
20. Deng, X., Papadimitriou, C., Safra, S.: On the complexity of equilibria. In Proceedings of the Thiry-Fourth Annual ACM Symposium on Theory of Computing, pp. 67–71. ACM (2002)
21. Deng, X., Papadimitriou, C., Safra, S.: On the complexity of price equilibria. J. Comput. Syst. Sci. **67**(2), 311–324 (2003)
22. Dierker, E.: Equilibrium analysis of exchange economies with indivisible commodities. Econ. J. Econ. Soc. **39**, 997–1008 (1971)
23. Edelman, P., Fishburn, P.: Fair division of indivisible items among people with similar preferences. Math. Soc. Sci. **41**(3), 327–347 (2001)
24. Eisenberg, E., Gale, D.: Consensus of subjective probabilities: the pari-mutuel method. Ann. Math. Stat. **30**(1), 165–168 (1959)
25. Gjerstad, S.: Multiple equilibria in exchange economies with homothetic, nearly identical preference. University of Minnesota, Center for Economic Research, Discussion Paper, 288 (1996)
26. Guruswami, V., Hartline, J.D., Karlin, A.R., Kempe, D., Kenyon, C., McSherry, F.: On profit-maximizing envy-free pricing. In: SODA, vol. 5, pp. 1164–1173. Citeseer (2005)
27. Jain, K.: A polynomial time algorithm for computing an Arrow-Debreu market equilibrium for linear utilities. SIAM J. Comput. **37**(1), 303–318 (2007)
28. Kirman, A.P., Koch, K.-J.: Market excess demand in exchange economies with identical preferences and collinear endowments. Rev. Econ. Stud. **53**(3), 457–463 (1986)
29. Mas-Colell, A., Whinston, M.D., Green, J.R., et al.: Microeconomic Theory, vol. 1. Oxford University Press, New York (1995)
30. Othman, A., Papadimitriou, C., Rubinstein, A.: The complexity of fairness through equilibrium. ACM Trans. Econ. Comput. (TEAC) **4**(4), 1–19 (2016)
31. Procaccia, A.D., Wang, J.: Fair enough: guaranteeing approximate maximin shares. In: Proceedings of the Fifteenth ACM Conference on Economics and Computation, pp. 675–692. ACM (2014)
32. Segal-Halevi, E.: Competitive equilibrium for almost all incomes: existence and fairness. arXiv preprint arXiv:1705.04212 (2017)
33. Weller, D.: Fair division of a measurable space. J. Math. Econ. **14**(1), 5–17 (1985)

Cost Sharing in Two-Sided Markets

Sreenivas Gollapudi[1], Kostas Kollias[1], and Ali Shameli[2(✉)]

[1] Google Research, Mountain View, USA
[2] Massachusetts Institute of Technology, Cambridge, USA

Abstract. Motivated by the emergence of popular service-based two-sided markets where sellers can serve multiple buyers at the same time, we formulate and study the *two-sided cost sharing* problem. In two-sided cost sharing, sellers incur different costs for serving different subsets of buyers and buyers have different values for being served by different sellers. Both buyers and sellers are self-interested agents whose values and costs are private information. We study the problem from the perspective of an intermediary platform that matches buyers to sellers and assigns prices and wages in an effort to maximize gains from trade (i.e., buyer values minus seller costs) subject to budget-balance in an incentive compatible manner. In our markets of interest, agents trade the (often same) services multiple times. Moreover, the value and cost for the same service differs based on the context (e.g., location, urgency, weather conditions, etc.). In this framework, we design mechanisms that are efficient, ex-ante budget-balanced, ex-ante individually rational, dominant strategy incentive compatible, and ex-ante in the core (a natural generalization of the core that we define here).

Keywords: Cost sharing · Mechanism design · Two-sided markets

1 Introduction

The recent emergence of sharing economy has brought renewed interest in the scientific community on studying two-sided markets where services are traded. One example of such markets are ride-sharing services like Uber and Lyft where one side of the market, i.e. drivers, provide a service to the other side of the market, namely riders. An important characteristic of such markets is the ability of a seller to offer service to *multiple* buyers at the same time. For example, Uber Pool and Lyft Line typically assign a driver to multiple riders at the same time; as long as the number of riders does not exceed the capacity of the car. This is in contrast to the one-to-one assignment that happens in other popular two-sided markets such as Amazon and Ebay. Central to the design of the above markets are the problems of price and wage computation as well as assignment of buyers to sellers.

Consider a simpler one sided case where we have multiple buyers and one service provider. In such a case, a service provider incurs a cost $c(S)$ for serving a subset S of its customers. In the case of ride-sharing, $c(S)$ is the cost incurred

I. Caragiannis and K. A. Hansen (Eds.): SAGT 2021, LNCS 12885, pp. 251–265, 2021.
https://doi.org/10.1007/978-3-030-85947-3_17

by a cab driver to serve the riders in S. Each rider i values the ride v_i which is known only to i. In this case, the utility derived by the rider is $v_i - p_i$ where p_i is the price charged to the rider for the ride. Depending on the pricing mechanism chosen by the ride-sharing platform, a rider might have an incentive to misreport her value to derive higher utility. The solution to this problem involves solving a *cost sharing* problem [14,16,20,22,26–28,30]. A cost sharing mechanism first asks each buyer to report their value for being served and then decides the assignment as well as the price each user pays on the buyer side in a way that the cost of the seller is covered by the payments of the buyers.

The reader may note that in the above *one-sided* setting, only the values of the users are private while the cost function $c(S)$ of the providers is known to the platform. In this study, we propose and study the *two-sided cost sharing* problem that generalizes the one-sided setting to the case where the costs are also private information to the sellers and the platform procures their services by offering wages. One challenge for such settings is designing a mechanism that can actually extract the true values and cost functions of buyers and sellers respectively.

In designing our mechanism, there are various objectives that we aim to achieve. A two-sided cost-sharing mechanism is efficient if it maximizes the sum of valuations of all buyers in the assignment minus the cost incurred by the sellers (which is equivalently called the *gains from trade*, a popular objective in the literature for designing mechanisms for two sided markets); It is dominant-strategy incentive compatible (DSIC) if for every buyer and seller, revealing their true value and cost respectively is a dominant strategy; it is weakly budget-balanced (BB) if, in the assignment, the price realized from all buyers is at least as large as the wages paid to all the sellers; it is individually rational (IR) if no agent incurs a loss participating in the mechanism; finally, a solution of the mechanism (which consists of an assignment and vectors of wages and prices) is in the *core* if the utilities of the agents are such that no subset of them can form a coalition and produce welfare higher than their collective utility in the proposed solution.

Two salient features of services in the sharing economy are - a) an agent participates many times in the market and b) the agent types tend to be dependent on environmental and circumstantial parameters (such as the current location, traffic volume in the surrounding area, weather conditions, urgency, etc.) and are not intrinsic to the agents. Therefore, our work focuses on designing two-sided cost sharing mechanisms that will satisfy the properties that pertain to agent utilities, namely IR and the core, in expectation. To be more precise, our mechanisms are efficient, dominant strategy IC (DSIC), ex-ante IR, ex-ante weakly BB, and ex-ante in the core.

We note that, on top of being suitable for our applications of interest, these properties are also tight from a technical perspective: Efficiency and IC are satisfied as their strongest possible versions and weakly BB is a platform constraint that we satisfy. Strengthening ex-ante IR is not possible even when relaxing IC to Bayesian IC (as given by the Myerson-Satterthwaite impossibility theorem [29] for the single buyer-single seller case) or even when relaxing efficiency (gains from trade) to approximate efficiency (as shown in [4,5,11], again for a single buyer and a single seller). Moreover, it is conjectured to be impossible even when relaxing both IC and efficiency, as supported by partial impossibility results and experimental evidence [5].

1.1 Results and Techniques

Table 1. Summary of our results.

Cost function	Result
1 submodular seller	Optimum welfare and core
Capacitated NGS sellers	Optimum welfare and core
General sellers with constant capacity constraints	Approx welfare and core
Super additive sellers	Approx welfare and core

As we explained above, our main contributions are mechanisms that are efficient, DSIC, ex-ante weakly BB, ex-ante IR, and ex-ante in the core. In Sect. 3 we study classes of cost functions that allow us to design an efficient mechanism, i.e. a mechanism that maximizes gains from trade. Subsequently in Sect. 4 we study general cost functions for which we devise an approximately efficient mechanism. The cases we study, are characterized by the cost functions of the sellers. We study 4 different scenarios. First the case where we only have one single seller, second, when we have multiple sellers with negative gross substitutes cost functions, third, when we have multiple sellers with general cost functions and constant capacity constraints, and lastly, when we have multiple sellers with superadditive cost functions (Table 1).

Given that our setting is multi-dimensional, it is known that the design space for truthful mechanisms is strongly restricted and the main tool in our disposal is the family of VCG-type mechanisms [9,21,33]. Our first technical contribution is designing a VCG-like mechanism that guarantees the DSIC property as well as the induced outcome being ex-ante in the core. To do so, we first give an algorithm that computes utilities in the core for the case with known values and costs, by means of a primal-dual pair of LPs. This result is of independent interest in itself as it generalizes a result of [3] to our various models. We then show that the utilities for different realizations can be combined point-wise to yield weakly BB wages and payments that are ex-ante in the core.

A second technical contribution is the proper use of sampling to achieve our properties of interest with high probability in polynomial time in certain submodels. We note that, in this sampling scenario, it is trickier to guarantee that the expected total utility (over sampled points) of a group of agents matches the expected welfare (over all points) that they could generate. However, with appropriate parameter selection and arguments, we show that we can approximate to arbitrary precision the utility *per agent*, which then yields the required properties.

Finally, for the case of multiple agents and general cost functions with constant capacity, since we can no longer achieve an assignment that maximizes gains from trade, it is more challenging to attain the DSIC property. However, we design a rounding scheme that rounds the optimal fractional solution of the LP corresponding to gains from trade, into an integral solution that achieves exactly a fixed fraction of the LP objective. This allows us to maintain the DSIC property for our mechanism. Subsequently, for the case of uncapacitated super additive cost functions, we present a transformation of the game to fit the framework of [23] and get a truthful mechanism, even in the absence of a welfare-maximizing algorithm. The framework of [23] is one-sided and requires utility functions that, among other properties, are monotone. Our setting is two-sided and the utility generated by each seller is not necessarily monotone in the set of buyers. However, we show how to get past these issues and design a mechanism that works for our model.

1.2 Related Work

Our work is related to two areas of literature: two-sided markets and (one-sided) cost sharing. In two-sided markets, a series of papers studies two-sided auctions that approximate efficiency with respect to the sum of values of the items' holders after trade, as opposed to the gains from trade version that we study here [4,10,12,17,24]. With respect to gains from trade, approximating the optimal gains from trade is impossible in many settings [4,5,11]. Nevertheless, there do exist results that impose assumptions on the distributions and/or approximate a weaker benchmark: the gains from trade achieved by the "second-best" mechanism of Myerson and Satterthwaite [29], which is known to be the one that maximizes gains from trade subject to interim IR and BIC [5,7,11,25]. All these results rely heavily on the fact that the studied settings are single-dimensional and break down even in minor departures from single-dimensionality. Approximating gains from trade even in simple multi-dimensional settings seems like a challenging problem. In a slightly different setting, other works [1,2] approximate gains from trade with non-atomic populations of agents.

On the cost sharing side of the literature, early work on cooperative games and the core can be found in [6,19,31,32]. Related to our results on computing outcomes in the core for known agent types is the work by Bateni et al. [3] who show how to compute a solution in the core of a game where suppliers deal with manufacturers and all information is public. With a simple transformation, one can show that the model in [3] is equivalent to our model with additive cost functions. In this sense, our results on computing solutions in the core for known agent types generalize the corresponding result of [3] to broader classes of cost functions. Considering mechanisms for one-sided cost sharing, most works focus on the version of the problem where there are no prior distributions on the values of the buyers. We remark that, in the absence of prior knowledge, the two-sided setting is hopeless in trading-off efficiency and budget balance. A simple example with a buyer with a large value v and a seller with 0 cost is enough to see that. The only price and wage that make this setting efficient and truthful are a price of 0 for the buyer and a wage of v for the seller. The only exception to this rule, is the work by Fu et al. [18], who consider a Bayesian setting and show that any approximation algorithm for the underlying problem can be transformed into a mechanism for the cost sharing problem with a logarithmic loss in efficiency.

Finally, we note that our mechanisms are related to the AGV mechanism [13] and the mechanisms in [8], which also achieve ex-ante IR guarantees. The AGV mechanism is efficient, BB, Bayesian IC, and ex-ante IR under certain conditions (such as no costs) which are very different from our setting. The work in [8] focuses on auction settings (including a two-sided auction with a single seller and a single item) and designs mechanisms that are efficient, BB, DSIC, and ex-ante IR. Our mechanisms focus on the richer two-sided cost-sharing setting, in which we provide solutions in the core (an important consideration in our applications of interest as, otherwise, agents are incentivized to deal outside of the market), and also address computational considerations (as they run in polynomial time for various settings).

2 Preliminaries

Our market model is comprised of a set of $m \geq 1$ sellers M and a set of $n \geq 1$ buyers N. Each buyer $i \in N$ is unit demand and has value v_{ij} for being served by seller $j \in M$. Each seller $j \in M$ is endowed with a cost function $c_j(S)$ which gives the cost of the seller to serve the buyers in $S \subseteq N$. We assume that $c_j(\emptyset) = 0$ for all $j \in M$. Optionally, the model can impose a capacity constraint on the sellers with each seller j being able to accept k_j buyers. We make the natural assumption that values are bounded and further, without loss of generality and for simplicity of exposition, that they are in $[0, 1]$. Buyers and sellers interact with an intermediary platform that determines the assignment of buyers to sellers as well as prices p_i for the buyers $i \in N$ and wages w_j for the sellers $j \in M$. The utility of a buyer i that is matched to a seller j is $u_i = v_{ij} - p_i$. The utility of a seller who is assigned buyers S is $u_j = w_j - c_j(S)$.

We assume the existence of discrete prior distributions over the types of each buyer and seller. The type of a seller specifies her cost function whereas the type of a buyer specifies her values. We assume agent types are drawn independently and that the prior distributions are common knowledge. Note that our discreteness assumption is a) natural, since these are distributions over possible payments which are by definition discrete, and b) without a major impact on the model since any continuous distribution can be replaced by a discrete version with an arbitrarily small approximation to the results. The solution that the platform needs to come up with is specified as an assignment of buyers to sellers, a price vector for the buyers, and a wage vector for the sellers. Throughout the paper we describe the assignment both as a collection of buyer subsets $S_1, S_2, \ldots, S_m$, with S_j being the set assigned to seller $j \in M$, and as a mapping function $\sigma(\cdot)$, with $\sigma(i)$ being the seller to which buyer $i \in N$ is assigned. Generally, our goal is to maximize gains from trade, i.e., the total value of matched buyers minus the total cost of sellers. Sometimes we refer to this objective as the social welfare.

We next describe the mechanism properties that appear in our results.

Efficiency. A mechanism is *efficient* if it maximizes the gain from trade, i.e., the total value of matched buyers minus the total cost of sellers. A mechanism is *α-efficient* if it achieves an α approximation to the optimal gains from trade.

Weak Budget Balance (BB). A mechanism is *ex-ante weakly BB* if the expected sum of prices extracted from the buyers is at least equal to the expected sum of wages paid to the sellers.

Individual Rationality (IR). A mechanism is *ex-ante IR* if every agent has non-negative expected utility before all types are drawn. We say a mechanism is ϵ ex-ante IR if the agent expected utilities are at least $-\epsilon$.

Incentive Compatibility. A solution is *dominant strategy incentive compatible* (DSIC) if an agent cannot improve her utility by misreporting her type, even after learning the types of other agents.

In addition to these standard properties, we are interested in obtaining solutions that are *in the core* of the cost-sharing game, something that would encourage agents to adhere to the platform's solution and stay in its market.

Cost-Sharing Core. A solution is *in the core* if the sum of buyer and seller utilities equals the welfare they produce and there does not exist a coalition of buyers and sellers who can generate welfare higher than the sum of their utilities. More generally, the *α-core* requests that every set of agents can't produce welfare higher than α times their total utilities. The α-core property guarantees that agents on platform achieve in expectation, at least α fraction of the maximum utility they can potentially gain in the market by forming a private coalition.

Before moving forward with presenting our results, we define the classes of *submodular* and *negative gross substitutes* seller cost functions which we will use as part of our results. For the latter, we begin with the standard gross substitutes definition and then present the negative gross substitutes definition and how it is placed in our framework. This class of functions is interesting to study since it represents the theoretical border of tractability for obtaining the optimal solution for this problem.

Submodular Cost Functions. A cost function c is *submodular* if for every subsets of buyers $S, S' \in 2^N$ it is the case that $c(S)+c(S') \geq c(S \cup S') + c(S \cap S')$.

Gross Substitutes Functions. A function f defined over the set of buyers N satisfies the *gross substitutes* condition if and only if the following holds. Let p be a vector of prices charged to the buyers and let $D(p) = \arg\max_{S \subseteq N}\{f(S) - \sum_{i \in S} p_i\}$ be the demand set. Then, for every price vector p, every $S \in D(p)$, and every $q \geq p$, there exists a set $T \subseteq N$ such that $(S \setminus A) \cup T \in D(q)$, where $A = \{i \in N : \ q_i \geq p_i\}$ is the set of items for which the prices increase from p to q.

Negative Gross Substitutes Cost Functions in Two-Sided Cost-Sharing. A seller cost function, which maps each subset of buyers $S \in 2^N$ to the real cost $c(S)$, satisfies the *negative gross substitutes* condition if and only if the following holds. Let p be a vector of prices charged to the buyers and let $D(p) = \arg\max_{S \subseteq N} c(S) - \sum_{i \in S} p_i$. Then, for every price vector p, every $S \in D(p)$, and every $q \leq p$, there exists a set $T \subseteq N$ such that $(S \setminus A) \cup T \in D(q)$, where $A = \{i \in N : q_i \leq p_i\}$.

Super Additive Cost Functions. A cost function c is *super additive* if for every pair of disjoint subsets of buyers $S, S' \in 2^N$ we have $c(S) + c(S') \leq c(S \cup S')$.

The class of functions satisfying the gross substitutes property contains, for example, all additive and unit-demand functions and is contained in the class of submodular functions. To abbreviate, we will say that a function is (negative) GS if it satisfies the (negative) gross substitutes condition.

In Sect. 3, we show how to design an efficient mechanism for special cases such as when we have one seller with a submodular cost function or the case where we have multiple sellers with negative gross substitutes cost functions. Note that, in general, this problem is very difficult, and as proved in Proposition 1, even for the case where the cost functions are constant over non-empty subsets, it is still NP-hard to design an efficient mechanism. That is why, in Sect. 4, we study approximately efficient mechanisms for more general cost functions.

This version of the paper does not include any proofs, however, we provide proofs of all the arguments in the full version of the paper available on arXiv[1].

Proposition 1. *Given multiple sellers with cost functions that are constant over non-empty subsets, it is NP-hard to find an assignment that maximizes gains from trade.*

3 Efficient Mechanism

In this section we describe the main mechanism for most of the settings we study. The mechanism requires access to two algorithms: a) an algorithm to compute a welfare-maximizing assignment of buyers to sellers with known values and costs and b) a deterministic algorithm to compute non-negative utilities that are in the (approximate) core of the cost-sharing game and sum up to the optimal welfare, again, with known values and costs. We discuss these algorithms further in Sect. 3.1 and in the sections that correspond to the different models we study. For the polynomial time version of our mechanism, both of these algorithms must run in polynomial time. Let WELFARE-ALG be the welfare maximizing algorithm and let CORE-ALG be the algorithm that computes utilities in the α-core. The exact value of α will depend on the exact model, i.e., on the number of sellers and the class of cost functions under consideration.

Mechanism 1. *Is defined as follows:*

- *Allocation Rule: Given the reported values v_{ij} for $i \in N, j \in M$, and cost functions $c_j : 2^N \to \mathbb{R}$ for $j \in M$, output the welfare-maximizing allocation computed by* WELFARE-ALG.

[1] https://arxiv.org/abs/1809.02718.

- *Pre-processing: For every realization of agent types r that has some probability q_r, compute buyer utilities $y_i^r, i \in N$, and seller utilities $z_j^r, j \in M$, that are non-negative, in the α-core, and sum up to the optimal welfare using* CORE-ALG. *Let $y_i = \sum_r q_r y_i^r$, be the expected utility of buyer i over all realizations and let $z_j = \sum_r q_r z_j^r$, be the expected utility of seller j over all realizations.*
- *Buyer prices: The price charged to buyer i is*

$$p_i = \sum_{j \in M} c_j(S_j) - \sum_{i' \in N, i' \neq i} v_{i'\sigma(i')} + \sum_{i' \in N, i' \neq i} y_{i'} + \sum_{j \in M} z_j,$$

where S_j is the set of buyers assigned to seller j and $\sigma(i')$ is the seller that i' is assigned to.
- *Seller wages: The wage paid to seller j is*

$$w_j = \sum_{i \in N} v_{i\sigma(i)} - \sum_{j' \in M, j' \neq j} c_{j'}(S_{j'}) - \sum_{i \in N} y_i - \sum_{j' \in M, j' \neq j} z_{j'},$$

where S_j is the set of buyers assigned to seller j and $\sigma(i)$ is the seller that i is assigned to.

Theorem 1. *Mechanism 1 is efficient, ex-ante weakly BB, DSIC, ex-ante IR, and ex-ante in the α-core.*

The second step of Mechanism 1 might not be feasible in polynomial time. We now modify our mechanism to make it run in polynomial time (assuming WELFARE-ALG and CORE-ALG run in polynomial time) as follows. Define *Mechanism 2* to be exactly like Mechanism 1, however we replace the second step with the following:

- Pre-processing: Sample a set C of $c = n^2(n+m)^5/\epsilon^3$ realizations of agent types, for some small parameter $\epsilon > 0$. For every sample $r \in C$, compute buyer utilities $y_i^r, i \in N$, and seller utilities $z_j^r, j \in M$, that are non-negative, in the α-core, and sum up to the optimal welfare using CORE-ALG. Let

$$y_i = \left(\sum_{r \in C} \frac{y_i^r}{c} \right) + \frac{\epsilon}{(n+m)^2},$$

be the slightly shifted average utility of buyer i over all sampled realizations and let

$$z_j = \left(\sum_{r \in C} \frac{z_j^r}{c} \right) + \frac{\epsilon}{(n+m)^2},$$

be the slightly shifted average utility of seller j over all sampled realizations.

Theorem 2. *For arbitrarily small $\epsilon > 0$, Mechanism 2 runs in time polynomial in n, m, and $1/\epsilon$, and is efficient, DSIC, and, with probability $1 - \epsilon$, ex-ante weakly BB, ϵ ex-ante IR, and ex-ante in the $\alpha(1+\delta)$-core, where $\delta = 2\epsilon/W$ and W is the expected optimal welfare.*

3.1 Welfare Maximization and Core Computation Algorithms

In this section we further discuss WELFARE-ALG and CORE-ALG. We begin with WELFARE-ALG and note that in Mechanism 1, for which no run-time guarantees are provided, WELFARE-ALG is simply exhaustive search and is available in all models. For Mechanism 2, WELFARE-ALG must be an algorithm that solves the optimization problem of assigning buyers to sellers in polynomial time. This can be done in several models (as we explain in upcoming sections), such as the case with negative GS cost functions and the case with a single uncapacitated seller with a submodular cost function.

Our core computation algorithm, CORE-ALG, relies on the following primal-dual pair of linear programs. The PRIMAL:

$$\begin{aligned}
\text{maximize} \quad & \sum_{j\in M}\sum_{S\subseteq N,|S|\leq k_j} x_{jS}\left(\sum_{i\in S} v_{ij} - c_j(S)\right) & \\
\text{subject to} \quad & \sum_{S\subseteq N} x_{jS} \leq 1 & \forall j \in M \\
& \sum_{j\in M}\sum_{S\ni i} x_{jS} \leq 1 & \forall i \in N \\
& x_{jS} \geq 0 & \forall j \in M, \forall S \subseteq N
\end{aligned}$$

and the DUAL:

$$\begin{aligned}
\text{minimize} \quad & \sum_{i\in N} y_i + \sum_{j\in M} z_j & \\
\text{subject to} \quad & \sum_{i\in S} y_i + z_j \geq \sum_{i\in S} v_{ij} - c_j(S) & \forall j \in M, \forall S \subseteq N, |S| \leq k_j \\
& y_i \geq 0 & \forall i \in N \\
& z_j \geq 0 & \forall j \in M
\end{aligned}$$

Let W^* be the optimal value of PRIMAL and let W be the optimal value among integral solutions to PRIMAL. The utilities that CORE-ALG outputs are precisely the dual variables scaled by W/W^*. The following theorem shows that these values are indeed in the approximate core.

Theorem 3. *Let (y^*, z^*) be the solution to* DUAL *and let $(y, z) = (y^*, z^*)W/W^*$, where W^* is the optimal value for* PRIMAL *and W the value of the integral optimal solution to* PRIMAL. *Then (y, z) gives utilities y_i for the buyers $i \in N$ and utilities z_j for the sellers $j \in M$ that are in the α-approximate core, with α the integrality gap of* PRIMAL.

With respect to running time considerations, we need to be able to solve DUAL in polynomial time.

For the case where we only have one submodular seller we can show that the integrality gap of PRIMAL is 1 and we can solve the primal optimally in polynomial time.

This also holds when we have multiple sellers with NGS cost functions. These functions, which are strictly more general than linear functions, represent the limit of tractability for solving PRIMAL optimally over the space of integral solutions. Once we leave the space of NGS functions, we have to rely on approximately efficient mechanisms to achieve our results.

This implies that for these two cases, we can utilize Mechanism 2 to achieve in polynomial time, a mechanism that is efficient, DSIC, and, with probability $1-\epsilon$, ex-ante weakly BB, ϵ ex-ante IR, and ex-ante in the $(1+\delta)$-core, where $\delta = 2\epsilon/W$ and W is the expected optimal welfare. Please refer to Theorem 1 and 2 for more details.

4 Approximately Efficient Mechanism

In this section, we present a polynomial time mechanism that addresses the case of intractable models, such as multiple submodular sellers. Our mechanism will achieve approximate efficiency and will be in the approximate core. The mechanism requires access to an algorithm that computes a convex combination of integral solutions that is equal to $1/\gamma$ times the fractional optimal solution of the PRIMAL of Sect. 3.1, for some given γ. We discuss this algorithm, which we call APPROX-WELFARE-ALG, further in Sect. 4.1. We now present the mechanism's specifics.

Mechanism 3. Specifics:

- Allocation Rule: Given the reported values v_{ij} for $i \in N, j \in M$, and cost functions $c_j : 2^N \to \mathbb{R}$ for $j \in M$, let x^* be the optimal solution to the PRIMAL linear program in Sect. 3.1. Our allocation is the lottery x that is output by APPROX-WELFARE-ALG.
- Pre-processing: Sample a set C of $c = n^2(n+m)^5/\epsilon^3$ realizations of agent types, for some small parameter $\epsilon > 0$. For every sample $r \in C$, compute buyer utilities $y_i^r, i \in N$, and seller utilities $z_j^r, j \in M$, by solving the DUAL of Sect. 3.1. Let $y_i = \left(\sum_{r\in C} \frac{y_i^r}{c}\right) + \frac{\epsilon}{(n+m)^2}$, be the slightly shifted average utility of buyer i over all sampled realizations and let $z_j = \left(\sum_{r\in C} \frac{z_j^r}{c}\right) + \frac{\epsilon}{(n+m)^2}$, be the slightly shifted average utility of seller j over all sampled realizations. Also, define

$$v_i(x^*) = \sum_{j\in M, S\ni i} x_{jS}^* v_{ij} \qquad \text{and} \qquad c_j(x^*) = \sum_{S\subseteq N} x_{jS}^* c_j(S),$$

which can be interpreted as the extracted value of buyer i and the incurred cost of seller j under fractional solution x^*.
- Buyer prices: The price charged to buyer i is

$$p_i = \frac{1}{\gamma}\left(\sum_{j\in M} c_j(x^*) - \sum_{i'\in N, i'\neq i} v_i(x^*) + \sum_{i'\in N, i'\neq i} y_{i'} + \sum_{j\in M} z_j\right).$$

– Seller wages: The wage paid to seller j is

$$w_j = \frac{1}{\gamma}\left(\sum_{i\in N} v_i(x^*) - \sum_{j'\in M, j'\neq j} c_{j'}(x^*) - \sum_{i\in N} y_i - \sum_{j'\in M, j'\neq j} z_{j'}\right).$$

Theorem 4. *For arbitrarily small $\epsilon > 0$, Mechanism 3 runs in time polynomial in n, m, and $1/\epsilon$, and is γ-efficient, DSIC, and, with probability $1-\epsilon$, ex-ante weakly BB, ϵ ex-ante IR, and ex-ante in the $\gamma(1+\delta)$-core, where $\delta = 2\epsilon/W$ and W is the expected welfare of the mechanism.*

4.1 Approximate Welfare Algorithm

In this section, we present APPROX-WELFARE-ALG, which outputs a convex combination of integral solutions that is precisely x^*/γ with x^* the fractional optimal solution to PRIMAL. We study two different settings. One is for general cost functions when we treat the capacity constraint of sellers as a constant. The other setting is when the cost functions are super additive.

General Cost Functions with Constant Capacity Constraints. In this section we study the case where sellers can have general cost functions and the capacity of all the sellers is a constant C. One can easily extend our result to the case where these constants are different for different sellers, however, for ease of exposition, we assume the same capacity constraint for all sellers.

Our approximate welfare algorithms consists of two parts. First we need to argue that we can find an optimal feasible solution to PRIMAL linear program in Sect. 3.1. And next we need to argue that we can in fact assign the buyers to sellers in a way that the assignment is a convex combination of integral solutions such that the expected assignment is exactly equal to $1/\gamma$ times the fractional optimal solution for some constant γ. This equality is important, since otherwise, the proof of DSIC property in Theorem 4 does not go through.

To optimally solve the primal, we first explain how we can design a separation oracle for the dual. This is not too hard, since for each seller $j \in M$, given the values of $\{y\}_i, \{z\}_j$, we have to be able to check the following constraint in polynomial time

$$\sum_{i\in S} y_i + z_j \geq \sum_{i\in S} v_{ij} - c_j(S) \qquad \forall S \subseteq N, |S| \leq C.$$

However, since C is just a constant, we can just enumerate over all such subsets and check them one by one. This allows us to solve the PRIMAL optimally to achieve solution x^*. Next we will show how to round this optimal fractional solution into an integral solution, such that for each $j \in M$ and $S \subseteq N$, the probability that seller j is assigned set S is exactly $x^*_{j,S}/\gamma$ for $\gamma = C+1$.

To do this, first pick an arbitrary order for all pairs (j, S) where $j \in M$ and $S \subseteq N$ and call them $(j_1, S_1), (j_2, S_2), \ldots, (j_k, S_k)$ where k is the total number of such pairs. Then run the following rounding algorithm.

Algorithm 1. Rounding Algorithm

1: **input**: An optimal solution x^* to PRIMAL.
2: **output**: A random integral and feasible assignment x such that $\mathbb{E}[x_{j,S}] = \frac{x^*_{j,S}}{\gamma+1}$ for all $j \in M$ and $S \subseteq N$.
3: **For** i=1...k **do**:
4: **If** none of the buyers in S_i have been assigned to a seller, and, no set of buyers have been assigned to seller j_i **then**:
5: Define set R as follows

$$R = \{l | l < i, (S_i \cap S_l \neq \emptyset \text{ or } j_l = j_i)\}$$

6: Assign buyers in S_i to seller j_i with probability $\frac{x_{j_i,S_i}}{\gamma(1-\sum_{l\in R} x_{j_l,S_l}/\gamma)}$.
7: **End**

We have the following result.

Theorem 5. *Given an optimal solution x^* to* PRIMAL, *Algorithm 1 will generate a random integral assignment x, such that*

$$\mathbb{E}[x_{j,S}] = \frac{1}{C+1} x^*_{j,S} \qquad \forall j \in M, S \subseteq N.$$

Super Additive Cost Functions. The idea in this section is to, similar to the previous section, first solve the PRIMAL linear program to obtain an optimal fractional solution x^*. This is in general a hard problem, however, we assume in this section, that we are given access to a demand oracle for the problem. The demand oracle allows us to design a separation oracle for the DUAL, which in turn allows us to obtain x^* in polynomial time. We will then use a method inspired by [15], that can round this fractional solution to an integral solution with a $\sqrt{n}$ approximation ratio in polynomial time. Finally, following the following lemma due to [23], we show how we can design an algorithm that round the solution of x^* into a random integral solution x, such that $\mathbb{E}[x] = x^*/\sqrt{n}$.

Lemma 1. (Lavi and Swamy [23]). *Let x^* be the fractional optimal solution to* PRIMAL *and γ be such that there exists a γ-approximation algorithm for the buyer to seller assignment problem and γ also bounds the integrality gap of* PRIMAL. *Then, there exists an algorithm which we call* LOTTERY-ALG *that can be used to obtain, in polynomial time, a convex combination of integral solutions that is equal to x^*/γ, under the following conditions on the welfare generated by each seller j and her matched buyers: a) it is a monotone function, b) it is 0 for an empty set of buyers, and c) we have a polynomial time demand oracle for it.*

Note that the utility function of the sellers in our setting does not satisfy the conditions presented in 1. Namely, the welfare of each seller $V_j(S) = \sum_{i\in S} v_{ij} - c_j(S)$ is not monotone. To fix this issue, we define another utility function for our sellers as follows,

$$\hat{V}_j(S) = \max_{S' \subseteq S} \sum_{i \in S'} v_{ij} - c_j(S').$$

Now we have the following result.

Theorem 6. *For the case of super additive cost functions, using* LOTTERY-ALG *of Lemma 1, with $\hat{V}_j(S)$ as the welfare functions, we can in polynomial time, achieve a random integral assignment x such that $x = x^*/\gamma$, with $\gamma = O(\sqrt{n})$.*

References

1. Alijani, R., Banerjee, S., Gollapudi, S., Kollias, K., Munagala, K.: Two-sided facility location. CoRR abs/1711.11392 (2017). http://arxiv.org/abs/1711.11392
2. Banerjee, S., Gollapudi, S., Kollias, K., Munagala, K.: Segmenting two-sided markets. In: Proceedings of the 26th International Conference on World Wide Web, WWW 2017, Perth, Australia, 3–7 April 2017, pp. 63–72 (2017). https://doi.org/10.1145/3038912.3052578. http://doi.acm.org/10.1145/3038912.3052578
3. Bateni, M.H., Hajiaghayi, M.T., Immorlica, N., Mahini, H.: The cooperative game theory foundations of network bargaining games. In: Abramsky, S., Gavoille, C., Kirchner, C., Meyer auf der Heide, F., Spirakis, P.G. (eds.) ICALP 2010. LNCS, vol. 6198, pp. 67–78. Springer, Heidelberg (2010). https://doi.org/10.1007/978-3-642-14165-2_7
4. Blumrosen, L., Dobzinski, S.: (Almost) efficient mechanisms for bilateral trading. CoRR abs/1604.04876 (2016). http://arxiv.org/abs/1604.04876
5. Blumrosen, L., Mizrahi, Y.: Approximating gains-from-trade in bilateral trading. In: Cai, Y., Vetta, A. (eds.) WINE 2016. LNCS, vol. 10123, pp. 400–413. Springer, Heidelberg (2016). https://doi.org/10.1007/978-3-662-54110-4_28
6. Bondareva, O.N.: Some applications of linear programming to cooperative games. In: Problemy Kibernetiki (1963)
7. Brustle, J., Cai, Y., Wu, F., Zhao, M.: Approximating gains from trade in two-sided markets via simple mechanisms. In: Proceedings of the 2017 ACM Conference on Economics and Computation, EC 2017, Cambridge, MA, USA, 26–30 June 2017, pp. 589–590 (2017). https://doi.org/10.1145/3033274.3085148. http://doi.acm.org/10.1145/3033274.3085148
8. Cavallo, R.: Efficient mechanisms with risky participation. In: Proceedings of the 22nd International Joint Conference on Artificial Intelligence, IJCAI 2011, Barcelona, Catalonia, Spain, 16–22 July 2011, pp. 133–138 (2011). https://doi.org/10.5591/978-1-57735-516-8/IJCAI11-034
9. Clarke, E.: Multipart pricing of public goods. Public Choice **11**(1), 17–33 (1971). https://EconPapers.repec.org/RePEc:kap:pubcho:v:11:y:1971:i:1:p:17-33
10. Colini-Baldeschi, R., Goldberg, P.W., de Keijzer, B., Leonardi, S., Roughgarden, T., Turchetta, S.: Approximately efficient two-sided combinatorial auctions. In: Proceedings of the 2017 ACM Conference on Economics and Computation, EC 2017, Cambridge, MA, USA, 26–30 June 2017, pp. 591–608 (2017). https://doi.org/10.1145/3033274.3085128. http://doi.acm.org/10.1145/3033274.3085128
11. Colini-Baldeschi, R., Goldberg, P., de Keijzer, B., Leonardi, S., Turchetta, S.: Fixed price approximability of the optimal gain from trade. In: Devanur, N.R., Lu, P. (eds.) WINE 2017. LNCS, vol. 10660, pp. 146–160. Springer, Cham (2017). https://doi.org/10.1007/978-3-319-71924-5_11

12. Colini-Baldeschi, R., de Keijzer, B., Leonardi, S., Turchetta, S.: Approximately efficient double auctions with strong budget balance. In: Proceedings of the Twenty-Seventh Annual ACM-SIAM Symposium on Discrete Algorithms, SODA 2016, Arlington, VA, USA, 10–12 January 2016, pp. 1424–1443 (2016). https://doi.org/10.1137/1.9781611974331.ch98
13. d'Aspremont, C., Gerard-Varet, L.A.: Incentives and incomplete information. J. Public Econ. **11**(1), 25–45 (1979). https://EconPapers.repec.org/RePEc:eee:pubeco:v:11:y:1979:i:1:p:25-45
14. Dobzinski, S., Mehta, A., Roughgarden, T., Sundararajan, M.: Is Shapley cost sharing optimal? Games Econ. Behav. **108**, 130–138 (2018). https://doi.org/10.1016/j.geb.2017.03.008
15. Dobzinski, S., Nisan, N., Schapira, M.: Approximation algorithms for combinatorial auctions with complement-free bidders. In: Proceedings of the 37th Annual ACM Symposium on Theory of Computing, Baltimore, MD, USA, 22–24 May 2005, pp. 610–618 (2005). https://doi.org/10.1145/1060590.1060681
16. Dobzinski, S., Ovadia, S.: Combinatorial cost sharing. In: Proceedings of the 2017 ACM Conference on Economics and Computation, EC 2017, Cambridge, MA, USA, 26–30 June 2017, pp. 387–404 (2017). https://doi.org/10.1145/3033274.3085141. http://doi.acm.org/10.1145/3033274.3085141
17. Dütting, P., Talgam-Cohen, I., Roughgarden, T.: Modularity and greed in double auctions. Games Econ. Behav. **105**, 59–83 (2017). https://doi.org/10.1016/j.geb.2017.06.008
18. Fu, H., Lucier, B., Sivan, B., Syrgkanis, V.: Cost-recovering Bayesian algorithmic mechanism design. In: ACM Conference on Electronic Commerce, EC 2013, Philadelphia, PA, USA, 16–20 June 2013, pp. 453–470 (2013). https://doi.org/10.1145/2482540.2482591. http://doi.acm.org/10.1145/2482540.2482591
19. Gillies, D.B.: Solutions to general non-zero-sum games. In: Tucker, A.W., Luce, R.D. (eds.) Contributions to the Theory of Games IV. No. 40 in Annals of Mathematics Studies, pp. 47–85. Princeton University Press, Princeton (1959)
20. Green, J., Kohlberg, E., Laffont, J.J.: Partial equilibrium approach to the free-rider problem. J. Public Econ. **6**(4), 375–394 (1976). https://EconPapers.repec.org/RePEc:eee:pubeco:v:6:y:1976:i:4:p:375-394
21. Groves, T.: Incentives in teams. Econometrica **41**(4), 617–31 (1973). https://EconPapers.repec.org/RePEc:ecm:emetrp:v:41:y:1973:i:4:p:617-31
22. Gupta, A., Könemann, J., Leonardi, S., Ravi, R., Schäfer, G.: An efficient cost-sharing mechanism for the prize-collecting Steiner forest problem. In: Proceedings of the Eighteenth Annual ACM-SIAM Symposium on Discrete Algorithms, SODA 2007, New Orleans, Louisiana, USA, 7–9 January 2007, pp. 1153–1162 (2007). http://dl.acm.org/citation.cfm?id=1283383.1283507
23. Lavi, R., Swamy, C.: Truthful and near-optimal mechanism design via linear programming. In: Proceedings of the 46th Annual IEEE Symposium on Foundations of Computer Science (FOCS 2005), Pittsburgh, PA, USA, 23–25 October 2005, pp. 595–604 (2005). https://doi.org/10.1109/SFCS.2005.76
24. McAfee, P.R.: A dominant strategy double auction. J. Econ. Theory **56**(2), 434–450 (1992). https://EconPapers.repec.org/RePEc:eee:jetheo:v:56:y:1992:i:2:p:434-450
25. McAfee, P.R.: The gains from trade under fixed price mechanisms. Appl. Econ. Res. Bull. **1**, 1–10 (2008)
26. Mehta, A., Roughgarden, T., Sundararajan, M.: Beyond Moulin mechanisms. Games Econ. Behav. **67**(1), 125–155 (2009). https://doi.org/10.1016/j.geb.2008.06.005

27. Moulin, H.: Incremental cost sharing: characterization by coalition strategy-proofness. Soc. Choice Welf. **16**(2), 279–320 (1999). https://EconPapers.repec.org/RePEc:spr:sochwe:v:16:y:1999:i:2:p:279-320
28. Moulin, H., Shenker, S.: Strategy proof sharing of submodular costs: budget balance versus efficiency. Econ. Theory **18**(3), 511–533 (2001). https://EconPapers.repec.org/RePEc:spr:joecth:v:18:y:2001:i:3:p:511-533
29. Myerson, R., Satterthwaite, M.A.: Efficient mechanisms for bilateral trading. J. Econ. Theory **29**(2), 265–281 (1983). https://EconPapers.repec.org/RePEc:eee:jetheo:v:29:y:1983:i:2:p:265-281
30. Roughgarden, T., Sundararajan, M.: Quantifying inefficiency in cost-sharing mechanisms. J. ACM **56**(4), 23:1–23:33 (2009). https://doi.org/10.1145/1538902.1538907. http://doi.acm.org/10.1145/1538902.1538907
31. Scarf, H.: The core of an N person game. Econometrica **35**(1), 50–69 (1965)
32. Shapley, L.S.: On balanced sets and cores. Naval Res. Logist. Q. **14**, 453–460 (1967)
33. Vickrey, W.: Counterspeculation, auctions, and competitive sealed tenders. J. Financ. **16**(1), 8–37 (1961). https://EconPapers.repec.org/RePEc:bla:jfinan:v:16:y:1961:i:1:p:8-37

The Three-Dimensional Stable Roommates Problem with Additively Separable Preferences

Michael McKay(✉) and David Manlove(✉)

School of Computing Science, University of Glasgow, Glasgow, UK
m.mckay.1@research.gla.ac.uk, david.manlove@glasgow.ac.uk

Abstract. The Stable Roommates problem involves matching a set of agents into pairs based on the agents' strict ordinal preference lists. The matching must be stable, meaning that no two agents strictly prefer each other to their assigned partners. A number of three-dimensional variants exist, in which agents are instead matched into triples. Both the original problem and these variants can also be viewed as hedonic games. We formalise a three-dimensional variant using general additively separable preferences, in which each agent provides an integer valuation of every other agent. In this variant, we show that a stable matching may not exist and that the related decision problem is NP-complete, even when the valuations are binary. In contrast, we show that if the valuations are binary and symmetric then a stable matching must exist and can be found in polynomial time. We also consider the related problem of finding a stable matching with maximum utilitarian welfare when valuations are binary and symmetric. We show that this optimisation problem is NP-hard and present a novel 2-approximation algorithm.

Keywords: Stable roommates · Stable matching · Three dimensional roommates · Hedonic games · Coalition formation · Complexity

1 Introduction

The Stable Roommates problem (SR) is a classical problem in the domain of matching under preferences. It involves a set of agents that must be matched into pairs. Each agent provides a *preference list*, ranking all other agents in strict order. We call a set of pairs in which each agent appears in exactly one pair a *matching*. The goal is to produce a matching M that admits no *blocking pair*, which comprises two agents, each of whom prefers the other to their assigned partner in M. Such a matching is called *stable*. This problem originates from a seminal paper of Gale and Shapley, published in 1962, as a generalisation of

This work was supported by the Engineering and Physical Sciences Research Council (Doctoral Training Partnership grant number EP/R513222/1 and grant number EP/P028306/1).

I. Caragiannis and K. A. Hansen (Eds.): SAGT 2021, LNCS 12885, pp. 266–280, 2021.
https://doi.org/10.1007/978-3-030-85947-3_18

the Stable Marriage problem [15]. They showed that an SR instance need not contain a stable matching. In 1985, Irving presented a polynomial-time algorithm to either find a stable matching or report that none exist, given an arbitrary SR instance [20]. Since then, many papers have explored extensions and variants of the fundamental SR problem model.

In this paper we consider the extension of SR to three dimensions (i.e., agents must be matched into triples rather than pairs). A number of different formalisms have already been proposed in the literature. The first, presented in 1991 by Ng and Hirschberg, was the *3-Person Stable Assignment Problem* (3PSA) [24]. In 3PSA, agents' preference lists are formed by ranking every pair of other agents in strict order. A matching M is a partition of the agents into unordered triples. A *blocking triple* t of M involves three agents that each prefer their two partners in t to their two assigned partners in M. Accordingly, a *stable matching* is one that admits no blocking triple. The authors showed that an instance of this model may not contain a stable matching and the associated decision problem is NP-complete [24]. In the instances constructed by their reduction, agents' preferences may be *inconsistent* [19], meaning that it is impossible to derive a logical order of individual agents from a preference list ranking pairs of agents.

In 2007, Huang considered the restriction of 3PSA to *consistent* preferences. He showed that a stable matching may still not exist and the decision problem remains NP-complete [18,19]. In his technical report, he also described another variant of 3PSA using *Precedence by Ordinal Number* (PON). PON involves each agent providing a preference list ranking all other agents individually. An agent's preference over pairs is then based on the sum of the ranks of the agents in each pair. Huang left open the problem of finding a stable matching, as defined here, in the PON variant. He also proposed another problem variant involving a more general system than PON, in which agents provide arbitrary numerical "ratings". It is this variant that we consider in this paper. He concluded his report by asking if there exist special cases of 3PSA in which stable matchings can be found using polynomial time algorithms. This question is another motivation for our paper.

The same year, Iwama, Miyazaki and Okamoto presented another variant of 3PSA [21]. In this model, agents rank individual agents in strict order of preference, and an ordering over pairs is inferred using a specific *set extension rule* [5,7]. The authors showed that a stable matching may not exist and that the decision problem remains NP-complete.

In 2009, Arkin et al. presented another variant of 3PSA called *Geometric 3D-SR* [1]. In this model, preference lists ranking pairs are derived from agents' relative positions in a metric space. Among other results, they showed that in this model a stable matching, as defined here, need not exist. In 2013, Deineko and Woeginger showed that the corresponding decision problem is NP-complete [14].

All of the problem models described thus far, including SR, can be viewed as *hedonic games* [6]. A hedonic game is a type of *coalition formation game*. In general, coalition formation games involve partitioning a set of agents into disjoint sets, or coalitions, based on agents' preferences. The term 'hedonic' refers to the fact that agents are only concerned with the coalition that they belong to. The study of hedonic games and coalition formation games is broad and many different problem models have been considered in the literature [17].

In particular, SR and its three-dimensional variants can be viewed as hedonic games with a constraint on permissible coalition sizes [26]. In the context of a hedonic game, the direct analogy of stability as described here is *core stability*. In a given hedonic game, a partition is *core stable* if there exists no set of agents S, of any size, where each agent in S prefers S to their assigned coalition [6].

Recently, Boehmer and Elkind considered a number of hedonic game variants, including 3PSA, which they described as *multidimensional roommate games* [8]. In their paper they supposed that the agents have *types*, and an agent's preference between two coalitions depends only on the proportion of agents of each type in each coalition. They showed that, for a number of different 'solution concepts', the related problems are NP-hard, although many problems are solvable in linear time when the room size is a fixed parameter. For stability in particular, they presented an integer linear programming formulation to find a stable matching in a given instance, if one exists, in linear time.

In 2020, Bredereck et al. considered another variation of multidimensional roommate games involving either a *master list* or *master poset*, a central list or poset from which all agents' preference lists are derived [10]. They presented two positive results relating to restrictions of the problem involving a master poset although they showed for either a master list or master poset that finding a stable matching in general remains NP-hard or W[1]-hard, for three very natural parameters.

Other research involving hedonic games with similar constraints has considered Pareto optimality rather than stability [13]; 'flatmate games', in which any coalition contains three or fewer agents [9]; and strategic aspects [27].

The template of a hedonic game helps us formalise the extension of SR to three dimensions. In this paper we apply the well-known system of *additively separable preferences* [2]. In a general hedonic game, additive separable preferences are derived from each agent α_i assigning a numerical valuation $val_{\alpha_i}(\alpha_j)$ to every other agent α_j. A preference between two sets is then obtained by comparing the sum of valuations of the agents in each set. This system formalises the system of "ratings" proposed by Huang [19]. In a general hedonic game with additively separable preferences, a core stable partition need not exist, and the associated decision problem is strongly NP-hard [25]. This result holds even when preferences are *symmetric*, meaning that $val_{\alpha_i}(\alpha_j) = val_{\alpha_j}(\alpha_i)$ for any two agents α_i, α_j [3].

The three-dimensional variant of SR that we consider in this paper can also be described as an additively separable hedonic game in which each coalition in a feasible partition has size three. To be consistent with previous research relating to three-dimensional variants of SR [19,21], in this paper we refer to a partition into triples as a *matching* rather than a partition and write *stable matching* rather than *core stable partition*. We finally remark that the usage of the terminology "three-dimensional" to refer to the coalition size rather than, say, the number of agent sets [24], is consistent with previous work in the literature [1,10,21,26].

Our Contribution. In this paper we use additively separable preferences to formalise the three-dimensional variant of SR first proposed by Huang in 2007 [19]. The problem model can be equally viewed as a modified hedonic game with additively separable preferences [3,25]. We show that deciding if a stable matching exists is NP-complete, even when valuations are binary (Sect. 3). In contrast, when valuations are binary and symmetric we show that a stable matching always exists and give an $O(|N|^3)$ algorithm for finding one, where N is the set of agents (Sects. 4.1–4.4). We believe that this restriction to binary and symmetric preferences has practical as well as theoretical significance. For example, this model could be applied to a social network graph involving a symmetric "friendship" relation between users. Alternatively, in a setting involving real people it might be reasonable for an administrator to remove all asymmetric valuations from the original preferences.

We also consider the notion of *utility* based on agents' valuations of their partners in a given matching. This leads us to the notion of *utilitarian welfare* [4, 11] which is the sum of the utilities of all agents in a given matching. We consider the problem of finding a stable matching with maximum utilitarian welfare given an instance in which valuations are binary and symmetric. We prove that this optimisation problem is NP-hard and provide a novel 2-approximation algorithm (Sect. 4.5).

We continue in the next section (Sect. 2) with some preliminary definitions and results.

2 Preliminary Definitions and Results

Let $N = \{\alpha_1, \ldots, \alpha_{|N|}\}$ be a set of *agents*. A *triple* is an unordered set of three agents. A *matching* M comprises a set of pairwise disjoint triples. For any agent α_i, if some triple in M contains α_i then we say that α_i is *matched* and use $M(\alpha_i)$ to refer to that triple. If no triple in M contains α_i then we say that α_i is *unmatched* and write $M(\alpha_i) = \varnothing$. Given a matching M and two distinct agents α_i, α_j, if $M(\alpha_i) = M(\alpha_j)$ then we say that α_j is a *partner* of α_i.

We define *additively separable preferences* as follows. Each agent α_i supplies a *valuation function* $val_{\alpha_i} : N \setminus \{\alpha_i\} \longrightarrow \mathbb{Z}$. Given agent α_i, let the *utility* of any set $S \subseteq N$ be $u_{\alpha_i}(S) = \sum_{\alpha_j \in S \setminus \{\alpha_i\}} val_{\alpha_i}(\alpha_j)$. We say that $\alpha_i \in N$ prefers some triple t_1 to another triple t_2 if $u_{\alpha_i}(t_1) > u_{\alpha_i}(t_2)$. An agent's preference between two distinct matchings depends only on that agent's partners in each matching, so given a matching M we write $u_{\alpha_i}(M)$ as shorthand for $u_{\alpha_i}(M(\alpha_i))$. Let $V = \bigcup_{\alpha_i \in N} val_{\alpha_i}$ be the collection of all valuation functions.

Suppose we have some pair (N, V) and a matching M involving the agents in N. We say that a triple $\{\alpha_{k_1}, \alpha_{k_2}, \alpha_{k_3}\}$ *blocks* M in (N, V) if $u_{\alpha_{k_1}}(\{\alpha_{k_2}, \alpha_{k_3}\}) > u_{\alpha_{k_1}}(M)$, $u_{\alpha_{k_2}}(\{\alpha_{k_1}, \alpha_{k_3}\}) > u_{\alpha_{k_2}}(M)$, and $u_{\alpha_{k_3}}(\{\alpha_{k_1}, \alpha_{k_2}\}) > u_{\alpha_{k_3}}(M)$. If no triple in N blocks M in (N, V) then we say that M is *stable* in (N, V). We say that (N, V) *contains* a stable matching if at least one matching exists in (N, V) that is stable.

We now define the *Three-Dimensional Stable Roommates problem with Additively Separable preferences* (3D-SR-AS). An instance of 3D-SR-AS is given by the pair (N, V). The problem is to either find a stable matching in (N, V) or report that no stable matching exists. In this paper we consider two different restrictions of this model. The first is when preferences are *binary*, meaning $val_{\alpha_i}(\alpha_j) \in \{0, 1\}$ for any $\alpha_i, \alpha_j \in N$. The second is when preferences are also *symmetric*, meaning $val_{\alpha_i}(\alpha_j) = val_{\alpha_j}(\alpha_i)$ for any $\alpha_i, \alpha_j \in N$.

Lemma 1 illustrates a fundamental property of matchings in instances of 3D-SR-AS. We shall use it extensively in the proofs. Throughout this paper the omitted proofs can be found in the full version [23].

Lemma 1. *Given an instance (N, V) of 3D-SR-AS, suppose that M and M' are matchings in (N, V). Any triple that blocks M' but does not block M contains at least one agent $\alpha_i \in N$ where $u_{\alpha_i}(M') < u_{\alpha_i}(M)$.*

We also make an observation that unmatched agents may be arbitrarily matched if required. The proof follows from Lemma 1.

Proposition 1. *Suppose we are given an instance (N, V) of 3D-SR-AS. Suppose $|N| = 3k + l$ where $k \geq 0$ and $0 \leq l < 3$. If a stable matching M exists in (N, V) then without loss of generality we may assume that $|M| = k$.*

Finally, some notes on notation: in this paper, we use $L = \langle \dots \rangle$ to construct an ordered list of elements L. If L and L' are lists then we write $L \cdot L'$ meaning the concatenation of L' to the end of L. We also write L_i to mean the ith element of list L, starting from $i = 1$, and $e \in L$ to describe membership of an element e in L. When working with sets of sets, we write $\bigcup S$ to mean $\bigcup_{T \in S} T$.

3 General Binary Preferences

Let 3D-SR-AS-BIN be the restriction of 3D-SR-AS in which preferences are binary but need not be symmetric. In this section we establish the NP-completeness of deciding whether a stable matching exists, given an instance (N, V) of 3D-SR-AS-BIN.

Theorem 1. *Given an instance of 3D-SR-AS-BIN, the problem of deciding whether a stable matching exists is* NP*-complete. The result holds even if each agent must be matched.*

Proof Sketch. Given an instance (N, V) of 3D-SR-AS-BIN and a matching M, it is straightforward to test in $O(|N|^3)$ time if M is stable in (N, V). This shows that the decision version of 3D-SR-AS-BIN belongs to the class NP.

We present a polynomial-time reduction from Partition Into Triangles (PIT), which is the following decision problem: "Given a simple undirected graph $G = (W, E)$ where $W = \{w_1, w_2, \dots, w_{3q}\}$ for some integer q, can the vertices of G

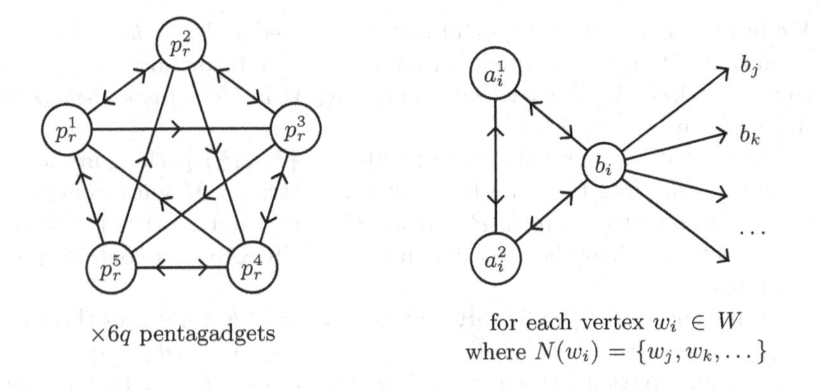

Fig. 1. The reduction from PIT to 3D-SR-AS-BIN. Each vertex represents an agent. An arc is present from agent α_i to agent α_j if $val_{\alpha_i}(\alpha_j) = 1$.

be partitioned into q disjoint sets $X = \{X_1, X_2, \dots, X_q\}$, each set containing exactly three vertices, such that for each $X_p = \{w_i, w_j, w_k\} \in X$ all three of the edges $\{w_i, w_j\}$, $\{w_i, w_k\}$, and $\{w_j, w_k\}$ belong to E?" PIT is NP-complete [16].

The reduction from PIT to 3D-SR-AS-BIN is as follows (see Fig. 1). Unless otherwise specified assume that $val_{\alpha_i}(\alpha_j) = 0$ for any $\alpha_i, \alpha_j \in N$. For each vertex $w_i \in W$ create agents a_i^1, a_i^2, b_i in N. Then set $val_{a_i^1}(a_i^2) = val_{a_i^1}(b_i) = 1$, $val_{a_i^2}(a_i^1) = val_{a_i^2}(b_i) = 1$, $val_{b_i}(a_i^1) = val_{b_i}(a_i^2) = 1$, and $val_{b_i}(b_j) = 1$ if $\{w_i, w_j\} \in E$ for any $w_j \in N \setminus \{w_i\}$. Next, for each r where $1 \leq r \leq 6q$ create $p_r^1, p_r^2, p_r^3, p_r^4, p_r^5$ in N. Then set $val_{p_r^1}(p_r^2) = val_{p_r^1}(p_r^3) = val_{p_r^1}(p_r^5) = 1$, $val_{p_r^2}(p_r^3) = val_{p_r^2}(p_r^4) = val_{p_r^2}(p_r^1) = 1$, $val_{p_r^3}(p_r^4) = val_{p_r^3}(p_r^5) = val_{p_r^3}(p_r^2) = 1$, $val_{p_r^4}(p_r^5) = val_{p_r^4}(p_r^1) = val_{p_r^4}(p_r^3) = 1$, and $val_{p_r^5}(p_r^1) = val_{p_r^5}(p_r^2) = val_{p_r^5}(p_r^4) = 1$. We shall refer to $\{p_r^1, \dots, p_r^5\}$ as the rth *pentagadget*. Note that $|N| = 39q$. In the full proof of this theorem, contained in [23], we show that a partition into triangles X exists in $G = (W, E)$ if and only if a stable matching M exists in (N, V) where $|M| = |N|/3$. □

4 Symmetric Binary Preferences

Consider the restriction of 3D-SR-AS in which preferences are binary and symmetric, which we call *3D-SR-SAS-BIN*. In this section we show that every instance of 3D-SR-SAS-BIN admits a stable matching. We give a step-by-step constructive proof of this result between Sects. 4.1–4.4, leading to an $O(|N|^3)$ algorithm for finding a stable matching. In Sect. 4.5 we consider an optimisation problem related to 3D-SR-SAS-BIN.

4.1 Preliminaries

An instance (N, V) of 3D-SR-SAS-BIN corresponds to a simple undirected graph $G = (N, E)$ where $\{\alpha_i, \alpha_j\} \in E$ if $val_{\alpha_i}(\alpha_j) = 1$, which we refer to as the *underlying graph*.

We introduce a restricted type of matching called a *P-matching*. Recall that by definition, $M(\alpha_p) = \varnothing$ implies that $u_{\alpha_p}(M) = 0$ for any $\alpha_p \in N$ in an arbitrary matching M. We say that a matching M in (N, V) is a *P-matching* if $M(\alpha_p) \neq \varnothing$ implies $u_{\alpha_p}(M) > 0$.

It follows that a P-matching corresponds to a $\{K_3, P_3\}$-packing in the underlying graph [22]. Note that any triple in a P-matching M must contain some agent with utility two. A *stable P-matching* is a P-matching that is also stable. We will eventually show that any instance of 3D-SR-SAS-BIN contains a stable P-matching.

In an instance (N, V) of 3D-SR-SAS-BIN, a *triangle* comprises three agents $\alpha_{m_1}, \alpha_{m_2}, \alpha_{m_3}$ such that $val_{\alpha_{m_1}}(\alpha_{m_2}) = val_{\alpha_{m_2}}(\alpha_{m_3}) = val_{\alpha_{m_3}}(\alpha_{m_1}) = 1$. If (N, V) contains no triangle then we say it is *triangle-free*. If (N, V) is not triangle-free then it can be reduced by successively removing three agents that belong to a triangle until it is triangle-free. This operation corresponds to removing a *maximal triangle packing* (see [12,22]) in the underlying graph and can be performed in $O(|N|^3)$ time. The resulting instance is triangle-free. We summarise this observation in the following lemma.

Lemma 2. *Given an instance (N, V) of 3D-SR-SAS-BIN, we can identify an instance (N', V') of 3D-SR-SAS-BIN and a set of triples $M_\triangle$ in $O(|N|^3)$ time such that (N', V') is triangle-free, $|N'| \leq |N|$, and if M is a stable P-matching in (N', V') then $M' = M \cup M_\triangle$ is a stable P-matching in (N, V).*

4.2 Repairing a *P*-Matching in a Triangle-Free Instance

In this section we consider an arbitrary triangle-free instance (N, V) of 3D-SR-SAS-BIN. Since the only instance referred to in this section is (N, V) so here we shorten "is stable in (N, V)" to "is stable", or similar.

We first define a special type of P-matching which is 'repairable'. We then present Algorithm `repair` (Algorithm 1), which, given (N, V) and a 'repairable' P-matching M, constructs a new P-matching M' that is stable. We shall see in the next section how this relates to a more general algorithm that, given a triangle-free instance, constructs a P-matching that is stable in that instance.

Given a triangle-free instance (N, V), we say a P-matching M is *repairable* if it is not stable and there exists exactly one $\alpha_i \in N$ where $u_{\alpha_i}(M) = 0$ and any triple that blocks M comprises $\{\alpha_i, \alpha_{j_1}, \alpha_{j_2}\}$ for some $\alpha_{j_1}, \alpha_{j_2} \in N$ where $u_{\alpha_{j_1}}(M) = 1$, $u_{\alpha_{j_2}}(M) = 0$, and $val_{\alpha_i}(\alpha_{j_1}) = val_{\alpha_{j_1}}(\alpha_{j_2}) = 1$.

We now provide some intuition behind Algorithm `repair` and refer the reader to Fig. 2. Recall that the overall goal of the algorithm is to construct a stable P-matching M'. Since the given P-matching M is repairable, our aim will be to modify M such that $u_{\alpha_i}(M') \geq 1$ while ensuring that no three agents that are ordered to different triples in M' block M'. The stability of the constructed P-matching M' then follows.

The algorithm begins by selecting some triple $\{\alpha_i, \alpha_{j_1}, \alpha_{j_2}\}$ that blocks M. The two agents in $M(\alpha_{j_1}) \setminus \{\alpha_{j_1}\}$ are labelled α_{j_3} and α_{j_4}. We present two example scenarios in which it is possible to construct a stable P-matching. First, suppose there exists some α_{z_1} where $val_{\alpha_{j_3}}(\alpha_{z_1}) = 1$ and $u_{\alpha_{z_1}}(M) = 0$. Construct M' from M by removing $\{\alpha_{j_1}, \alpha_{j_2}, \alpha_{j_3}\}$ and adding $\{\alpha_i, \alpha_{j_1}, \alpha_{j_2}\}$ and $\{\alpha_{j_3}, \alpha_{j_4}, \alpha_{z_1}\}$. Now, $u_{\alpha_i}(M') = 1$ and $u_{\alpha_p}(M') \geq u_{\alpha_p}(M)$ for any $\alpha_p \in N \backslash \{\alpha_i\}$. It follows by Lemma 1 that M' is stable. Second, suppose there exists no such α_{z_1} but there exists some α_{z_2} where $val_{\alpha_{j_4}}(\alpha_{z_2}) = 1$ and $u_{\alpha_{z_2}}(M) = 0$. Now construct M' from M by removing $\{\alpha_{j_1}, \alpha_{j_2}, \alpha_{j_3}\}$ and adding $\{\alpha_i, \alpha_{j_1}, \alpha_{j_2}\}$ and $\{\alpha_{j_3}, \alpha_{j_4}, \alpha_{z_2}\}$. Note that $u_{\alpha_i}(M') = 1$ and $u_{\alpha_p}(M') \geq u_{\alpha_p}(M)$ for any $\alpha_p \in N \setminus \{\alpha_i, \alpha_{j_3}\}$. It can be shown that α_{j_3} does not belong to a triple that blocks M' since no α_{z_1} exists as described. It follows again by Lemma 1 that M' is stable. Generalising the approach in the two example scenarios, the algorithm constructs a list S of agents, which initially comprises $\langle \alpha_{j_1}, \alpha_{j_3}, \alpha_{j_4} \rangle$. The list S has length $3c$ for some $c \geq 1$, where $\{S_{3c-2}, S_{3c-1}, S_{3c}\} \in M$ and $val_{S_p}(S_{p+1}) = 1$ for each p ($1 \leq p < 3c$). The list S therefore corresponds to a path in the underlying graph. In each iteration of the main loop, three agents belonging to some

---

Algorithm 1. Algorithm `repair`

Input: a triangle-free instance (N, V) of 3D-SR-SAS-BIN, repairable P-matching M in (N, V) (Section 4.2) with some such $\alpha_i \in N$.
Output: stable P-matching M' in (N, V)

$\{\alpha_{j_1}, \alpha_{j_2}\} \leftarrow$ some $\alpha_{j_1}, \alpha_{j_2} \in N$ where $\{\alpha_i, \alpha_{j_1}, \alpha_{j_2}\}$ blocks M and $u_{\alpha_{j_1}}(M) = 1$
$\{\alpha_{j_3}, \alpha_{j_4}\} \leftarrow M(\alpha_{j_1}) \setminus \{\alpha_{j_1}\}$ where $u_{\alpha_{j_3}}(M) = 2$
$S \leftarrow \langle \alpha_{j_1}, \alpha_{j_3}, \alpha_{j_4} \rangle$
$c \leftarrow 1$
$b \leftarrow 0$
$\alpha_{z_1}, \alpha_{z_2}, \alpha_{y_1}, \alpha_{y_2}, \alpha_{w_1} \leftarrow \perp$
while true
  $\alpha_{z_1} \leftarrow$ some $\alpha_{z_1} \in N \setminus \{\alpha_i\}$ where $val_{\alpha_{z_1}}(S_{3c-1}) = 1$ and $u_{\alpha_{z_1}}(M) = 0$, else $\perp$
  $\alpha_{z_2} \leftarrow$ some $\alpha_{z_2} \in N \setminus \{\alpha_i, \alpha_{j_2}\}$ where $val_{\alpha_{z_2}}(S_{3c}) = 1$ and $u_{\alpha_{z_2}}(M) = 0$, else $\perp$
  $\alpha_{y_1} \leftarrow$ some $\alpha_{y_1} \in N$ where $val_{S_{3c}}(\alpha_i) = val_{\alpha_{y_1}}(\alpha_i) = 1$ and $u_{\alpha_{y_1}}(M) = 0$, else $\perp$
  $\alpha_{y_2} \leftarrow$ some $\alpha_{y_2} \in N$ where $val_{S_{3c}}(\alpha_{j_2}) = val_{\alpha_{y_2}}(\alpha_{j_2}) = 1$ and $u_{\alpha_{y_2}}(M) = 0$, else $\perp$
  $b \leftarrow$ some $1 \leq b < c$ where $val_{S_{3b}}(\alpha_{j_2}) = val_{S_{3c}}(S_{3b}) = 1$, else 0
  $\alpha_{w_1} \leftarrow$ some $\alpha_{w_1} \in N$ where $val_{S_{3c}}(\alpha_{w_1}) = 1$, $u_{\alpha_{w_1}}(M) = 1$ and $\alpha_{w_1} \notin S$
    and there exists some $\alpha_{z_3} \in N \setminus \{\alpha_i\}$ where $val_{\alpha_{w_1}}(\alpha_{z_3}) = 1$ and $u_{\alpha_{z_3}}(M) = 0$,
    else $\perp$
  if $\alpha_{z_1} \neq \perp$ or $\alpha_{z_2} \neq \perp$ or $\alpha_{y_1} \neq \perp$ or $\alpha_{y_2} \neq \perp$ or $b > 0$ or $\alpha_{w_1} = \perp$ **then**
    break
  else
    $\{\alpha_{w_2}, \alpha_{w_3}\} \leftarrow M(\alpha_{w_1}) \setminus \{\alpha_{w_1}\}$ where $u_{\alpha_{w_2}}(M) = 2$
    $S \leftarrow S \cdot \langle \alpha_{w_1}, \alpha_{w_2}, \alpha_{w_3} \rangle$
    $c \leftarrow c + 1$
  end if
end while

continued overleaf

Algorithm 1. Algorithm **repair**

continued from previous page

if $\alpha_{z_1} \neq \perp$ and $\alpha_{z_1} \neq \alpha_{j_2}$ **then**
 ▷ Case 1
 $M_S \leftarrow \{\{\alpha_i, \alpha_{j_1}, \alpha_{j_2}\}\} \cup \bigcup_{1 \leq d < c} \{\{S_{3d-1}, S_{3d}, S_{3d+1}\}\} \cup \{\{\alpha_{z_1}, S_{3c-1}, S_{3c}\}\}$
else if $\alpha_{z_2} \neq \perp$ **then**
 ▷ Case 2
 $M_S \leftarrow \{\{\alpha_i, \alpha_{j_1}, \alpha_{j_2}\}\} \cup \bigcup_{1 \leq d < c} \{\{S_{3d-1}, S_{3d}, S_{3d+1}\}\} \cup \{\{S_{3c-1}, S_{3c}, \alpha_{z_2}\}\}$
else if $\alpha_{z_1} \neq \perp$ and $\alpha_{z_1} = \alpha_{j_2}$ **then**
 ▷ Case 3
 $\alpha_{z_4} \leftarrow$ some $\alpha_{z_4} \in N \setminus \{\alpha_i, \alpha_{j_2}\}$ where $val_{S_{3c-2}}(\alpha_{z_4}) = 1$ and $u_{\alpha_{z_4}}(M) = 0$
 $M_S \leftarrow \{\{\alpha_i, \alpha_{j_1}, \alpha_{j_3}\}\} \cup \bigcup_{1 \leq d < c-1} \{\{S_{3d}, S_{3d+1}, S_{3d+2}\}\} \cup \{\{S_{3c-3}, S_{3c-2}, \alpha_{z_4}\}\}$
 $\cup \{\{S_{3c-1}, S_{3c}, \alpha_{j_2}\}\}$
else if $\alpha_{y_1} \neq \perp$ **then**
 ▷ Case 4
 $M_S \leftarrow \{\{\alpha_{j_2}, \alpha_{j_1}, \alpha_{j_3}\}\} \cup \bigcup_{1 \leq d < c} \{\{S_{3d}, S_{3d+1}, S_{3d+2}\}\} \cup \{\{S_{3c}, \alpha_i, \alpha_{y_1}\}\}$
else if $\alpha_{y_2} \neq \perp$ **then**
 ▷ Case 5
 $M_S \leftarrow \{\{\alpha_i, \alpha_{j_1}, \alpha_{j_3}\}\} \cup \bigcup_{1 \leq d < c} \{\{S_{3d}, S_{3d+1}, S_{3d+2}\}\} \cup \{\{S_{3c}, \alpha_{j_2}, \alpha_{y_2}\}\}$
else if $b > 0$ **then**
 ▷ Case 6
 $\alpha_{z_5} \leftarrow$ some $\alpha_{z_5} \in N \setminus \{\alpha_i, \alpha_{j_2}\}$ where $val_{S_{3b+1}}(\alpha_{z_3}) = 1$ and $u_{\alpha_{z_3}}(M) = 0$
 $M_S \leftarrow \{\{\alpha_i, \alpha_{j_1}, \alpha_{j_3}\}\} \cup \bigcup_{1 \leq d < b} \{\{S_{3d}, S_{3d+1}, S_{3d+2}\}\} \cup \{\{\alpha_{z_4}, S_{3b+1}, S_{3b+2}\}\}$
 $\cup \bigcup_{b+1 \leq d < c} \{\{S_{3d}, S_{3d+1}, S_{3d+2}\}\} \cup \{\{S_{3c}, S_{3b}, \alpha_{j_2}\}\}$
else
 ▷ Case 7. Note that $\alpha_{w_1} = \perp$.
 $M_S \leftarrow \{\{\alpha_i, \alpha_{j_1}, \alpha_{j_3}\}\} \cup \bigcup_{1 \leq d < c} \{\{S_{3d}, S_{3d+1}, S_{3d+2}\}\}$
end if
return $M' = M_S \cup \{r \in M \mid r \cap S = \varnothing\}$

triple in M are appended to the end of S. The loop continues until S satisfies at least one of six specific conditions. We show that eventually at least one of these conditions must hold.

These six stopping conditions correspond to seven different cases, labelled Case 1–Case 7, in which a stable P-matching M' may be constructed. The exact construction of M' depends on which condition(s) caused the main loop to terminate. Cases 1 and 3 generalise the first example scenario, in which some α_{z_1} exists as described. Case 2 generalises the second example scenario, in which no such α_{z_1} exists but some α_{z_2} exists as described. Cases 4–7 correspond to similar scenarios. The six stopping conditions and seven corresponding constructions of

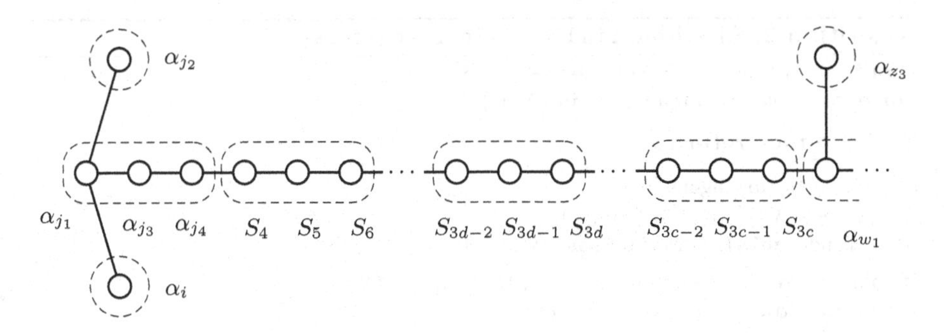

Fig. 2. Players and triples in M before a new iteration of the `while` loop

M' are somewhat hierarchical. For example, the proof that M' is stable in Case 4 relies on the fact that in no iteration did the condition for Cases 1 and 3 hold. A similar reliance exists in the proofs of each of the other cases. The proof that M' is stable in Case 7 is the most complex. It relies on the fact that no condition relating to any of the previous six cases held in the final or some previous iteration of the main loop. Further intuition for the different cases is given in the full version of this paper [23].

Algorithm `repair` is presented in Algorithm 1 in two parts. The first part involves the construction of S and exploration of the instance. The second part involves the construction of M'. The following lemma establishes the correctness and complexity of this algorithm.

Lemma 3. *Algorithm* `repair` *returns a stable P-matching in $O(|N|^2)$ time.*

4.3 Finding a Stable P-Matching in a Triangle-Free Instance

In the previous section we supposed that (N, V) was a triangle-free instance of 3D-SR-SAS-BIN and considered a P-matching M that was repairable (Sect. 4.2). We presented Algorithm `repair`, which can be used to construct a stable P-matching M' in $O(|N|^2)$ time (Lemma 3). In this section we present Algorithm `findStableInTriangleFree` (Algorithm 2), which, given a triangle-free instance (N, V), constructs a P-matching M' that is stable in (N, V). Algorithm `findStableInTriangleFree` is recursive. The algorithm first removes an arbitrary agent α_i to construct a smaller instance (N', V'). It then uses a recursive call to construct a P-matching M that is stable in (N', V'). By Lemma 1, any triple that blocks M in the larger instance (N, V) must contain α_i or block M in (N', V'). There are then three cases involving types of triple that block M in (N', V'). In two out of three cases, M' can be constructed by adding to M a new triple containing α_i and two players unmatched in M. In the third case, M is not stable in (N, V) but, by design, is repairable (see Sect. 4.2). It follows that Algorithm `repair` can be used to construct a P-matching that is stable in (N, V) (Lemma 3). It is relatively straightforward to show that the running time of Algorithm `findStableInTriangleFree` is $O(|N|^3)$.

Algorithm 2. Algorithm `findStableInTriangleFree`

Input: an instance (N, V) of 3D-SR-SAS-BIN
Output: stable P-matching M' in (N, V)

if $|N| = 2$ **then return** $\varnothing$

$\alpha_i \leftarrow$ an arbitrary agent in N
$(N', V') \leftarrow (N \setminus \{\alpha_i\}, V \setminus \{val_{\alpha_i}\})$
$M \leftarrow$ `findStableInTriangleFree`$((N', V'))$
if some $\alpha_{l_1}, \alpha_{l_2} \in N$ exist where $u_{\alpha_{l_1}}(M) = u_{\alpha_{l_2}}(M) = 0$
  and $val_{\alpha_i}(\alpha_{l_1}) = val_{\alpha_i}(\alpha_{l_2}) = 1$ **then**
  return $M \cup \{\{\alpha_i, \alpha_{l_1}, \alpha_{l_2}\}\}$
else if some $\alpha_{l_3}, \alpha_{l_4} \in N$ exist where $u_{\alpha_{l_3}}(M) = u_{\alpha_{l_4}}(M) = 0$
  and $val_{\alpha_i}(\alpha_{l_3}) = val_{\alpha_{l_3}}(\alpha_{l_4}) = 1$ **then**
  return $M \cup \{\{\alpha_i, \alpha_{l_3}, \alpha_{l_4}\}\}$
else if some $\alpha_{l_5}, \alpha_{l_6} \in N$ exist where $u_{\alpha_{l_5}}(M) = 1$, $u_{\alpha_{l_6}}(M) = 0$
  and $val_{\alpha_i}(\alpha_{l_5}) = val_{\alpha_{l_5}}(\alpha_{l_6}) = 1$ **then**
  ▷ M is repairable in (N, V) (see Section 4.2). Note that $\alpha_{j_1} = \alpha_{l_5}$ and $\alpha_{j_2} = \alpha_{l_6}$.
  return `repair`$((N, V), M, \alpha_i)$
else
  return M
end if

Lemma 4. *Algorithm* `findStableInTriangleFree` *returns a stable P-matching in (N, V) in $O(|N|^3)$ time.*

4.4 Finding a Stable P-Matching in an Arbitrary Instance

In the previous section we considered instances of 3D-SR-SAS-BIN that are triangle-free. We showed that, given such an instance, Algorithm `findStableInTriangleFree` can be used to find a stable P-matching in $O(|N|^3)$ time (Lemma 4). In Sect. 4.1, we showed that an arbitrary instance can be reduced in $O(|N|^3)$ time to construct a corresponding triangle-free instance (Lemma 2). Algorithm `findStable` therefore comprises two steps. First, the instance is reduced by removing a maximal set of triangles. Call this set $M_\triangle$. Then, Algorithm `findStableInTriangleFree` is called to construct a P-matching M' that is stable in the reduced, triangle-free instance. It is straightforward to show that $M_\triangle \cup M'$ is a stable P-matching. The running time of Algorithm `findStable` is thus $O(|N|^3)$. A pseudocode description of Algorithm `findStable` can be found in the full version of this paper [23]. We arrive at the following result.

Theorem 2. *Given an instance (N, V) of 3D-SR-SAS-BIN, a stable P-matching, and hence a stable matching, must exist and can be found in $O(|N|^3)$ time. Moreover, if $|N|$ is a multiple of three then, if required, every agent can be matched in the returned stable matching.*

4.5 Stability and Utilitarian Welfare

Given an instance (N, V) of 3D-SR-SAS-BIN and matching M, let the *utilitarian welfare* [4,11] of a set $S \subseteq N$, denoted $u_S(M)$, be $\sum_{\alpha_i \in S} u_{\alpha_i}(M)$. Let $u(M)$ be short for $u_N(M)$. Given a matching M in an arbitrary instance (N, V) of 3D-SR-SAS-BIN, it follows that $0 \leq u(M) \leq 2|N|$. It is natural to then consider the optimisation problem of finding a stable matching with maximum utilitarian welfare, which we refer to as *3D-SR-SAS-BIN-MAXUW*. This problem is closely related to Partition Into Triangles (PIT, defined in Sect. 3), which we reduce from in the proof that 3D-SR-SAS-BIN-MAXUW is NP-hard.

Theorem 3. *3D-SR-SAS-BIN-MAXUW is* NP*-hard.*

We note that the reduction from PIT to 3D-SR-SAS-BIN-MAXUW also shows that the problem of finding a (not-necessarily stable) matching with maximum utilitarian welfare, given an instance of 3D-SR-SAS-BIN, is also NP-hard.

In Sect. 4.4 we showed that, given an arbitrary instance (N, V) of 3D-SR-SAS-BIN, a stable P-matching exists and can be found in $O(|N|^3)$ time. We now present Algorithm `findStableUW` (Algorithm 3) as an approximation algorithm for 3D-SR-SAS-BIN-MAXUW.

This algorithm first calls Algorithm `findStable` to construct a stable P-matching. It then orders the unmatched agents into triples such that the produced matching is still stable in (N, V) (by Lemma 1) but is not necessarily a P-matching. The pseudocode description of Algorithm `findStableUW` includes

---

Algorithm 3. Algorithm `findStableUW`

Input: an instance (N, V) of 3D-SR-SAS-BIN
Output: stable matching M_A in (N, V)

$M_1 \leftarrow$ `findStable`$((N, V))$
$U \leftarrow$ agents in N unmatched in M_1
$Y \leftarrow$ `maximum2DMatching`$((N, V),\ U)$
if $|Y| \geq \lfloor |U|/3 \rfloor$ **then**
  $X \leftarrow$ any $\lfloor |U|/3 \rfloor$ elements of Y
else
  ▷ Note that since Y is a set of disjoint pairs, it follows that
   $|U \setminus \bigcup Y| = |U| - 2|Y| \geq \lfloor |U|/3 \rfloor - |Y|$.
  $W \leftarrow$ an arbitrary set of $\lfloor |U|/3 \rfloor - |Y|$ pairs of elements in $U \setminus \bigcup Y$
  $X \leftarrow Y \cup W$
end if
$Z \leftarrow U \setminus \bigcup X$
▷ Suppose $X = \{x_1, x_2, \ldots, x_{\lfloor |U|/3 \rfloor}\}$ and $Z = \{z_1, z_2, \ldots, z_{\lfloor |U|/3 \rfloor}\}$.
▷ Note that x_i is a pair of agents and z_i is a single agent for each $1 \leq i \leq \lfloor |U|/3 \rfloor$.
$M_2 \leftarrow \{x_i \cup \{z_i\}$ for each $1 \leq i \leq \lfloor |U|/3 \rfloor\}$
return $M_1 \cup M_2$

---

a call to `maximum2DMatching`. Given an instance (N, V) and some set $U \subseteq N$, this subroutine returns a (two-dimensional) *maximum cardinality matching* Y in the subgraph of G, the underlying graph of (N, V), induced by U. From Y, Algorithm `findStableUW` constructs a set X of pairs with cardinality $\lfloor |U|/3 \rfloor$. It also constructs a set Z from the remaining agents, also with cardinality $\lfloor |U|/3 \rfloor$. Finally, it constructs the matching M_2 such that each triple in M_2 is union of a pair of agents in X and a single agent in Z. Let M_{A} be an arbitrary matching returned by Algorithm `findStableUW` given (N, V). Suppose M_{opt} is a stable matching in (N, V) with maximum utilitarian welfare. To prove the performance guarantee of Algorithm `findStableUW` we show that $2u(M_{\mathrm{A}}) \geq u(M_{\mathrm{opt}})$. The proof involves apportioning the welfare of agents in M_{A} by the triples of those agents in M_{opt}.

Theorem 4. *Algorithm* `findStableUW` *is a 2-approximation algorithm for 3D-SR-SAS-BIN-MAXUW.*

In the instance of 3D-SR-SAS-BIN shown in Fig. 3, Algorithm `findStableUW` always returns $M_{\mathrm{A}} = \{\{\alpha_3, \alpha_5, \alpha_6\}\}$ while $M_{\mathrm{opt}} = \{\{\alpha_1, \alpha_2, \alpha_3\}, \{\alpha_4, \alpha_5, \alpha_8\}, \{\alpha_6, \alpha_7, \alpha_9\}\}$. Since $u(M_{\mathrm{A}}) = 6$ and $u(M_{\mathrm{opt}}) = 12$ it follows that $u(M_{\mathrm{opt}}) = 2u(M_{\mathrm{A}})$. This shows that the analysis of Algorithm `findStableUW` is tight. Moreover, this particular instance shows that any approximation algorithm with a better performance ratio than 2 should not always begin, like Algorithm `findStableUW` does, by selecting a maximal set of triangles.

5 Open Questions

In this paper we have considered the three-dimensional stable roommates problem with additively separable preferences. We considered the special cases in which preferences are binary but not necessarily symmetric, and both binary and symmetric. There are several interesting directions for future research.

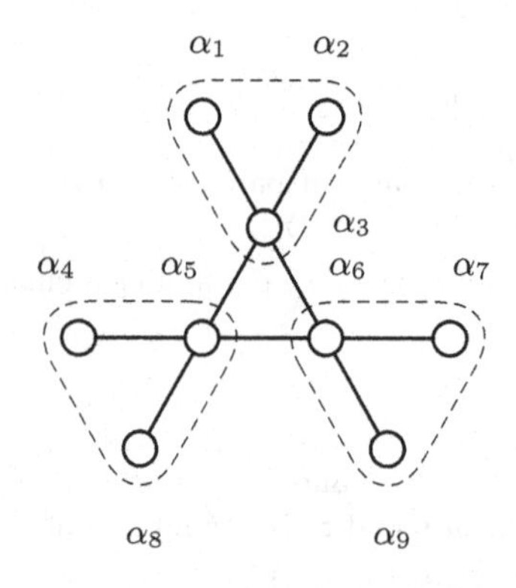

Fig. 3. An instance in which $u(M_{\mathrm{opt}}) = 2u(M_{\mathrm{A}})$.

- Does there exist an approximation algorithm for 3D-SR-SAS-BIN-MAXUW (Sect. 4.5) with a better performance guarantee than 2?
- In 3D-SR-AS, there are numerous possible restrictions besides symmetric and binary preferences. Do any other restrictions ensure that a stable matching exists? For example, we could consider the restriction in which preferences are symmetric and $val_{\alpha_i} \in \{0, 1, 2\}$ for each $\alpha_i \in N$.
- Additively separable preferences are one possible structure of agents' preferences that can be applied in a model of three-dimensional SR. Are there other systems of preferences that result in new models in which a stable matching can be found in polynomial time?
- The 3D-SR-AS problem model can be generalised to higher dimensions. It would be natural to ask if the algorithm for 3D-SR-SAS-BIN can be generalised to the same problem in $k \geq 3$ dimensions, in which a k-set of agents S is blocking if, for each of the k agents in S, the utility of S is strictly greater than that agent's utility in the matching. We conjecture that when $k \geq 4$, a stable matching need not exist, and that the associated decision problem is NP-complete, even when preferences are both binary and symmetric.

References

1. Arkin, E., Bae, S., Efrat, A., Mitchell, J., Okamoto, K., Polishchuk, V.: Geometric stable roommates. Inf. Process. Lett. **109**, 219–224 (2009)
2. Aziz, H., Brandt, F., Seedig, H.G.: Optimal partitions in additively separable hedonic games. In: Proceedings of IJCAI 2011: The 22nd International Joint Conference on Artificial Intelligence, vol. 1, pp. 43–48. AAAI Press (2011)
3. Aziz, H., Brandt, F., Seedig, H.G.: Computing desirable partitions in additively separable hedonic games. Artif. Intell. **195**, 316–334 (2013)
4. Aziz, H., Gaspers, S., Gudmundsson, J., Mestre, J., Täubig, H.: Welfare maximization in fractional hedonic games. In: Proceedings of IJCAI 2015: The 24th International Joint Conference on Artificial Intelligence, pp. 461–467. AAAI Press (2015)
5. Aziz, H., Lang, J., Monnot, J.: Computing Pareto optimal committees. In: Proceedings of IJCAI 2016: The 25th International Joint Conference on Artificial Intelligence, pp. 60–66. AAAI Press (2016)
6. Aziz, H., Savani, R., Moulin, H.: Hedonic games. In: Brandt, F., Conitzer, V., Endriss, U., Lang, J., Procaccia, A.D. (eds.) Handbook of Computational Social Choice, pp. 356–376. Cambridge University Press (2016)
7. Barberà, S., Bossert, W., Pattanaik, P.: Ranking sets of objects. In: Barberà, S., Hammond, P., Seidl, C. (eds.) Handbook of Utility Theory, vol. 2, chap. 17, pp. 893–977. Kluwer Academic Publishers (2004)
8. Boehmer, N., Elkind, E.: Stable roommate problem with diversity preferences. In: Proceedings of IJCAI 2020: The 29th International Joint Conference on Artificial Intelligence, pp. 96–102. IJCAI Organization (2020)
9. Brandt, F., Bullinger, M.: Finding and recognizing popular coalition structures. In: Proceedings of AAMAS 2020: The 19th International Conference on Autonomous Agents and Multiagent Systems, pp. 195–203. IFAAMAS (2020)

10. Bredereck, R., Heeger, K., Knop, D., Niedermeier, R.: Multidimensional stable roommates with master list. In: Chen, X., Gravin, N., Hoefer, M., Mehta, R. (eds.) WINE 2020. LNCS, vol. 12495, pp. 59–73. Springer, Cham (2020). https://doi.org/10.1007/978-3-030-64946-3_5
11. Bullinger, M.: Pareto-optimality in cardinal hedonic games. In: Proceedings of AAMAS 2020: The 19th International Conference on Autonomous Agents and Multiagent Systems, pp. 213–221. IFAAMAS (2020)
12. Chataigner, F., Manić, G., Wakabayashi, Y., Yuster, R.: Approximation algorithms and hardness results for the clique packing problem. Discret. Appl. Math. **157**(7), 1396–1406 (2009)
13. Cseh, Á., Fleiner, T., Harján, P.: Pareto optimal coalitions of fixed size. J. Mech. Inst. Des. **4**(1), 87–108 (2019)
14. Deineko, V.G., Woeginger, G.J.: Two hardness results for core stability in hedonic coalition formation games. Discret. Appl. Math. **161**(13), 1837–1842 (2013)
15. Gale, D., Shapley, L.: College admissions and the stability of marriage. Amer. Math. Monthly **69**, 9–15 (1962)
16. Garey, M., Johnson, D.: Computers and Intractability. Freeman, San Francisco (1979)
17. Hajduková, J.: Coalition formation games: a survey. Int. Game Theory Rev. **8**(4), 613–641 (2006)
18. Huang, C.-C.: Two's company, three's a crowd: stable family and threesome roommates problems. In: Arge, L., Hoffmann, M., Welzl, E. (eds.) ESA 2007. LNCS, vol. 4698, pp. 558–569. Springer, Heidelberg (2007). https://doi.org/10.1007/978-3-540-75520-3_50
19. Huang, C.-C.: Two's company, three's a crowd: stable family and threesome roommates problems. Computer Science Technical Report TR2007-598, Dartmouth College (2007)
20. Irving, R.: An efficient algorithm for the "stable roommates" problem. J. Algorithms **6**, 577–595 (1985)
21. Iwama, K., Miyazaki, S., Okamoto, K.: Stable roommates problem with triple rooms. In: Proceedings of WAAC 2007: The 10th Korea-Japan Workshop on Algorithms and Computation, pp. 105–112 (2007)
22. Kirkpatrick, D.G., Hell, P.: On the complexity of general graph factor problems. SIAM J. Comput. **12**(3), 601–609 (1983)
23. McKay, M., Manlove, D.: The three-dimensional stable roommates problem with additively separable preferences (preprint). arXiv:2107.04368 [cs.GT]
24. Ng, C., Hirschberg, D.: Three-dimensional stable matching problems. SIAM J. Discret. Math. **4**(2), 245–252 (1991)
25. Sung, S.C., Dimitrov, D.: Computational complexity in additive hedonic games. Eur. J. Oper. Res. **203**(3), 635–639 (2010)
26. Woeginger, G.J.: Core stability in hedonic coalition formation. In: van Emde Boas, P., Groen, F.C.A., Italiano, G.F., Nawrocki, J., Sack, H. (eds.) SOFSEM 2013. LNCS, vol. 7741, pp. 33–50. Springer, Heidelberg (2013). https://doi.org/10.1007/978-3-642-35843-2_4
27. Wright, M., Vorobeychik, Y.: Mechanism design for team formation. In: Proceedings of the 29th Conference on Artificial Intelligence, AAAI 2015, pp. 1050–1056. AAAI Press (2015)

Descending the Stable Matching Lattice: How Many Strategic Agents Are Required to Turn Pessimality to Optimality?

Ndiamé Ndiaye(✉), Sergey Norin, and Adrian Vetta

McGill University, Montreal, Canada
ndiame.ndiaye@mcgill.ca

Abstract. The set of stable matchings induces a distributive lattice. The supremum of the stable matching lattice is the boy-optimal (girl-pessimal) stable matching and the infimum is the girl-optimal (boy-pessimal) stable matching. The classical boy-proposal deferred-acceptance algorithm returns the supremum of the lattice, that is, the boy-optimal stable matching. In this paper, we study the smallest group of girls, called the *minimum winning coalition of girls*, that can act strategically, but independently, to force the boy-proposal deferred-acceptance algorithm to output the girl-optimal stable matching. We characterize the minimum winning coalition in terms of stable matching rotations. Our two main results concern the random matching model. First, the expected cardinality of the minimum winning coalition is small, specifically $(\frac{1}{2}+o(1))\log n$. This resolves a conjecture of Kupfer [13]. Second, in contrast, a randomly selected coalition must contain nearly every girl to ensure it is a winning coalition asymptotically almost surely. Equivalently, for any $\varepsilon > 0$, the probability a random group of $(1-\varepsilon)n$ girls is *not* a winning coalition is at least $\delta(\varepsilon) > 0$.

1 Introduction

We study the stable matching problem with n boys and n girls. Each boy has a preference ranking over the girls and vice versa. A matching is *stable* if there is no boy-girl pair that prefer each other over their current partners in the matching. A stable matching always exists and can be found by the deferred-acceptance algorithm [5]. Furthermore, the set of stable matchings forms a lattice whose supremum matches each boy to his *best* stable-partner and each girl to her *worst* stable-partner. This matching is called the *boy-optimal* (girl-pessimal) stable matching. Conversely, the infimum of the lattice matches each boy to his worst stable-partner and each girl to her best stable-partner. Consequently this matching is called the *girl-optimal* (boy-pessimal) stable matching.

Interestingly, the deferred-acceptance algorithm outputs the optimal stable matching for the proposing side. Perhaps surprisingly, the choice of which side makes the proposal can make a significant difference. For example, for the random matching model, where the preference list of each boy and girl is sampled

I. Caragiannis and K. A. Hansen (Eds.): SAGT 2021, LNCS 12885, pp. 281–295, 2021.
https://doi.org/10.1007/978-3-030-85947-3_19

uniformly and independently, Pittel [15] showed the boy-proposal deferred acceptance algorithm assigns the boys with much better ranking partners than the girls. Specifically, with high probability, the sum of the partner ranks is close to $n \log n$ for the boys and close to $\frac{n^2}{\log n}$ for the girls. Hence, on average, each boy ranks his partner at position $\log n$ at the boy-optimal stable matching while each girl only ranks her partner at position $\frac{n}{\log n}$. Consequently, collectively the girls may have a much higher preference for the infimum (girl-optimal) stable matching than the supremum (girl-pessimal) stable matching output by the boy-proposal deferred-acceptance algorithm.

Remarkably, Ashlagi et al. [1] proved that in an *unbalanced market* with one fewer girls than boys this advantage to the boys is reversed. In the random matching model, with high probability, each girl is matched to a boy she ranks at $\log n$ on average and each boy is matched to a girl he ranks at $\frac{n}{\log n}$ on average, even using the boy-proposal deferred-acceptance algorithm.[1] Kupfer [13] then showed a similar effect arises in a balanced market in which exactly one girl acts strategically. The expected rank of the partner of each girl improves to $O(\log^4 n)$ while the expected rank of the partner of each boy deteriorates to $\Omega(\frac{n}{\log^{2+\epsilon} n})$. Thus, just one strategic girl suffices for the stable matching output by the boy-proposal deferred-acceptance algorithm to change from the supremum of the lattice to a stable matching "close" to the infimum. But how many strategic girls are required to guarantee the infimum itself is output? Kupfer [13] conjectured that $O(\log n)$ girls suffice in expectation. In this paper we prove this conjecture. More precisely, we show that the minimum number of strategic girls required is $\frac{1}{2} \log n + O(\log \log n) = (\frac{1}{2} + o(1)) \log n$ in expectation. Consequently, the expected cardinality of the optimal winning coalition of girls is relatively small. Conversely, a random coalition of girls must be extremely large, namely of cardinality $n - o(n)$, if it is to be a winning coalition asymptotically almost surely. We prove that, for any $\varepsilon > 0$, the probability a random group of $(1 - \varepsilon)n$ girls is *not* a winning coalition is at least a constant.

1.1 Overview

In Sect. 2, we present the relevant background on the stable matching problem, in particular, concerning the stable matching lattice and the rotation poset. In Sect. 3 we provide a characterization of winning coalitions of girls in terms of minimal rotations in the rotation poset. In Sect. 4, we present the random matching model studied for the main results of the paper. Our first main result is given in Sect. 5 and shows that in random instances the cardinality of the minimum winning coalition is much closer to the lower bound than the upper bound. Specifically, in the random matching model, the expected cardinality of the minimum winning coalition is $\frac{1}{2} \log n + O(\log \log n)$. Our second main result is presented in Sect. 6 and shows that for a randomly selected coalition to be a winning coalition with probability $1 - o(1)$, it must have cardinality $n - o(n)$.

[1] In fact, an unbalanced market essentially contains a unique stable matching; see [1] for details.

An example illustrating the concepts along with the proofs of the lemmas and theorems are deferred to the full version of the paper.

2 The Stable Matching Problem

Here we review the stable matching problem and the concepts and results relevant to this paper. The reader is referred to the book [9] by Gusfield and Irving for a comprehensive introduction to stable matchings.

We are given a set $B = \{b_1, b_2, \ldots, b_n\}$ of boys and a set $G = \{g_1, g_2, \ldots, g_n\}$ of girls. Every boy $b \in B$ has a preference ranking $\succ_b$ over the girls; similarly, every girl $g \in G$ has a preference ranking $\succ_g$ over the boys. Now let μ be a (perfect) matching between the boys and girls. We say that boy b is matched to girl $\mu(b)$ in the matching μ; similarly, girl g is matched to boy $\mu(g)$. Boy b and girl g form a *blocking pair* $\{b, g\}$ if they prefer each other to their partners in the matching μ; that is $g \succ_b \mu(b)$ **and** $b \succ_g \mu(g)$. A matching μ that contains no blocking pair is called *stable*; otherwise it is unstable. In the *stable matching problem*, the task is to find a stable matching.

2.1 The Deferred-Acceptance Algorithm

The first question to answer is whether or not a stable matching is guaranteed to exist. Indeed a stable matching always exists, as shown in the seminal work of Gale and Shapley [5]. Their proof was constructive; the *deferred-acceptance algorithm*, described in Algorithm 1, outputs a stable matching.

Algorithm 1: Deferred-Acceptance (Boy-Proposal Version)

```
while there is an unmatched boy b do
    Let b propose to his favourite girl g who has not yet rejected him;
    if g is unmatched then
        g provisionally matches with b;
    else if g is provisionally matched to b̂ then
        g provisionally matches to her favourite of b and b̂, and rejects the
        other;
```

The key observation here is that only a girl can reject a provisional match. Thus, from a girl's perspective, her provisional match can only improve as the algorithm runs. It follows that the deferred-acceptance algorithm terminates when every girl has received at least one proposal. In addition, from a boy's perspective, his provisional match can only get worse as the algorithm runs. Indeed, it would be pointless for a boy to propose to girl who has already rejected him. Thus, each boy will make at most n proposals. Furthermore, because each boy makes proposals in decreasing order of preference, every girl must eventually receive a proposal. Thus the deferred-acceptance algorithm must terminate with a perfect matching μ.

Theorem 2.1 (Gale and Shapley 1962 [5]**).** *The deferred-acceptance algorithm outputs a stable matching.*

2.2 The Stable Matching Lattice

So a stable matching always exists. In fact, there may be an exponential number of stable matchings [11]. The set $\mathcal{M}$ of all stable matchings forms a poset $(\mathcal{M}, \geqslant)$ whose order $\geqslant$ is defined via the preference lists of the boys. Specifically, $\mu_1 \geqslant \mu_2$ if and only if every boy weakly prefers their partner in the stable matching μ_1 to their partner in the stable matching μ_2; that is $\mu_1(b) \succeq_b \mu_2(b)$, for every boy b.

Conway (see Knuth [11]) observed that the poset $(\mathcal{M}, \geqslant)$ is in fact a *distributive lattice*. Thus, by the lattice property, each pair of stable matchings μ_1 and μ_2 has a *join* (least upper bound) and a *meet* (greatest lower bound) in the lattice. Moreover, the join $\hat{\mu} = \mu_1 \vee \mu_2$ has the remarkable property that each boy b is matched to his *most preferred* partner amongst the girls $\mu_1(b)$ and $\mu_2(b)$. Similarly, in the meet $\check{\mu} = \mu_1 \wedge \mu_2$ each boy is matched to his *least preferred* partner amongst the girls $\mu_1(b)$ and $\mu_2(b)$. In particular, in the *supremum* $\mathbf{1} = \bigvee_{\mu \in \mathcal{M}} \mu$ of the lattice each boy is matched to his most preferred partner from any stable matching (called his *best stable-partner*). Accordingly, the matching $\mathbf{1}$ is called the *boy-optimal* stable matching. On the other hand, in the *infimum* $\mathbf{0} = \bigwedge_{\mu \in \mathcal{M}} \mu$ of the lattice each boy is matched to his least preferred partner from any stable matching (called his *worst stable-partner*). Accordingly, the matching $\mathbf{0}$ is called the *boy-pessimal* stable matching.

Theorem 2.2 *[5]. The deferred-acceptance algorithm outputs the boy-optimal stable matching.*

The reader may have observed that the description of the deferred-acceptance algorithm given in Algorithm 1 is ill-specified. In particular, which unmatched boy is selected to make the next proposal? Theorem 2.2 explains the laxity of our description. It is irrelevant which unmatched boy is chosen in each step, the final outcome is guaranteed to be the boy-optimal stable matching! In fact, the original description of the algorithm by Gale and Shapley [5] allowed for simultaneous proposals by unmatched boys – again this has no effect on the stable matching output.

The inverse poset $(\mathcal{M}, \leqslant)$ is also of fundamental interest. Indeed, McVitie and Wilson [14] made the surprising observation that $(\mathcal{M}, \leqslant)$ is the lattice defined using the preference lists of the girls rather than the boys. That is, every boy weakly prefers their partner in the stable matching μ_1 to their partner in the stable matching μ_2 if and only if every girl weakly prefers their partner in the stable matching μ_2 to their partner in the stable matching μ_1.

Theorem 2.3 *[14]. If $\mu_1 \geqslant \mu_2$ in the lattice $(\mathcal{M}, \geqslant)$ then every girl weakly prefers μ_2 over μ_1.*

Consequently, the boy-optimal stable matching **1** is also the *girl-pessimal* stable matching and the boy-pessimal stable matching **0** is the *girl-optimal* stable matching.

2.3 The Rotation Poset

Recall that the lattice $(\mathcal{M}, \geqslant)$ is a *distributive* lattice. This is important because the *fundamental theorem for finite distributive lattices* of Birkhoff [2] states that associated with any distributive lattice $\mathcal{L}$ is a unique *auxiliary poset* $\mathcal{P}(\mathcal{L})$. Specifically, the order ideals (or down-sets) of the auxiliary poset $\mathcal{P}$, ordered by inclusion, form the lattice $\mathcal{L}$. We refer the reader to the book of Stanley [18] for details on the fundamental theorem for finite distributive lattices. For our purposes, however, it is sufficient to note that the auxiliary poset $\mathcal{P}$ for the stable matching lattice $(\mathcal{M}, \geqslant)$ has an elegant combinatorial description that is very amenable in studying stable matchings.

In particular, the auxiliary poset for the stable matching lattice is called the *rotation poset* $\mathcal{P} = (\mathcal{R}, \geq)$ and was first discovered by Irving and Leather [10]. The elements of the auxiliary poset are *rotations*. Informally, given a stable matching μ, a rotation will rearrange the partners of a suitably chosen subset of the boys in a circular fashion to produce another stable matching. Formally, a rotation $R \in \mathcal{R}$ is a subset of the pairs in the stable matching μ, $R = [(b_0, g_0), (b_1, g_1), \dots, (b_k, g_k)]$, such that for each boy b_i, the girl $g_{i+1\,(\mathrm{mod}\,k+1)}$ is the first girl *after* his current stable-partner g_i on his preference list who would accept a proposal from him. That is, g_{i+1} prefers boy b_i over her current partner boy b_{i+1} *and* every girl g that boy b_i ranks on his list between g_i and g_{i+1} prefers her current partner in μ over b_i.

In this case, we say that R is a *rotation exposed* by the stable matching μ. Let $\hat{\mu} = \mu \otimes R$ be the perfect matching obtained by matching boy b_i with the girl $g_{i+1\,(\mathrm{mod}\,k+1)}$, for each $0 \leq i \leq k$, with all other matches the same as in μ. Irving and Leather [10] showed that $\hat{\mu}$ is also a stable matching. More importantly they proved:

Theorem 2.4 *[10]. The matching $\hat{\mu}$ is covered*[2] *by μ in the Hasse diagram of the stable matching lattice if and only if $\hat{\mu} = \mu \otimes R$ for some rotation R exposed by μ.*

Theorem 2.4 implies we may traverse the stable matching lattice $(\mathcal{M}, \geqslant)$ using rotations. As stated, we may also derive a poset $\mathcal{P} = (\mathcal{R}, \geq)$ whose elements are rotations. Let $\mathcal{R}_\mu$ be the set of all rotations exposed in μ. Then $\mathcal{R} = \bigcup_{\mu \in \mathcal{M}} \mathcal{R}_\mu$ is the set of all rotations. We then define the partial order $\geq$ as follows. Let $R_1 \geq R_2$ in $\mathcal{P}$ *if and only if* for any stable matching $\mu_1 \in \{\mu \in \mathcal{M} : R_1 \in \mathcal{R}_\mu\}$ and any stable matching $\mu_2 \in \{\mu \in \mathcal{M} : R_2 \in \mathcal{R}_\mu\}$, either μ_1 and μ_2 are incomparable or $\mu_1 \geqslant \mu_2$ in $(\mathcal{M}, \geqslant)$. This rotation poset $\mathcal{P} = (\mathcal{R}, \geq)$ is the auxiliary poset for the

---
[2] We say y is *covered* by x in a poset if $x \geqslant y$ and there is no element z such that $x \geqslant z \geqslant y$.

stable matching lattice $(\mathcal{M}, \geqslant)$; see Gusfield and Irving [9]. In particular, there is a bijection between stable matchings and *antichains* of the rotation poset.

For any stable matching $\mu = \{(b_1, g_1), (b_2, g_2), \ldots, (b_n, g_n)\}$ we define an auxiliary directed graph $H(\mu)$. This graph, which we call the *(exposed) rotation graph*, has a vertex i for each boy b_i. There is an arc from i to j if the next girl on b_i's list to prefer b_i over her current partner is g_j. If for some b_i, no such girl exists, then i has out-degree 0; otherwise it has out-degree 1. By definition, the rotations exposed in μ are exactly the cycles of $H(\mu)$. For example, if $\mu = \mathbf{1}$ then $H(\mathbf{1})$ consists of the set of rotations exposed in the boy-optimal stable matching. We call these the *maximal rotations*.

A rotation R exposed in μ is *minimal* if $\mu \otimes R = \mathbf{0}$. Equivalently, the *minimal rotations* are the set of rotations exposed in the girl-optimal stable matching $\mathbf{0}$ when ordering using the preferences of the girls rather than the boys.

3 Incentives in the Stable Matchings Problem

Intuitively, because the deferred-acceptance algorithm outputs the boy-optimal stable matching, there is no incentive for a boy not to propose to the girls in order of preference. This fact was formally proven by Dubins and Freedman [3]. On the other hand, because the stable matching is girl-pessimal, it can be beneficial for a girl to strategize. Indeed, Roth [17] showed that no stable matching mechanism exists that is incentive compatible for every participant.

3.1 The Minimum Winning Coalition of Girls

The structure of the stable matching lattice $\mathcal{L}$ is extremely useful in understanding the incentives that arise in the stable matching problem. For example, the following structure will be of importance in this paper. Let $F \subseteq G$ be a group of girls and let $\mathcal{M}_F$ be the collection of stable matchings where every girl in F is matched to their best stable-partner. Given the aforementioned properties of the join and meet operation in the stable matching lattice, it is easy to verify that $\mathcal{L}_F = (\mathcal{M}_F, \geqslant)$ is also a lattice. Thus, $\mathcal{L}_F$ has a supremum $\mathbf{1}_F$ which is the boy-optimal stable matching given that every girl in F is matched to their best stable-partner. Similarly, $\mathcal{L}_F$ has a infimum $\mathbf{0}_F$ which is the boy-pessimal stable matching given that every girl in F is matched to their best stable-partner. Observe that $\mathbf{0}_F$ is the girl-optimal stable-matching $\mathbf{0}$, for any subset F of the girls.

Why is this useful here? Well, imagine that each girl in F rejects anyone who is not their best stable-partner. Then the deferred-acceptance algorithm will output the stable matching $\mathbf{1}_F$; see also the works of Gale and Sotomayor [6] on strong equilibria and of Gonczarowski [7] on blacklisting. Of course, if $F = G$ then both $\mathbf{1}_G$ and $\mathbf{0}_G$ must match every girl to their optimal stable partner so $\mathbf{1}_G = \mathbf{0}_G = \mathbf{0}$.

We will call any $F \subseteq G$ such that $\mathbf{1}_F = \mathbf{0}$ a *winning coalition* and the smallest such group is called a *minimum winning coalition*. Winning coalitions can be found using the rotation poset.

Theorem 3.1. *A set of girls is a winning coalition if and only if it contains at least one girl from each minimal rotation in the rotation poset* $(\mathcal{R}, \geq)$

Theorem 3.1 allows us to find a minimum winning coalition.

Corollary 3.2. *The cardinality of the minimum winning coalition is equal to the cardinality of the set of minimal rotations in the rotation poset* $(\mathcal{R}, \geq)$.

3.2 Efficiency and Extremal Properties

From the structure inherent in Theorem 3.1 and Corollary 3.2 we can make several straight-forward deductions regarding winning coalitions.

First, Theorem 3.1 implies that we have a polynomial algorithm to verify winning coalitions. Likewise Corollary 3.2 implies that we have a polynomial time algorithm to compute the minimum winning coalition. In fact, the techniques of Gusfield [8] (see also [9]) can now be used to solve both problems in $O(n^2)$ time.

Second, we can upper bound the cardinality of the minimum winning coalition.

Lemma 3.3. *In any stable matching problem the minimum winning coalition has cardinality at most* $\lfloor \frac{n}{2} \rfloor$.

Can this upper bound on the cardinality of the minimum winning coalition ever be obtained? The answer is yes. In fact, every integer between 0 and $\lfloor \frac{n}{2} \rfloor$ can be the cardinality of the smallest winning coalition.

Theorem 3.4. *For each* $0 \leq k \leq \lfloor \frac{n}{2} \rfloor$ *there exists a stable matching instance where the minimum winning coalition has cardinality exactly* k.

We remark that the instances constructed in the proof of Theorem 3.4 have 2^k stable matchings. As k can be as large as $\lfloor \frac{n}{2} \rfloor$, this gives a simple proof of the well known fact that the number of stable matchings may be exponential in the number of participants [11].

We now have all the tools required to address the main questions in this paper.

4 The Random Matching Model

For the rest of the paper we use the *random matching model* which was first studied by Wilson [19] and subsequently examined in detail by Knuth, Pittel and coauthors [11,12,15,16]. Here the preference ranking of each boy and each girl is drawn uniformly and independently from the symmetric group $\mathbf{S_n}$. Specifically, each preference ranking is a random permutation of the set $[n] = \{1, 2, \ldots, n\}$.

We may now state the two main results of the paper. First, in the random matching model, the expected cardinality of the minimum winning coalition is $O(\log n)$.

Theorem 4.1. *In the random matching model, the expected cardinality of the minimum winning coalition F is* $\mathbb{E}(|F|) = \log(\sqrt{n}) + O(\log \log n)$

So the minimum winning coalition is small. Surprisingly, in sharp contrast, our second result states that a random coalition must contain nearly **every** girl if it is to form a winning coalition with high probability. Equivalently:

Theorem 4.2. *In the random matching model, $\forall \varepsilon > 0$, $\exists \delta(\varepsilon) > 0$ such that for a random coalition F of cardinality $(1 - \varepsilon) \cdot n$ the probability that F is* **not** *a winning coalition is at least $\delta(\varepsilon)$.*

To prove these results, recall Theorem 3.1 which states that a winning coalition F must intersect each *minimal rotation* in the rotation poset $(\mathcal{R}, \geq)$. Thus, for Theorem 4.1 it suffices to show that the expected number of minimal rotations is $O(\log n)$. To show Theorem 4.2 we must lower bound the probability that a randomly chosen coalition of girls contains at least one girl in each minimal rotation. Our approach is to show the likelihood of a small cardinality minimal rotation is high. In particular, we prove there is a minimal rotation containing exactly two girls with constant probability. It immediately follows that a random coalition must contain nearly all the girls if it is to be a winning coalition with high probability.

So our proofs require that we study the set of minimal rotations in the random matching model. The following two "tricks" will be useful in performing our analyses. First, instead of minimal rotations we may, in fact, study the set $\mathcal{R}^{\max}$ of *maximal rotations*, that is the rotations that are exposed at the boy-optimal stable matching $\mathbf{1}$. This is equivalent because Theorem 2.3 tells us that the inverse lattice $(\mathcal{M}, \leqslant)$ is the stable matching lattice ordered according to the preferences of the girls. This symmetry implies that the behaviour of minimal rotations is identical to the behaviour of maximal rotations as the maximal rotations of one lattice are the minimal rotations of the other. But why is the switch to maximal rotations from minimal rotations helpful? Simply put, as we are using the boy-proposal version of the deferred acceptance algorithm, we obtain the boy-optimal stable matching and, consequently, it is more convenient to reason about the rotations exposed at $\mathbf{1}$, that is, the maximal rotations.

Second, it will be convenient to view the deferred acceptance algorithm with random preferences in an alternative manner. In particular, instead of generating the preference rankings in advance, we may generate them dynamically. Specifically, when a boy b is selected to make a proposal he asks a girl g chosen uniformly at random. If b has already proposed to g then this proposal is immediately rejected; such a proposal is termed *redundant*. Meanwhile, g maintains a preference ranking only for the boys that have proposed to her. Thus if this is the kth distinct proposal made to girl g then she assigns to b a rank chosen uniformly at random among $\{1, \ldots k\}$. In particular, in the deferred acceptance algorithm g accepts the proposal with probability $1/k$. As explained by Knuth et al. [12], this process is equivalent to randomly generating the preference rankings independently in advance. Furthermore, recall from Theorem 2.2 that the deferred acceptance algorithm will output the boy-optimal stable matching regardless of the order of proposals. It follows that, for the purposes of analysis, we may assume the algorithm selects the unmatched boy with the lowest index to make the next proposal.

So our task now is to investigate the properties of maximal rotations, that is directed cycles in the rotation graph $H(\mathbf{1})$. Intuitively, this relates to the study of directed cycles in *random graphs* with out-degrees exactly one. But there is one major problem. In random graphs the choice of out-neighbour is independent for each vertex. But in the rotation graph $H(\mathbf{1})$ this independence is lost. In particular, the arcs in $H(\mathbf{1})$ share intricate dependencies and specifically depend on who made and who received each proposal in obtaining the boy-optimal stable matching $\mathbf{1}$. Moreover, a vertex may even have out-degree zero in $H(\mathbf{1})$. Essentially, the remainder of paper is devoted to showing that the myriad of dependencies that arise are collectively of small total consequence. It will then follow that the expected number or maximal rotations and the minimum cardinality of a maximal rotation both behave in a predictable manner, similar to that of directed cycles in random graphs with out-degrees exactly one. Namely, the expected number of cycles is close to $\frac{\log n}{2}$ and the existence a cycle of size two with constant probability [4].

Consequently, to study maximal rotations we must consider $H(\mathbf{1})$. We do this via a two-phase approach. In the *first phase* we calculate the boy-optimal stable matching $\mathbf{1}$, without loss of generality, $\mathbf{1} = \{(b_1, g_1), (b_2, g_2), \ldots, (b_n, g_n)\}$. This of course can be found by running the boy-proposal deferred acceptance algorithm. In the *second phase*, we calculate the rotation graph $H(\mathbf{1})$. But, as explained in Sect. 2.3, we can find the rotations by running the boy-proposal deferred acceptance algorithm longer.

In fact, to calculate (i) the expected number of maximal rotations and (ii) the probability that there is a maximum rotation of cardinality 2, we will not need the entire rotation graph $H(\mathbf{1})$ only subgraphs of it. Moreover, the subgraphs we require will be different in each case. Consequently, the second phases required to prove Theorem 4.1 and Theorem 4.2 will each be slightly different. These distinct second phases will be described in detail in Sect. 5 and Sect. 6, respectively. They both, however, share fundamental properties which will be exploited in shortening the subsequent proofs.

4.1 A Technical Tool for Counters

Before describing the two algorithms, we present a technical lemma that we will use repeated in analyzing the deviations that arise in their application. To formalize the lemma, we require the notion of a *state*. The state of the algorithm at any point is the record of all the (random) choices made so far: the sequence of proposals and the preference rankings generated by the girls. Thus we are working in the probability space (Ω, P) of all possible states Ω of the algorithm and the probabilities of reaching them.

We index the intermediate states of the algorithm by the number of proposals made to reach it. Let Ω_t denote the set of all possible states of the procedure after t proposals. A random variable X_t is *Ω_t-measurable* if X_t is determined by the algorithm state after t proposals, that is X is constant on each part of Ω_t. We say that a sequence $(X_t)_{t\geq 0}$ of random variables is a *counter* if X_t is Ω_t-measurable and $X_t - X_{t-1} \in \{0, 1\}$. Thus counters count the number of certain events occurring over the course of the algorithm. As an example, the number of successful proposals among the first t proposals is a counter.

Our main tool is Lemma 4.3 below which is used to control large deviations of counters. Let $B_{k,p}$ be a random variable which follows a binomial distribution with parameters k and p. We say that a collection of states $\mathcal{G}$ is *monotone* if for every state $S \notin \mathcal{G}$ we have $S' \notin \mathcal{G}$ for every state S' that can be reached from S. For example, the collection of states in which every girl received at most one proposal is monotone. Let $\{\mathcal{S}_t | t \in \mathbb{N}\}$ be the sequence of random variables corresponding to the state of the algorithm at time t

Lemma 4.3. *Let $\mathcal{G}$ be a monotone collection of states and let $(X_t)_{t\geq 0}$ be a counter. If $P(X_{t'+1} - X_{t'} = 1 | \mathcal{S}_{t'} = S_{t'}) \geq p$ for every state $S_{t'} \in \Omega_{t'} \cap \mathcal{G}$, for any $t' \in [t, t+k]$, then, for any $\lambda \geq 0$ and any $k \geq 1$,*

$$P\left((X_{t+k} - X_t \leq \lambda) \wedge (\mathcal{S}_{t+k-1} \in \mathcal{G}) | \mathcal{S}_t = S_t\right) \;\leq\; P\left(B_{k,p} \leq \lambda\right).$$

We also use a version of this lemma in which we give an upper bound to the probability that a counter is bounded below.

5 Minimum Winning Coalitions

In this section, we will evaluate the expected cardinality of the minimum winning coalition. Recall, it suffices is to find the expected number of directed cycles, $\mathcal{R}^{\max}$, in the rotation graph $H(\mathbf{1})$. To do this, it will be useful to describe the cardinality of $\mathcal{R}^{\max}$ in a more manipulable form. Specifically, for any boy b_i define a variable Z_i to be $\frac{1}{|R|}$ if b_i is in a maximal rotation R and 0 otherwise. Then we obtain that $|\mathcal{R}^{\max}| = \sum_{R\in\mathcal{R}^{\max}} 1 = \sum_{R\in\mathcal{R}^{\max}} \sum_{(b,g)\in R} \frac{1}{|R|} = \sum_{i=1}^{n} Z_i$.

By linearity of expectation, the expected cardinality of the minimum winning coalition F is $\mathbb{E}(|F|) = \mathbb{E}(|\mathcal{R}^{\max}|) = \mathbb{E}\left(\sum_{i=1}^{n} Z_i\right) = \sum_{i=1}^{n} \mathbb{E}(Z_i)$. This equality has an important consequence. Recall from Sect. 4 that the difficulty in computing $\mathbb{E}(|F|)$ is the myriad of dependencies that arise in the formation of the rotations in $\mathcal{R}^{\max}$. But we can now infer that, to quantify the dependency effects, rather than count expected rotations directly, it suffices to focus simply on computing $\mathbb{E}(Z_i)$.

5.1 Generating Maximal Rotations from the Rotation Graph

Ergo, our task now is to evaluate $\mathbb{E}(Z_i)$. For this we study a two-phase randomized algorithm, henceforth referred to as the *algorithm*, for generating the potential maximal rotation containing a given boy. The first phase computes the boy-optimal stable matching $\mathbf{1} = \{(b_1, g_1), (b_2, g_2), \ldots, (b_n, g_n)\}$. In the *second phase* we use a variation of the deferred acceptance algorithm to generate arcs in (a subgraph of) the rotation graph and generate a random variable Z.

The second phase starts with a randomly selected boy i_1 who makes uniformly random proposals until the first time he proposes to a girl g_j who prefers him over her partner b_j in the boy-optimal stable matching. The boy b_j will make the next sequence of proposals. The process terminates if we find a maximal rotation. Moreover, if this rotation is completed because girl g_{i_1} receives and accepts a proposal then we have found a maximal rotation containing boy i_1. In this case we also update Z. Formally, we initialize the second-phase by:

- Choose i_1 from $\{1, 2, \ldots, n\}$ uniformly at random.
- Initialize the potential cycle in the rotation digraph containing i_1 by setting $R = [i_1]$.

Once $R = [i_1, \ldots, i_k]$ is found, we generate the arc of the rotation digraph emanating from i_k, as follows.

- Let boy b_{i_k} make uniformly random proposals until the first time he proposes to a girl g_j such that g_j ranks b_{i_k} higher than b_j. That is, g_j ranks b_{i_k} higher than her pessimal stable partner.
  - If $j \notin R$ then we set $i_{k+1} = j$, $R = [i_1, \ldots, i_k, i_{k+1}]$, and recurse.
  - If $j \in R$ then we terminate the procedure. We set $Z = \frac{1}{|R|}$, if $j = i_1$, and $Z = 0$, otherwise.
- If, instead, boy b_{i_k} gets rejected by all the girls then the vertex i_k has no-outgoing arcs in the rotation graph. Thus, b_{i_1} belongs to no maximal rotation, so we terminate the procedure and set $Z = 0$.

We emphasize that as the second phase runs, we do not change any assigned partnerships. Specifically, when a girl receives a proposal we always compare her rank for the proposing boy to the rank of her pessimal partner, regardless of any other proposals she may have received during the second phase. Note $Z = Z_{i_1}$, where i_1 was chosen uniformly at random. The next lemma is then implied by noting that the expectation of Z is the average of the expectations of the Z_i.

Lemma 5.1. $\mathbb{E}(|F|) = n \cdot \mathbb{E}(Z)$ *where Z is the random variable generated by the algorithm.*

Recall, b_{i_1} is in a maximal rotation if and only if the rotation graph of the boy optimal stable matching has a cycle containing b_{i_1}. Observe that every connected component of a directed graph in which each vertex has out-degree 1 contains exactly one cycle. Hence, if we find a cycle in the same connected component as b_i but which does not contain him then b_{i_1} is not in a maximal rotation. Then, since $|F| = \sum_{j=1}^{n} Z_{i_j}$, we get $\mathbb{E}(|F|) = \sum_{j=1}^{n} \mathbb{E}(Z_{i_j}) = n \cdot \mathbb{E}(Z)$.

5.2 Properties of the Two-Phase Algorithm

We now present a series of properties that arise with high enough probability during the two-phase process. In particular, the process does not deviate too far from its expected behaviour. For example, the running time of each phase is not much longer than expected, no girl receives too many proposals, and no boy makes too many proposals. To formalize this, let T_1 and T_2 be the number of proposals made in the first and second phases, respectively, and let $T = T_1 + T_2$. Further, let a *run* be a sequence of consecutive proposals made by the same boy in the same phase. Now consider the following properties that may apply to a state:

I. The algorithm has not terminated.
II. If the algorithm is in the first phase then $t \leq 5n \log n$. If the algorithm is in the second phase then $T_1 \leq 5n \log n$
III. If the algorithm has not found a rotation yet then $t \leq T_1 + \sqrt{n} \log^3 n$.
IV. Each girl has received at most $21 \log n$ proposals.
V. Each boy started at most $21 \log n$ runs.
VI. Each run contained at most $111 \log^2 n$ proposals.
VII. Each boy has made at most $\log^4 n$ proposals.

Let $\mathcal{G}$ be the set of all states that satisfy properties I to VII. We call these *good* states. Any state that is not good is *bad*. Let G_* denote the event that the algorithm is in a good state the step before it terminates. Let $\overline{G_*}$ be the complement of G_*.

We remark that, for technical reasons, we will assume the second-phase terminates if $n \log n$ proposals are made during that phase. This assumption is superfluous here by conditon III, which states the second phase has at most $\sqrt{n} \log^3 n$ proposals. However, the assumption is useful as it will allow the following lemma to also apply for the modified second-phase algorithm that we use in Sect. 6.

Lemma 5.2. *For n sufficiently large,* $P(G_*) \geq 1 - O(n^{-4})$.

So, we are in a good state the period before the algorithm terminates with high probability. It follows that the magnitude of the expected number of maximal rotations can be evaluated by consideration of good states.

Now, to calculate the expected number of maximal rotations we must analyze in more detail the second phase of the algorithm. In particular, this section is devoted to the proof of the following lemma.

Lemma 5.3. *Let* S_* *be the terminal state of the first phase.*
If $P\left(\overline{G_*}|S_*\right) \leq \frac{1}{n^3}$ *then*

$$\mathbb{E}\left(Z|S_*\right) = \frac{\log n}{2n} + O\left(\frac{\log \log n}{n}\right).$$

We remark that that our first main result, Theorem 4.1, readily follows from Lemma 5.3 via Lemmas 5.1 and 5.2. It is also worth noting that III implies that the second phase has at most $\sqrt{n} \log^3 n$ proposals when G_* occurs, due to the fact we stop once we find our first cycle.

6 Random Winning Coalitions

In this section, we consider the case where the girls in the coalition are themselves randomly selected. Our task now is to prove that almost every girl must be selected if we wish to obtain a winning coalition with high probability To do this, it will suffice to prove that there is a maximal rotation of cardinality two with constant probability.

6.1 Generating Maximal Rotations from the Rotation Graph

Let Z' be a random variable counting the number of maximal rotations of cardinality two. Again, to analyze Z' we use a two-phase algorithm. The *first phase* is the same as before. We simply generate the boy-optimal stable matching $\mathbf{1} = \{(b_1, g_1), (b_2, g_2), \ldots, (b_n, g_n)\}$. But the *second phase* is slightly different. Previously we had to evaluate the expected number of maximal rotations and, to achieve that, it sufficed to end the second phase once we had found one rotation. Now, because we are interested in maximal rotations or cardinality two we will extend the second phase and terminate only when and if we find rotation of cardinality two.

So now in the second phase we use the following algorithm to generate the random variable Z', initialized at 0:

- Choose i_1 from $\{1, 2, \ldots, n\}$ uniformly at random.
- Initialize the set of indices of boys who have made proposals in the second phase with $\mathscr{I} = \{i_1\}$.
- Set $\texttt{tar} = \infty$.

For motivation, at any step, girl $g_{\texttt{tar}}$ can be viewed as the target girl. If she accepts the next proposal then this will complete a rotation of cardinality two. Observe that we intitialize $\texttt{tar} = \infty$ as it is impossible to complete a rotation in the fist step.

To complete the description of the second-phase, assume we have $\mathscr{I} = \{i_1, \ldots, i_k\}$. If $k < \frac{n}{2}$ *and* less than $n \log n$ proposals in total have been made then we generate the next arc of the rotation digraph starting at i_k, as follows:

- Let boy b_{i_k} make uniformly random proposals until the first time he proposes to a girl g_j such that g_j ranks b_{i_k} higher than b_j.
  - If $j = \texttt{tar}$ then increment Z' by 1. Recurse.
  - If $j \in \mathscr{I} \setminus \{\texttt{tar}\}$ then pick i_{k+1} from $\{1, 2, \ldots, n\} \setminus \mathscr{I}$ uniformly at random. Set $\mathscr{I} = \{i_1, \ldots, i_k, i_{k+1}\}$ and $\texttt{tar} = \infty$. Recurse.
  - If $j \notin \mathscr{I}$ then set $i_{k+1} = j$, $\texttt{tar} = i_k$, $\mathscr{I} = \{i_1, \ldots, i_k, i_{k+1}\}$. Recurse.
- If, instead, boy b_{i_k} gets rejected by all the girls then return $Z' = 0$

Lemma 6.1. *The probability of the existence of a maximal rotation of size two is lower bounded by* $P(Z' \geq 1)$.

Therefore, our aim is to prove that $P(Z' \geq 1) = \Omega(1)$, where Z' is the random variable generated by the algorithm.

6.2 Bounding the Probability of Missing a Rotation

Our objective now is to show that the behaviour of this new two-phase algorithm does not deviate too much from its expected behaviour. Specifically, we show it satisfies a series of properties with sufficiently high probability. Properties I to VII as well as T_1 and T_2 are as defined in Sect. 5. But now we require several more properties. To describe these, let p_{S_t} denote the probability of the next proposal being accepted when in state $\mathcal{S}_t$. We are interested in the states satisfying the following properties:

VIII. $t \geq \frac{1}{2} n \log n$
IX. No more than $n^{\frac{9}{10}}$ girls have received less than $\frac{1}{4} \log n$ proposals.
X. No more than $\sqrt{n}$ girls have received a redundant proposal.
XI. $\{p_\tau | T_1 \leq \tau \leq t\} \subseteq \left[\frac{1}{22 \log n}, \frac{5}{\log n}\right]$
XII. $T_2 \geq \frac{1}{20} n \log n$

Lemma 6.2. *For n sufficiently large, the algorithm in a good state satisfying these conditions the period before it terminates with high probability.*

We can complete the proof of our second main result in two steps. First, we show that the probability of a maximal rotation of cardinality two existing is at least a constant, namely $P(Z' \geq 1) = \Omega(1)$. The second step is then easy. If there is a maximal rotation of cardinality two then a random coalition of cardinality at most $(1 - \epsilon) \cdot n$ will not be a winning coalition with constant probability.

7 Conclusion

We have evaluated the expected cardinality of the minimum winning coalition. We believe this result is of theoretical interest and that the techniques applied may have broader applications for stable matching problems. In terms of practical value it is worth discussing the assumptions inherent in the model. The assumption of uniform and independent random preferences, while ubiquitous in the theoretical literature, is somewhat unrealistic in real-world stable matching instances. Furthermore, as presented, the model assumes full information, which is clearly not realistic in practice. However, to implement the behavioural strategy presented in this paper, the assumption of full information is **not** required. It needs only that a girl has a good approximation of the rank of her best stable partner. But, by the results of Pittel [15], she does know this with high probability. Consequently, a near-optimal implementation of her behavioural strategy requires knowledge only of her own preference list! This allows for a risk-free method to output a matching close in the lattice to the girl-optimal stable matching. Similarly, as discussed, although our presentation has been in terms of a coalition of girls, each girl is able to implement a near-optimal behavioural strategy independent of who the other girls in the coalition may be or what their preferences are.

Acknowledgements. We would like to thank the various reviewers whose comments have helped us improve this paper.

References

1. Ashlagi, I., Kanoria, Y., Leshno, J.: Unbalanced random matching markets: the stark effect of competition. J. Polit. Econ. **125**(1), 69–98 (2017)
2. Birkhoff, G.: Rings of sets. Duke Math. J. **3**(3), 443–454 (1937)
3. Dubins, L., Freedman, D.: Machiavelli and the Gale-Shapley algorithm. Amer. Math. Mon. **88**(7), 485–494 (1981)
4. Flajolet, P., Odlyzko, A.M.: Random mapping statistics. In: Quisquater, J.J., Vandewalle, J. (eds.) Advances in Cryptology – EUROCRYPT '89, pp. 329–354. Springer, Berlin Heidelberg, Berlin, Heidelberg (1990). https://doi.org/10.1007/3-540-46885-4_34
5. Gale, D., Shapley, L.: College admissions and the stability of marriage. Amer. Math. Mon. **69**(1), 9–15 (1962)
6. Gale, D., Sotomayor, M.: Ms. Machiavelli and the stable matching problem. Am. Math. Mon. **92**(4), 261–268 (1985)
7. Gonczarowski, Y.: Manipulation of stable matchings using minimal blacklists. In: Proceedings of the Fifteenth ACM Conference on Economics and Computation, EC 2014, Association for Computing Machinery, p. 449 (2014)
8. Gusfield, D.: Three fast algorithms for four problems in stable marriage. SIAM J. Comput. **16**(1), 111–128 (1987)
9. Gusfield, D., Irving, R.: The Stable Marriage Problem: Structure and Algorithms. MIT Press, Cambridge (1989)
10. Irving, R., Leather, P.: The complexity of counting stable marriages. SIAM J. Comput. **15**(3), 655–667 (1986)
11. Knuth, D.: Mariages stables et leurs relations avec d'autres problèmes combinatoires. Les Presses de l'Université de Montréal (1982)
12. Knuth, D., Motwani, R., Pittel, B.: Stable husbands. Random Struct. Algorithms **1**(1), 1–14 (1990)
13. Kupfer, R.: The instability of stable matchings: the influence of one strategic agent on the matching market. In: Proceedings of the 16th Conference on Web and Internet Economics (WINE) (2020)
14. McVitie, D., Wilson, L.: The stable marriage problem. Commun. ACM **14**, 486–490 (1971)
15. Pittel, B.: The average number of stable matchings. SIAM J. Discret. Math. **2**(4), 530–549 (1989)
16. Pittel, B.: On likely solutions of a stable matching problem. Ann. Appl. Probab. **2**(2), 358–501 (1992)
17. Roth, A.: The economics of matching: stability and incentives. Math. Oper. Res. **7**(4), 617–628 (1982)
18. Stanley, R.: Enumerative Combinatorics, Volume I. Cambridge University Press (1997)
19. Wilson, L.: An analysis of the stable marriage assignment algorithm. BIT Numer. Math. **12**, 569–575 (1972)

References

Social Choice and Cooperative Games

Metric-Distortion Bounds Under Limited Information

Ioannis Anagnostides(✉), Dimitris Fotakis, and Panagiotis Patsilinakos

National Technical University, Athens, Greece

Abstract. In this work we study the metric distortion problem in voting theory under a limited amount of ordinal information. Our primary contribution is threefold. First, we consider mechanisms which perform a sequence of pairwise comparisons between candidates. We show that a widely-popular deterministic mechanism employed in most knockout phases yields distortion $\mathcal{O}(\log m)$ while eliciting only $m-1$ out of $\Theta(m^2)$ possible pairwise comparisons, where m represents the number of candidates. We also provide a matching lower bound on its distortion. In contrast, any mechanism which performs fewer than $m - 1$ pairwise comparisons has unbounded distortion. Moreover, we study the power of deterministic mechanisms under incomplete rankings. Most notably, when every agent provides her k-top preferences we show an upper bound of $6m/k + 1$ on the distortion, for any $k \in \{1, 2, \ldots, m\}$, substantially improving over the previous bound of $12m/k$ recently established by Kempe [25,26]. Finally, we are concerned with the sample complexity required to ensure near-optimal distortion with high probability. Our main contribution is to show that a random sample of $\Theta(m/\epsilon^2)$ voters suffices to guarantee distortion $3 + \epsilon$ with high probability, for any sufficiently small $\epsilon > 0$. This result is based on analyzing the sensitivity of the deterministic mechanism introduced by Gkatzelis, Halpern, and Shah [22].

Keywords: Social choice · Metric distortion · Pairwise comparisons · Incomplete rankings · Sampling

1 Introduction

Aggregating the preferences of individuals into a collective decision lies at the heart of social choice. According to the classic theory of Von Neumann and Morgenstern [31] individual preferences are captured through a *utility function*, which assigns numerical (or *cardinal*) values to each alternative. Yet, in voting theory, as well as in most practical applications, mechanisms typically elicit only *ordinal* information from the voters, indicating an order of preferences over the candidates. Although this might seem at odds with a utilitarian framework, it has been recognized that it might be hard for a voter to specify a precise numerical value for an alternative, and providing only ordinal information substantially reduces the cognitive burden. This begs the question: What is the loss

I. Caragiannis and K. A. Hansen (Eds.): SAGT 2021, LNCS 12885, pp. 299–313, 2021.
https://doi.org/10.1007/978-3-030-85947-3_20

in efficiency of a mechanism extracting only ordinal information with respect to the *utilitarian social welfare*, i.e. the sum of individual utilities over a chosen candidate? The framework of *distortion* introduced by Procaccia and Rosenschein [32] measures exactly this loss, and has received considerable attention in recent years.

As it turns out, the guarantees we can hope for crucially depend on the assumptions we make on the utility functions. For example, in the absence of any structure Procaccia and Rosenschein observed that *no* ordinal mechanism can obtain bounded distortion [32]. In this work we focus on the *metric distortion* framework, introduced by Anshelevich et al. [3], wherein voters and candidates are thought of as points in some arbitrary metric space; this is akin to models in spatial voting theory [17]. In this context, the voters' preferences are measured by means of their "proximity" from each candidate, and the goal is to output a candidate who (approximately) minimizes the *social cost*, i.e. the cumulative distances from the voters.

A common assumption made in this line of work is that the algorithm has access to the entire total rankings of the voters. However, there are many practical scenarios in which it might be desirable to truncate the ordinal information elicited by the mechanism. For example, requesting only the top preferences could further relieve the cognitive burden since it might be hard for a voter to compare alternatives which lie on the bottom of her list (for additional motivation for considering incomplete or partial orderings see [16,21], and references therein), while any truncation in the elicited information would also translate to more efficient communication. These reasons have driven several authors to study the decay of distortion under missing information [5,18,24,26], potentially allowing some randomization (see our related work subsection). In this work we follow this line of research, offering several new insights and improved bounds over prior results.

1.1 Overview of Results

First, we study voting rules which perform a sequence of pairwise comparisons between two candidates, with the result of each comparison being determined by the majority rule over the entire population of voters. This class includes many common mechanisms such as Copeland's rule [33], and has received considerable attention in the literature of social choice; cf., see [27], and references therein. Within the framework of (metric) distortion the following fundamental question arises:

> *How many pairwise comparisons between two candidates are needed to guarantee non-trivial bounds on the distortion?*

For example, Copeland's rule elicits all possible pairwise comparisons, i.e. $\binom{m}{2} = \Theta(m^2)$, and guarantees distortion at most 5 [3]. Thus, it is natural to ask whether we can substantially truncate the number of elicited pairwise comparisons without sacrificing too much the efficiency of the mechanism. In this context, we provide the following strong positive result:

Theorem 1. *There exists a deterministic mechanism which elicits only* $m - 1$ *pairwise comparisons and guarantees distortion* $\mathcal{O}(\log m)$.

Our mechanism is particularly simple and natural: In every round we arbitrarily pair the remaining candidates and we only extract the corresponding comparisons. Next, we eliminate all the candidates who lost and we continue recursively until a single candidate emerges victorious. Interestingly, this mechanism is widely employed in practical applications, for example in the knockout phases of most competitions. The main technical ingredient of the analysis is a powerful lemma developed by Kempe via an LP duality argument [25]. Specifically, Kempe characterized the social cost ratio between two candidates when there exists a sequence of intermediate alternatives such that every candidate in the chain pairwise-defeats the next one. We also supplement our analysis for this mechanism with a matching lower bound on a carefully constructed instance. Moreover, we show that any mechanism which performs (strictly) fewer than $m-1$ pairwise comparisons has *unbounded* distortion. This limitation applies even if we allow randomization either during the elicitation or the winner-determination phase.

Next, we study deterministic mechanisms which only receive an incomplete order of preferences from every voter, instead of the entire rankings as it is usually assumed. This setting has already received attention in the literature, most notably by Kempe [26], and has numerous applications in real-life voting systems. Arguably the most important such consideration arises when every voter provides her k-top preferences, for some parameter $k \in [m]$. Kempe showed [26] that there exists a deterministic mechanism which elicits only the k-top preferences and whose distortion is upper-bounded by $79\,m/k$; using a tool developed in [25] this bound can be improved all the way down to $12\,m/k$. However, this still leaves a substantial gap with respect to the best-known lower bound, which is $2\,m/k$ if we ignore some additive constant factors. Thus, Kempe [26] left as an open question whether the aforementioned upper bound can be improved. In our work we make substantial progress towards bridging this gap, proving the following:

Theorem 2. *There exists a deterministic mechanism which only elicits the* k*-top preferences and yields distortion at most* $6m/k + 1$.

We should stress that the constant factors are of particular importance in this framework; indeed, closing the gap even for the special case of $k = m$ has received intense scrutiny in recent years [3,22,25,30]. From a technical standpoint the main technique for proving such upper bounds consists of identifying a candidate for which there exists a path to any other node such that every candidate in the path pairwise-defeats the next one by a sufficiently large margin (which depends on k). Importantly, the derived upper bound crucially depends on the length of the path. Our main technical contribution is to show that there always exists a path of length 2 with the aforedescribed property, while the previous best result by Kempe established the claim only for paths of length 3.

We also provide some other interesting bounds for deterministic mechanisms under missing information. Most notably, if the voting rule performs well on an arbitrary (potentially adversarially selected) subset of the voters can we quantify its distortion over the entire population? We answer this question with a sharp upper bound in Theorem 5. In fact, we use this result as a tool for some of our other proofs, but nonetheless we consider it to be of independent interest.

Finally, we consider mechanisms which receive information from only a "small" *random sample* of voters; that is, we are concerned with the *sample complexity* required to ensure efficiency, which boils down to the following fundamental question:

How large should the size of the sample be in order to guarantee near-optimal distortion with high probability?

More precisely, we are interested in deriving sample-complexity bounds which are *independent* on the number of voters n. This endeavor is particularly motivated given that in most applications $n \gg m$. Naturally, sampling approximations are particularly standard in the literature of social choice. Indeed, in many scenarios one wishes to predict the outcome of an election based on small sample (e.g. in polls or exit polls), while in many other applications it is considered even infeasible to elicit the entire input (e.g. in online surveys). In this context, we will be content with obtaining near-optimal distortion *with high probability* (e.g. 99%). This immediately deviates from the line of research studying randomized mechanisms (cf. see [5]) wherein it suffices to obtain a guarantee in expectation. We point out that it has been well-understood that a guarantee only in expectation might be insufficient in many cases (e.g. see [18]). In this context, we analyze the sample complexity of PLURALITYMATCHING, the mechanism of Gkatzelis et al. [22] which recovers the optimal distortion bound of 3 (among deterministic mechanisms), establishing the following result:

Theorem 3. *For any sufficiently small $\epsilon > 0$ there exists a mechanism which takes a sample of size $\Theta(m/\epsilon^2)$ voters and yields distortion at most $3 + \epsilon$ with probability* 0.99.

More precisely, the main ingredient of PLURALITYMATCHING is a maximum-matching subroutine for a certain bipartite graph. Our first observation is that the size of the maximum matching can be determined through a much smaller graph which satisfies a "proportionality" condition with respect to a maximum-matching decomposition. Although this condition cannot be explicitly met since the algorithm is agnostic to the decomposition, our observation is that sampling (with sufficiently many samples) will approximately satisfy this requirement, eventually leading to the desired conclusion. It should be noted that our approach is *distribution-independent*.

To conclude, we provide several experimental findings in real-life voting applications from the standpoint of the (metric) distortion framework. We are mostly concerned with comparing the results of the scoring systems used in practice against a mechanism which explicitly attempts to minimize the distortion; the latter is realized with the linear programming mechanism of Goel et al. [23]. Specifically, we analyze the efficiency of the scoring rule used in the Eurovision song contest. Interestingly, we find that the winner in the actual competition is the candidate who minimizes the distortion. Our implementation is publicly available at https://github.com/ioannisAnagno/Voting-MetricDistortion.

We should remark that due to space constraints most of our proofs, as well as some additional results are presented in the full version of our paper [2].

1.2 Related Work

Research in the metric distortion framework was initiated by Anshelevich et al. [3]. Specifically, they analyzed the distortion of several common voting rules, most notably establishing that Copeland's rule has distortion at most 5, with the bound being tight for certain instances. They also conjectured that the *ranked pairs* mechanism always achieves distortion at most 3, which is also the lower bound for any deterministic mechanism. This conjecture was disproved by Goel et al. [23], while they also studied *fairness* properties of certain voting rules. The barrier of 5 set out by Copeland was broken by Munagala and Wang [30], presenting a novel deterministic rule with distortion $2+\sqrt{5}$. The same bound was obtained by Kempe [25] through an LP duality framework, who also articulated sufficient conditions for proving the existence of a deterministic mechanism with distortion 3. This conjecture was only recently confirmed by Gkatzelis et al. [22], introducing the *plurality matching* mechanism. Closely related to our study is also the work of Gross et al. [24], wherein the authors provide a near-optimal mechanism which only asks $m+1$ voters for their top-ranked alternatives. One of the main differences with our setting is that we require an efficiency-guarantee with high probability, and not in expectation.

Broader Context. Beyond the metric case most focus has been on analyzing distortion under a *unit-sum* assumption on the utility function. In particular, Boutilier et al. [11] provide several upper and lower bounds, while they also study learning-theoretic aspects under the premise that every agent's utility is drawn from a distribution. Moreover, several multi-winner extensions have been studied in the literature. Caragiannis et al. [15] studied the *committee selection* problem, which consists of selecting k alternatives that maximize the social welfare, assuming that the value of each agent is defined as the maximum value derived from the committee's members. We also refer to [8] for the *participatory budgeting* problem, and to [9] when the output of the mechanism should be a total order over alternatives (instead of a single winner). The trade-off between efficiency and communication has been addressed in [28,29] (see also [1]). We should also note a series of works analyzing the power of ordinal preferences for some fundamental graph-theoretic problems [6,7,19]. Finally, we point out that strategic issues are typically ignored within this line of work. We will also posit that agents provide truthfully their preferences, but we refer to [10,14] for rigorous considerations on the strategic issues that arise. We refer the interested reader to the survey of Anshelevich et al. [4], as we have certainly not exhausted the literature.

2 Preliminaries

A *metric space* is a pair $(\mathcal{M}, d)$, where $d : \mathcal{M} \times \mathcal{M} \mapsto \mathbb{R}$ constitutes a *metric* on $\mathcal{M}$, i.e., (i) $\forall x, y \in \mathcal{M}, d(x, y) = 0 \iff x = y$ (identity of indiscernibles), (ii) $\forall x, y \in d(x, y) = d(y, x)$ (symmetry), and (iii) $\forall x, y, z \in \mathcal{M}, d(x, y) \leq d(x, z) + d(z, y)$ (triangle inequality). Consider a set of n voters $V = \{1, 2, \ldots, n\}$ and a set of m candidates $C = \{a, b, \ldots,\}$; candidates will be typically represented with lowercase letters such as a, b, w, x. We assume that every voter $i \in V$ is associated with a point $v_i \in \mathcal{M}$, and every candidate $a \in C$ to a point $c_a \in \mathcal{M}$. Our goal is to select some candidate x in order to minimize the *social cost*: $\text{SC}(x) = \sum_{i=1}^{n} d(v_i, c_x)$. This task would be trivial if we had access to the agents' distances from all the candidates. However, in the standard *metric distortion* framework, every agent i provides only a *ranking* (a total order) σ_i over the points in C according to the *order* of i's distances from the candidates. We assume that ties are broken arbitrarily, subject to transitivity.

In this work we are considering a substantially more general setting, wherein every agent provides a subset of σ_i. More precisely, we assume that agent i provides as input a set $\mathcal{P}_i$ of ordered pairs of distinct candidates, such that $(a, b) \in \mathcal{P}_i \implies a \succ_i b$, where $a, b \in C$; it will always be assumed that $\mathcal{P}_i$ corresponds to the *transitive closure* of the input. We will allow $\mathcal{P}_i$ to be the empty set, in which case i does not provide any information to the mechanism; with a slight abuse of notation we will let $\mathcal{P}_i \equiv \sigma_i$ when i provides the entire order of preferences. We will say that the input $\mathcal{P} = (\mathcal{P}_1, \ldots, \mathcal{P}_n)$ is consistent with the metric d if $(a, b) \in \mathcal{P}_i \implies d(v_i, c_a) \leq d(v_i, c_b), \forall i \in V$, and this will be denoted with $d \rhd \mathcal{P}$. We will also represent with $\text{top}(i)$ i's most preferred candidates.

A *deterministic social choice rule* is a function that maps an *election* in the form of a 3-tuple $\mathcal{E} = (V, C, \mathcal{P})$ to a single candidate $a \in C$. We will measure the performance of f for a given input of preferences $\mathcal{P}$ in terms of its *distortion*, namely, the worst-case approximation ratio it provides with respect to the social cost:

$$\text{distortion}(f; \mathcal{P}) = \sup \frac{\text{SC}(f(\mathcal{P}))}{\min_{a \in C} \text{SC}(a)}, \tag{1}$$

where the supremum is taken over all metrics such that $d \rhd \mathcal{P}$. The distortion of a social choice rule f is the maximum of $\text{distortion}(f; \mathcal{P})$ over all possible input preferences $\mathcal{P}$. In other words, once the mechanism selects a candidate (or a distribution over candidates if the social choice rule is *randomized*) an adversary can select any metric space subject to being consistent with the input preferences. Similarly, in Sect. 3 where we study mechanisms performing pairwise comparisons, the adversary can select any metric space consistent with the elicited comparisons.

3 Sequence of Pairwise Comparisons

Consider the tournament graph $T = (C, E)$ where $(a, b) \in E$ if and only if candidate a pairwise-defeats candidate b; it will be tacitly assumed—without any loss of generality—that ties are broken arbitrarily so that T is indeed a tournament. In this section we study mechanisms which elicit edges from T, and we are interested in establishing a trade-off between the number of elicited edges and the distortion of the mechanism. We commence with the following lower bound:

Proposition 1. *There are instances for which any deterministic mechanism which elicits (strictly) fewer than $m-1$ edges from T has unbounded distortion.*

In fact, the same limitation applies even if we allow randomization, either during the elicitation or the winner-determination phase. Importantly, we will show that $m-1$ edges from T suffice to obtain near-optimal distortion. To this end, we will employ a powerful technical lemma by Kempe [25], proved via an LP-duality argument.

Lemma 1. ([25]). *Let $a_1, a_2, \dots a_\ell$ be a sequence of distinct candidates such that for every $i = 2, \dots, \ell$ at least half of the agents prefer candidate a_{i-1} over candidate a_i. Then, $\mathrm{SC}(a_1) \leq (2\ell - 1)\,\mathrm{SC}(a_\ell)$.*

Armed with this important lemma we introduce the DominationRoot mechanism, which operates with access only to a pairwise comparison oracle; namely, $\mathfrak{O}$ takes as input two distinct candidates $a, b \in C$ and returns the *losing* candidate based on the voters' preferences.

Mechanism 1: DominationRoot

Input: Set of candidates C, Pairwise comparison oracle $\mathfrak{O}$;
Output: Winner $w \in C$;
1. Initialize $S := C$;
2. Construct arbitrarily a set Π of $\lfloor S/2 \rfloor$ pairings from S;
3. For every $\{a, b\} \in \Pi$ remove $\mathfrak{O}(a, b)$ from S;
4. If $|S| = 1$ **return** $w \in S$; otherwise, continue from step 2;

The performance of the DominationRoot mechanism can be understood using Lemma 1, leading to the following conclusion:

Theorem 4. DominationRoot *elicits only $m-1$ edges from T and guarantees distortion at most $2\lceil \log m \rceil + 1$.*

This theorem along with Proposition 1 imply a remarkable gap depending on whether the mechanism is able to elicit at least $m-1$ pairwise comparisons. We also provide a matching lower bound for DominationRoot:

Proposition 2. *There exist instances for which* DominationRoot *yields distortion at least $2 \log m + 1$.*

This lower bound is shown in two steps. First, we prove that Lemma 1 is tight; the pattern of this construction is illustrated in Fig. 1a. Then, we devise an instance and a sequence of pairing (Fig. 1b) for which Proposition 2 reduces to the tightness of Lemma 1.

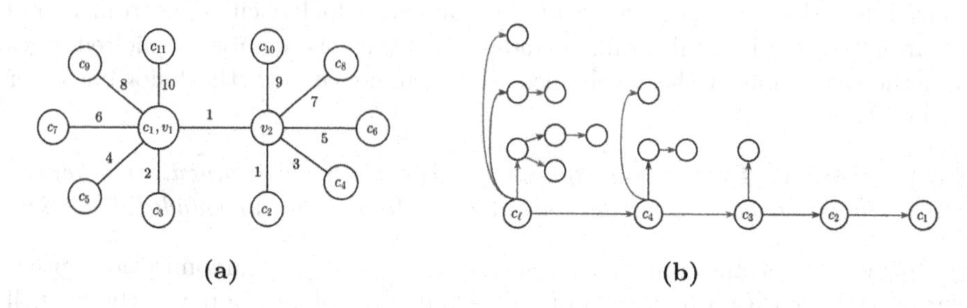

Fig. 1. (a) A metric embedding of voters and candidates establishing that Lemma 1 is tight. (b) A sequence of pairings such that c_ℓ emerges victorious. We have highlighted with different colors pairings that correspond to different rounds.

4 Distortion of Deterministic Rules Under Incomplete Rankings

In this section we study the performance of deterministic voting rules under incomplete rankings. We commence this section with another useful lemma by Kempe [25].

Lemma 2. ([25]). *Consider three distinct candidates $w, y, x \in C$ so that at least $\alpha \cdot n$ voters prefer w over y, and y over x. Then,*

$$\frac{\text{SC}(w)}{\text{SC}(x)} \leq \frac{2}{\alpha} + 1. \tag{2}$$

As a warm-up, we first employ this lemma to characterize the distortion when for all pairs of candidates at least a small fraction of voters has provided their pairwise preferences.

Proposition 3. *Consider an election $\mathcal{E} = (V, C, \mathcal{P})$ such that for every pair of distinct candidates $a, b \in C$ it holds that $\sum_{i=1}^{n} \mathbb{1}\{(a,b) \in \mathcal{P}_i \vee (b,a) \in \mathcal{P}_i\} \geq \alpha \cdot n$. Then, there exists a voting rule which obtains distortion at most $4/\alpha + 1$.*

We should remark that this upper bound is tight (up to constant factors), at least for certain instances. Interestingly, Proposition 3 suggests one possible preference elicitation strategy: collect the information about the preferences in a "balanced" manner.

4.1 Missing Voters

Next, consider a mechanism which has access to the votes of only a subset $V \setminus Q$ of voters, where $Q \subset V$ is the set of *missing voters* such that $|Q| = \epsilon \cdot n$. If the mechanism performs well on $V \setminus Q$ can we characterize the distortion over the entire set of voters as ϵ increases? Observe that this setting is tantamount to $\mathcal{P}_i = \emptyset$ for all $i \in Q$. In the following theorem we provide a sharp bound:

Theorem 5. *Consider a mechanism with distortion at most ℓ w.r.t. an arbitrary subset with $(1 - \epsilon)$ fraction of all the voters, for some $\epsilon \in (0, 1)$. Then, the distortion of the mechanism w.r.t. the entire population is upper-bounded by*

$$\ell + \frac{\epsilon}{1 - \epsilon}(\ell + 1). \tag{3}$$

The proof of this theorem uses some standard techniques to identify and characterize the worst-case scenario. It should also be noted that the derived bound is tight, for example in the presence of 2 candidates. In the sequel, we will use this bound as a tool for establishing some of our results.

4.2 Top Preferences

In this subsection we investigate how the distortion increases when every voter provides only her k-top preferences, for some parameter $k \in [m]$. It should be noted that the two extreme cases are well understood. Specifically, when $k = m$ the mechanism has access to the entire rankings and we know that any deterministic mechanism has distortion at least 3, which is also the upper bound established in [22]. On the other end of the spectrum, when $k = 1$ the plurality rule—which incidentally is the optimal deterministic mechanism when only the top preference is given—yields distortion at most $2m - 1$ [3]. Consequently, the question is to quantify the decay of distortion as we gradually increase k. In this context, we should point out that Kempe [26] obtained the lower bound of $2m/k$ (ignoring some additive constant factors) for any deterministic mechanism which elicits only the k-top preferences. In the following theorem we come closer to matching this lower bound.

Theorem 6. *There exists a deterministic mechanism which elicits only the k-top preferences from every voter out of m candidates and has distortion at most $6m/k + 1$.*

For the proof of this theorem we analyze the directed graph $\widehat{G} = (C, \widehat{E})$, where $(a, b) \in \widehat{E}$ if and only if at least a fraction of $k/(3m)$ voters prefer a over b. In particular, we manage to show that this graph always admits a *king* vertex, and then our claim follows by Lemma 2. Notably, this implies that if $k = \gamma \cdot m$ for some $\gamma \in (0, 1)$, the distortion is at most $6/\gamma + 1$. Our analysis substantially improves over the previous best-known bound which was $12m/k$ [25,26], but nonetheless there is still a gap between the aforementioned lower bound. Before we conclude this section we elaborate on how one can further improve upon our the bound of Theorem 6.

Conjecture 1. If we assume that every agent provides her k-top preferences for some $k \in [m]$, there is a candidate $a \in C$ and a subset $S \subseteq V$ such that

(i) There exists a perfect matching $M : S \mapsto S$ in the integral domination graph of a (see Definition 1 in the next section);
(ii) $|S| \geq n \times k/m$.

When $k = m$ this conjecture was shown to be true by Gkatzelis et al. [22]. On the other end of the spectrum, when $k = 1$ it is easy to verify that the plurality winner establishes this conjecture. In this context, we observe that if this conjecture holds we would immediately obtain a substantial improvement over the bound of Theorem 6.

Proposition 4. *If Conjecture 1 holds, then there exists a deterministic mechanism which elicits only the k-top preferences and yields distortion at most* $4m/k - 1$.

5 Randomized Preference Elicitation and Sampling

Previously we characterized the distortion when only a deterministically (and potentially adversarially) selected subset of voters has provided information to the mechanism. This raises the question of bounding the distortion when the mechanism elicits information from only a small *random sample* of voters. Here a single sample corresponds to the *entire* ranking of a voter. We stress that randomization is only allowed during the preference elicitation process; for any given random sample as input the mechanism has to select a candidate *deterministically.* We commence this section with a simple lower bound, which essentially follows from a standard result by Canetti et al. [12].

Proposition 5. *Any mechanism which yields distortion at most* $3+\epsilon$ *with probability at least* $1 - \delta$ *requires* $\Omega(\log(1/\delta)/\epsilon^2)$ *samples, even for* $m = 2$.

5.1 Approximating PluralityMatching

In light of Proposition 5 the main question that arises is whether we can asymptotically reach the optimal distortion bound of 3. To this end, we will analyze a sampling approximation of PluralityMatching, a deterministic mechanism introduced by Gkatzelis et al. [22] which obtains the optimal distortion bound of 3. To keep the exposition reasonably self-contained we recall some basic facts about PluralityMatching.

Definition 1. *For an election* $\mathcal{E} = (V, C, \sigma)$ *and a candidate* $a \in C$*, the integral domination graph of candidate a is the bipartite graph* $G(a) = (V, V, E_a)$*, where* $(i, j) \in E_a$ *if and only if* $a \succeq_i \text{top}(j)$.

Proposition 6. ([22]). *There exists a candidate* $a \in C$ *whose integral domination graph* $G(a)$ *admits a perfect matching.*

Before we proceed let us first introduce some notation. For this subsection it will be convenient to use numerical values in the set $\{1, 2, \ldots, m\}$ to represent the candidates. We let $\Pi_j = \sum_{i \in V} \mathbb{1}\{\text{top}(i) = j\}$, i.e. the number of voters for which $j \in C$ is the top candidate. For candidate $j \in C$ we let $G(j)$ be the integral domination graph of j, and M_j be a maximum matching in $G(j)$. In the sequel, it will be useful to "decompose" M_j as follows. We consider the partition of V into $V_j^0, V_j^1, \ldots, V_j^m$ such that $V_j^k = \{i \in V : M_j(i) = k\}$ for all $k \in [m]$, while V_j^0 represents the subset of voters which remained unmatched under M_j.

Moreover, consider a set $\mathcal{S} = S_j^0 \cup S_j^1 \cup \cdots \cup S_j^m$ such that $S_j^k \subseteq V_j^k$ for all k; we also let $c = |\mathcal{S}|$, and $\Pi_j' = c/n \times \Pi_j$. For now let us assume that $\Pi_j' \in \mathbb{N}$ for all j. We let $G^{\mathcal{S}}(j)$ represent the induced subgraph of $G(j)$ w.r.t. the subset $\mathcal{S} \subseteq V$ and the new plurality scores Π_j'. We start our analysis with the following observation:

Lemma 3. *Assume that* $\mathcal{S}$ *is such that* $|S_j^k|/c = |V_j^k|/n$ *for all* k. *Then, if* $M_j^{\mathcal{S}}$ *represents the maximum matching in* $G^{\mathcal{S}}(j)$, *it follows that* $|M_j^{\mathcal{S}}|/c = |M_j|/n$.

Let us denote with $\Phi_j = M_j/n$; roughly speaking, we know from [22] that Φ_j is a good indicator of the "quality" of candidate j. Importantly, Lemma 3 tells us that we can determine Φ_j in a much smaller graph, if only we had a decomposition that satisfied the "proportionality" condition of the claim. Of course, determining explicitly such a decomposition makes little sense given that we do not know the sets V_j^k, but the main observation is that we can approximately satisfy the condition of Lemma 3 through sampling. It should be noted that we previously assumed that $\Pi_j' \in \mathbb{N}$, i.e. we ignored rounding errors. However, in the worst-case rounding errors can only induce an error of at most m/c in the value of Φ_j; thus, we remark that our subsequent selection of c will be such that this error will be innocuous, in the sense that it will be subsumed by the "sampling error" (see Lemma 5). Before we proceed, recall that for $\mathbf{p}, \widehat{\mathbf{p}} \in \Delta([k])$,

$$\mathrm{d_{TV}}(\mathbf{p}, \widehat{\mathbf{p}}) \stackrel{\text{def}}{=} \sup_{S \subseteq [k]} |\mathbf{p}(S) - \widehat{\mathbf{p}}(S)| = \frac{1}{2} ||\mathbf{p} - \widehat{\mathbf{p}}||_1, \tag{4}$$

where $|| \cdot ||_1$ represents the ℓ_1 norm. In this context, we will use the following standard fact (e.g., see [13]):

Lemma 4. *Consider a discrete distribution* $\mathbf{p} \in \Delta([k])$ *and let* $\widehat{\mathbf{p}}$ *be the empirical distribution derived from* N *independent samples. For any* $\epsilon > 0$ *and* $\delta \in (0, 1)$, *if* $N = \Theta((k + \log(1/\delta))/\epsilon^2)$ *it follows that* $\mathrm{d_{TV}}(\mathbf{p}, \widehat{\mathbf{p}}) \leq \epsilon$ *with probability at least* $1 - \delta$.

As a result, if we draw a set $\mathcal{S}$ with $|\mathcal{S}| = c = \Theta((m + \log(1/\delta))/\epsilon^2)$ samples (*without* replacement[1]) we can guarantee that

[1] Although the samples are not independent since we are not replacing them, observe that the induced bias is negligible for n substantially larger than m.

$$\sum_{k=0}^{m} \left| \frac{|S_j^k|}{c} - \frac{|V_j^k|}{n} \right| \leq 2\epsilon; \tag{5}$$

$$\sum_{k=1}^{m} \left| \frac{\widehat{\Pi}_k}{c} - \frac{\Pi_k}{n} \right| \leq 2\epsilon, \tag{6}$$

where S_j^k represents the subset of $\mathcal{S}$ which intersects V_j^k, and $\widehat{\Pi}_k$ is the empirical plurality score of candidate k. Thus, the following lemma follows directly from Lemma 3 and Lemma 4.

Lemma 5. *Let $\widehat{\Phi}_j = |\widehat{M}_j|/c$, where $\widehat{M}_j$ is the maximum matching in the graph $G^{\mathcal{S}}(j)$. Then, if $|\mathcal{S}| = \Theta((m + \log(1/\delta)/\epsilon^2)$ for some $\epsilon, \delta \in (0, 1)$, it follows that $(1 - \epsilon)\Phi_j \leq \widehat{\Phi}_j \leq (1 + \epsilon)\Phi_j$ with probability at least $1 - \delta$.*

As a result, if we combine all of these ingredients we can establish the main result of this section:

Theorem 7. *For any $\epsilon \in (0, 4]$ and $\delta \in (0, 1)$ there exists a mechanism which takes a sample of size $\Theta((m + \log(m/\delta))/\epsilon^2)$ voters and yields distortion at most $3 + \epsilon$ with probability at least $1 - \delta$.*

6 Experiments

Finally, we analyze the performance of the scoring system used in the Eurovision song contest, so let us first give a basic overview of the competition and the voting rule employed. Fist of all, we will only focus on the final stage of the competition, wherein a set of m countries compete amongst each other and a set of n countries—which is a strict superset of the contenders—provide their preferences over the finalists. Eurovision employs a specific *positional scoring* system which works as follows. Every country assigns 12 points to its highest preference, 10 points to its second-highest preference, and from $8 - 1$ points to each of its next 8 preferences; note that no country can vote for itself. This scoring system shall be referred to as the SCORING rule. It should be noted that the authors in [34] quantify the distortion for some specific scoring rules (e.g. the *harmonic rule*). We will make the working hypothesis that for every country the assigned scores correspond to its actual order of preferences. Nonetheless, we stress that the assigned scores of every country have been themselves obtained by *preference aggregation*, and as such they are themselves subject to *distortion*, but we will tacitly suppress this issue.[2]

We will focus on the competitions held between 2004 and 2008; during these years the number of finalists (or candidates) m was 24, with the exception of 2008 where 25 countries were represented in the final. We should note that for our experiments we used a dataset from `Kaggle`. Observe that every "voter" only

[2] We refer the interested reader to the work of Filos-Ratsikas and Voudouris [20].

provides its top $k = 10$ preferences, while the countries which are represented in the final are 0-*decisive* (see [5]). The main question that concerns us is whether the SCORING rule employed for the competition yields very different results from the *instance-optimal* mechanism (which we refer to as MINIMAX-LP) of Goel et al. [23]—which is based on linear programming. Our results are summarized in Table 1.

Perhaps surprisingly, on all occasions the winners in the two mechanisms coincide; on the other hand, there are generally substantial differences below the first position. It is also interesting to note that on all occasions the winner has a remarkably small distortion, at least compared to the theoretical bounds.

Table 1. Summary of our findings for the Eurovision song contests held between 2004 and 2008. For every year we have indicated the top three countries according to the MINIMAX-LP rule and the SCORING system employed in the actual contest.

Year	MINIMAX-LP rule		SCORING rule		# of Countries
	Country	Distortion	Country	Score	
2004	Ukraine	**1.1786**	Ukraine	**280**	36
	Serbia & Montenegro	1.4444	Serbia & Montenegro	263	
	Turkey	1.4746	Greece	252	
2005	Greece	**1.4068**	Greece	**230**	39
	Switzerland	1.4127	Malta	192	
	Moldova	1.4194	Romania	158	
2006	Finland	**1.3000**	Finland	**292**	38
	Romania	1.4262	Russia	248	
	Russia	1.4407	Bosnia & Herzegovina	229	
2007	Serbia	**1.3235**	Serbia	**268**	42
	Ukraine	1.3667	Ukraine	235	
	Russia	1.5231	Russia	207	
2008	Russia	**1.3562**	Russia	**272**	43
	Greece	1.4507	Ukraine	230	
	Ukraine	1.4923	Greece	218	

References

1. Amanatidis, G., Birmpas, G., Filos-Ratsikas, A., Voudouris, A.A.: Peeking behind the ordinal curtain: improving distortion via cardinal queries. In: The Thirty-Fourth AAAI Conference on Artificial Intelligence, AAAI 2020, pp. 1782–1789. AAAI Press (2020)
2. Anagnostides, I., Fotakis, D., Patsilinakos, P.: Metric-distortion bounds under limited information. CoRR abs/2107.02489 (2021)
3. Anshelevich, E., Bhardwaj, O., Postl, J.: Approximating optimal social choice under metric preferences. In: Bonet, B., Koenig, S. (eds.) Proceedings of the Twenty-Ninth AAAI Conference on Artificial Intelligence, 2015, pp. 777–783. AAAI Press (2015)
4. Anshelevich, E., Filos-Ratsikas, A., Shah, N., Voudouris, A.A.: Distortion in social choice problems: the first 15 years and beyond (2021)

5. Anshelevich, E., Postl, J.: Randomized social choice functions under metric preferences. In: Kambhampati, S. (ed.) Proceedings of the Twenty-Fifth International Joint Conference on Artificial Intelligence, IJCAI 2016, pp. 46–59. IJCAI/AAAI Press (2016)
6. Anshelevich, E., Sekar, S.: Blind, greedy, and random: algorithms for matching and clustering using only ordinal information. In: Schuurmans, D., Wellman, M.P. (eds.) Proceedings of the Thirtieth AAAI Conference on Artificial Intelligence, pp. 390–396. AAAI Press (2016)
7. Anshelevich, E., Zhu, W.: Ordinal approximation for social choice, matching, and facility location problems given candidate positions. In: Christodoulou, G., Harks, T. (eds.) WINE 2018. LNCS, vol. 11316, pp. 3–20. Springer, Cham (2018). https://doi.org/10.1007/978-3-030-04612-5_1
8. Benade, G., Nath, S., Procaccia, A.D., Shah, N.: Preference elicitation for participatory budgeting. In: Singh, S.P., Markovitch, S. (eds.) Proceedings of the Thirty-First AAAI Conference on Artificial Intelligence, pp. 376–382. AAAI Press (2017)
9. Benadè, G., Procaccia, A.D., Qiao, M.: Low-distortion social welfare functions. In: The Thirty-Third AAAI Conference on Artificial Intelligence, AAAI 2019, pp. 1788–1795. AAAI Press (2019)
10. Bhaskar, U., Dani, V., Ghosh, A.: Truthful and near-optimal mechanisms for welfare maximization in multi-winner elections. In: McIlraith, S.A., Weinberger, K.Q. (eds.) Proceedings of the Thirty-Second AAAI Conference on Artificial Intelligence, (AAAI-18), pp. 925–932. AAAI Press (2018)
11. Boutilier, C., Caragiannis, I., Haber, S., Lu, T., Procaccia, A.D., Sheffet, O.: Optimal social choice functions: a utilitarian view. Artif. Intell. **227**, 190–213 (2015)
12. Canetti, R., Even, G., Goldreich, O.: Lower bounds for sampling algorithms for estimating the average. Inf. Process. Lett. **53**(1), 17–25 (1995)
13. Canonne, C.L.: A short note on learning discrete distributions (2020)
14. Caragiannis, I., Filos-Ratsikas, A., Frederiksen, S.K.S., Hansen, K.A., Tan, Z.: Truthful facility assignment with resource augmentation: an exact analysis of serial dictatorship. In: Cai, Y., Vetta, A. (eds.) WINE 2016. LNCS, vol. 10123, pp. 236–250. Springer, Heidelberg (2016). https://doi.org/10.1007/978-3-662-54110-4_17
15. Caragiannis, I., Nath, S., Procaccia, A.D., Shah, N.: Subset selection via implicit utilitarian voting. In: Kambhampati, S. (ed.) Proceedings of the Twenty-Fifth International Joint Conference on Artificial Intelligence, IJCAI 2016, pp. 151–157. IJCAI/AAAI Press (2016)
16. Chen, S., Liu, J., Wang, H., Augusto, J.C.: Ordering based decision making - a survey. Inf. Fusion **14**(4), 521–531 (2013)
17. Cho, S., Endersby, J.W.: Issues, the spatial theory of voting, and British general elections: a comparison of proximity and directional models. Public Choice **114**(3/4), 275–293 (2003)
18. Fain, B., Goel, A., Munagala, K., Prabhu, N.: Random dictators with a random referee: constant sample complexity mechanisms for social choice. In: The Thirty-Third AAAI Conference on Artificial Intelligence, AAAI 2019, pp. 1893–1900. AAAI Press (2019)
19. Filos-Ratsikas, A., Frederiksen, S.K.S., Zhang, J.: Social welfare in one-sided matchings: random priority and beyond. In: Lavi, R. (ed.) SAGT 2014. LNCS, vol. 8768, pp. 1–12. Springer, Heidelberg (2014). https://doi.org/10.1007/978-3-662-44803-8_1
20. Filos-Ratsikas, A., Voudouris, A.A.: Approximate mechanism design for distributed facility location. CoRR abs/2007.06304 (2020)

21. Fotakis, D., Kalavasis, A., Stavropoulos, K.: Aggregating incomplete and noisy rankings. In: Banerjee, A., Fukumizu, K. (eds.) The 24th International Conference on Artificial Intelligence and Statistics, AISTATS 2021. Proceedings of Machine Learning Research, vol. 130, pp. 2278–2286. PMLR (2021)
22. Gkatzelis, V., Halpern, D., Shah, N.: Resolving the optimal metric distortion conjecture. In: 61st IEEE Annual Symposium on Foundations of Computer Science, FOCS 2020, pp. 1427–1438. IEEE (2020)
23. Goel, A., Krishnaswamy, A.K., Munagala, K.: Metric distortion of social choice rules: lower bounds and fairness properties. In: Proceedings of the 2017 ACM Conference on Economics and Computation, p. 287–304. EC 2017, Association for Computing Machinery (2017)
24. Gross, S., Anshelevich, E., Xia, L.: Vote until two of you agree: mechanisms with small distortion and sample complexity. In: Singh, S.P., Markovitch, S. (eds.) Proceedings of the Thirty-First AAAI Conference on Artificial Intelligence, 2017, pp. 544–550. AAAI Press (2017)
25. Kempe, D.: An analysis framework for metric voting based on LP duality. In: The Thirty-Fourth AAAI Conference on Artificial Intelligence, AAAI 2020, pp. 2079–2086. AAAI Press (2020)
26. Kempe, D.: Communication, distortion, and randomness in metric voting. In: The Thirty-Fourth AAAI Conference on Artificial Intelligence, AAAI 2020, pp. 2087–2094. AAAI Press (2020)
27. Lang, J., Pini, M.S., Rossi, F., Venable, K.B., Walsh, T.: Winner determination in sequential majority voting. In: Veloso, M.M. (ed.) IJCAI 2007, Proceedings of the 20th International Joint Conference on Artificial Intelligence, pp. 1372–1377 (2007)
28. Mandal, D., Procaccia, A.D., Shah, N., Woodruff, D.P.: Efficient and thrifty voting by any means necessary. In: Advances in Neural Information Processing Systems 32: Annual Conference on Neural Information Processing Systems 2019, pp. 7178–7189 (2019)
29. Mandal, D., Shah, N., Woodruff, D.P.: Optimal communication-distortion tradeoff in voting. In: Proceedings of the 21st ACM Conference on Economics and Computation, p. 795–813. EC 2020, Association for Computing Machinery (2020)
30. Munagala, K., Wang, K.: Improved metric distortion for deterministic social choice rules. In: Karlin, A., Immorlica, N., Johari, R. (eds.) Proceedings of the 2019 ACM Conference on Economics and Computation, EC 2019, pp. 245–262. ACM (2019)
31. von Neumann, J., Morgenstern, O.: Theory of Games and Economic Behavior. Princeton University Press, Princeton (1944)
32. Procaccia, A.D., Rosenschein, J.S.: The distortion of cardinal preferences in voting. In: Klusch, M., Rovatsos, M., Payne, T.R. (eds.) CIA 2006. LNCS (LNAI), vol. 4149, pp. 317–331. Springer, Heidelberg (2006). https://doi.org/10.1007/11839354_23
33. Saari, D.G., Merlin, V.R.: The Copeland method: I.: relationships and the dictionary. Econ. Theor. **8**(1), 51–76 (1996)
34. Skowron, P.K., Elkind, E.: Social choice under metric preferences: Scoring rules and STV. In: Singh, S.P., Markovitch, S. (eds.) Proceedings of the Thirty-First AAAI Conference on Artificial Intelligence, February 4–9, 2017, San Francisco, California, USA, pp. 706–712. AAAI Press (2017)

Hedonic Expertise Games

Bugra Caskurlu(✉), Fatih Erdem Kizilkaya, and Berkehan Ozen

TOBB University of Economics and Technology, Ankara, Turkey
{bcaskurlu,f.kizilkaya,b.ozen}@etu.edu.tr

Abstract. We consider a team formation setting where agents have varying levels of expertise in a global set of required skills, and teams are ranked with respect to how well the expertise of teammates complement each other. We model this setting as a hedonic game, and we show that this class of games possesses many desirable properties, some of which are as follows: A partition that is Nash stable, core stable and Pareto optimal is always guaranteed to exist. A contractually individually stable partition (and a Nash stable partition in a restricted setting) can be found in polynomial-time. A core stable partition can be approximated within a factor of $1 - \frac{1}{e}$ and this bound is tight. We discover a larger and relatively general class of hedonic games, where the above existence guarantee holds. For this larger class, we present simple dynamics that converge to a Nash stable partition in a relatively low number of moves.

Keywords: Team formation · Hedonic games · Common ranking property

1 Introduction

Hedonic games provide a simple formal model for numerous problems, where a set of agents is required to be partitioned into stable coalitions [11], such as research group formation [1], group activity selection [10] or task allocation [17] problems. In this paper, we follow this line of research by introducing a model for the formation of stable teams. For the ease of understanding, we define our model below using a simple example where students in a classroom need to form teams for a project assignment.

Model. In our model, a global set of skills and for each agent a level of expertise in each of these skills are given. For instance, the required skills for a class project assignment may be `(Python, Java, SQL)` where the expertise of two students, say Alice and Bob, in these skills are `(1, 3, 3)` and `(3, 3, 1)` respectively. We measure the success of a team by how well the expertise of teammates complement each other. For instance, notice that Alice may compensate the lack of expertise of Bob in `SQL`, just as Bob may compensate Alice in `Python`. We say that a coalition's *joint expertise* in some skill is the maximum level of expertise of its members in that skill, and its *joint utility* is the sum of its joint

I. Caragiannis and K. A. Hansen (Eds.): SAGT 2021, LNCS 12885, pp. 314–328, 2021.
https://doi.org/10.1007/978-3-030-85947-3_21

expertise in each skill. For instance, the team formed by Alice and Bob would have a joint expertise of 3 in each skill, and thus a joint utility of 9.

We next define the utility of agents. In our classroom example, even if some students do not contribute to their teams as much as their teammates, note that they will still receive the same grade as their teammates, which is the typical case in a significant number of scenarios involving teams. Therefore, we define the utility of agents simply as the joint utility of their coalition. For instance, if Alice and Bob form a coalition, they both will have a utility of 9. Under this assumption, notice that all agents are better off in the grand coalition. However, in most real-life scenarios there exists a limit on the sizes of coalitions that can be formed due to inherent constraints and/or coordination problems. For instance, it would not make sense if the whole class formed a single team, in the classroom example above. Therefore, we additionally have an upper bound on the sizes of coalitions that are possible to form.

The above setting can be modeled as a subclass of hedonic game, which we refer to as *hedonic expertise games* (HEGs). HEGs naturally model a variety of team contests such as hackathons in which software developers, graphic designers, project managers, and other domain experts collaborate on software projects. Various other team formation settings are studied in the literature, which are similar to HEGs in that agents are endowed with a set of skills, or some other kind of resource (see, for example [3,21]). What differs between HEGs and these models is that, in HEGs, instead of completion of a set of tasks or goals, agents are solely motivated to form teams that best complement each other's strengths, which captures the main concern in many team contests and group projects. Moreover, they are not modeled as non-transferable utility coalitional games, such as hedonic games (see the related work section for more details).

HEGs also have the useful property that the joint utility function (defined above for possible coalitions) is monotone and submodular, which is discussed below in more detail.

Common Ranking Property. A hedonic game where all agents in a coalition receive the same utility is said to possess the common ranking property [13], and the class of those games is referred to as *hedonic games with common ranking property* (HGCRP). We may represent an HGCRP instance with a single joint utility function U defined for each possible coalition C, i.e., each member of C receives a utility of $U(C)$ if C is formed. It is clear that HEGs are a subclass of HGCRP. Moreover, it turns out that the joint utility function defined in HEGs is monotone and submodular, which gives rise to some desirable complexity results.

To the best of our knowledge, a monotonicity restriction has not yet been considered in HGCRP. Therefore, we also study this restriction on HGCRP, and obtain some interesting results which we discuss below.

Monotonicity Restriction. We show that a monotonicity restriction in HGCRP ensures the existence of a Nash stable, core stable and Pareto optimal partition. We also study better response dynamics (based on Nash stability) in this setting. We identify a large set of initial partitions for which there exists a sequence of better responses that reaches a Nash stable partition in a low number of moves. We also give a simple decentralized algorithm that finds such a sequence.

Related Work. We now discuss some notable classes of games that are similar to HEGs in their motivation and/or formulation. In coalitional skill games [3], each agent has a set of skills that are required to complete various tasks. Each task requires a set of skills in order to be completed, and a coalition can accomplish the task only if its members cover the set of required skills for the task. The game is modeled as a transferable utility coalitional game where the characteristic function maps the achieved set of tasks to a real value. That is, the authors are concerned with how the value of a coalition is distributed among its members, depending on their contribution to the accomplishment of tasks. In the hedonic games literature, the focus is on which coalitions are formed instead.

Another similar model is coalitional resource games [21], where agents are interested in achieving a single goal among their set of goals. Each agent has different amounts of various resources, which are required to reach these goals. A goal set is said to be satisfying for a coalition, if for every agent in that coalition it contains a goal desired by that agent; and a goal set is said to be feasible for a coalition, if its members collectively have sufficient resources to achieve all the goals in that set. A solution in this model is simply a goal set that is both feasible and satisfying for a coalition, i.e., this model does not technically define a coalitional game.

Unlike HEGs, a naive representation of HGCRP and coalitional skill games can be exponential, respectively, in the number of agents and tasks. There has been a considerable amount of effort put in the literature to design succinctly representable hedonic games (see, for example [4,5,12]). Among those, HEGs are somewhat related to $\mathcal{B}$-hedonic games [9], in which each agent ranks all other agents and prefers a coalition over another if it contains an agent that she ranked higher. HEGs might be thought of as a multidimensional generalization of $\mathcal{B}$-hedonic games where each agent has multiple rankings over other agents, which corresponds to different skills. However, each agent's ranking over other agents is identical in our setting, which is not necessarily the case in $\mathcal{B}$-hedonic games.

Research on hedonic games has been mainly focused on the existence and computational complexity of deciding the existence of partitions under various stability criteria. The actual process of forming coalitions based on individual behavior has received little attention until very recently. Brandt et al. initiated the study of better response dynamics (based on individual stability) leading to stable partitions in a variety of classes of hedonic games [6]. In an earlier study, better response dynamics (based on contractually individual stability) have been shown to converge in symmetric additively separable hedonic games [15].

For a detailed discussion of hedonic games literature, we refer to [7,16].

Contributions and Organization. In Sect. 2, we introduce HEGs and monotone HGCRP formally, and define the stability and optimality concepts we use.

In Sect. 3, we prove that a partition that is Nash stable, core stable and Pareto optimal is guaranteed to exist in monotone HGCRP (and thus in HEGs). The economical interpretation of this existence guarantee is that efficiency need not be sacrificed for the sake of stability with respect to both individual and group based deviations.

In Sect. 4, we introduce a decentralized algorithm that finds a Nash stable partition of a given HEG instance. Our procedure terminates in a linear number of moves, if the number of levels of expertise is bounded above by a constant.[1] There exists such a bound for most practical purposes because the expertise in some real-life skill is most commonly measured by a small number of levels such as `(0: None, 1: Beginner, 2: Intermediate, 3: Advanced)`. We also show that finding a contractually individually stable partition is polynomial-time solvable in HEGs, even if expertise levels are not bounded.

In Sect. 5, we show that a $(1-\frac{1}{e})$-approximate core stable partition of a given HEG instance can be found in polynomial-time and that this is the best approximation ratio achievable, unless $\mathsf{P} = \mathsf{NP}$. We also show that verifying a core stable partition is intractable, unless $\mathsf{NP} = \mathsf{co\text{-}NP}$.

In Sect. 6, we show that finding a perfect partition, or a socially optimal partition, or a Pareto optimal partition is NP-HARD. We also show that verifying a Pareto optimal partition is coNP-COMPLETE.

In Sect. 7, we conclude the paper and discuss future directions.

Lastly, the overall picture for our results on HEGs is given in Fig. 1.

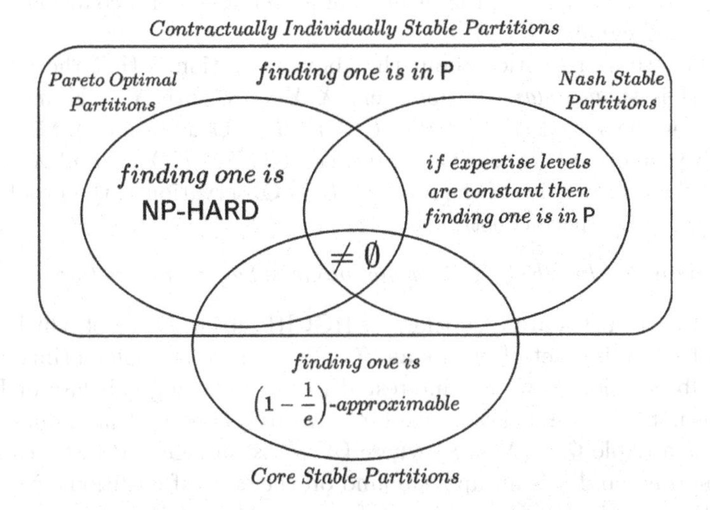

Fig. 1. The set of core stable, Nash stable, contractually individually stable and Pareto optimal partitions of HEGs are drawn in the above Venn diagram. The intersection of all of these sets of partitions are guaranteed to be nonempty. The computational complexity of finding one such partition in these sets of partitions are stated above.

[1] This follows from our investigation on better response dynamics in monotone HGCRP, which we discussed above.

2 Model and Background

We now formally define *hedonic expertise games* (HEGs). We have a set of *agents* N and a set of *skills* S. For each agent $i \in N$, we have a non-negative integer-valued *expertise function* $e_i : S \to \mathbb{Z}_{\geq 0}$, where $e_i(s)$ denotes the expertise that agent $i \in N$ has in skill $s \in S$. Lastly, we have an upper bound of κ on the sizes of coalitions, i.e., no coalition of size greater than κ can be formed. We denote an HEG instance by $\mathcal{G} = (N, S, e, \kappa)$. Moreover, we refer to the subclass of HEGs in which $e_i : S \to \{0, 1, \ldots, \beta\}$ for all $i \in N$, where β is a constant, as $(0, \beta)$-HEGs.

For each coalition C, we define the *joint expertise function* $E_C : S \to \mathbb{Z}_{\geq 0}$ as $E_C(s) = \max_{i \in C} e_i(s)$, i.e., $E_C(s)$ is the maximum expertise level that a member of coalition C has in skill s. Lastly, we define the *joint utility* of a coalition C as $U(C) = \sum_{s \in S} E_C(s)$, i.e., $U(C)$ is the sum of the maximum expertise that a member of coalition C has in each skill $s \in S$. We define $U(\emptyset) = 0$ as a convention.

A solution of an HEG instance is a partition π over the set of agents N where for each coalition $C \in \pi$ we have $|C| \leq \kappa$. (Throughout the rest of the paper, when we refer to a partition π, it is implicitly assumed that $|C| \leq \kappa$ for all $C \in \pi$ for the sake of briefness.) We use $\pi(i)$ to denote the coalition containing agent $i \in N$ in partition π. We use $u_i(\pi)$ to denote the *utility* of agent i in partition π where $u_i(\pi) = U(\pi(i))$, i.e., the utilities of all members of a coalition $C \in \pi$ are the same, and equal to $U(C)$.

The first thing to notice about the above definition is that the joint utility function U is *submodular*, i.e., for every $X, Y \subseteq N$ with $X \subseteq Y$ and for every $x \in N \backslash Y$, we have that $U(X \cup \{x\}) - U(X) \geq U(Y \cup \{x\}) - U(Y)$. Moreover, the joint utility function U is also *monotone*, i.e., $U(X) \leq U(Y)$ for all $X \subseteq Y \subseteq N$. We state these properties of the function U in Observation 1, the proof of which is omitted due to space constraints.

Observation 1. *In HEGs, U is a monotone submodular function.*

Recall that HEGs are a subclass of HGCRP, an instance of which is a pair (N, U) where N is a set of agents and $U : 2^N \to \mathbb{Z}_{\geq 0}$ is a joint utility function. Due to Observation 1, we are interested in the following subclass of HGCRP, which has not been studied earlier as far as we are aware of. A *monotone HGCRP* instance is a triple $\mathcal{G} = (N, U, \kappa)$ where (N, U) is an HGCRP instance in which U is monotone, and κ is an upper bound on the sizes of coalitions. Notice that, as it is the case in HEGs, this game form would be trivial without an upper bound on the size of the coalitions since otherwise all agents would be better off in the grand coalition. It is clear that HEGs are a subclass of monotone HGCRP by Observation 1.

2.1 Stability and Optimality

We now formally define the stability and optimality concepts that we study, in the context of monotone HGCRP.

The main stability concepts based on individual deviations [5] are as follows:

- A partition π is *Nash stable (NS)* if no agent $i \in N$ can benefit from moving to an existing[2] coalition $C \in \pi$ such that $|C| < \kappa$, i.e., $U(\pi(i)) \geq U(C \cup \{i\})$ for all $C \in \pi$ where $|C| < \kappa$.
- A partition π is *individually stable (IS)* if no agent $i \in N$ can benefit from moving to an existing coalition $C \in \pi$ such that $|C| < \kappa$, without making an agent in C worse off, i.e., $U(\pi(i)) \geq U(C \cup \{i\})$ or $U(C) > U(C \cup \{i\})$ for all $C \in \pi$ where $|C| < \kappa$.
- A partition π is *contractually individually stable (CIS)* if no agent $i \in N$ can benefit from moving to an existing coalition $C \in \pi$ such that $|C| < \kappa$, without making an agent in neither C nor $\pi(i)$ worse off, i.e., $U(\pi(i)) \geq U(C \cup \{i\})$ or $U(C) > U(C \cup \{i\})$ or $U(\pi(i)) > U(\pi(i)\backslash\{i\})$ for all $C \in \pi$ where $|C| < \kappa$.

It is clear that an NS partition is IS, and an IS partition is CIS by definition. However, an IS partition is also NS in monotone HGCRP since $U(C) \leq U(C')$ for all $C \subseteq C' \subseteq N$. In other words, the sets of NS partitions and IS partitions are identical in monotone HGCRP.[3]

The main stability concept based on group deviations [20], and its approximate adaptation is as follows:

- A coalition C is said to *block* π, if $U(C) > u_i(\pi)$ for all agents $i \in C$, i.e., any agent $i \in C$ is better off in C than she is in her coalition $\pi(i)$. A partition π is *core stable (CS)* if no coalition blocks π.
- Similarly, a coalition C is said to *α-approximately block* π where $\alpha \leq 1$, if $\alpha \cdot U(C) > u_i(\pi)$ for all agents $i \in C$. Similarly, a partition π is *α-approximate CS* if no coalition α-approximately blocks π. Note that a 1-approximate CS partition is simply a CS partition.

We now introduce the following notation which comes in handy with the above stability concepts. For a partition π and a coalition $C \notin \pi$, we define π_C as the partition induced on π by C, i.e., π_C is the partition that would arise if the agents in C collectively deviated from π to form coalition C, i.e., $\pi_C(i) = C$ for all $i \in C$, and $\pi_C(j) = \pi(j)\backslash C$ for all $j \in N\backslash C$.

Notice that if a partition π is not NS then there exists an agent $i \in N$ and a coalition $C \in \pi$ such that $u_i(\pi_{C \cup \{i\}}) > u_i(\pi)$ and $|C| < \kappa$. Also notice that if coalition C blocks partition π then $u_i(\pi_C) > u_i(\pi)$, for all agents $i \in C$.

The main optimality concepts are as follows:

- A partition π is *perfect* if all agents are in their most preferred coalition.
- The *social welfare* $W(\pi)$ of a partition π is defined as the sum of the utilities of all the agents, i.e., $W(\pi) = \sum_{i \in N} u_i(\pi)$. A *socially optimal (SO)* solution is a partition for which the social welfare is maximized.

---

[2] Moving to an empty coalition is also permissible, but we can omit this case w.l.o.g. due to monotonicity.

[3] Similarly, the sets of CIS partitions and contractually Nash stable [19] partitions are identical in monotone HGCRP.

- A partition π' *Pareto dominates* a partition π if $u_i(\pi') \geq u_i(\pi)$ for all agents $i \in N$, and there exists some agent i for which the inequality is strict. A partition π is said to be *Pareto optimal (PO)* if no partition π' Pareto dominates π.

Observe that a perfect partition is necessarily SO, and a SO partition is necessarily PO. Though a SO partition (and thus a PO partition) is guaranteed to always exist, a perfect partition does not necessarily exist in HEGs (and thus in monotone HGCRP). Moreover, observe that a PO partition is necessarily CIS.

3 Existence Guarantees

The common ranking property has long been known for guaranteeing the existence of a CS partition in hedonic games via a simple greedy algorithm [13]. Moreover, existence of a partition which is both CS and PO in HGCRP has recently been proven by giving an asymmetric and transitive relation ψ defined over the set of partitions, where a maximal partition with respect to ψ is both CS and PO [8]. This is established by applying two potential function arguments in a successive manner as described below.

Given a partition π, $\psi(\pi)$ is defined as the sequence of the utilities of all the agents in a *non-increasing order* (if two agents have the same utility then that value is repeated in the sequence, i.e., the length of $\psi(\pi)$ is exactly $|N|$). It is shown that **(CS)** and **(PO)** given below hold for any partition π, where we use $\rhd$ to denote "lexicographically greater than".

(CS) If there exists a coalition C that blocks π then $\psi(\pi_C) \rhd \psi(\pi)$.
(PO) If there exists a partition π' that Pareto dominates π then $\psi(\pi') \rhd \psi(\pi)$.

On the other hand, an HGCRP instance does not necessarily possess an NS partition as can be seen from Example 1.

Example 1. Consider the HGCRP instance $\mathcal{G} = (N, U)$, where $N = \{1, 2\}$, and U is defined as $U(\{1\}) = 1$, $U(\{1, 2\}) = 2$ and $U(\{2\}) = 3$. Notice that HGCRP instance $\mathcal{G}$ does not possess an NS partition.

In contrast, monotone HGCRP do not only admit an NS partition but also an NS, CS and PO partition as stated in Theorem 1, which improves upon the aforementioned existence guarantee in HGCRP.

Theorem 1. *In monotone HGCRP, a partition that is NS, CS and PO is always guaranteed to exist.*

In order to prove Theorem 1, notice that we only need to include a third potential function argument which establishes that a maximal partition with respect to ψ is also NS, due to **(CS)** and **(PO)**. This is done in Lemma 1 whose proof is omitted due to space constraints.

Lemma 1. *Given a partition π of a monotone HGCRP instance, if there exists an agent i which benefits from moving to an existing coalition $C \in \pi$ such that $|C| < \kappa$ (i.e., $u_i(\pi_{C\cup\{i\}}) > u_i(\pi)$), then $\psi(\pi_{C\cup\{i\}}) \rhd \psi(\pi)$.*

Recall that HEGs are a subclass of monotone HGCRP by Observation 1. Therefore, Theorem 1 has the following corollary.

Corollary 1. *In HEGs, a partition that is NS, CS and PO is always guaranteed to exist.*

4 Efficiently Computable Stable Partitions

We first present a decentralized algorithm for finding an NS partition of a given monotone HGCRP instance. Our algorithm accompanies a restricted version of *better response dynamics*, i.e., while the current partition π is not NS, an agent i moves to an existing coalition $C \in \pi$ such that $|C| < \kappa$ and $u_i(\pi_{C\cup\{i\}}) > u_i(\pi)$, which we refer to as a *better response* of agent i in partition π. Note that better response dynamics is guaranteed to converge to an NS partition in monotone HGCRP by Lemma 1.

For the ease of analysis, we force a natural restriction on better response dynamics. We refer to this restricted class as *imitative better response dynamics*, which we describe below.

Imitative Better Response Dynamics. Given a partition π, suppose that an agent i benefits from moving to an existing coalition $C \in \pi$ such that $|C| < \kappa$, i.e., $U(C\cup\{i\}) > U(\pi(i))$. Suppose that agent i takes the above better response. If $|C \cup \{i\}| < \kappa$, then notice that another agent $i' \in \pi(i)\backslash\{i\}$ also benefits from moving to $C\cup\{i\}$, since $U(C\cup\{i\}) > U(\pi(i)) \geq U(\pi(i)\backslash\{i\})$. That is, if the size of the coalition that the last agent i has moved to did not reach the upper bound of κ, then an agent $i' \in \pi(i)\backslash\{i\}$ simply *"imitates"* the last agent i by moving to the same coalition. Otherwise, an arbitrary agent takes a better response.

We refer to a partition π where each coalition $C \in \pi$ has a size of exactly κ (except maybe for one coalition) as a *complete partition.* We now show that the above procedure starting from any complete partition π converges to an NS partition in $O(|N| \cdot U(N))$ moves.

Theorem 2. *In monotone HGCRP, imitative better response dynamics starting from any complete partition converges to an NS partition in $O(|N|\cdot U(N))$ moves.*

Proof. Suppose that a monotone HGCRP instance $\mathcal{G} = (N, U, \kappa)$ is given. Let π be a complete partition of $\mathcal{G}$. We refer to the coalitions in π whose size is exactly κ as $C_1, \ldots, C_{\lfloor |N|/\kappa \rfloor}$. Notice that if these are the only coalitions in π then we are done since no agent can move to another coalition in π. Hence, we assume w.l.o.g. that there is another coalition $L \in \pi$ such that $|L| < \kappa$ which consists of the leftover agents from those coalitions whose sizes are exactly κ.

Suppose that π is not NS. Then, there exists an agent $j \in C_i$ which benefits from moving to coalition L, i.e., $U(L\cup\{j\}) > U(C_i)$. Suppose that agent j takes this better response. Note that then $\kappa - |L| - 1$ agents in $C_i \backslash \{j\}$ will imitate agent j by moving to the same coalition. Notice that after these $\kappa - |L|$ moves, the resulting partition will still consist of $\lfloor |N|/\kappa \rfloor$ coalitions whose sizes are exactly κ, and an additional coalition that consists of the leftover agents. We exploit this structure as follows:

- Let C_i' denote the resulting coalition after agent j and other $\kappa - |L| - 1$ agents in $C_i \backslash \{j\}$ move to coalition L, i.e., $C_i' = L \cup \{j\} \cup K$ where $K \subseteq C_i \backslash \{j\}$ is an arbitrary subset of agents of size $\kappa - |L| - 1$.
- Let L' denote the remaining coalition after agent j and other $\kappa - |L| - 1$ agents in $C_i \backslash \{j\}$ move to coalition L, i.e., $L' = C_i \backslash (K \cup \{j\})$.

Notice that we can obtain the resulting partition, say π', after these $\kappa - |L|$ moves by updating C_i as C_i' and L as L' in partition π. Moreover, notice that $U(C_i') = U(L \cup \{j\} \cup K) \geq U(L \cup \{j\}) > U(C_i)$, which means the joint utility of the coalition which we refer to as C_i is strictly greater in π' than in π. Since $U(C_i)$ is an integer between 0 and $U(N)$, this means the number of moves is bounded by $U(N) \cdot \lfloor |N|/\kappa \rfloor \cdot (\kappa - |L|) = O(|N| \cdot U(N))$, which finishes our proof. □

In $(0, \beta)$-HEGs, notice that $U(N)$ is at most $\beta \cdot |S|$. Therefore, Theorem 2 has the following corollary.

Corollary 2. *In $(0, \beta)$-HEGs, an NS partition can be found in polynomial-time via imitative better response dynamics in $O(|N| \cdot |S|)$ moves.*

Corollary 2 is significant since we can say that even boundedly-rational agents that imitate the last agent if possible, and take the first beneficial move otherwise, will converge to an NS partition quickly.

We next present a polynomial-time algorithm for finding a CIS partition in HEGs, not only in $(0, \beta)$-HEGs unlike our previous result.

Theorem 3. *In HEGs, a CIS partition can be found in polynomial-time.*

Proof. Suppose that an HEG instance $\mathcal{G} = (N, S, e, \kappa)$ is given. Exactly as in Theorem 2, we begin with a partition $\pi = (C_1, \ldots, C_{\lfloor |N|/\kappa \rfloor}, L)$ where $|C_i| = \kappa$ for all $C_i \in \pi$ and $|L| < \kappa$. Note that if $L = \emptyset$ (i.e. κ divides $|N|$) then we are already done. Hence, we assume w.l.o.g. that $L \neq \emptyset$.

Recognizing which agents of a coalition are (or would be) "critical" lies in the heart of our proof. We say that an agent i is *critical* for a coalition C if there exists a skill $s \in S$ such that $e_i(s) > E_{C \backslash \{i\}}(s)$. Note that an agent $i \in C$ is critical for coalition C if and only if $U(C) > U(C \backslash \{i\})$. This means that if each agent $i \in C$ is critical for C, then no agent in C can leave coalition C without making an agent in C worse off.

Notice that π is not CIS if and only if there exists an agent $j \in C_i$ such that[4]:

[4] And also $U(L \cup \{j\}) \geq U(L)$ which trivially holds in HEGs due to monotonicity.

- $U(L \cup \{j\}) > U(C_i)$
(which means there exists a skill $s \in S$ such that $E_L(s) > E_{C_i}(s)$, which then implies that there exists a critical agent $j' \in L$ for coalition C_i),
- $U(C_i \backslash \{j\}) \geq U(C_i)$
(which means that agent j is not critical for C_i),

If this is the case, then we update C_i and L as follows:

- Let $C_i' = (C_i \backslash \{j\}) \cup \{j'\}$.
- Let $L' = (L \backslash \{j'\}) \cup \{j\}$.

Let $\gamma(C)$ denote the number of critical agents for coalition C that are also in coalition C. Since j' is critical for C_i whereas j is not, we have $\gamma(C_i') > \gamma(C_i)$. Therefore, if we update C_i as C_i', L as L' and repeat the above procedure, we will eventually reach a CIS partition. We now only need to show that the number of iterations will be polynomial.

Notice that $E_{C_i'}(s) \geq E_{C_i}(s)$ for all $s \in S$. Therefore, no matter how many iterations have passed, agent j cannot ever become critical for coalition C_i. However, for j to be able to return to C_i, she must be critical for C_i. Hence, j cannot ever return to C_i. This means that the number of iterations is bounded by $\lfloor |N|/\kappa \rfloor \cdot |N|$, which finishes our proof. □

5 Approximating Core Stable Partitions

We devote this section to show that finding a $(1 - \frac{1}{e})$-approximate CS partition of an HEG instance is polynomial-time solvable, and this is the best possible approximation ratio achievable, unless $\mathsf{P} = \mathsf{NP}$.

Initially, we study the problem of finding a coalition with maximum joint utility, which we formally specify as follows.

MAXIMUM-JOINT-UTILITY = *"Given an HEG instance $\mathcal{G} = (N, S, e, \kappa)$, find a subset of agents $C^* \subseteq N$ which maximizes $U(C^*)$ such that $|C^*| \leq \kappa$."*

In the lemma below, we show that this problem is inapproximable within better than a ratio of $1 - \frac{1}{e}$, even for (0, 1)-HEGs. We then use this lemma to show that a $(1 - \frac{1}{e} + \epsilon)$-approximate CS partition cannot be found in polynomial-time for any $\epsilon > 0$.

Lemma 2. *In* $(0,1)$*-HEGs, the above problem of* MAXIMUM-JOINT-UTILITY *is inapproximable within better than a ratio of* $1 - \frac{1}{e}$, *unless* $\mathsf{P} = \mathsf{NP}$.

Proof. We will give an approximation preserving S-reduction from MAXIMUM-COVERAGE problem, which is known to be inapproximable within better than $1 - \frac{1}{e}$, unless $\mathsf{P} = \mathsf{NP}$ [14]. An instance of MAXIMUM-COVERAGE problem consists of a universe $\mathcal{U} = \{1, \ldots, m\}$, a family $\mathcal{S} = \{\mathcal{S}_1, \ldots, \mathcal{S}_n\}$ of subsets of $\mathcal{U}$, an integer k, and the objective is finding a subset $\mathcal{C} \subseteq \mathcal{S}$ such that $|\mathcal{C}| \leq k$ which maximizes $\text{cov}(\mathcal{C}) = |\cup_{\mathcal{S}_i \in \mathcal{C}} \mathcal{S}_i|$. Given a MAXIMUM-COVERAGE instance $\mathcal{I} = (\mathcal{U}, \mathcal{S}, k)$, we build a (0, 1)-HEG instance $\mathcal{G}_{\mathcal{I}} = (N, S, e, \kappa)$ as follows:

- Universe $\mathcal{U}$ corresponds to the set of skills S.
- Each subset $\mathcal{S}_i \in \mathcal{S}$ corresponds to an agent $i \in N$, whose expertise function is defined for each skill $s \in S$ as $e_i(s) = 1$ if $s \in \mathcal{S}_i$, and $e_i(s) = 0$ otherwise.
- Finally, k corresponds to the upper bound κ on the sizes of coalitions.

Notice that a coalition C (i.e., a subset of the set of agents N such that $|C| \leq \kappa$) corresponds to a subset $\mathcal{C} \subseteq \mathcal{S}$ such that $U(C) = \text{cov}(\mathcal{C})$ and $|\mathcal{C}| \leq k$; and the reverse also holds, which completes our S-reduction. □

Theorem 4. *In* $(0,1)$*-HEGs, a* $(1 - \frac{1}{e} + \epsilon)$*-approximate CS partition cannot be found in polynomial-time for any constant* $\epsilon > 0$*, unless* $\mathsf{P} = \mathsf{NP}$.

Proof. For some $\epsilon > 0$, suppose that a $(1 - \frac{1}{e} + \epsilon)$-approximate CS partition π of a given $(0,1)$-HEG instance $\mathcal{G} = (N, S, e, \kappa)$ can be found in polynomial-time. Let C^* be a coalition in $\mathcal{G}$ with maximum joint utility. Then, there exists an agent $i^* \in C^*$ such that $(1 - \frac{1}{e} + \epsilon) \cdot U(C^*) \leq u_{i^*}(\pi)$, because otherwise C^* would $(1 - \frac{1}{e} + \epsilon)$-approximately block π.

Let $C \in \pi$ be a coalition such that $U(C) \geq U(C')$ for all $C' \in \pi$. Note that C can be found in polynomial-time. Notice that $(1 - \frac{1}{e} + \epsilon) \cdot U(C^*) \leq u_{i^*}(\pi) \leq U(C)$. This means that we could devise a $(1 - \frac{1}{e} + \epsilon)$-approximation algorithm for MAXIMUM-JOINT-UTILITY problem by simply returning C. Unless $\mathsf{P} = \mathsf{NP}$, this creates a contradiction by Lemma 2, which finishes our proof. □

Theorem 4 also has the following interesting implication, the proof of which is omitted due to space constraints.

Theorem 5. *In* $(0,1)$*-HEGs, it is not possible to verify whether a given partition is CS or not in polynomial-time, unless* $\mathsf{NP} = \mathsf{co\text{-}NP}$.

We next show that MAXIMUM-JOINT-UTILITY problem is approximable within a ratio of $1 - \frac{1}{e}$ by the so-called *standard greedy algorithm*, which is described below in the context of HEGs.

> *"Begin with an empty coalition* C*, and greedily add the agent that increase the joint utility of* C *the most, until reaching the upper bound of* κ*".*

We then use the above algorithm as a subroutine to find a $(1 - \frac{1}{e})$-approximate CS partition in HEGs.

Lemma 3. *In HEGs,* MAXIMUM-JOINT-UTILITY *problem is* $(1 - \frac{1}{e})$*-approximable by the standard greedy algorithm.*

Proof. Recall that the joint utility function U is monotone and submodular as given in Observation 1. Due to the upper bound on the sizes of coalitions, this means that MAXIMUM-JOINT-UTILITY is simply a problem of maximizing a monotone submodular function subject to a cardinality constraint, which is known to be $(1 - \frac{1}{e})$-approximable by the standard greedy algorithm [18]. □

Theorem 6. *In HEGs, a* $(1 - \frac{1}{e})$*-approximate CS partition can be found in polynomial-time.*

Proof. Consider an HEG instance $\mathcal{G} = (N, S, e, \kappa)$. Let C^* be a coalition of $\mathcal{G}$ with maximum joint utility. Note that running the standard greedy algorithm will return a coalition C such that $U(C) \geq (1 - \frac{1}{e}) \cdot U(C^*)$ due to Lemma 3. Since no agent can have a strictly greater utility than $U(C^*)$, notice that if π is a partition where $C \in \pi$ then no agent $i \in C$ can participate in a coalition that $(1 - \frac{1}{e})$-approximately blocks π. Thus, by assuming coalition C is formed, we can reduce the problem of finding a $(1 - \frac{1}{e})$-approximate CS partition in $\mathcal{G}$ into one of finding a $(1 - \frac{1}{e})$-approximate CS partition in $\mathcal{G}' = (N \backslash C, S, e, \kappa)$. This means that we can build a $(1 - \frac{1}{e})$-approximate CS partition in polynomial-time by repeatedly running the standard greedy algorithm, which finishes our proof. □

6 Intractability of Computing Optimal Partitions

This section is devoted to show that finding a perfect (if exists), SO or PO partition in $(0,1)$-HEGs is intractable, and so is verifying if a partition is PO. For any subclass of hedonic games, if deciding whether there exists a perfect partition is NP-HARD, and if a given partition can be verified to be perfect in polynomial-time, then it is known that finding a PO or SO partition is also NP-HARD in the same subclass of hedonic games [2]. However, we cannot directly use this method since it is not clear how we can check efficiently whether a given partition is perfect in HEGs. The trick is to construct special HEG instances where a partition can be verified to be perfect easily, throughout the reductions.

Theorem 7. *In $(0,1)$-HEGs: (i) deciding if a perfect partition exists is* NP-HARD, *(ii) finding a SO partition is* NP-HARD, *(iii) finding a PO partition is* NP-HARD *and (iv) verifying whether a partition is PO is* coNP-COMPLETE.

Proof. All of the proofs for (i), (ii), (iii) and (iv) are via a polynomial-time mapping reduction from SET-COVER problem, in which we are given a universe $\mathcal{U} = \{1, \ldots, m\}$ and a family $\mathcal{S} = \{\mathcal{S}_1, \ldots, \mathcal{S}_n\}$ of subsets of $\mathcal{U}$ along with a positive integer k; and then, we are required to decide whether there exists a *set cover* $\mathcal{C} \subseteq \mathcal{S}$ whose size is at most k, i.e., we need to have $|\mathcal{C}| \leq k$ and $\cup_{\mathcal{S}_i \in \mathcal{C}} \mathcal{S}_i = \mathcal{U}$. Given an instance $\mathcal{I} = (\mathcal{U}, \mathcal{S}, k)$ of SET-COVER, we build a $(0,1)$-HEG instance $\mathcal{G}_\mathcal{I} = (N, S, e, \kappa)$ as follows:

- Universe $\mathcal{U}$ corresponds to the set of skills S.
- Each subset $\mathcal{S}_i \in \mathcal{S}$ corresponds to an agent $i \in N$, whose expertise function is defined for each skill $s \in S$ as $e_i(s) = 1$ if $s \in \mathcal{S}_i$, and $e_i(s) = 0$ otherwise.
- Let $x = \lceil \frac{n-k}{k-1} \rceil$ and $X = \{n+1, \ldots, n+x\}$ where each $i \in X$ corresponds to an agent $i \in N$ such that $e_i(s) = 1$ for all $s \in S$.
- Finally, k corresponds to the upper bound κ on the sizes of coalitions.

Notice that $|N| = n + x$ and $\lceil \frac{n+x}{k} \rceil = x + 1$. Therefore, a partition π of $\mathcal{G}_\mathcal{I}$ contains at least $x + 1$ coalitions. Hence, there exists a coalition $C \in \pi$ which does not contain any agent in X. Thus, we can map back any partition π of $\mathcal{G}_\mathcal{I}$ to a feasible solution $\mathcal{C}$ of $\mathcal{I}$ by returning coalition C. Notice that coalition C corresponds to a set cover $\mathcal{C}$ of $\mathcal{I}$ if and only if $U(C) = m$.

(*i*) Since all the agents can attain the utility of m by participating in a coalition with an agent in X, a partition π is perfect if and only if $u_i(\pi) = m$ for all $i \in N$. Notice that such an allocation π exists if and only if there exists a set cover $\mathcal{C}$ of $\mathcal{I}$ whose size is less than k. Therefore, deciding whether there exists a perfect partition in $(0,1)$-HEGs is NP-HARD.

The rest of our results are proven via the same construction as above but by exploiting a reduction from the problem given in (*i*).

(*ii*) Notice that if there exists a perfect partition, then any SO partition needs to be also perfect. If we could find a SO partition π of $\mathcal{G}_\mathcal{I}$ in polynomial-time, then we could decide if there exists a perfect partition in polynomial-time by simply checking whether $u_i(\pi) = m$ for all $i \in N$. However, this would be contradictory. Therefore, finding a SO partition is also NP-HARD in $(0,1)$-HEGs.

(*iii*) Assume for the sake of contradiction that we can find a PO partition π of $\mathcal{G}_\mathcal{I}$ in polynomial-time. Suppose that π is not a perfect partition. Then, there cannot exist a perfect partition π^* of $\mathcal{G}_\mathcal{I}$ because otherwise π^* would Pareto dominate π by definition. Then, we could decide if there exists a perfect partition by checking whether $u_i(\pi) = m$ for all $i \in N$, which would be contradictory. Therefore, finding a PO partition is also NP-HARD in $(0,1)$-HEGs.

(*iv*) Let $\pi = (X_1, \ldots, X_x, C)$ be a partition of $\mathcal{G}_\mathcal{I}$ such that $X \cap C = \emptyset$, $U(C) < m$ and $n + i \in X_i$ for all i. We show that there exists a perfect partition of $\mathcal{G}_\mathcal{I}$ if and only if π is not PO. Notice that π is not a perfect partition, since $U(C) < m$. Therefore, it is clear that if π is PO then there does not exist a perfect partition.

Now suppose that π is not PO. Then, there exists a partition, say π', that Pareto dominates π. Recall that there must be a coalition $C' \in \pi'$ such that $C' \cap X = \emptyset$. Note that $C' \neq C$ since otherwise no agent would be better off in π' with respect to π. Moreover, since all agents except those in C had a utility of m in π, some agents would get worse off in π', unless $U(C') = m$. Recall that this implies the existence of a perfect partition, and thus, we are done.

Finally, note that verifying a whether a given partition π is PO is in coNP, since a partition π' that Pareto dominates π is a counterexample that is verifiable in polynomial-time. Therefore, verifying a PO partition is coNP-COMPLETE. □

7 Conclusion

In this paper, we investigated computational aspects of HEGs and we concluded that stable solutions based on individual deviations (namely NS partitions if the level of expertise is bounded by a constant, and CIS partitions in general) can be computed efficiently, whereas stable solutions based on group deviations (namely CS partitions) can be approximated within a factor of $1 - \frac{1}{e} \approx 0.632$. On the other hand, we showed that finding a perfect, SO or PO partition, and verifying a CS or PO partition is intractable. Yet the computational complexity of finding an NS partition in HEGs remains open.

Moreover, we showed that the existence guarantees given in HEGs arise from the fact that HEGs is a subclass of a much more general class of hedonic games, which we referred to as monotone HGCRP. For this larger class, we introduced imitative better response dynamics, which demonstrate how boundedly-rational agents playing this game can naturally converge to an NS partition in a relatively low number of moves. Using the upper bound we obtained for monotone HGCRP, we showed that, in $(0, \beta)$-HEGs, imitative better response dynamics converge to an NS partition in a polynomial number of moves. But the convergence rate of (usual) better response dynamics remains open.

Future Directions. A strong assumption in our model is that skills are additive. One might, for instance, use L^2-norm or L^∞-norm (instead of L^1-norm) to define the joint utility of a coalition C from the joint expertise values of C in each skill. One might also define the joint expertise of a coalition C in a skill s as the sum, average or geometric mean (instead of maximum) of expertise levels of members of C have in s. All of these alternative definitions might drastically change the properties of the game, some of which might be worth studying. On the other hand, note that the definition used in this paper desirably favor agents that complement each other's abilities, and reflect the diminishing returns of each new member (due to submodularity).

Another strong assumption in our model is that the utility functions of agents are uniform. One might, for instance, consider a setting where each agent only cares about a subset of skills, which would violate the uniformity. One might be also concerned by fairness considerations since the utility of an agent does not depend on the contribution of the expertise that she provides on the skill. First, note that HEGs are best thought of as a group of contestants who try to form teams in order to increase their chances of winning. However, suppose now that a team actually wins and needs to share a given amount of prize money. In order to address this problem, our model can be simply reformulated as a transferable utility coalition game where the characteristic function corresponds to the joint utility function (since U is monotone). Such a research direction (similar to [3]) could be also interesting.

References

1. Alcalde, J., Revilla, P.: Researching with whom? Stability and manipulation. J. Math. Econ. **40**(8), 869–887 (2004)
2. Aziz, H., Brandt, F., Harrenstein, P.: Pareto optimality in coalition formation. Games Econom. Behav. **82**, 562–581 (2013)
3. Bachrach, Y., Parkes, D.C., Rosenschein, J.S.: Computing cooperative solution concepts in coalitional skill games. Artif. Intell. **204**, 1–21 (2013)
4. Banerjee, S., Konishi, H., Sönmez, T.: Core in a simple coalition formation game. Soc. Choice Welfare **18**(1), 135–153 (2001)
5. Bogomolnaia, A., Jackson, M.O.: The stability of hedonic coalition structures. Games Econom. Behav. **38**(2), 201–230 (2002)
6. Brandt, F., Bullinger, M., Wilczynski, A.: Reaching individually stable coalition structures in hedonic games. In: Proceedings of the 35th Conference on Artificial Intelligence (AAAI), vol. 35, pp. 5211–5218 (2021)

7. Brandt, F., Conitzer, V., Endriss, U., Lang, J., Procaccia, A.D.: Handbook of Computational Social Choice. Cambridge University Press, USA (2016)
8. Caskurlu, B., Kizilkaya, F.E.: On hedonic games with common ranking property. In: Heggernes, P. (ed.) CIAC 2019. LNCS, vol. 11485, pp. 137–148. Springer, Cham (2019). https://doi.org/10.1007/978-3-030-17402-6_12
9. Cechlárová, K., Romero-Medina, A.: Stability in coalition formation games. Internat. J. Game Theory **29**(4), 487–494 (2001)
10. Darmann, A., Elkind, E., Kurz, S., Lang, J., Schauer, J., Woeginger, G.: Group activity selection problem. In: Goldberg, P.W. (ed.) WINE 2012. LNCS, vol. 7695, pp. 156–169. Springer, Heidelberg (2012). https://doi.org/10.1007/978-3-642-35311-6_12
11. Drèze, J.H., Greenberg, J.: Hedonic coalitions: optimality and stability. Econometrica **48**(4), 987–1003 (1980)
12. Elkind, E., Wooldridge, M.: Hedonic coalition nets. In: Proceedings of The 8th International Conference on Autonomous Agents and Multiagent Systems (AAMAS), vol. 1, pp. 417–424 (2009)
13. Farrell, J., Scotchmer, S.: Partnerships. Q. J. Econ. **103**(2), 279–297 (1988)
14. Feige, U.: A threshold of ln n for approximating set cover. J. ACM **45**(4), 634–652 (1998)
15. Gairing, M., Savani, R.: Computing stable outcomes in hedonic games. In: Kontogiannis, S., Koutsoupias, E., Spirakis, P.G. (eds.) SAGT 2010. LNCS, vol. 6386, pp. 174–185. Springer, Heidelberg (2010). https://doi.org/10.1007/978-3-642-16170-4_16
16. Hajdukova, J.: Coalition formation games: a survey. Int. Game Theory Rev. **8**(4), 613–641 (2006)
17. Jang, I., Shin, H.-S., Tsourdos, A.: Anonymous hedonic game for task allocation in a large-scale multiple agent system. IEEE Trans. Rob. **34**(6), 1534–1548 (2018)
18. Nemhauser, G.L., Wolsey, L.A., Fisher, M.L.: An analysis of approximations for maximizing submodular set functions-i. Math. Program. **14**(1), 265–294 (1978)
19. Sung, S.C., Dimitrov, D.: On myopic stability concepts for hedonic games. Theor. Decis. **62**(1), 31–45 (2006)
20. Woeginger, G.J.: Core stability in hedonic coalition formation. In: van Emde Boas, P., Groen, F.C.A., Italiano, G.F., Nawrocki, J., Sack, H. (eds.) SOFSEM 2013. LNCS, vol. 7741, pp. 33–50. Springer, Heidelberg (2013). https://doi.org/10.1007/978-3-642-35843-2_4
21. Wooldridge, M., Dunne, P.E.: On the computational complexity of coalitional resource games. Artif. Intell. **170**(10), 835–871 (2006)

When Dividing Mixed Manna Is Easier Than Dividing Goods: Competitive Equilibria with a Constant Number of Chores

Jugal Garg[1], Martin Hoefer[2], Peter McGlaughlin[1], and Marco Schmalhofer[2](✉)

[1] University of Illinois at Urbana-Champaign, Urbana-Champaign, USA
{jugal,mcglghl2}@illinois.edu
[2] Goethe University Frankfurt, Frankfurt am Main, Germany
{mhoefer,schmalhofer}@em.uni-frankfurt.de

Abstract. We study markets with mixed manna, where m divisible goods and chores shall be divided among n agents to obtain a competitive equilibrium. Equilibrium allocations are known to satisfy many fairness and efficiency conditions. While a lot of recent work in fair division is restricted to linear utilities, we focus on a substantial generalization to separable piecewise-linear and concave (SPLC) utilities. We first derive polynomial-time algorithms for markets with a constant number of items or a constant number of agents. Our main result is a polynomial-time algorithm for instances with a constant number of chores (as well as any number of goods and agents) under the condition that chores dominate the utility of the agents. Interestingly, this stands in contrast to the case when the goods dominate the agents utility in equilibrium, where the problem is known to be PPAD-hard even without chores.

1 Introduction

The allocation of a set of items to a set of agents in a *fair* and *efficient* manner is the main challenge in fair division, a prominent field in economics with a variety of well-established concepts and techniques [22]. *Algorithms* for fair division have recently prompted a large amount of research interest in AI, due to many important applications arising from computer-aided decision making in various parts of society [10, Part II]. Standard criteria for fair and efficient allocation in markets include envy-freeness (EF; no agent prefers the bundle of goods from another agent), proportionality (PROP; every agent gets a bundle that has at least her "average" value), and Pareto-optimality (PO). Interestingly, all these criteria are achieved in a *competitive equilibrium from equal incomes (CEEI)*, an equilibrium allocation in a market when every agent has $1 of (fake) money.

For more than two decades, the computation of competitive equilibria (with and without equal incomes) has been a main line of research in fair division

J. Garg and P. McGlaughlin—Supported by NSF grant CCF-1942321 (CAREER).
M. Hoefer and M. Schmalhofer—Supported by DFG grant Ho 3831/5-1, 6-1 and 7-1.

I. Caragiannis and K. A. Hansen (Eds.): SAGT 2021, LNCS 12885, pp. 329–344, 2021.
https://doi.org/10.1007/978-3-030-85947-3_22

and, more broadly, at the intersection of economics and computer science [23, Chapters 5+6]. An intriguing recent development in this area is the consideration of *chores* and, more generally, *mixed manna*. In an allocation problem with mixed manna there are goods and chores. Goods are desired by at least one of the agents (e.g., cake), chores are undesirable for all agents (e.g., job shifts, cleaning tasks). In particular, chores are not disposable. All goods can and chores must be allocated to the agents. The goal again is to satisfy fairness criteria such as EF, PROP, and/or PO. The consideration of mixed manna substantially generalizes our understanding of fair division and represents an intriguing challenge for algorithms to computing such allocations when they exist.

In a seminal contribution [7] the existence of competitive equilibria under general conditions for instances with mixed manna were established. Moreover, even for mixed manna, CEEI retain their attractive fairness properties. Clearly, this raises a natural question from a computational perspective, which we study in this paper: *Under which conditions can competitive equilibria be computed in polynomial time for markets with mixed manna?*

The answers depend on whether we consider instances with only goods, only chores or, more generally, true mixed manna. For only goods, markets with linear utilities allow even strongly polynomial-time algorithms [19,24]. For additively separable piecewise-linear concave (SPLC) utilities, the problem is PPAD-hard [14]. For only chores, the problem is PPAD-hard for linear utilities when we allow agents to have infinitely negative utility for some chores [12]. For mixed manna, an equilibrium can be computed efficiently for linear utilities, when we have a constant number of agents or a constant number of items [17].

1.1 Contribution and Outline

In this paper, we provide polynomial-time algorithms for computing competitive equilibria in markets with mixed manna. The introduction of the formal model and preliminary results are given in Sect. 2. As a first set of results, we show a polynomial-time algorithm to compute equilibria in markets with SPLC utilities when the number of agents or items (i.e., goods and bads) is constant. This substantially generalizes the results in [17] where only linear utilities are considered. SPLC utilities are quite more general and applicable as they model natural properties like decreasing marginals while maintaining (piecewise) linearity; see, e.g., [18]. The discussion of these results is given in Sect. 3. We note that this is the first polynomial time algorithm to compute a competitive equilibrium of mixed manna with SPLC utilities under any assumptions. Our main result is then presented in Sect. 4 – an efficient algorithm for computing competitive equilibria in *negative* instances with arbitrary many agents, goods, and a constant number of chores. The agents can have SPLC utilities for goods, but we assume linear utilities for chores. Negativity is a condition that implies that chores dominate the utility of the agents (for a formal definition see Sect. 2). This is a notable contrast to *positive* instances with SPLC utilities for goods, where computation of an equilibrium is PPAD-hard, even without chores.

Finally, in Sect. 5 we discuss an algorithm that rounds any equilibrium for markets with divisible mixed manna to an allocation for the same market with indivisible mixed manna. The resulting allocation guarantees Pareto-optimality (PO) and a notion of proportionality[1].

1.2 Further Related Work

The literature on competitive equilibrium in markets with only goods is vast, and a complete review is beyond the scope of this paper. Instead, we refer the reader the books [9,22,25], and focus on the case of mixed manna.

While most of the work in fair division focuses on goods, there are a few works for the case of bads [4,9,25,26]. The study of competitive division with a mixed manna was initiated by [7]. They establish equilibrium existence and show further properties, e.g., that multiple, disconnected equilibria may exist, and polynomial-time computation is possible if there are either two agents or two items with linear utility functions [8].

On the algorithmic side, an algorithm to compute a competitive allocation of bads with linear utility functions was recently proposed in [11]. The algorithm runs in strongly polynomial time if either the number of agents or bads is constant. This result was generalized in [17] to a mixed manna. Our work generalizes this further to the case of SPLC utilities.

Chaudhury et al. [13] provided an algorithm to compute an equilibrium of mixed manna with SPLC utility functions. However, our work differs from theirs in two important ways. First, our approach allows for computing *all* equilibria in an instance, while in [13] only one is found. In negative instances where 'bads overwhelm the goods', there are generally multiple equilibria in which agents receive significantly different utilities, i.e., an agent might prefer one equilibrium over another. Thus, finding *all* equilibria might enable a social planner to offer an allocation that is more 'fair' to all agents. Second, the algorithm in [13] has polynomial runtime when the number of agents or items is constant for instances with only bads. Our algorithm runs in polynomial time for a more general setting of mixed manna under the same conditions.

Fair allocation of *indivisible* items is an intensely studied problem. Recently, attention has shifted to the case of all bads or mixed manna, see e.g. [1,3,6,20]. Most closely related to our work is a recent contribution [2] presenting an algorithm to compute an indivisible allocation that is PO and PROP1 in markets with mixed manna. Our algorithm has a number of similarities with the approach in [2]. A notable difference is that in our case the divisible allocation constitutes a competitive equilibrium in the divisible market. Hence, our algorithm comes with the additional benefit of strengthening the algorithmic connection between competitive equilibrium and fair indivisible allocations.

---

[1] More precisely, the allocation satisfies an adaptation of proportionality up to one good (PROP1) to mixed manna.

2 Preliminaries

2.1 Fair Division with Mixed Manna

We consider fair division of mixed manna, in which there is a set $N = [n]$ of n agents and a set $M = [m]$ of m divisible items. We strive to divide the items among the agents. W.l.o.g. we may assume that there is a unit amount of each item. A fractional allocation $x = \{x_1, \ldots, x_n\}$ assigns each agent $i \in N$ a bundle of items $x_i = (x_{i1}, \ldots, x_{im})$, where $x_{ij} \in [0,1]$ is the amount of item j agent i receives. An allocation is feasible if all items are fully assigned, i.e., $\forall j \in M$, $\sum_{i \in N} x_{ij} = 1$. For the rest of the paper, we assume all allocations are feasible, unless otherwise explicitly stated.

Each agent $i \in N$ has a utility function u_i that maps the received bundle to a numerical value. In this paper, we assume all utility functions are additively separable over items, piecewise linear, and concave (SPLC). Formally, agent i's utility for receiving x_{ij} amount of item $j \in M$ is given by the piecewise linear and concave function $f_{ij}(x_{ij})$, and the total utility for the bundle x_i is given by $u_i(x_i) = \sum_{j \in M} f_{ij}(x_{ij})$. Let $\{u_{ij1}, \ldots, u_{ijk}\}$ be the slopes of each linear segment of f_{ij} with lengths $\{l_{ij1}, \ldots, l_{ijk}\}$. In contrast to the familiar case of disposable goods where $f_{ij} \geq 0$, $\forall i \in N$, $\forall j \in M$, a mixed manna allows $f_{ij} \in \mathbb{R}$, i.e., an agent may get *positive or negative* utility for an item. We assume each agent labels each item either an (individual) *good* or *bad*. If item j is a good for agent i, then $f_{ij} > 0$ and $u_{ij1} > u_{ij2} > \cdots > u_{ijk} > 0$, which implies concavity and captures the classical condition of decreasing marginal returns. Otherwise, j is a bad for agent i, then $f_{ij} < 0$ and $0 > u_{ij1} > u_{ij2} > \cdots > u_{ijk}$. Note that two agents i, i' might disagree the label of a given item j, e.g., j can be a good for i and a bad for i'. For simplicity of the technical exposition, we assume that $u_{ijk} \neq 0$ for all segments.[2] Let $|f_{ij}|$ denote the number of linear segments of f_{ij}. Also, we sometimes write (i, j, k) to refer to the k-th segment of the function f_{ij}.

Instance Types. In [7], the authors show that every fair division instance with mixed manna falls into one of three *types*: positive, negative, or null. The type roughly indicates whether there is a 'surplus' of goods or bads.

More formally, let $N^+ = \{i \in N : \max_{j \in M} u_{ij1} > 0\}$ be the set of *attracted* agents, i.e., agents that each have at least one good, whereas $N^- = N \setminus N^+$ is the set of *repulsed* agents that have only bads. We use $\mathcal{X}$ to denote the set of feasible allocations, and $\mathcal{U}$ for the set of agent utilities over all feasible allocations. If $u \in \mathcal{U}$, then $u = (u_1(x_1), \ldots, u_n(x_n))$ for some $x \in \mathcal{X}$. Next, we define $\Gamma_+ = \mathbb{R}_+^{N^+} \times \{0\}^{N^-}$. Note that in Γ_+ attracted agents benefit (the $\mathbb{R}_+^{N^+}$ portion), without harming any repulsed agents (the $\{0\}^{N^-}$ portion). Also, let $\Gamma_{++} = \mathbb{R}_{++}^{N^+} \times \{0\}^{N^-}$ be the relative interior of Γ_+.

[2] While we conjecture that conceptually all our ideas can be applied also when $u_{ijk} = 0$ is allowed, the analysis of such segments generates a lot of technicalities, which we leave for future work.

Definition 1. *A fair division instance is called*

- *positive if* $\mathcal{U} \cap \Gamma_{++} \neq \emptyset$
- *null if* $\mathcal{U} \cap \Gamma_{+} = \{\mathbf{0}\}$
- *negative if* $\mathcal{U} \cap \Gamma_{+} = \emptyset$.

For an interpretation of a positive instance, we can ensure all attracted agents receive strictly positive utility without harming any repulsed agents (who dislike all items). Conversely, in a negative instance, no feasible allocation gives *all* attracted agents non-negative utility. Finally, in null instances, the only feasible allocations which give all agents non-negative utility satisfy $u_i(x_i) = 0$, $\forall i \in N$.

Determining the Instance Type. We can determine the type of a given instance with SPLC utilities in polynomial time by solving the following LP. The approach extends results of [17] for linear utilities.

$$\begin{aligned} \max \quad & t \\ s.t. \quad & \sum_{j,k} u_{ijk} x_{ijk} \geq t, \ \forall i \in N^+ \\ & \sum_{i \in N^+, k} x_{ijk} = 1, \ \forall j \in M \\ & 0 \leq x_{ijk} \leq l_{ijk}, \ \forall i \in N^+, j \in M \end{aligned} \tag{1}$$

The solution t gives a lower bound on any attracted agent's utilities by the first set of constraints. The second set of constraints simply requires that all items are fully allocated among attracted agents, and the third set of constraints ensures that segments aren't overallocated.

Proposition 1. *Let* (t^*, x^*) *be a solution to* (1). *The sign of* t^* *determines the instance type:*

- *If* $t^* > 0$, *then the instance is positive.*
- *If* $t^* = 0$, *then the instance is null.*
- *If* $t^* < 0$, *then the instance is negative.*

Proof. First suppose that $t^* > 0$. Then all attracted agents receive strictly positive utility, while repulsed agents receive no allocation. Hence the instance is positive by Definition 1.

Next suppose that $t^* = 0$. We want to show that the only feasible allocations which give all agents non-negative utility satisfy $u_i(x_i) = 0$, $\forall i \in N$. For contradiction suppose not. Then at least one agent $k \in N^+$ receives strictly positive utility $u_k(x_k) > 0$, and some other agent $i \in N^+$ receives a total utility of $u_i(x_i) = 0$. We now construct an alternate allocation y so that $u_i(y_i) > 0$, $\forall i \in N^+$, contradicting the optimality of $t^* = 0$.

Let $M^+ = \{j \in M : \max_{i \in N} u_{ij1} > 0\}$ and $M^- = M \setminus M^+ = \{j \in M : \max_{i \in N} u_{ij1} < 0\}$. First observe that for any good $j \in M^+$, there is $i \in N^+$ such that $u_{ij1} > 0$. Therefore, we may assume that no agent $i' \in N^+$ with $u_{i'j1} < 0$ receives any part $x_{i'j} > 0$ of j. This is valid since reallocating $x_{i'j}$ to i, i.e., $y_{ij} = x_{ij} + x_{i'j}$ and $y_{i'j} = 0$ improves both agents utilities.

Next consider any agent i with a non-zero allocation, i.e., $x_i \neq 0$, such that $u_i(x_i) = 0$. Since $x_i \neq 0$, we must have $x_{ij}, x_{ij'} > 0$, for some $j \in M^+$ and $j' \in M^-$. If $u_k(x_k) = \epsilon > 0$ for some $k \in N^+$, then we can reallocate some portion of $x_{ij'}$ to agent k to make both agents utilities strictly positive, i.e., transfer a fraction of bad j' from agent i to agent k. Let d be the final segment of j' with positive allocation $x_{kj'd} > 0$. If there is none, then $d = 1$. Now set

$$y_{ij'} = x_{ij'} - \min\left(\frac{\min(\epsilon, l_{kj'd} - x_{kj'd})}{2|u_{kj'd}|}, x_{ij'}\right)$$

and

$$y_{kj'} = x_{kj'} + \min\left(\frac{\min(\epsilon, l_{kj'd} - x_{kj'd})}{2|u_{kj'd}|}, x_{ij'}\right).$$

Then $u_k(y_k) = \epsilon/2 > 0$, and $u_i(y_i) > 0$.

After the steps above, either $u_i(y_i) > 0$ or $y_i = 0$, for all $i \in N^+$. If $u_i(y_i) > 0$ for all $i \in N^+$, then we reach a contradiction to that $t^* = 0$ is optimal. Therefore, assume that $y_i = 0$ for some $i \in N^+$. By definition of N^+, there is $j \in M^+$ such that $u_{ij1} > 0$. Further, all items are fully allocated in x, so there is $z \in N^+$ with $x_{zj} > 0$, and $u_z(y_z) = \epsilon > 0$. Let d be the last segment with $x_{zjd} > 0$. Suppose we reallocate a portion of x_{zj} to agent i:

$$y_{ij} = \min\left(\frac{\min(\epsilon, l_{ij1})}{2|u_{zjd}|}, x_{zjd}\right)$$

and

$$y_{zj} = x_{zj} - \min\left(\frac{\min(\epsilon, l_{ij1})}{2|u_{zjd}|}, x_{zjd}\right).$$

Then $u_z(y_z) \geq \epsilon/2 > 0$ and $u_i(y_i) > 0$. Repeating this step for all $i \in N^+$ with $x_i = 0$ ensures that $u_i(y_i) > 0$ for all $i \in N^+$, which contradicts that t^* maximizes (1).

The above argument shows that if $t^* = 0$, then x^* must satisfy $u_i(x_i^*) = 0, \forall i \in N^+$, so the instance is null. Finally, repeating the above arguments in case $t^* < 0$ shows that the instance must be negative. □

2.2 Competitive Equilibrium

We are interested in computing *competitive equilibria (CE)*. To define this notion, we turn a fair division instance into a market. We endow each agent $i \in N$ with a budget e_i of (virtual) currency. We assume that all agents' budgets are restricted to have the same sign, $\text{sign}(e_i) = \text{sign}(e_j), \forall i, j \in N$. The sign of agents' budgets corresponds to the instance type. In positive instances, we assume strictly positive budgets with $e_i > 0$, while in negative instances all budgets are strictly negative. In null instances, there is no money, and computing a CE is easy.[3] Hence, for the rest of the paper, we concentrate on positive and negative instances.

[3] Any feasible allocation that gives all agents non-negative utility can be seen as CE. We can compute such an allocation when solving the LP (1) to determine the instance type.

A competitive equilibrium consists of an allocation x and a vector of prices $p = (p_1, \ldots, p_m)$ for the items. In markets with mixed manna, the prices of an item can also be positive or negative. A price $p_j > 0$ represents a payment to receive a fraction of an item an agent enjoys, while $p_j \leq 0$ means that agents are paid to receive a fraction of a bad item they dislike. Nevertheless, we say i *buys* item j whenever $x_{ij} > 0$.

Definition 2 (Competitive Equilibrium). *A pair (x^*, p^*) of allocation and prices is a* competitive equilibrium *if:*

1.) Items are fully allocated: $\sum_{i \in N} x^*_{ij} = 1, \ \forall j \in M$.
2.) Budgets are fully spent: $\sum_{j \in M} x^*_{ij} p^*_j = e_i, \ \forall i \in N$.
3.) Each agent $i \in N$ buys a utility-maximizing bundle:

$$x^*_i \in \arg \max_{x_i \in \mathbb{R}^m} u_i(x_i), \quad s.t. \sum_{j \in M} x_{ij} p^*_j \leq e_i, \ x_{ij} \geq 0. \tag{2}$$

Our algorithms for computing CE apply even to scenarios with different budgets, where agents have different entitlements to the items (e.g., when dissolving a business partnership where one partner is more senior than another). The prominent special case of equal budgets, i.e., $e_i = e_j, \ \forall i, j \in N$, is a *competitive equilibrium from equal incomes (CEEI)*.

Bogomolnaia et al. [7] show that CE exist under very general conditions and satisfy a number of fairness criteria. The following theorem summarizes the result in our context.

Theorem 1. *If agents' utility functions are SPLC, then a competitive equilibrium always exists. The allocation is Pareto-optimal, satisfies envy-freeness and proportionality.*

Global Goods and Bads. It is easy to see that any item j either has $p^*_j > 0$ in every CE or $p^*_j \leq 0$. If $u_{ij1} > 0$ for some agent $i \in N$, then $p^*_j > 0$, since otherwise agent i has infinite demand for j in (2) regardless of the budget e_i. Then p^*_j cannot be an equilibrium price. If $\max_i u_{ij1} \leq 0$, then $p^*_j < 0$, since otherwise no agent chooses to purchase j in (2). Therefore item j is not allocated at all.

Hence, in addition to *individual* goods and bads for each agent, we define a *global set of goods* $M^+ = \{j \in M : \ \max_{i \in N} u_{ij1} > 0\}$ and the complement, a global set of *bads* $M^- = M \setminus M^+ = \{j \in M : \ \max_{i \in N} u_{ij1} < 0\}$.

2.3 Optimal Bundles

Let us analyze the structure of an agent's optimal bundle in a CE. Note that for SPLC utilities, the optimization problem in (2) is an LP. We use variables x_{ijk} as agent i's allocation on the k-th segment of item j. Since the segment (i, j, k)

has length l_{ijk}, we have $0 \leq x_{ijk} \leq l_{ijk}$. Given a vector of prices p, agent i then solves the LP $\max_{x_i} \left\{ \sum_{j,k} u_{ijk} x_{ijk} \middle| \sum_{j,k} x_{ijk} p_j \leq e_i, 0 \leq x_{ijk} \leq l_{ijk} \right\}$.

Bang and Pain Per Buck. Given prices p, we define agent i's *bang per buck* for the k-th segment of good $j \in M^+$ as $bpb_{ijk} = u_{ijk}/p_j$, and the *pain per buck* for the k-th segment of bad $j \in M^-$ as $ppb_{ijk} = u_{ijk}/p_j$. Note that *bpb* (*ppb*) gives the utility (disutility) per unit spending on a good (bad). Next, we partition the segments of i's utility function into the equivalence classes $\{G_1^i, \ldots, G_k^i\}$ with the same *bpb*, where the G_j^i are labeled in decreasing order of *bpb*. Similarly, we define $\{B_1^i, \ldots, B_{k'}^i\}$ as the equivalence classes of segments with the same *ppb* labeled in increasing order. Intuitively, agent i must buy the segments of the G_j^i's in increasing order, i.e., all of G_1^i, then all of G_2^i and so on, since they provide the highest utility per unit spending. Similarly, i buys the segments of the B_j^i's in increasing order since they provide the minimum disutility per unit spending. These facts are easy consequences of KKT conditions applied to the above LP.

Forced and Flexible Segments. If agent i exhausts her budget in the segments G_r^i and B_s^i, then she buys all the segments in G_1^i through G_{r-1}^i, and B_1^i through B_{s-1}^i. We call these *forced segments* since i must buy them to maximize her utility. We call the segments of G_r^i and B_s^i *flexible segments*, since i can buy a fraction of any of the segments, but she need not buy the entire (or even any part) of these segments. Finally, we call segments of a class *undesirable* when they have lower *bpb* than G_r^i or higher *ppb* than B_s^i.

The following proposition shows a structural condition on the bang and pain per buck of flexible segments for goods and bads in a CE.

Proposition 2. *Let (x, p) be a CE, and let G_r^i and B_s^i be flexible segments of agent i. If $(i, j, k) \in G_r^i$ and $(i, j', k') \in B_s^i$, then, $bpb_{ijk} = ppb_{ij'k'}$.*

Proof. Clearly, $bpb_{ijk} \geq ppb_{ij'k'}$ otherwise disutility per unit earning exceeds utility gained per unit spending. For contradiction, suppose that $bpb_{ijk} > ppb_{ij'k'}$. Recall that this means i's utility gained per unit spending on the segment (i, j, k) is higher than her utility lost per unit earning on segment (i, j', k'). We want to show that i can increase her utility by purchasing a small additional amount of each item.

Formally, suppose i purchases δ_{ijk} and $\delta_{ij'k'}$ more of segments (i, j, k) and (i, j', k') respectively, and let y_i be her new bundle: $y_{ijk} = x_{ijk} + \delta_{ijk}$, $y_{ij'k'} = x_{ij'k'} + \delta_{ij'k'}$, and $y_{ilt} = x_{ilt}$ otherwise. By purchasing in the ratio $\delta_{ij'k'} = -\delta_{ijk} p_j / p_{j'}$, i's spending remains unchanged. Further, choosing $\max(\delta_{ijk}, \delta_{ij'k'}) \leq \max(l_{ijk} - x_{ijk}, l_{ij'k'} - x_{ij'k'})$ ensures that her new bundle y_i remains on the segments (i, j, k) and (i, j', k'). Therefore, y_i is a feasible bundle with the same total spending as x_i. Now observe that

$$u_i(y_i) - u_i(x_i) = \delta_{ijk} u_{ijk} + \delta_{ij'k'} u_{ij'k'} = \delta_{ijk} p_j \left(\frac{u_{ijk}}{p_j} - \frac{u_{ij'k'}}{p_{j'}} \right) > 0,$$

since $u_{ijk}/p_j = bpb_{ijk} > ppb_{ij'k'} = u_{ij'k'}/p_{j'}$. Therefore, i's bundle x_i is not optimal for prices p, contradicting that (x, p) is a CE. $\square$

UPB Graph. Given prices p, we define the following bipartite graph $G(p) = (V, E)$ that we refer to as the *utility per buck graph (UPB)*. We drop the price argument when the meaning is clear. We create a vertex for each agent $i \in N$ on one side and a vertex each item $j \in M$ on the other side. Let $G^i_k \cup B^i_{k'}$ be the flexible segments for agent i. We create the following edges: (i, j), $\forall j \in G^i_k \cup B^i_{k'}$, $\forall i \in N$.

3 Constant Number of Agents or Items

In this section, we discuss an algorithm for computing all CE for instances with SPLC utilities when there is a constant number of agents or a constant number of items. Our approach represents an extension of algorithms for linear utilities [17]. The treatment of SPLC utilities creates a number of technical challenges in the correct handling of forced and flexible segments.

We assume that the input is a market with agents, items, utilities, and budgets[4] (in accordance with the instance type). Our algorithm is based on the 'cell' decomposition technique pioneered by [15]. It rests on the fact that k hyperplanes separate $\mathbb{R}^d$ into $O(k^d)$ non-empty regions or *cells*. If d is constant, then this creates only polynomially many cells. We choose hyperplanes so that each cell corresponds to a unique set of forced and a unique set of flexible segments for each agent. We call such a set system a *UPB configuration*. Since agents only purchase segments from a UPB configuration in a CE, each cell uniquely determines which items an agent might purchase. Hence, for a cell it remains to check the conditions for a CE: 1.) all items are fully sold, and 2.) all agents spend their budget. Note that the optimal bundles condition will get automatically satisfied by consistent selection of forced and flexible segments.

Overall, our algorithm proceeds as follows: 1.) Enumerate the polynomially many UPB configurations via cell decomposition. Then, for each UPB configuration: 2.) Check whether there are feasible prices. 3.) Check whether for these prices there is a CE allocation consistent with the UPB configuration. We here concentrate on step 1, polynomial-time algorithms for steps 2 and 3 are described in the full version.

3.1 Finding UPB Configurations

We present a cell decomposition to determine all meaningful UPB configurations. We show that if the number of agents or items is constant, we obtain only poly(n, m) cells. Using polynomial-time algorithms for finding prices and allocations (full version), we get a polynomial-time algorithm to compute all CE.

Constant Number of Agents. Let $n = |N| = d$ be a constant. Suppose for a given set of prices, B^i_r and G^i_s are agent i's flexible segments. Then, for any

[4] Alternatively, if the goal is to compute CEEIs for a fair division instance, we can determine the instance type in polynomial time by solving the LP (1) and then assign appropriate budgets $e_i = 1$ or $e_i = -1$ for all $i \in N$.

$(i, j, k) \in B_r^i \cup G_s^i$, we have $u_{ijk}/p_j = \alpha_i > 0$. Also, any segment with $u_{ijk}/p_j > \alpha_i$ must be forced for good $j \in M^+$, and any segment with $u_{ijk}/p_j < \alpha_i$ must be forced for a bad $j \in M^-$. Note that if (i, j, k) is a flexible segment for i, then $u_{ijk}/\alpha_i = p_j$.

Let $\lambda_i = 1/\alpha_i$, and consider $\mathbb{R}^n$ with coordinates $\lambda_1, \ldots, \lambda_d$. For each tuple (a, b, j, r, s) where $a, b \in N$, $j \in M$, $r \leq |f_{aj}|$ and $s \leq |f_{bj}|$, we create a hyperplane $u_{ajr}\lambda_a - u_{bjs}\lambda_b = 0$. In the $>$ half-space, we have $u_{ajr}\lambda_a > u_{bjs}\lambda_b$. If (b, j, s) is flexible segment of good $j \in M^+$ for agent b, then $u_{ajr}\lambda_a > u_{bjs}\lambda_b = p_j$, or $u_{ajr}/p_j > 1/\lambda_a = \alpha_a$, i.e., the segment (a, j, r) is forced for agent a. Similarly, (b, j, s) is flexible segment of bad $j \in M^-$ for agent b, then in the $>$ half-space $u_{ajr}/p_j < \alpha_a$, i.e., (a, j, r) is forced for agent a.

A *cell* is the intersection of these half-spaces, which gives a partial ordering on the $u_{ijk}\lambda_i$'s. We sort the segments of good $j \in M^+$ in the decreasing order of $u_{ijk}\lambda_i$, and partition them into equivalence classes $G_1^j, \ldots, G_s^j$ with the same $u_{ijk}\lambda_i$ value. Similarly, we create equivalence classes $B_1^j, \ldots, B_s^j$ for bad j by sorting the $u_{ijk}\lambda_i$ in increasing order. By the above discussion, if flexible segments of good $j \in M^+$ are in G_t^j, then all segments in $G_{t'}^j$ with $t' < t$ are forced. Let $B_{<k}^j = \cup_{z=1}^{k-1} B_z^i$ and define $G_{<k}^i$ similarly. Now the flexible segment, say s, of good j is the largest integer such that $\sum_{(i,j,k) \in G_{<s}^j} l_{ijk} < 1$, since the last spending by any agent on good j before it is fully sold happens on G_s^j. The same holds for any bad $j \in M^-$. Then, each cell corresponds to a unique UPB configuration. Let $S = \max_{i,j} |f_{ij}|$. Observe that the total number of hyperplanes created is $m\binom{nS}{2} = O(mn^2S^2)$, and they divide $\mathbb{R}^n$ into at most $O((mn^2S^2)^n) = O((mS^2)^d)$ many cells.

Constant Number of Items. We concentrate on negative instances, i.e., $e_i < 0$. One can adapt the argument to positive instances by swapping the roles of goods and bads. Due to space constraints we discuss a high-level overview here. Let $m = |M| = d$, a constant. Consider $\mathbb{R}^d$ with coordinates $p_1, \ldots, p_d$. For each tuple (i, j, k, r, s) where $i \in N$, $j \neq k \in M$, $r \leq |f_{ij}|$ and $s \leq |f_{ik}|$, we create a hyperplane $u_{ijr}p_k - u_{iks}p_j = 0$. Each hyperplane divides $\mathbb{R}^d$ into regions with signs $>$, $=$, or $<$, where the sign of determines whether i prefers the segment (i, j, r) or (i, k, s), e.g., if $j, k \in M^+$ then $u_{ijr}/p_j \geq u_{iks}/p_k$ in the $\geq$ region. A cell is the intersection of these half-spaces, so that a cell gives a partial ordering of bpb_{ijr} and ppb_{ijr} for each agent $i \in N$. Sort the segments (i, j, r) of goods in decreasing order of *bpb* for agent i and create the equivalence classes $G_1^i, \ldots, G_c^i$ with the same *bpb*. Similarly create the equivalence classes $B_1^i, \ldots, B_{c'}^i$ of segments of bads with the same *ppb*, sorted in increasing order. We let ppb_j be the *ppb* of B_j^i, and bpb_j be the *bpb* of G_j^i.

Let $B_{<j}^i = \cup_{z=1}^{j-1} B_z^i$ and define $G_{<j}^i$ similarly. Also let $B_{<1}^i = G_{<1}^i = \emptyset$. If B_j^i and G_k^i are i's flexible segments, then $B_{<j}^i$ and $G_{<k}^i$ are her forced segments. Thus, each choice of flexible segments B_j^i and G_k^i for each agent yields a unique UPB configuration.

To find agent i's flexible segments we add another set of hyperplanes $\sum_{(i,j,k) \in B_{<r}^i \cup G_{<s}^i} l_{ijk}p_j - e_i = 0$ to partition cells into sub-cells. The sign of

sub-cell >, =, or < determines whether an agent over- or underspends her budget. For example, in a negative instance where $e_i = -1$, then in the > region $\sum_{(i,j,k)\in B^i_{<r}\cup G^i_{<s}} l_{ijk}p_j > e_i$, so if agent i purchases all segments of $B^i_{<r} \cup G^i_{<s}$, then she still needs to purchase more bads to reach her budget. From this information we can ultimately determine i's flexible segments. This aspect is the most significant challenge by SPLC utilities over the linear case in [17].

4 Constant Number of Bads

In this section, we show that in a negative instance when agents have linear utility functions for bads we can relax the requirement for a constant number of items, and instead only ask for a constant number of bads. To be clear, we still allow any number of goods with SPLC utilities. This result improves on [17] by using a weaker set of assumptions to obtain a polynomial time algorithm to compute a CE of mixed manna.

Note that linear utility function is SPLC with a single segment, i.e., $f_{ij}(x_{ij}) = u_{ij}x_{ij}$. For a set of prices p, define the minimum pain per buck bads as $mpb_i = \arg\min_{j\in M^-} u_{ij}/p_j$ and let $\alpha_i = \min_{j\in M^-} u_{ij}/p_j$. In a negative instance where all agents must purchase some bads, α_i is well defined. Let G^i_k be agent i's flexible segment for goods with bang per buck bpb_k. Then $bpb_k = \alpha_i$, and any segments G^i_j with $bpb_j > bpb_k$ are forced.

Finding UPB Configurations. The algorithm has the same basic structure as in Sect. 3.1: we use a cell decomposition to enumerate UPB configurations, then determine prices and check if an equilibrium allocation exists. The difference lies in the cell decomposition. It can be seen as a hybrid of the techniques used in the two scenarios in Sect. 3.1.

In a negative instance, agents have negative budgets $e_i = -1$, and must earn on some bads. First, we determine the mpb_i bads for each agent in a cell using a similar approach as the constant number of items case. This gives the set of bads each agent might purchase and determines the value of $\alpha_i = \min_{j\in M^-} u_{ij}/p_j$. In the constant number of agents case, we used the variables $\lambda_i = 1/\alpha_i$ to determine mbb_i goods and mpb_i bads for each agent. Now we adapt the approach using the variables $p_j/u_{ij} = 1/\alpha_i = \lambda_i$, for bad $j \in mpb_i$.

Theorem 2. *Suppose the instance is negative and that agents have linear utility functions for bads and SPLC utility functions for goods. If the number of bads is constant, then we can compute all CE in polynomial time.*

Proof. Let $d = |M^-|$ be a constant. Consider $\mathbb{R}^d$ with coordinates $p_1, \ldots, p_d$. For each agent $i \in N$ and each pair of bads $j, k \in M^-$ we introduce the hyperplane $u_{ij}p_k - u_{ik}p_j = 0$, which partitions $\mathbb{R}^d$ into regions with sign >, =, or <. Thus, a cell gives a partial ordering on the terms $u_{ij}p_k$. Sort the bads by $u_{ij}p_k$ values under this ordering, i.e., $j < k$ if $u_{ij}p_k < u_{ik}p_j$, and let $B^i_1, \ldots, B^i_c$ be the equivalence classes listed in increasing order. Then B^i_1 are the mbb_i goods for agent i in the cell. To see this, suppose $(i, j) \in B^i_1$ and $(i, k) \in B^i_z$ for some $z > 1$.

Then, $u_{ij}p_k < u_{ik}p_j$, or $u_{ij}/p_j < u_{ik}/p_k$, i.e., $j \in mbb_i$. We use $\binom{d}{2}$ hyperplanes for each agent, giving $O(nd^2)$ in total. Therefore there are at most $O(n^d)$ cells.

Note that all bads $j \in mpb_i$ have *ppb* of $\alpha_i = \min_{j \in M^-} u_{ij}/p_j$. Recall that we used $\lambda_i = 1/\alpha_i$ to determine the forced and flexible segments when the number of agents is constant. We follow a similar procedure, this time using $p_j/u_{ij} = 1/\alpha_i = \lambda_i$, for a $j \in mpb_i$. To simplify notation, for each agent i pick a bad $k \in mpb_i$, and let $c(i) = 1/u_{ik}$ and define $p(i) = p_k$. Then $p(i)c(i) = \lambda_i = 1/\alpha$.

We now determine the flexible segments of goods for each agent in a given cell. For each tuple (i, i', j, k, k') where $i \neq i' \in N$, $j \in M^+$, $k < |f_{ij}|$, $k' < |f_{i'j}|$, we create a hyperplane $u_{ijk}c(i)p(i) - u_{i'jk'}c(i')p(i') = 0$, if $p(i) \neq p(i')$. Otherwise, we compare the values $u_{ijk}|c(i)|$ and $u_{i'jk'}|c(i')|$ directly, since $p(i)$, $c(i) < 0$. This further divides a cell into sub-cells where we have a partial ordering on the agents' segments for each good $j \in M^+$, i.e., $(i, j, k) > (i', j, k')$ if $u_{ijk}c(i)p(i) > u_{i'jk'}c(i')p(i')$ since c_i, $p(i) < 0$. For each good $j \in M^+$, define $G_1^j, \ldots, G_c^j$ as the equivalence classes with the same $u_{ijk}c(i)p(i)$ value, sorted in decreasing order.

Since each good must be fully sold, let r be largest integer such that $\sum_{(i,j,k) \in G_{<r}^j} l_{ijk} < 1$, i.e., j becomes fully sold once agents purchase the segments of $G_{\leq r}^j$. Then, G_r^j are the flexible segments. Indeed, let $(i, j, k) \in G_r^j$ be a flexible segment for agent i. This means that $u_{ijk}/p_j = \alpha_i$, or $p_j = u_{ijk}/\alpha_i = u_{ijk}c(i)p(i)$, by our choice of $c(i)$ and $p(i)$. Consider the segment $(i', j, k') \in G_q^j$, for some $q < r$. Then, $\frac{u_{i'jk'}}{\alpha_{i'}} = u_{i'jk'}c(i')p(i') > u_{ijk}c(i)p(i) = p_j$, i.e., $\frac{u_{i'jk'}}{p_j} > \alpha_{i'}$, so that (i', j, k') is a forced segment for agent i. Also, by our choice of r, the final segments of j that agents purchase are G_r^j.

Let S be the maximum number of segments of any agents' utility functions. We formed sub-cells by adding hyperplanes for each tuple (i, i', j, k, k') where $i \neq i' \in N$, $j \in M^+$, $k < |f_{ij}|$, $k' < |f_{i'j}|$. We created $|M^+|\binom{nS}{2} = O(mn^2S^2)$ overall in any given cell, which partitions the cell into at most $O(m^d(nS)^{2d})$ sub-cells. As previously calculated, there are $O(n^d)$ cells. The total number of sub-cells is $O(m^d n^{3d} S^{2d})$, which is poly(n, m, S) for constant d. □

Remark: If both goods and (constantly many) bads have SPLC utilities, we need to find agent i's flexible segments of bads B_s^i. The *ppb* of theses segments is α_i. However, flexible segments are determined by ensuring an agent spends her entire budget, which obviously depends on both goods and bads. Thus, we cannot consider goods and bads separately as we have done in this proof. Finding a polynomial-time algorithm in this case is an interesting open problem.

5 Indivisible Manna

Finally, we turn to fair division with *indivisible* mixed manna. We assume that there are m indivisible items. Each agent $i \in N$ has a utility value u_{ij} for each item $j \in M$. In this section, we assume that the utilities for the agents are additive, i.e., $u_i(S_i) = \sum_{j \in S_i} u_{ij}$ for every subset $S_i \subseteq M$ of items assigned to agent i. Item j is a *good for agent* i if $u_{ij} > 0$. If $u_{ij} < 0$, then j is a *bad for*

i.[5] More globally, we define sets of (global) goods and bads as in Sect. 2, i.e., $M^+ = \{j \in M : \max_{i \in N} u_{ij} > 0\}$ and $M^- = M \setminus M^+$.

In a *feasible allocation*, we *can* assign the goods but *must* assign all bads to the agents. Clearly, in a *Pareto-optimal (PO)* feasible allocation, we assign all items; in particular, goods only get assigned to agents that have positive value for them. While finding a feasible allocation is trivial, our goal is to satisfy a natural fairness criterion that we term *proportional up to a single item*. Our definition is a direct extension of the version for goods to mixed manna.

Definition 3. *A feasible allocation $S = (S_1, \ldots, S_n)$ for an instance with mixed manna is* proportional up to a single item (PROP1) *if for every agent i there is $j \in M^+$ such that $u_i(S_i \cup \{j\}) \geq \frac{1}{n} \cdot u_i(M)$ or $j \in S_i \cap M^-$ such that $u_i(S_i \setminus \{j\}) \geq \frac{1}{n} \cdot u_i(M)$.*

Our main result is a polynomial-time rounding algorithm that yields a feasible PROP1 allocation. Our algorithm is inspired by algorithms for markets with only goods [5,21]. We pretend the instance is divisible with linear utilities, compute a CEEI based on the instance type (all budgets are 1, 0, or -1 respectively), and then use our algorithm to round the CEEI to an indivisible allocation that is feasible and PROP1. For positive and negative instances it is also PO.

Theorem 3. *There is a polynomial-time algorithm to round any CEEI for a divisible instance with mixed manna and linear utilities to a feasible indivisible PROP1 allocation. For positive/negative instances the rounded allocation is PO.*

Proof. Due to space constraints, we show the result for positive instances. Consider a CEEI (x, p) in a positive instance. Agent i only buys from a subset of goods that give the maximum bang per buck $mbb_i = \max_{k \in M^+} u_{ik}/p_k$ and/or a subset of bads that give minimum pain per buck $mpb_i = \min_{k \in M^-} u_{ik}/p_k$. If i buys both goods and bads, then $mbb_i = mpb_i$. The sets of mbb_i goods and mpb_i bads are invariant to scaling all utility values u_{ij} by a common factor $\gamma_i > 0$. Further, properties feasibility, PO, and PROP1 are also invariant to such a scaling. Hence, we assume w.l.o.g. that the utilities are scaled such that whenever $x_{ij} > 0$, this implies $u_{ij}/p_j = 1$. As a consequence, since all budgets are 1, we have $u_i(x_i) = 1$ for all $i \in N$. Further, by market clearing, $\sum_j p_j = \sum_i e_i = n$. Hence, with a budget of n, any agent i would be able to buy all goods and bads. However, when doing so, every good delivers at most a utility per unit spending of 1, and every bad at least a pain per utility of earning of 1. As a consequence, $u_i(M) \leq n$, and $u_i(x_i) \geq \frac{1}{n} u_i(M)$.

Now consider the *allocation graph* G, i.e., the bipartite graph composed of agents, items, and edges $E = \{\{i, j\} \in N \times M \mid x_{ij} > 0\}$. Because the allocation x is fractional PO (i.e., no other allocation makes an agent better off without making someone else worse off), we can use standard arguments for linear markets and assume that the allocation graph is a forest [8,16]. Moreover, for the

[5] For consistency with previous sections, we assume that $u_{ij} \neq 0$ throughout. Our arguments can be adapted easily by assuming that when $u_{ij} = 0$, j is a good for i.

same reason, it holds that $x_{ij} > 0$ and $u_{ij} < 0$ only when $j \in M^-$ is a (global) bad. Thus, for every agent, the set of incident goods in G fulfills

$$\sum_{j \in M^+ : x_{ij} > 0} u_{ij} \geq 1 - \sum_{j \in M^- : x_{ij} > 0} x_{ij} u_{ij}. \tag{3}$$

For our rounding algorithm, we consider every tree T of G separately. We root T in an arbitrary agent r and initialize $S_r = \emptyset$. Now apply a greedy algorithm: First add all incident leaf items to S_r, since these items are already assigned fully to r in the CEEI. Then go through the remaining children goods of r in non-increasing order of u_{rj}, and add good j to S_r as long as $u_r(S_r \cup \{j\}) \leq 1$. Due to (3) and since we include only leaf bads of r into S_r, there is either child good j' with $u_r(S_r \cup \{j'\}) > 1$ or we add all child items to S_r resulting in $u_r(S_r) = 1$. In both cases, S_r fulfills PROP1.

We recursively apply the greedy approach. Remove r from T and all children goods assigned to r. Assign each remaining child item of r to its respective child agent. This splits T into a number of subtrees of $T_1, T_2, \ldots$. The new roots are the grandchildren $r_1, r_2, \ldots$ of r. We label each new root whether it received its parent good (RG), its parent bad (RB), or did not receive its parent good (NG). Note that, recursively, a parent bad is always assigned to the child agent.

If r_i is (RG), it is easy to see that the greedy procedure and the arguments for r can be applied directly to assign a subset to r_i that is PROP1. If r_i is (RB), let j_i be the parent bad. We apply the greedy procedure, but stop only after the first child good that yields $u_{r_i}(S_{r_i} \setminus \{j_i\}) \geq 1$. Such a good exists, since r_i buys a fraction of j_i in the CEEI, and the set of all child goods of r_i fulfills (3). Clearly, the resulting set S_{r_i} is PROP1. Finally, if r_i is (NG), let j_i be the parent good. Hence, we apply the greedy algorithm, but stop only after the first child good gives $u_{r_i}(S_{r_i} \cup \{j_i\}) \geq 1$. Again, such a good exists due to (3). Again, the resulting S_{r_i} is PROP1. Using these arguments, we can proceed recursively top-down through the entire tree. The resulting allocation is PROP1.

For PO, recall that we scale utilities based on mbb_i and mpb_i values. We allocate only mbb_i goods and mpb_i bads in the CEEI, so $x_{ij} > 0$ only if $u_{ij}/p_j = 1$, i.e., if $x_{ij} > 0$, then $u_{ij} = p_j$. The other items have less value, i.e., if $x_{ij} = 0$, then $u_{ij} \leq p_j$. The algorithm assigns item j to agent i only if $x_{ij} > 0$ and, thus, $u_{ij} = p_j$. As a consequence, the algorithm gives each item to an agent with maximum scaled utility for that item. Hence, the allocation maximizes the sum of all scaled utilities of the agents. This proves that it is PO.

References

1. Aziz, H., Caragiannis, I., Igarashi, A., Walsh, T.: Fair allocation of indivisible goods and chores. In: Proceedings of the 28th International Joint Conference on Artificial Intelligence (IJCAI) (2019)
2. Aziz, H., Moulin, H., Sandomirskiy, F.: A polynomial-time algorithm for computing a Pareto optimal and almost proportional allocation. Oper. Res. Lett. **48**(5), 573–578 (2020)

3. Aziz, H., Rey, S.: Almost group envy-free allocation of indivisible goods and chores. IJCAI (2020)
4. Azrieli, Y., Shmaya, E.: Rental harmony with roommates. J. Econ. Theory **153**, 128–137 (2014)
5. Barman, S., Krishnamurthy, S.K.: On the proximity of markets with integral equilibria. In: Proceedings of the 33rd Conference on Artificial Intelligence (AAAI) (2019)
6. Bhaskar, U., Sricharan, A., Vaish, R.: On approximate envy-freeness for indivisible chores and mixed resources (2020). arxiv:2012.06788
7. Bogomolnaia, A., Moulin, H., Sandomirskiy, F., Yanovskaia, E.: Competitive division of a mixed manna. Econometrica **85**(6), 1847–1871 (2017)
8. Bogomolnaia, A., Moulin, H., Sandomirskiy, F., Yanovskaia, E.: Dividing bads under additive utilities. Soc. Choice Welf. **52**(3), 395–417 (2018). https://doi.org/10.1007/s00355-018-1157-x
9. Brams, S.J., Taylor, A.D.: Fair Division - From Cake-Cutting to Dispute Resolution. Cambridge University Press, Cambridge (1996)
10. Brandt, F., Conitzer, V., Endriss, U., Lang, J., Procaccia, A. (eds.): Handbook of Computational Social Choice. Cambridge University Press, Cambridge (2016)
11. Branzei, S., Sandomirskiy, F.: Algorithms for competitive division of chores (2019). arXiv:1907.01766
12. Chaudhury, B.R., Garg, J., McGlaughlin, P., Mehta, R.: Dividing bads is harder than dividing goods: on the complexity of fair and efficient division of chores (2020). arxiv:2008.00285
13. Chaudhury, B.R., Garg, J., McGlaughlin, P., Mehta, R.: Competitive allocation of a mixed manna. In: Proceedings of the 31st Symposium on Discrete Algorithms (SODA) (2021)
14. Chen, X., Teng, S.: Spending is not easier than trading: on the computational equivalence of fisher and Arrow-Debreu equilibria. In: Proceedings of the 20th International Symposium on Algorithms and Computation (ISAAC), pp. 647–656 (2009)
15. Devanur, N., Kannan, R.: Market equilibria in polynomial time for fixed number of goods or agents. In: Proceedings of the 49th Symposium on Foundations of Computer Science (FOCS), pp. 45–53 (2008)
16. Devanur, N., Papadimitriou, C., Saberi, A., Vazirani, V.: Market equilibrium via a primal-dual algorithm for a convex program. J. ACM **55**(5), 1–18 (2008)
17. Garg, J., McGlaughlin, P.: Computing competitive equilibria with mixed manna. In: Proceedings of the 19th Conference on Autonomous Agents and Multi-Agent Systems (AAMAS), pp. 420–428 (2020)
18. Garg, J., Mehta, R., Sohoni, M., Vazirani, V.V.: A complementary pivot algorithm for market equilibrium under separable, piecewise-linear concave utilities. SIAM J. Comput. **44**(6), 1820–1847 (2015)
19. Garg, J., Végh, L.A.: A strongly polynomial algorithm for linear exchange markets. In: Proceedings of the 51st Symposium on Theory of Computing (STOC) (2019)
20. Huang, X., Lu, P.: An algorithmic framework for approximating maximin share allocation of chores (2019). arXiv:1907.04505
21. McGlaughlin, P., Garg, J.: Improving Nash social welfare approximations. J. Artif. Intell. Res. **68**, 225–245 (2020)
22. Moulin, H.: Fair Division and Collective Welfare. MIT Press, Cambridge (2003)
23. Nisan, N., Tardos, É., Roughgarden, T., Vazirani, V. (eds.): Algorithmic Game Theory. Cambridge University Press, Cambridge (2007)

24. Orlin, J.: Improved algorithms for computing Fisher's market clearing prices. In: Proceedings of the 42nd Symposium on Theory of Computing (STOC), pp. 291–300 (2010)
25. Robertson, J., Webb, W.: Cake-Cutting Algorithms: Be Fair If You Can. AK Peters, MA (1998)
26. Su, F.E.: Rental harmony: sperner's lemma in fair division. Am. Math. Mon. **106**(10), 930–942 (1999)

Computing Fair and Efficient Allocations with Few Utility Values

Jugal Garg and Aniket Murhekar(✉)

University of Illinois at Urbana-Champaign, Champaign, USA
{jugal,aniket2}@illinois.edu

Abstract. We study the problem of allocating indivisible goods among agents with additive valuations in a fair and efficient manner, when agents have *few* utility values for the goods. We consider the compelling fairness notion of envy-freeness up to any good (EFX) in conjunction with Pareto-optimality (PO). Amanatidis et al. [1] showed that when there are at most two utility values, an EFX allocation can be computed in polynomial-time. We improve this result by showing that for such instances an allocation that is EFX and PO can be computed in polynomial-time. This is the first class apart from identical or binary valuations, for which EFX+PO allocations are shown to exist and are polynomial-time computable. In contrast, we show that when there are three utility values, EFX+PO allocations need not exist, and even deciding if EFX+PO allocations exist is NP-hard.

Our techniques allow us to obtain similar results for the fairness notion of equitability up to any good (EQX) together with PO. We show that for instances with two positive values an EQX+PO allocation can be computed in polynomial-time, and deciding if an EQX+PO allocation exists is NP-hard when there are three utility values.

We also study the problem of maximizing Nash welfare (MNW), and show that our EFX+PO algorithm returns an allocation that approximates the MNW to a factor of 1.061 for two valued instances, in addition to being EFX+PO. In contrast, we show that for three valued instances, computing an MNW allocation is APX-hard.

Keywords: Fair and efficient allocation · EFX · Nash welfare · EQX

1 Introduction

The problem of fair division was formally introduced by Steinhaus [36], and has since been extensively studied in various fields, including economics and computer science [10,32]. It concerns allocating resources (goods) to agents in a *fair* and *efficient* manner, and has various practical applications such as rent division, division of inheritance, course allocation, and government auctions.

Supported by NSF Grant CCF-1942321 (CAREER).

I. Caragiannis and K. A. Hansen (Eds.): SAGT 2021, LNCS 12885, pp. 345–359, 2021.
https://doi.org/10.1007/978-3-030-85947-3_23

Arguably, the most popular notion of fairness is *envy-freeness* (EF) [19,37], which requires that every agent prefers their own bundle of goods to that of any other. However in the case of *indivisible* goods, EF allocations need not even exist (consider allocating 1 good among 2 agents). This motivated the study of its relaxations. One such relaxation is *envy-freeness up to one good* (EF1) allocation, defined by Budish [11], where every agent prefers their own bundle to the bundle of any other agent after removing *some* good from the other agent's bundle. It is well-known that an EF1 allocation always exists and is polynomial-time computable [29]. However, an EF1 allocation may be unsatisfactory because it allows the removal of the *most valuable* good from the other agent's bundle, which might be the main reason for huge envy to exist in the first place. Therefore, stronger fairness notions are desirable in many settings.

A stronger notion is called *envy-free up to any good* (EFX), defined by Caragiannis et al. [12], which requires every agent to prefer their bundle over the bundle of any other agent after removing *any* good from the other agent's bundle. Clearly, any allocation that is EFX is also EF1, but not vice-versa. The existence of EFX allocations is known for identical valuations [34], and was recently shown for 3 agents with additive valuations [15].[1] At the same time, we want the output allocation to be efficient because a fair allocation by itself may be highly inefficient. Consider for example two agents A_1 and A_2 and 2 goods g_1 and g_2 where A_i values only g_i and does not value the other good. The allocation in which g_1 is assigned to A_2 and g_2 is assigned to A_1 is clearly EFX. However both agents get zero utility, which is highly inefficient. The allocation in which g_i is assigned to A_i is more desirable since it is both fair as well as efficient.

The standard notion of economic efficiency is Pareto optimality (PO). An allocation is said to be PO if no other allocation makes an agent better off without making someone else worse off. A stronger notion called fractional Pareto optimality (fPO) requires that no other fractional allocation makes an agent better off without making someone else worse off. Every fPO allocation is therefore PO, but not vice-versa (see the appendix for an example). Another reason to prefer fPO allocations over PO allocations is that the former admit efficient verification while the latter do not: given an allocation, it can be checked in polynomial time if it is fPO [5], whereas checking if an allocation is PO is coNP-complete [27]. Hence if a centralized entity responsible for allocating resources claims the allocation is fPO, each agent can individually verify that this is indeed the case; in contrast such a check is not efficiently possible if the guarantee is only PO.

An important question is whether the notions of fairness (EF1 or EFX) can be achieved in conjunction with the efficiency notions (PO or fPO). Further, if yes, then whether they can be computed in polynomial-time. For this, the concept of *Nash welfare* provides a partial answer. The Nash welfare is defined as the geometric mean of the agents' utilities, and by maximizing it we achieve

[1] Settling the (non-)existence of EFX allocations is considered the biggest open question in fair division [35]; see [16] and references therein for recent progress on this problem.

a tradeoff between efficiency and fairness. Caragiannis et al. [12] showed that the maximum Nash welfare (MNW) allocations are EF1 and PO under additive valuations. However, the problem of computing an MNW allocation is APX-hard [28] (hard to approximate). Bypassing this barrier, Barman et al. [5] devised a pseudo-polynomial-time algorithm that computes an EF1+PO allocation. In a recent paper, Garg *et al.* [23] showed that an EF1+fPO allocation can be computed in pseudo-polynomial time. For the special case of binary additive valuations an MNW allocation is EFX+fPO, and is known to be polynomial-time computable [6,18].

1.1 Our Contributions

In this work, we obtain several novel results on the notions of EFX, EQX, PO, and MNW, especially for instances in which agents have few values for the goods. A fair division instance is called k-valued if values that agents have for the goods belong a set of size k.

EFX. Recently, Amanatidis *et al.* [1] showed that for 2-valued instances any MNW allocation is EFX+PO, but left open the question of whether it can be computed in polynomial-time. They presented a polynomial-time algorithm which computes an EFX allocation for 2-valued instances, however, the outcome of their algorithm need not be PO (see the appendix for an example). In this work, we show EFX+fPO allocations always exist for 2-valued instances and can be computed in polynomial-time.[2] Further, apart from the classes of identical valuations and binary valuations, this is the first class for which EFX+PO allocations exist and can be computed in polynomial-time.

In general, EFX+PO allocations are not guaranteed to exist [34]. We therefore ask the natural question: what is the complexity of *checking* if an instance admits an EFX+PO allocation? We show that this problem is NP-hard, somewhat surprisingly, even for 3-valued instances.

EQX. Our techniques allow us to obtain similar results for the fairness notion of *equitability up to any good* [20,26]. An allocation is said to be EQX (resp. EQ1) if the utility an agent gets from her bundle is no less than the utility any other agent gets after removing *any* (resp. some) good from their bundle. We show that for positive 2-valued instances, an EQX+PO allocation can be computed in polynomial-time, and in contrast, even checking existence of EQX+PO allocations for 3-valued instances is NP-hard.

MNW. Our EFX+PO algorithm returns an allocation that approximates the maximum Nash welfare to a factor of 1.061 in addition to being EFX and PO. This guarantee is better than the best known 1.45-approximation algorithm of [5] for the MNW problem.

Amanatidis *et al.* [1] showed that computing an MNW allocation is NP-hard for 3-values instances, which, as they remark "extends the hardness aspect, but

[2] Our results extend to the much broader class where there are two values $\{a_i, b_i\}$ per agent, but a_i/b_i is the same across agents.

not the inapproximability, of the result of Lee [28] for 5-valued instances", who had shown that MNW is NP-hard to approximate within a factor of 1.00008. In our work, we extend the inapproximability aspect too, and show that it is NP-hard to approximate the MNW to a factor of 1.00019, even for 3-valued instances, which is better than Lee's result.

Thus, for the problems of computing (i) EFX+PO, (ii) EQX+PO, and (iii) MNW allocations, our work improves the state-of-the-art and also crucially pinpoints the boundary between tractable and intractable cases.

1.2 Other Related Work

Barman *et al.* [5] showed that for n agents and m goods, an EF1+PO allocation can be computed in time $\mathsf{poly}(n, m, v_{max})$, where v_{max} is the maximum utility value. Their algorithm first perturbs the values to a desirable form, and then computes an EF1+fPO allocation for the perturbed instance, which for a small-enough perturbation is EF1+PO for the original instance. Their approach is via *integral market-equilibria*, which guarantees fPO at every step. Our algorithm uses a similar approach, with one main difference being that we do not need to consider any approximate instance and can work directly with the given values. The outcome of our algorithm is EFX+fPO, which beats the guarantee of EF1+PO.

Another key difference is the *run-time analysis*: our arguments show termination in $\mathsf{poly}(n, m)$ time for 2-valued instances, even when $v_{max} = 2^{\Omega(n+m)}$, whereas the analysis of Barman *et al.* only shows a $\mathsf{poly}(n, m, v_{max})$ time bound, even for 2-valued instances.

Recently, Garg and Murhekar [33] showed that an EF1+fPO allocation can be computed in $\mathsf{poly}(n, m, v_{max})$-time, by using integral market-equilibria. They also showed that an EF1+fPO allocation can be computed in $\mathsf{poly}(n, m)$-time for k-valued instances where k is a constant, however they do not show that the allocation returned by their algorithm is EFX for 2-valued instances.

Freeman *et al.* [20] showed that EQ.1+PO allocations can be computed in pseudo-polynomial-time for instances with positive values. They also show that the leximin solution, i.e., the allocation that maximizes the minimum utility, and subject to this, maximizes the second minimum utility, and so on; is EQX+PO. However, as remarked in [34], computing a leximin solution is intractable.

Barman *et al.* [6] showed that for identical valuations, any EFX allocation provides a 1.061-approximation to the MNW. Garg *et al.* [21] show a 1.069-hardness of approximating MNW, although for 4-valued instances.

Instances with few values have been widely considered in the fair division literature: for instance Golovin [25] presents approximation algorithms and hardness results for computing max-min fair allocations in 3-valued instances; Aziz *et al.* [3] show PO is efficiently verifiable for 2-valued instances and coNP-hard for 3-valued instances; Aziz [2], and Vazirani and Yannakakis [38] study the Hylland-Zeckhauser scheme for probabilistic assignment of goods in 2-valued instances; Bogomolnaia and Moulin [9] study matching problems with 2-valued (dichotomous) preferences; Bliem *et al.* [8] study fixed-parameter tractability for computing EF+PO allocations with parameter $n + z$, where z is the number of values; and Garg *et al.* [24] study leximin assignments of papers ranked by reviewers on a small scale, in particular they present an efficient algorithm for 2 ranks, i.e., "high or low interest" and show NP-hardness for 3 ranks. More generally, such instances have been studied in resource allocation contexts, including makespan minimization with 2 or 3 job sizes [13,39].

2 Preliminaries

For $t \in \mathbb{N}$, let $[t]$ denote $\{1, \ldots, t\}$.

Problem Setting. A *fair division instance* is a tuple (N, M, V), where $N = [n]$ is a set of $n \in \mathbb{N}$ agents, $M = [m]$ is the set of $m \in \mathbb{N}$ indivisible items, and $V = \{v_1, \ldots, v_n\}$ is a set of utility functions, one for each agent $i \in N$. Each utility function $v_i : M \to \mathbb{Z}_{\geq 0}$ is specified by m numbers $v_{ij} \in \mathbb{Z}_{\geq 0}$, one for each good $j \in M$, which denotes the value agent i has for good j. We assume that the valuation functions are additive, that is, for every agent $i \in N$, and for $S \subseteq M$, $v_i(S) = \sum_{j \in S} v_{ij}$. Further, we assume that for every good j, there is some agent i such that $v_{ij} > 0$. Note that we can in general work with rational values without loss of generality, since they can be scaled to make them integral, and the efficiency and fairness guarantees we consider are scale-invariant.[3]

We call a fair division instance (N, M, V) a t-valued instance if $|\{v_{ij} : i \in N, j \in M\}| = t$. The class of 2-valued instances is made up of two disjoint fragments: binary instances, where all values $v_{ij} \in \{0, 1\}$; and $\{a, b\}$-instances, where all values $v_{ij} \in \{a, b\}$ for $a, b \in \mathbb{Z}_{>0}$. An important subclass of 3-valued instances is the $\{0, a, b\}$ class, wherein all values $v_{ij} \in \{0, a, b\}$ for $a, b \in \mathbb{Z}_{>0}$.

Allocation. An (integral) *allocation* $\mathbf{x}$ of goods to agents is a n-partition $(\mathbf{x}_1, \ldots, \mathbf{x}_n)$ of the goods, where $\mathbf{x}_i \subseteq M$ is the bundle of goods allotted to agent i, who gets a total value of $v_i(\mathbf{x}_i)$. A *fractional allocation* $\mathbf{x} \in [0, 1]^{n \times m}$ is a fractional assignment of the goods to agents such that for each good $j \in M$, $\sum_{i \in N} x_{ij} = 1$. Here, $x_{ij} \in [0, 1]$ denotes the fraction of good j allotted to agent i. In a fractional allocation $\mathbf{x}$, an agent i receives a value of $v_i(\mathbf{x}_i) = \sum_{j \in M} v_{ij} x_{ij}$.

Fairness Notions. An allocation $\mathbf{x}$ is said to be:

1. *Envy-free up to one good* (EF1) if for all $i, h \in N$, there exists a good $j \in \mathbf{x}_h$ s.t. $v_i(\mathbf{x}_i) \geq v_i(\mathbf{x}_h \setminus \{j\})$.
2. *Envy-free up to any good* (EFX) if for all $i, h \in N$ and for all goods $j \in \mathbf{x}_h$ we have $v_i(\mathbf{x}_i) \geq v_i(\mathbf{x}_h \setminus \{j\})$.
3. *Equitable up to one good* (EQ.1) if for all $i, h \in N$, there exists a good $j \in \mathbf{x}_h$ s.t. $v_i(\mathbf{x}_i) \geq v_h(\mathbf{x}_h \setminus \{j\})$.
4. *Equitable up to any good* (EQX) if for all $i, h \in N$ and for all goods $j \in \mathbf{x}_h$ we have $v_i(\mathbf{x}_i) \geq v_h(\mathbf{x}_h \setminus \{j\})$.

[3] The properties of EFX, PO, and Nash welfare are invariant under scaling, while EQX is not scale-invariant in general. However, in our algorithms this is not an issue since we only uniformly scale the valuations of all agents, which preserves EQX.

Pareto-optimality. An allocation $\mathbf{y}$ dominates an allocation $\mathbf{x}$ if for all $i \in N$, $v_i(\mathbf{y}_i) \geq v_i(\mathbf{x}_i)$ and there exists $h \in N$ s.t. $v_h(\mathbf{y}_h) > v_h(\mathbf{x}_h)$. An allocation is said to be *Pareto-optimal* (PO) if no allocation dominates it. Further, an allocation is said to be *fractionally* Pareto-optimal (fPO) if no fractional allocation dominates it. Thus, any fPO allocation is PO, but not vice-versa (see the appendix for an example).

Nash Welfare. The Nash welfare of an allocation $\mathbf{x}$ is given by $\mathsf{NW}(\mathbf{x}) = \left(\Pi_{i\in N} v_i(\mathbf{x}_i)\right)^{1/n}$. An allocation that maximizes the NW is called an MNW allocation or Nash optimal allocation. An allocation $\mathbf{x}$ *approximates* the MNW to a factor α if $\alpha \cdot \mathsf{NW}(\mathbf{x}) \geq \mathsf{NW}(\mathbf{x}^*)$, where $\mathbf{x}^*$ is an MNW allocation.

Fisher Markets. A Fisher market or a *market instance* is a tuple (N, M, V, e), where $N = [n]$ is a set of $n \in \mathbb{N}$ agents, $M = [m]$ is a set of $m \in \mathbb{N}$ divisible goods, $V = \{v_1, \ldots, v_n\}$ is a set of additive (linear) utility functions, and $e = \{e_1, \ldots, e_n\}$ is the set of agents' budgets, where each $e_i \geq 0$. In this model, agents can fractionally share goods. Each agent aims to obtain a bundle of goods that maximizes her total value subject to her budget constraint.

A *market outcome* is a fractional allocation $\mathbf{x}$ of the goods to the agents and a set of prices $\mathbf{p} = (p_1, \ldots, p_m)$ of the goods, where $p_j \geq 0$ for every $j \in M$. The spending of an agent i under the market outcome $(\mathbf{x}, \mathbf{p})$ is given by $\mathbf{p}(\mathbf{x}_i) = \sum_{j\in M} p_j x_{ij}$. For an agent i, we define the *bang-per-buck* ratio α_{ij} of good j as v_{ij}/p_j, and the *maximum bang-per-buck* (MBB) ratio $\alpha_i = \max_j \alpha_{ij}$. We define $mbb_i = \{j \in M : \alpha_i = v_{ij}/p_j\}$, called the *MBB-set*, to be the set of goods that give MBB to agent i at prices $\mathbf{p}$. A market outcome $(\mathbf{x}, \mathbf{p})$ is said to be *'on MBB'* if for all agents i and goods j, $x_{ij} > 0 \Rightarrow j \in mbb_i$. For integral $\mathbf{x}$, this means $\mathbf{x}_i \subseteq mbb_i$.

A market outcome $(\mathbf{x}, \mathbf{p})$ is said to be a *market equilibrium* if (i) the market clears, i.e., all goods are fully allocated. Thus, for all j, $\sum_{i\in N} x_{ij} = 1$, (ii) budget of all agents is exhausted, for all $i \in N$, $\sum_{j\in M} x_{ij} p_j = e_i$, and (iii) agents only spend money on MBB goods, i.e., $(\mathbf{x}, \mathbf{p})$ is on MBB.

Market equilibria are an important tool in computing fair and efficient allocations because of their remarkable fairness and efficiency properties; see e.g., [4,5,14,17,20,22,31]. The First Welfare Theorem [30] shows that for a market equilibrium $(\mathbf{x}, \mathbf{p})$ of a Fisher market instance $\mathcal{M}$, the allocation $\mathbf{x}$ is fPO. We include a proof in the appendix for completeness.

Theorem 1. *(First Welfare Theorem [30]) Let* $(\mathbf{x}, \mathbf{p})$ *be a equilibrium of a Fisher market* $\mathcal{M}$*. Then* $\mathbf{x}$ *is fractionally Pareto-optimal.*

Given an allocation $\mathbf{x}$ for a fair division instance (N, M, V) and a vector of prices $\mathbf{p}$ for the goods such that $(\mathbf{x}, \mathbf{p})$ is on MBB, one can define an associated Fisher market instance $\mathcal{M} = (N, M, V, e)$ by setting $e_i = \mathbf{p}(\mathbf{x}_i)$. It is easy to see that $(\mathbf{x}, \mathbf{p})$ is a market equilibrium of $\mathcal{M}$. Hence Theorem 1 implies:

Corollary 1. *Given a market outcome* $(\mathbf{x}, \mathbf{p})$ *on MBB, the allocation* $\mathbf{x}$ *is fPO.*

3 Computing EFX+PO Allocations

We first study the problem of computing an EFX+PO allocation for t-valued instances when $t \in \{2, 3\}$. We show that EFX+PO allocations can be computed in polynomial-time for 2-valued instances, and in contrast, computing such allocations for 3-valued instances is NP-hard.

3.1 EFX+PO Allocations for 2-Valued Instances

We consider $\{a, b\}$-instances, as it is known EFX+PO allocations can be efficiently computed for binary instances. We remark that while the allocation returned by the algorithm Match&Freeze of Amanatidis *et al.*, [1] for $\{a, b\}$-instances is EFX, it need not be PO (example in appendix). We improve this result by showing that:

Theorem 2. *Given a fair division $\{a, b\}$-instance $I = (N, M, V)$, an allocation that is EFX, fPO and approximates the maximum Nash welfare to a factor of 1.061 can be computed in polynomial-time.*

We prove this theorem by showing that Algorithm 1 computes such an allocation. We first define some relevant terms, including the concept of price envy-freeness introduced by Barman *et al.* [5]. A market outcome $(\mathbf{x}, \mathbf{p})$ is said to be *price envy-free up to one good* (pEF1) if for all agents $i, h \in N$ there is a good $j \in \mathbf{x}_h$ such that $\mathbf{p}(\mathbf{x}_i) \geq \mathbf{p}(\mathbf{x}_h \setminus \{j\})$. Similarly, we say it is pEFX if for all agents $i, h \in N$, and for all goods $j \in \mathbf{x}_h$ it holds that $\mathbf{p}(\mathbf{x}_i) \geq \mathbf{p}(\mathbf{x}_h \setminus \{j\})$. For market outcomes on MBB, the pEFX condition implies the EFX condition:

Lemma 1. *Let $(\mathbf{x}, \mathbf{p})$ be an integral market outcome on MBB. Then $\mathbf{x}$ is fPO. If $(\mathbf{x}, \mathbf{p})$ is pEFX, then $\mathbf{x}$ is EFX.*

Proof. The fact that $\mathbf{x}$ is fPO follows from Corollary 1. Since $(\mathbf{x}, \mathbf{p})$ is pEFX, for all pairs of agents $i, h \in N$, and all goods $j \in \mathbf{x}_h$ it holds that $\mathbf{p}(\mathbf{x}_i) \geq \mathbf{p}(\mathbf{x}_h \setminus \{j\})$. Since $(\mathbf{x}, \mathbf{p})$ is on MBB, $\mathbf{x}_i \subseteq mbb_i$. Let α_i be the MBB-ratio of i at the prices $\mathbf{p}$. By definition of MBB, $v_i(\mathbf{x}_i) = \alpha_i \mathbf{p}(\mathbf{x}_i)$, and $v_i(\mathbf{x}_h \setminus \{j\}) \leq \alpha_i \mathbf{p}(\mathbf{x}_h \setminus \{j\})$, for every $j \in \mathbf{x}_h$. Combining these, we get that $\mathbf{x}$ is EFX. □

Given a price vector $\mathbf{p}$, we define the MBB graph to be the bipartite graph $G = (N, M, E)$ where for an agent i and good j, $(i, j) \in E$ iff $j \in mbb_i$. Such edges are called *MBB edges*. Given an accompanying allocation $\mathbf{x}$, we supplement G to include *allocation edges*, an edge between agent i and good j if $j \in \mathbf{x}_i$.

We call the agent i with minimum $\mathbf{p}(\mathbf{x}_i)$ a *least spender* (LS), where ties are broken lexicographically. For agents $i_0, \dots, i_\ell$ and goods $j_1, \dots, j_\ell$, consider a path $P = (i_0, j_1, i_1, j_2, \dots, j_\ell, i_\ell)$ in the supplemented MBB graph, where for all $1 \leq \ell' \leq \ell$, $j_{\ell'} \in mbb_{(\ell'-1)} \cap \mathbf{x}_{\ell'}$. Define the *level* of an agent h to be the length of the shortest such path from the LS to h, and to be n if no such path exists. Define *alternating paths* to be such paths beginning with agents at a lower level and ending with agents at a strictly higher level. The edges in an alternating path alternate between MBB edges and allocation edges.

Algorithm 1. EFX+fPO allocation for $\{a, b\}$-instances

Input: Fair division $\{a, b\}$-instance (N, M, V)
Output: An integral allocation $\mathbf{x}$
1: Scale values to $\{1, k\}$, where $k = a/b > 1$.
2: $(\mathbf{x}, \mathbf{p}) \leftarrow$ Integral welfare-maximizing market allocation, where $p_j = v_{ij}$ for $j \in \mathbf{x}_i$.
3: Let $i \in \mathrm{argmin}_{h \in N} \mathbf{p}(\mathbf{x}_h)$ be the least spender
4: **if** there is an alternating path $(i, j_1, i_1, \ldots, j_\ell, i_\ell)$, s.t. $\mathbf{p}(\mathbf{x}_{i_\ell} \setminus \{j_\ell\}) > \mathbf{p}(\mathbf{x}_i)$ **then**
5: Transfer j_ℓ from i_ℓ to $i_{\ell-1}$
6: Repeat from Line 3
7: **if** $\forall$ agents $h \notin C_i$, and $\forall j \in \mathbf{x}_h : \mathbf{p}(\mathbf{x}_h \setminus \{j\}) \leq \mathbf{p}(\mathbf{x}_i)$ **then return x** ▷ pEFX condition satisfied for all agents not in component of LS, defined in Def.1
8: **else**
9: Raise prices of goods in C_i by a multiplicative factor of k
10: Repeat from Line 3

Definition 1 (Component C_i of a least spender i). *For a least spender i, define C_i^ℓ to be the set of all goods and agents which lie on alternating paths of length ℓ. Call $C_i = \bigcup_\ell C_i^\ell$ the component of i, the set of all goods and agents reachable from the least spender i through alternating paths.*

We now describe Algorithm 1. Let $k = a/b > 1$. Let us first scale the valuations to $\{1, k\}$ since both properties EFX and fPO are scale-invariant. The algorithm starts with a welfare maximizing integral allocation $(\mathbf{x}, \mathbf{p})$, where $p_j = v_{ij}$ if $j \in \mathbf{x}_i$. The algorithm then explores if there is an alternating path $P = (i = i_0, j_1, i_1, \cdots, j_\ell, i_\ell = h)$, where i is the LS agent, such that $\mathbf{p}(\mathbf{x}_h \setminus \{j_\ell\}) > \mathbf{p}(\mathbf{x}_i)$, i.e., an alternating path along which the pEF1 condition is violated for the LS agent. We call any such agent h who owns some good j such that the pEF1 condition is not satisfied by the LS with respect to good j, a pEF1-violator. When such a path is encountered, the algorithm *transfers* j_ℓ from h to $i_{\ell-1}$. This process is repeated from Line 3 to account for a possible change in the LS, until there is no such path in the component C_i of the LS agent. Suppose there is some agent $h \notin C_i$ for which the pEFX condition is not satisfied with respect to the LS, then the algorithm *raises the prices* of all goods in the component of the LS agent by a factor of k, and the algorithm proceeds once again from Line 3.

The proof of Theorem 2 relies on Lemmas 1-6. We first show that we can re-scale prices to $\{1, k\}$.

Lemma 2. *For every outcome $(\mathbf{x}, \mathbf{p})$ constructed during the run of Algorithm 1, there exists a set of prices $\mathbf{q}$ such that $(\mathbf{x}, \mathbf{q})$ is also on MBB, and for every $j \in M$, $q_j \in \{1, k\}$.*

Proof. Note that initially all prices are either 1 or k. Since all price rises are by a factor of k (Line 9), final prices are of the form $p_j = k^{s_j}$, for $s_j \in \mathbb{Z}_{\geq 0}$. Let j_0 be the smallest priced good with $p_{j_0} = k^s$, and let $j_0 \in \mathbf{x}_i$, for some agent $i \in N$. Then $\forall j \in \mathbf{x}_i : p_j \in \{k^s, k^{s+1}\}$. By the MBB condition for any agent $h \neq i$ for

$j' \in \mathbf{x}_h$ and $j \in \mathbf{x}_i$:

$$\frac{v_{hj'}}{p_{j'}} \geq \frac{v_{hj}}{p_j},$$

which gives:

$$p_{j'} \leq \frac{v_{hj'}}{v_{hj}} p_j \leq k^{s+2} \ .$$

Thus all $p_j \in \{k^s, k^{s+1}, k^{s+2}\}$. Either all $p_j \in \{k^s, k^{s+1}\}$, or $\exists j \in \mathbf{x}_h$ with $p_j = k^{s+2}$, for some agent $h \in N$. Then by the MBB condition for any good j':

$$\frac{v_{hj}}{p_j} \geq \frac{v_{hj'}}{p_{j'}},$$

which gives:

$$p_{j'} \geq \frac{v_{hj'}}{v_{hj}} p_j \geq k^{s+1} \ .$$

Thus either all $p_j \in \{k^s, k^{s+1}\}$ or all $p_j \in \{k^{s+1}, k^{s+2}\}$. In either case we can scale the prices to belong to $\{1, k\}$. □

This in fact shows that at any stage of Algorithm 1, the prices of goods are in $\{k^s, k^{s+1}\}$ for some $s \in \mathbb{Z}_{\geq 0}$. This, along with the fact that goods are always transferred along MBB edges, and the prices are raised only by factor of k, leads us to conclude that the MBB condition is never violated for any agent and the allocation is always on MBB throughout the run of the algorithm. Hence the allocation is fPO.

Lemma 3. *The allocation* $\mathbf{x}$ *returned by Algorithm 1 is on MBB w.r.t. prices* $\mathbf{p}$ *upon termination. Thus,* $\mathbf{x}$ *is fPO.*

The full proof of the above Lemma appears in the appendix. We now show: correctness:

Lemma 4. *The allocation* $\mathbf{x}$ *returned by Algorithm 1, together with the prices* $\mathbf{p}$ *on termination is pEFX.*

Proof. To see why $(\mathbf{x}, \mathbf{p})$ is pEFX, first note that by Lemma 2, we can assume the prices are in $\{1, k\}$. Suppose $(\mathbf{x}, \mathbf{p})$ is not pEFX. Then there must be an agent h and some good $j \in \mathbf{x}_h$ s.t. $\mathbf{p}(\mathbf{x}_h \setminus \{j\}) > \mathbf{p}(\mathbf{x}_i)$, where i is the least spender. If $h \notin C_i$, the algorithm would not have halted (negation of condition in line 8 holds). Therefore h is in C_i. Since the algorithm has halted, this means that along all alternating paths $(i, j_1, i_1, \ldots, h', j, h)$, it is the case that $\mathbf{p}(\mathbf{x}_h \setminus \{j\}) \leq \mathbf{p}(\mathbf{x}_i)$. Suppose there is some alternating path s.t. $p_j = 1$. We know for all $j' \in \mathbf{x}_h$, $p_{j'} \geq 1$. Thus:

$$\mathbf{p}(\mathbf{x}_i) \geq \mathbf{p}(\mathbf{x}_h \setminus \{j\}) = \mathbf{p}(\mathbf{x}_h) - 1 \geq \mathbf{p}(\mathbf{x}_h \setminus \{j'\}),$$

which means that i is pEFX towards h. Now suppose along all alternating paths $(i, j_1, i_1, \ldots, h', j, h)$, it holds that $p_j = k$. Since $(\mathbf{x}, \mathbf{p})$ is not pEFX, it must be the case that there is some good $j' \in \mathbf{x}_h$ that is not reachable from i via any alternating path, with $p_{j'} = 1$. This means that $j' \notin mbb_{h'}$. Since $j \in mbb_{h'}$, comparing the bang-per-buck ratios gives $v_{h'j}/p_j > v_{h'j'}/p_{j'}$. This implies $v_{h'j} > kv_{h'j'}$ which is not possible when $v_{h'j}, v_{h'j'} \in \{1, k\}$, thus leading to a contradiction. Hence we conclude that $(\mathbf{x}, \mathbf{p})$ is pEFX. □

Lemma 5. *Algorithm 1 terminates in polynomial-time.*

Proof. (Sketch) We first note that the number of alternating paths from an agent i to an agent h who owns a good j which is then transferred to an agent h' is at most $n \cdot n \cdot m$. Thus there are at most $\mathsf{poly}(n, m)$ transfers with the same LS and no price rise step.

Next, we argue that the number of identity changes of the least spender without a price rise step is $\mathsf{poly}(n, m)$. Suppose an agent i ceases to be the LS at iteration t, and subsequently (without price-rise steps) becomes the LS again for the first time at time t'. We show that the spending of i is strictly larger at t' than at t, and hence has strictly larger utility. Since all utility values are integers, the increase in i's utility is by at least 1. In any allocation $\mathbf{x}$, if s_i (resp. t_i) is the number of goods in $\mathbf{x}_i$ that are valued at b (resp. a) by i, the utility of i is $u_i = s_i b + t_i a$. Since $0 \leq s_i, t_i \leq m$, the number of different utility values i can get in any allocation is at most $O(m^2)$. Thus, for any agent i, the number of times her utility increases is at most $O(m^2)$. This is our key insight. It implies that without price rises, any agent can become the least spender only $O(m^2)$ times. Hence, the number of identity changes of the LS in the absence of price rise steps is at most $O(nm^2)$.

For polynomial run-time, it remains to be shown that the number of price-rises is $\mathsf{poly}(n, m)$. We do this via a potential function argument similar to [20]. The full proof is present in the appendix. □

Finally, we show that the allocation returned by our algorithm also provides a good approximation to the MNW, and defer the proof to the appendix.

Lemma 6. *Let* $\mathbf{x}$ *be the allocation output by Algorithm 1. If* $\mathbf{x}^*$ *is an MNW allocation, then* $\mathsf{NW}(\mathbf{x}) \geq \frac{1}{1.061}\mathsf{NW}(\mathbf{x}^*)$.

Proof. (Sketch) Let $\mathbf{p}$ be the price vector on termination. Consider a scaled fair division instance $I' = (N, M, V')$ with identical valuations, where $v'_{ij} = p_j$ for each $i \in N, j \in M$. Since $(\mathbf{x}, \mathbf{p})$ is pEFX for the instance I (Lemma 4), $\mathbf{x}$ is EFX for the instance I'. Barman et al., [6] showed that for identical valuations, any EFX allocation provides a 1.061-approximation to the maximum Nash welfare. Hence $\mathbf{x}$ provides a 1.061-approximation to the MNW of I', and we can show that because $(\mathbf{x}, \mathbf{p})$ is on MBB (from Lemma 3), $\mathbf{x}$ gives the same guarantee for the MNW of the instance I. □

Lemmas 1, 3, 4, 5, and 6 together prove Theorem 2.

3.2 EFX+PO for 3-Valued Instances

On generalizing the class of valuations slightly to $\{0, a, b\}$, EFX+PO allocations are no longer guaranteed to exist [34] (see the appendix for an example).

Therefore we investigate the complexity of checking if an EFX and PO allocation exists or not, and show that this problem is NP-hard.

Theorem 3. *Given a fair division instance $I = (N, M, V)$, checking if I admits an allocation that is both EFX and PO is NP-hard, even when I is a $\{0, a, b\}$-instance.*

We reduce from 2P2N3SAT, an instance of which consists of a 3SAT formula over n variables $\{x_i\}_{i\in[n]}$ in conjunctive normal form, and m distinct clauses $\{C_j\}_{j\in[m]}$, with three literals per clause. Additionally, each variable x_i appears exactly twice negated and exactly twice unnegated in the formula. Deciding if there exists a satisfying assignment for such a formula is known to be NP-complete [7]. Given a 2P2N3SAT-instance $(\{x_i\}_{i\in[n]}, \{C_j\}_{j\in[m]})$, we construct a fair division instance with $2n + m$ agents and $7n + m$ goods, with all values in $\{0, 1, 3\}$ as follows:

1. For every variable x_i, create two agents T_i and F_i. Also create 7 goods: $d_i^T, d_i^F, g_i, y_i^T, z_i^T, y_i^F, z_i^F$. Both T_i and F_i value g_i at 3. T_i values d_i^T, y_i^T, z_i^T at 1, and F_i values d_i^F, y_i^F, z_i^F at 1. T_i and F_i value all other goods at 0.
2. For every clause $C_j = \ell_1 \vee \ell_2 \vee \ell_3$, create one agent D_j and a good e_j. D_j values e_j at 1. If for any $k \in [3]$, $\ell_k = x_i$ for some $i \in [n]$ then D_j values y_i^T, z_i^T at 1; and if for any $k \in [3]$, $\ell_k = \neg x_i$ for some $i \in [n]$ then D_j values y_i^F, z_i^F at 1. D_j values all other goods at 0.

We show that this instance admits an EFX+PO allocation iff the formula has a satisfying assignment. We illustrate the correspondence between PO allocations and assignments, and how our construction enforces EFX allocations to give rise to satisfying assignments (and vice versa). In any PO allocation, for every $i \in [n]$, d_i^A must be assigned to A_i, for $A \in \{T, F\}$; and g_i must be assigned to T_i or F_i. Consider the assignment $x_i = A$, if g_i is allotted to A_i, for $A \in \{T, F\}$, for all $i \in [n]$. Suppose for some $i \in [n]$, g_i is allocated to T_i. Then in an EFX allocation, because F_i must not envy T_i after removing d_i^T from the bundle of T_i, F_i must get utility at least 3. This is only possible if both y_i^F and z_i^F are allocated to F_i. This leaves y_i^T, z_i^T for the clause agents, when x_i is True. Thus if there is a satisfying assignment, the remaining goods can be allocated to the clause agents in an EFX+PO manner. Also, if all assignments are unsatisfying, some clause agent will end up not being EFX towards another agent in any PO allocation.

We defer the full proof to the appendix, and also show:

Lemma 7. *Given a fair division instance $I = (N, M, V)$, checking if I admits an allocation that is both EFX and fPO is NP-complete, even when I is a $\{0, a, b\}$-instance.*

4 Computing EQX+PO Allocations

We now consider relaxations of the fairness notion of equitability, which demands that all agents receive roughly the same utility. An allocation is said to be

Algorithm 2. EQX+fPO allocation for $\{a,b\}$-instances

Input: Fair division $\{a,b\}$-instance (N,M,V)
Output: An integral allocation $\mathbf{x}$

1: $(\mathbf{x},\mathbf{p}) \leftarrow$ Integral welfare-maximizing market allocation, where $p_j = v_{ij}$ for $j \in \mathbf{x}_i$.
2: Let $i \in \mathsf{argmin}_{h\in N} v_h(\mathbf{x}_h)$ be the least utility (LU) agent
3: **if** there is an alternating path $(i, j_1, i_1, \dots, j_\ell, i_\ell)$, s.t. $v_{i_\ell}(\mathbf{x}_{i_\ell} \setminus \{j_\ell\}) > v_i(\mathbf{x}_i)$ **then**
4: Transfer j_ℓ from i_ℓ to $i_{\ell-1}$
5: Repeat from Line 2
6: **if** $\forall$ agents $h \notin C_i$, and $\forall j \in \mathbf{x}_h : v_h(\mathbf{x}_h \setminus \{j\}) \leq v_i(\mathbf{x}_i)$ **then return** $\mathbf{x}$ ▷ EQX condition satisfied for all agents not reachable through alt. paths from LU agent; C_i is defined in Def. 1
7: **else**
8: Raise prices of goods in C_i by a multiplicative factor of a/b
9: Repeat from Line 2

equitable up to any good (EQX) if for all $i, h \in N$, for all $j \in \mathbf{x}_h$ we have $v_i(\mathbf{x}_i) \geq v_h(\mathbf{x}_h \setminus \{j\})$. It is known that for binary instances, EQX+PO allocations can be computed in polynomial-time, whenever they exist [20]. Hence we first consider $\{a,b\}$-instances. We show that:

Theorem 4. *Given a fair division $\{a,b\}$-instance $I = (N,M,V)$, an allocation that is both EQX and fPO exists and can be computed in polynomial-time.*

We prove this by showing that Algorithm 2 terminates in polynomial-time with an EQX+fPO allocation. Since we are interested in EQX as opposed to EFX, we need not construct a pEFX allocation and can instead work directly with the values. Since the techniques used in the analysis of Algorithm 2 are similar to the analysis of Algorithm 1, we defer the full proof to the appendix. We remark here that our techniques also enable us to show that EQ.1+fPO allocations can be computed in polynomial-time for $\{a,b\}$-instances of *chores*.

For $\{0,a,b\}$-instances, EQX+PO allocations need not exist (example in appendix). Therefore, we study the complexity of checking if an EQX+PO allocation exists or not, and show that this problem too is NP-hard. The full proof is deferred to the appendix.

Theorem 5. *Given a fair division instance $I = (N,M,V)$, checking if I admits an allocation that is both EQX and PO is NP-hard, even when I is a $\{0,a,b\}$-instance.*

5 Maximizing Nash Welfare

We turn to the problem of maximizing Nash welfare for t-valued instances when $t \in \{2,3\}$. Recall that for $\{a,b\}$-instances, we showed in Lemma 6 that Algorithm 1 approximates the MNW to a 1.061-factor.

Turning to 3-valued instances, our final result shows APX-hardness for the MNW problem with we slighly generalize the class of allowed values to $\{0,a,b\}$.

This rules out the existence of a polynomial-time approximation scheme (PTAS) for the MNW problem even $\{0, a, b\}$-instances, thus strengthening the result of [1], who showed NP-hardness for the same. The full proof is present in the appendix.

Theorem 6. *Given a fair division instance $I = (N, M, V)$, it is NP-hard to approximate the MNW to a factor better than 1.00019, even for $\{0, a, b\}$-instances.*

We present the reduction and defer the full proof to the appendix. We consider a 2P2N3SAT-instance: $\{x_i\}_{i \in [n]}, \{C_j\}_{j \in [m]}$, where $3m = 4n$. For each variable x_i, we create two agents T_i, F_i and one good g_i which is valued at 2 by both T_i, F_i. For each clause C_j, we create a good h_j which is valued at 1 by agent A_i if setting $x_i = A$ makes C_j true, for $A \in \{T, F\}$. We also create $2n - m$ dummy goods $\{d_j\}_{j \in [2n-m]}$ which are valued at 1 by all agents. All other values are 0. We show that if we can approximate the MNW to a factor better than 1.00019, we can decide if there is an assignment with $\geq \rho_1 m$ clauses, or all assignments satisfy at most $\leq \rho_2 m$ clauses, for specific constants ρ_1, ρ_2. The latter problem is known to be NP-complete [7].

6 Conclusion

In this paper, we push the boundary between tractable and intractable cases for the problems of fair and efficient allocations. We presented positive algorithmic results for computing EFX+PO, EQX+PO, and 1.061-approximate MNW allocations for 2-valued instances. In contrast, we showed that for 3-valued instances, checking existence of EFX+PO (or EQX+PO) allocations is NP-complete, and computing MNW is APX-hard. Our techniques can be adapted to compute EQ.1+PO allocations for 2-valued instances of *chores*, and an interesting direction for future work is to see if we can compute EF1+PO allocations in the chores setting, even for 2-valued instances. We also leave open the problem of computing an MNW allocation for general 2-valued instances.

References

1. Amanatidis, G., Birmpas, G., Filos-Ratsikas, A., Hollender, A., Voudouris, A.A.: Maximum Nash welfare and other stories about EFX. In: Proceedings of the Twenty-Ninth International Joint Conference on Artificial Intelligence (IJCAI), pp. 24–30 (2020)
2. Aziz, H.: The Hylland-Zeckhauser rule under bi-valued utilities. CoRR abs/2006.15747 (2020)
3. Aziz, H., Biró, P., Lang, J., Lesca, J., Monnot, J.: Efficient reallocation under additive and responsive preferences. Theor. Comput. Sci. **790**, 1–15 (2019)
4. Barman, S., Krishnamurthy, S.: On the proximity of markets with integral equilibria. In: Proceedings of the 33rd AAAI Conference on Artificial Intelligence, pp. 1748–1755 (2019)

5. Barman, S., Krishnamurthy, S.K., Vaish, R.: Finding fair and efficient allocations. In: Proceedings of the 19th ACM Conference on Economics and Computation (EC), pp. 557–574 (2018)
6. Barman, S., Krishnamurthy, S.K., Vaish, R.: Greedy algorithms for maximizing Nash social welfare. In: Proceedings of the 17th International Conference on Autonomous Agents and MultiAgent Systems (AAMAS), pp. 7–13 (2018)
7. Berman, P., Karpinski, M., Scott, A.: Approximation hardness of short symmetric instances of MAX-3SAT. Electronic Colloquium on Computational Complexity (ECCC) (Jan 2003)
8. Bliem, B., Bredereck, R., Niedermeier, R.: Complexity of efficient and envy-free resource allocation: few agents, resources, or utility levels. In: Proceedings of the 25th International Joint Conference on Artificial Intelligence (IJCAI), pp. 102–108 (2016)
9. Bogomolnaia, A., Moulin, H.: Random matching under dichotomous preferences. Econometrica **72**(1), 257–279 (2004)
10. Brams, S., Taylor, A.: Fair Division: From Cake-Cutting to Dispute Resolution. Cambridge University Press, Cambridge (1996)
11. Budish, E.: The combinatorial assignment problem: Approximate competitive equilibrium from equal incomes. J. Polit. Econ. **119**(6), 1061–1103 (2011)
12. Caragiannis, I., Kurokawa, D., Moulin, H., Procaccia, A.D., Shah, N., Wang, J.: The unreasonable fairness of maximum Nash welfare. In: Proceedings of the 17th ACM Conference on Economics and Computation (EC), pp. 305–322 (2016)
13. Chakrabarty, D., Khanna, S., Li, S.: On (1, e)-restricted assignment makespan minimization. In: Proceedings of the 26th Annual ACM-SIAM Symposium on Discrete Algorithms (SODA), pp. 1087–1101 (2015)
14. Chaudhury, B.R., Cheung, Y.K., Garg, J., Garg, N., Hoefer, M., Mehlhorn, K.: On fair division for indivisible items. In: 38th IARCS Annual Conference on Foundations of Software Technology and Theoretical Computer Science (FSTTCS), pp. 1–17 (2018)
15. Chaudhury, B.R., Garg, J., Mehlhorn, K.: EFX exists for three agents. In: Proceedings of the 21st ACM Conference on Economics and Computation (EC), pp. 1–19 (2020)
16. Chaudhury, B.R., Garg, J., Mehlhorn, K., Mehta, R., Misra, P.: Improving EFX guarantees through rainbow cycle number. CoRR abs/2103.01628 (2021). To appear in ACM EC 2021
17. Cole, R., Gkatzelis, V.: Approximating the Nash social welfare with indivisible items. In: Proceedings of the Forty-Seventh Annual ACM Symposium on Theory of Computing (STOC), pp. 371–380 (2015)
18. Darmann, A., Schauer, J.: Maximizing Nash product social welfare in allocating indivisible goods. SSRN Electron. J. **247**(2), 548–559 (2014)
19. Foley, D.: Resource allocation and the public sector. Yale Econ. Essays **7**(1), 45–98 (1967)
20. Freeman, R., Sikdar, S., Vaish, R., Xia, L.: Equitable allocations of indivisible goods. In: Proceedings of the 28th International Joint Conference on Artificial Intelligence (IJCAI), pp. 280–286 (2019)
21. Garg, J., Hoefer, M., Mehlhorn, K.: Satiation in Fisher markets and approximation of Nash social welfare. CoRR abs/1707.04428 (2017)
22. Garg, J., Hoefer, M., Mehlhorn, K.: Approximating the Nash social welfare with budget-additive valuations. In: Proceedings of the 29th Annual ACM-SIAM Symposium on Discrete Algorithms (SODA), pp. 2326–2340 (2018)

23. Garg, J., Murhekar, A.: On fair and efficient allocations of indivisible goods. In: Proceedings of the 35th AAAI Conference on Artificial Intelligence (2021)
24. Mehlhorn, K.: Assigning papers to referees. In: Albers, S., Marchetti-Spaccamela, A., Matias, Y., Nikoletseas, S., Thomas, W. (eds.) ICALP 2009. LNCS, vol. 5555, pp. 1–2. Springer, Heidelberg (2009). https://doi.org/10.1007/978-3-642-02927-1_1
25. Golovin, D.: Max-min fair allocation of indivisible goods. Technical report, CMU-CS-05-144 (2005)
26. Gourvès, L., Monnot, J., Tlilane, L.: Near fairness in matroids. In: Proceedings of the 21st European Conference on Artificial Intelligence (ECAI), pp. 393–398 (2014)
27. de Keijzer, B., Bouveret, S., Klos, T., Zhang, Y.: On the complexity of efficiency and envy-freeness in fair division of indivisible goods with additive preferences. In: Rossi, F., Tsoukias, A. (eds.) ADT 2009. LNCS (LNAI), vol. 5783, pp. 98–110. Springer, Heidelberg (2009). https://doi.org/10.1007/978-3-642-04428-1_9
28. Lee, E.: APX-hardness of maximizing Nash social welfare with indivisible items. Inf. Process. Lett. **122**, 17–20 (07 2015)
29. Lipton, R.J., Markakis, E., Mossel, E., Saberi, A.: On approximately fair allocations of indivisible goods. In: Proceedings of the 5th ACM Conference on Electronic Commerce (EC), pp. 125–131 (2004)
30. Mas-Colell, A., et al.: Microeconomic Theory. Oxford University Press, Oxford (1995)
31. McGlaughlin, P., Garg, J.: Improving Nash social welfare approximations. J. Artif. Intell. Res. **68**, 225–245 (2020)
32. Moulin, H.: Fair Division and Collective Welfare. MIT Press, Cambridge (2004)
33. Murhekar, A., Garg, J.: On fair and efficient allocations of indivisible goods. In: Proceedings of the AAAI Conference on Artificial Intelligence, vol. 35, no. 6, pp. 5595–5602 (2021)
34. Plaut, B., Roughgarden, T.: Almost envy-freeness with general valuations. In: Proceedings of the Twenty-Ninth Annual ACM-SIAM Symposium on Discrete Algorithms (SODA), pp. 2584–2603 (2018)
35. Procaccia, A.D.: Technical perspective: an answer to fair division's most enigmatic question. Commun. ACM **63**(4), 118 (2020)
36. Steinhaus, H.: Sur la division pragmatique. Econometrica **17**(1), 315–319 (1949)
37. Varian, H.R.: Equity, envy, and efficiency. J. Econ. Theory **9**(1), 63–91 (1974)
38. Vazirani, V.V., Yannakakis, M.: Computational complexity of the Hylland-Zeckhauser scheme for one-sided matching markets. In: Proceedings of the 12th Innovations in Theoretical Computer Science Conference (ITCS) (2021)
39. Woeginger, G.J.: A polynomial-time approximation scheme for maximizing the minimum machine completion time. Oper. Res. Lett. **20**(4), 149–154 (1997)

An Approval-Based Model for Single-Step Liquid Democracy

Evangelos Markakis(✉) and Georgios Papasotiropoulos

Athens University of Economics and Business, Athens, Greece

Abstract. We study a Liquid Democracy framework where voters can express preferences in an approval form, regarding being represented by a subset of voters, casting a ballot themselves, or abstaining from the election. We examine, from a computational perspective, the problems of minimizing (resp. maximizing) the number of dissatisfied (resp. satisfied) voters. We first show that these problems are intractable even when each voter approves only a small subset of other voters. On the positive side, we establish constant factor approximation algorithms for that case, and exact algorithms under bounded treewidth of a convenient graph-theoretic representation, even when certain secondary objectives are also present. The results related to the treewidth are based on the powerful methodology of expressing graph properties via Monadic Second Order logic. We believe that this approach can turn out to be fruitful for other graph related questions that appear in Computational Social Choice.

1 Introduction

Liquid Democracy (LD) is a voting paradigm that has emerged as a flexible model for enhancing engagement in decision-making. The main idea in LD models is that a voter can choose either to vote herself or to delegate to another voter that she trusts to be more knowledgeable or reliable on the topic under consideration. A delegation under LD, is a transitive action, meaning that the voter not only transfers her own vote but also the voting power that has been delegated to her by others. Experimentations and real deployments have already taken place using platforms that support decision-making under Liquid Democracy. One of the first such systems that was put to real use was Župa, intended for a student union for the University of Novo Mesto in Slovenia. Another example is LiquidFeedback that was used by the German Pirate party (among others). Other political parties (such as the Flux Party in Australia) or regional organisations have also attempted to use or experiment with LD, leading to a growing practical appeal. Even further examples include the experiment run by Google via Google Votes, as well as Civicracy and, the more recently developed, Sovereign. We refer to [23] for an informative survey on these systems.

The interest generated by these attempts, has also led to theoretical studies on relevant voting models and has enriched the research agenda of the community. The goals of these works have been to provide more rigorous foundations

I. Caragiannis and K. A. Hansen (Eds.): SAGT 2021, LNCS 12885, pp. 360–375, 2021.
https://doi.org/10.1007/978-3-030-85947-3_24

and highlight the advantages and the negative aspects of LD models. Starting with the positive side, LD definitely has the potential to incentivize civic participation, both for expert voters on a certain topic, but also for users who feel less confident and can delegate to some other trusted voter. At the same time, it also forms a flexible means of participation, since there are no restrictions for physical presence, and usually there is also an option of instant recall of a delegation, whenever a voter no longer feels well represented.

Coming to the critique that has been made on LD, an issue that can become worrying is the formation of large delegation paths. Such paths tend to be undesirable since a voter who gets to cast a ballot may have a rather different opinion with the first voters of the path, who are being represented by her [16]. Secondly, LD faces the risk of having users accumulating excessive voting power, if no control action is taken [5]. Furthermore, another undesirable phenomenon is the creation of delegation cycles, which could result to a waste of participation for the involved voters. Despite the criticism, LD is still a young and promising field, for promoting novel methods of participation and decision-making, generating an increasing interest in the community. We therefore feel that several aspects have not yet been thoroughly studied, and new models and ideas are worth further investigation. Such efforts can help both in tackling some of the existing criticism but also in identifying additional inherent problems.

Contribution. We focus on a model, where voters can have approval-based preferences on the available actions. Each voter can have a set of approved delegations, and may also approve voting herself or even abstaining. Our main goal is the study of centralized algorithms for optimizing the overall satisfaction of the voters. For this objective, under our model, it turns out that it suffices to focus only on delegations to actual voters (i.e., delegation paths of unit length). Even with this simpler solution space, the problems we study turn out to be computationally hard. In Sect. 3, we start with the natural problem of minimizing the number of dissatisfied voters, where we establish a connection with classic combinatorial optimization problems, such as SET COVER and DOMINATING SET. we present approximation preserving reductions which allow us to obtain almost tight approximability and hardness results. The main conclusion from these is that one can have a small constant factor approximation when each voter approves a small number of possible representatives. A constant factor approximation can also be obtained for the variant of maximizing the number of satisfied voters, through a different approach of modeling this as a constraint satisfaction problem. Moving on, in Sect. 4, we consider the design of exact algorithms for the same problems. Our major highlight is the use of a logic-based technique, where it suffices to express properties by formulas in Monadic Second Order logic. In a nutshell, this approach yields an FPT algorithm, whenever the treewidth of an appropriate graph-theoretic representation of our problem is constant. Under the same restriction, polynomial time algorithms also exist when adding certain secondary objectives on top of minimizing (resp. maximizing) dissatisfaction (resp. satisfaction). To our knowledge, this framework has not received much attention in the social choice community and we expect that it could have further applicability for related problems.

1.1 Related Work

To position our work with respect to existing literature, we note that the works most related to ours are [15] and [12]. In terms of the model, we are mostly based on [15], which studies centralized algorithms and where voters specify possible delegations in an approval format. Coming to the differences, their model does not allow abstainers (which we do), but more importantly, [15] studies a different objective and no notion of satisfaction needs to be introduced (in Sect. 4 we also examine a related question). Our main optimization criteria are inspired mostly by [12], which among others, tries to quantify voters' dissatisfaction. Our differences with [12] is that they have voters with rank-based preferences and their optimization is w.r.t. equilibrium profiles and not over all possible delegations (in Sect. 5, we also provide a game-theoretic direction with some initial findings). We note also that these works, like ours, are agnostic to the final election outcome (preferences are w.r.t. delegations and not on actual votes).

More generally, the LD-related literature within computational social choice concerns *(i)* comparisons with direct democracy models, *(ii)* game-theoretic stability of delegations, *(iii)* axiomatic approaches. Concerning the first topic, local delegation mechanisms, under which every voter independently is making a choice, have been explored in [8,18]. For the second direction, one can view an LD framework as a game in which the voters can make a choice according to some given preference profile. Such games have been considered in [12,13]. At the same time, game-theoretic aspects have also been studied in [4] and, for the case of weighted delegations, in [27]. Concerning the third direction, a range of delegation schemes have been proposed to avoid delegation cycles [19], accumulation of high power in the election procedure [15] and existence of inconsistent outcomes [9]. Related paradigms to LD have also been considered, e.g. in [1,7,10].

2 Preliminaries

2.1 Approval Single-Step Liquid Model

We denote by $V = \{1, \ldots, n\}$ the set of agents who participate in the election process and we will refer to all members of V as voters (even though some of them may eventually not vote themselves). In the suggested model, which we refer to as Approval Single-Step Liquid model (ASSL), each voter $i \in V$ needs to express her preferences, in an approval-based format, on the options of *(i)* casting a ballot herself, *(ii)* abstaining from the election, *(iii)* delegating her vote to voter $j \in V \setminus \{i\}$. Namely, a voter may approve any combination of

- casting a ballot herself. We let C denote the set of all such voters.
- abstaining from the voting procedure (e.g., because she feels not well-informed on the topic). We let A denote the set of all such voters.
- delegating her vote to some other voter she trusts. For every $v \in V$, we denote by $N(v)$ the set of approved delegatees of v.

Note that we place no restriction on whether a voter accepts one or more of the above options (or even none of them). Hence, in a given instance it may be true that $C \cap A \neq \emptyset$ or that $C \cup A = \emptyset$ or that $v \in C$ and at the same time $N(v) \neq \emptyset$, etc. It is often natural and convenient to think of a graph-theoretic representation of the approved delegations. Hence, for every instance, we associate a directed graph $G = (V, E)$, such that $N(v)$ is the set of out-neighbors of v, i.e., $deg^+(v) = |N(v)|$, where $deg^+(v)$ is the out-degree of v. This will be particularly useful in Sect. 4.

Let a delegation function $d : V \rightarrow V \cup \{\bot\}$ express the final decision for each voter. We say that $d(v) = v$, if voter v votes, $d(v) = \bot$ if she abstains, and $d(v) = u \in N(v)$, if she delegates to voter u. Given a delegation function $d(\cdot)$, we refer to a voter who casts a ballot as a *guru*. The guru of a voter $v \in V$, denoted by $gu(v)$, can be found by following the successive delegations, as given by a delegation function $d(\cdot)$, starting from v until reaching a guru (if possible). Formally, $gu(v) = u$ if there exists a sequence of voters $u_1, \ldots, u_\ell$ such that $d(u_k) = u_{k+1}$ for every $k \in \{1, 2, \ldots, \ell - 1\}, u_1 = v, u_\ell = u$ and $d(u) = u$. Obviously, $gu(v) = v$ if v votes. In case the delegation path starting from v ends up in a voter u for which $d(u) = \bot$ then we say that v does not have a guru and we set $gu(v) = \infty$. Additionally, we do the same for the case where the successive delegations starting from some voter v form or end up in a cycle.

We say that a voter v is satisfied with the delegation function $d(\cdot)$ if v approves the outcome regarding her participation or her representation by another guru-voter. This means that either $d(v) = v$ and $v \in C$ or that $d(v) = \bot$ and $v \in A$ or that $d(v) = u$, with $u \neq v$, $u \neq \bot$ and $gu(u) \in N(v)$. In all other cases, the voter is dissatisfied. Our work mainly deals with the problem of finding centralized mechanisms for the following computational problems:

MINIMUM SOCIAL COST (MIN-SC)/MAXIMUM SOCIAL GOOD (MAX-SG)

Given: An instance of ASSL, i.e., the approval preferences of n voters regarding their intention to vote, abstain and delegate

Output: A delegation function that minimizes the number of dissatisfied voters/maximizes the number of satisfied voters

---

2.2 Warm-Up Observations

We start with some observations that will help us tackle the algorithmic problems under consideration. Given an instance of ASSL, let G be the corresponding graph with the approved delegations, as described in the previous subsection. A delegation function $d(\cdot)$, induces a subgraph of G that we denote by $G(d)$, so that (u, v) is an edge in $G(d)$ if and only if $d(u) = v$. Clearly, the out-degree of every vertex in $G(d)$ is at most one and thus it can contain isolated vertices, directed trees oriented towards the gurus, but in general it can also contain cycles, the presence of which can only deteriorate the solution. The next claim shows that we can significantly reduce our solution space. Its proof together with any other missing proof are deferred to the full version of the work.

Claim 1. *Consider a solution given by a delegation function $d(\cdot)$. There always exists a solution $d'(\cdot)$ which is at least as good (i.e., the number of satisfied voters is at least as high) as $d(\cdot)$, so that $G(d')$ is a collection of disjoint directed (towards the central vertex) stars, and voters that abstain form isolated vertices.*

Claim 1 justifies the name ASSL. One may discern similarities with proxy voting models (see e.g. [2]), under which every voter is being represented by her delegator, since no transitivity of votes is taken into account. Nevertheless, we still like to think of our model as a Liquid Democracy variant, because it is precisely the objectives that we study together with the centralized approach that enforce Claim 1. When discussing decentralized scenarios or game-theoretic questions (as we do in Sect. 5), longer delegation paths may also appear.

The next claim shows that for certain voters, we can a priori determine their action, when looking for an optimal solution and that we can be sure about the action of any voter who is dissatisfied under a given delegation function.

Claim 2. *Consider a solution given by a delegation function $d(\cdot)$. There always exists a solution which is at least as good (i.e., the number of satisfied voters is at least as high) as $d(\cdot)$, in which (i) every voter in C casts a ballot and (ii) if any voter is dissatisfied, it is because she is casting a ballot without approving it.*

Claim 2 takes care of voters in C. We cannot state something similar for the rest of the voters, since it might be socially better to dissatisfy a certain voter by asking her to cast a ballot so as to make other people (pointing to her) satisfied. In practice, this can also occur in cases where voting may be costly (in time or effort to become more informed) and one member of a community may need to act in favor of the common good, outweighing her cost.

3 Approximation Algorithms and Hardness Results

In this section we will mainly focus on MIN-SC, but we will also examine MAX-SG in Sect. 3.2 and further related questions in Sect. 3.3. We pay particular attention to instances where every voter approves only a constant number of other voters, i.e., $\Delta = \max_v |N(v)| = O(1)$. We find this to be a realistic case, as it is rather expected that voters cannot easily trust a big subset of the electorate.

3.1 Social Cost Minimization

We start by showing that the problem is intractable even when each voter approves at most 2 other voters. In fact, we show that our problem encodes a directed version of the DOMINATING SET problem, hence, beyond NP-hardness, we also inherit known results concerning hardness of approximation.

Theorem 1. *Let* $\Delta = \max_{v \in V} |N(v)| \geq 2$*. When* Δ *is constant, it is NP-hard to approximate* MIN-SC *with a ratio smaller than* $\max\{1.36, \Delta - 1\}$*. For general instances, it is NP-hard to achieve an approximation better than* $\ln n - \Theta(1) \ln \ln n$*.*

Since hardness results have been established for $\Delta \geq 2$, it is natural to question whether an optimal algorithm could be found for the case of $\Delta \leq 1$. This scenario is far from unexciting. Consider for instance a spatial model where voters are represented by points in some Euclidean space, interpreted as opinions on the outcomes of some issues. If each voter approves for delegation only the nearest located voter to her, we have precisely that $\Delta = 1$. The following theorem provides an affirmative answer in the above stated question (its proof is actually a direct Corollary of Theorem 5 from Sect. 4).

Theorem 2. *When* $\Delta \leq 1$*,* MIN-SC *can be solved in polynomial time.*

For higher values of Δ, we can only hope for approximation algorithms. As we show next, we complement Theorem 1 with asymptotically tight approximation guarantees by reducing MIN-SC to the SET COVER problem.

Theorem 3. *Let* $\Delta = \max_{v \in V} |N(v)| \geq 2$*. When* Δ *is constant there is a polynomial time algorithm for* MIN-SC *with a constant approximation ratio of* $(\Delta + 1)$*. For general instances, the problem is* $(\ln n - \ln \ln n + \Theta(1))$*-approximable.*

Proof. We will present a reduction that preserves approximability to the SET COVER problem. In an unweighted SET COVER instance, we are given a universe U and a collection $\mathcal{F}$ of subsets of U, and ask to find a cover of the universe with the minimum possible number of sets from $\mathcal{F}$. From an instance I of MIN-SC we create an instance I' of SET COVER as follows: We create a universe of elements U by adding one element for every voter, except for certain voters for which there is no such need. In particular, U contains one element for every $v \in V \setminus (C \cup A \cup \{u : \exists u' \in N(u) \cap C\})$. This means that in U we have excluded voters who can be satisfied without delegating to someone else as well as voters who can be satisfied by delegating to members of C (observe that because of Claim 2 (part (*i*)), all voters of C will be assigned to vote). Furthermore, to describe the collection $\mathcal{F}$ of sets in I', for every voter $v \in V \setminus C$ we add the set $S_v = U \cap (\{u : v \in N(u)\} \cup \{v\})$. If some S_v turns out to be the empty set, it can be simply disregarded (e.g. for a voter v with $N(v) \cap C \neq \emptyset$).

Lemma 1. *Let* $OPT(I), OPT(I')$ *be the costs of the optimal solutions in the instances* I *and* I' *respectively. Then* $OPT(I') \leq OPT(I)$*.*

Proof of Lemma 1. Let there be k dissatisfied voters in the optimal solution of the MIN-SC instance I. By making use of Claim 2, we can assume that these are members of $V \setminus C$ who are assigned to cast a ballot. Hence, for a dissatisfied voter v there exists a corresponding set S_v in I'. We will argue that by selecting these k sets that correspond to dissatisfied voters, we have a feasible solution for the SET COVER problem. Towards contradiction, assume that there is an element in I' that has not been covered by any of these sets. Because of the definition of U, there must exist a voter v in I who corresponds to that element and who only accepts to delegate to some voters who are not in C, i.e., each $u \in N(v)$ has a corresponding set S_u in $\mathcal{F}$ since $u \in V \setminus C$. Moreover, v should be satisfied in I, otherwise the set S_v would have been selected in the solution we constructed for I' and v would have been covered. Therefore, at least one of her approved voters, say u, is a guru, and the set S_u covers v, which is a contradiction. □

Lemma 2. *Given a feasible solution with cost $SOL(I')$ of the produced instance I', we can create a feasible solution of I, with cost $SOL(I) \leq SOL(I')$.*

Proof of Lemma 2. Say that we are given a solution for I' with cost $SOL(I') = k$, which means that by selecting a number of k sets, it is possible to cover every element of U. Consider a delegation function $d(\cdot)$ which asks every voter from $V \setminus C$ whose corresponding set has been selected in the cover, to cast a ballot. Following Claim 2, it also asks every voter from C to cast a ballot. From these, only the former k voters are dissatisfied, who vote but do not belong to C. We will argue that we can make all the remaining voters satisfied and hence we will have a solution with k dissatisfied voters.

Consider a voter $v \in V \setminus C$, whose set S_v was not included in the SET COVER solution. If $v \in A$, then v is assigned to abstain and she is satisfied. So, suppose that $v \in V \setminus (A \cup C)$ and also that $N(v) \neq \emptyset$ (otherwise, with $N(v) = \emptyset$, then S_v would have been selected in the cover). There are now two cases to consider:

Case 1: $N(v) \cap C \neq \emptyset$. Then v can delegate to a member of C and be satisfied.
Case 2: $N(v) \cap C = \emptyset$. Then by the construction of the universe U, we have that $v \in U$. Since we have selected a cover for U, v is covered by some set. Additionally, we have assumed that S_v was not picked in the cover, hence v is covered by some other set, say S_u, which means that u is a voter who is assigned to cast a vote and $v \in S_u$. But then v can delegate to u and be satisfied. □

By combining Lemma 1 and Lemma 2, we have that if we run any α-approximation algorithm for the SET COVER instance I', we can find a solution for the MIN-SC instance I, with the same guarantee since $SOL(I) \leq SOL(I') \leq \alpha OPT(I') \leq \alpha OPT(I)$. Recall that there exists a well known f-approximation algorithm for SET COVER, where f is the maximum number of sets that contain any element. Note also that in our construction, each element of I' that corresponds to a voter v of I, belongs to at most $|N(v)| + 1$ sets. This directly yields a $(\Delta + 1)$-approximation for our problem. Alternatively, when Δ is not bounded, we can use the best currently known approximation algorithm for the SET COVER problem, presented in [25], to obtain the desired result. □

3.2 Social Good Maximization

In all voting problems that involve a notion of dissatisfaction, one can study either minimization of dissatisfactions or maximization of satisfactions. The minimization version is slightly more popular, see e.g., [12] (also, in approval voting elections, it is more common to minimize the sum of distances from the optimal solution than to maximize the satisfaction score). Clearly, for ASSL, if we can solve optimally MIN-SC, the same holds for MAX-SG. The problems however can differ on their approximability properties.

Looking back on our findings for MIN-SC, we note that the results from Theorem 1 immediately yield NP-hardness for MAX-SG. The hardness of approximation however does not transfer. The result of Theorem 2 also applies.

Corollary 1. *Let* $\Delta = \max_{v \in V} |N(v)|$. *Then* MAX-SG *is NP-hard even when* $\Delta = 2$, *and it is efficiently solvable when* $\Delta \leq 1$.

Next, we also provide a constant factor approximation for constant Δ, albeit with a worse constant than the results for MIN-SC. The main insight for the next theorem is that we can exploit results from the rich domain of Constraint Satisfaction Problems (CSPs) and model MAX-SG as such.

Theorem 4. *Let* $\Delta = \max_{v \in V} |N(v)| \geq 2$. *When* Δ *is constant there is a polynomial time algorithm for* MAX-SG *with an approximation ratio of* $\frac{1}{(\Delta+2)^{\Delta+2}}$.

We leave as an open problem the question of whether there exist better approximations or whether one can establish hardness of approximation results.

3.3 Further Implications: Instances with Bounded Social Cost

We conclude this section by discussing some implications that can be derived by the reductions presented in Sect. 3.1, on relevant questions to MIN-SC. Let us start with the special case where the optimal cost of an instance I is zero, i.e., it is possible to satisfy all voters. Can we have an algorithm that detects this? It would be ideal to compute a delegation function that does not cause any dissatisfactions, and this is indeed possible. If $OPT(I) = 0$, then any approximation algorithm for MIN-SC of finite ratio will necessarily return an optimal solution. If $OPT(I) > 0$, the approximation algorithm will also return a positive cost. Hence, by using Theorem 3 we have the following:

Corollary 2. *Given an instance of* ASSL, *there exists a polynomial time algorithm that decides if it is possible to satisfy all voters, in which case it can also construct an optimal delegation function.*

Taking it a step further, suppose now we ask: Given an instance I, is it true that $OPT(I) \leq k$, for some positive constant k? This time, we can construct a SET COVER instance I' using the reduction presented in the proof of Theorem 3, and then we can enumerate all possible collections of subsets of size at most k. If a solution is found, it corresponds to a set of at most k dissatisfied voters. Hence, we can solve the problem in time $n^{O(k)}$. But now we can question whether there is hope for a substantially better running time. To answer this, we exploit the reduction used in Theorem 1 from DIRECTED DOMINATING SET. In particular, it is known by [14] that DOMINATING SET is $W[2]$-hard when parameterized by the solution cost, even in graphs of bounded average degree. Given that the directed version of DOMINATING SET inherits the hardness results of the undirected version in combination with the proof of Theorem 1, we get:

Corollary 3. *Unless $W[2] = FPT$, MIN-SC cannot be solved in time $f(k)n^{O(1)}$ even for the case where Δ is constant, where $f(k)$ is a computable function depending only on the minimum possible number k of dissatisfied voters.*

4 Exact Algorithms via Monadic Second Order Logic

The goal of this section is to focus on special cases that admit exact polynomial time algorithms. Our major highlight is the use of a logic-based technique for obtaining such algorithms. To our knowledge, this framework has not received much attention (if at all) from the computational social choice community despite its wide applicability on graph-theoretic problems. We therefore expect that this has the potential of further deployments for other related problems.

4.1 Optimization Under Bounded Treewidth

The general methodology involves the use of an algorithmic meta-theorem (for related surveys see [17] and [21]) to check the satisfiability of a formula that expresses a graph property, defined over an input graph of bounded treewidth. Roughly speaking, the treewidth is a graph parameter that indicates the "tree-likeness" of a graph. It was introduced independently by various authors mainly for undirected graphs (see [24] for an extended exposition of the origin of the notion) but its definition and intuition can be extended to directed graphs as well [3]. In our case, we will require bounded treewidth for the directed graph associated to an instance of ASSL.

The approach presented here was initiated by Courcelle [11], who used Monadic Second Order (MSO) logic to define graph properties. These, typically ask for some set of vertices or edges subject to certain constraints. For expressing a property in MSO, we can make use of variables for edges, vertices as well as for subsets of them. Apart from the variables, we can also have the usual[1] boolean connectives $\neg, \wedge, \vee, \Rightarrow$, quantifiers $\forall, \exists$, and the membership operator $\in$. The resulting running time for deciding properties expressible in MSO turns out to be exponentially dependent on the treewidth and the size of the formula.

[1] For ease of presentation, we will also use some set operations that although they are not explicitly allowed, they can be easily replaced by equivalent MSO expressions. For instance, $x \notin A \setminus B \equiv \neg((x \in A) \wedge \neg(x \in B))$ and $A \subseteq B \equiv (\forall x \in A \Rightarrow x \in B)$.

After Courcelle's theorem, there have been several works that extend the algorithmic implications of MSO logic. Most importantly, and most relevant to us, the framework of [3] can handle some types of optimization problems. Consider a formula $\phi(X_1, \ldots, X_r)$ in MSO, having $X_1, \ldots, X_r$ as free set variables, so that a property is true if there exists an assignment to the free variables that make ϕ satisfied. Then, we can optimize a weighted sum over elements that belong to any such set variable, subject to the formula ϕ being true (one needs to be careful though as the weights are taken in unary form). A representative example presented in [3] (see Theorem 3.6 therein for a wide variety of tractable problems w.r.t. treewidth) is MINIMUM DOMINATING SET in which we want to minimize $|X|$ subject to a formula that enforces the set X to be a dominating set.

We note that the results we use here require to have a representation of the tree decomposition of the input graph. But even if this is not readily available, its computation is in FPT w.r.t. the treewidth [6].

Our first result in this section shows that MIN-SC and MAX-SG are tractable when the treewidth of the associated graph is constant.

Theorem 5. *Consider an instance of* ASSL, *and let G be its corresponding graph. Then* MIN-SC *and* MAX-SG *are in FPT w.r.t. the treewidth of G.*

Proof. It suffices to solve MIN-SC since this yields an optimal solution to MAX-SG as well. In order to apply a framework of MSO logic, we first make a small modification to the graph G. We add a special vertex denoted by a and we add a directed edge (v, a) for every v for which $v \in A$. In this manner, abstainers will be encoded by "delegating" their vote to a. Let $G' = (V', E')$ be the resulting graph, where $V' = V \cup \{a\}$ and $E' = E \cup \{(v, a) : v \in A\}$. We observe that these additions do not affect the boundedness of the treewidth.

Lemma 3. *If G has bounded treewidth, so does G'.*

We will create an MSO formula $\phi(D, X)$ with 2 free variables, D and X, encoding an edge-set and a vertex set respectively. The rationale is that $\phi(D, X)$ becomes true when the edges of D encode a delegation function and X denotes the set of voters who are dissatisfied by the delegations of D. To write the formula, we also exploit the fact that the framework of [3] allows the use of a constant number of "distinguished" sets so that we can quantify over them as well (apart from quantification over V' and E'). We will use V, along with C and A (of voters who approve casting a ballot or abstaining respectively), as these special sets here. To proceed, $\phi(D, X)$ is the following formula:

$$
\begin{aligned}
& D \subseteq E' \wedge X \subseteq V \setminus C \wedge \\
& (\forall v \in V' \ (deg_D^+(v) \leq 1)) \wedge (\forall u, v, w \in V'((u, v) \in D \Rightarrow (v, w) \notin D)) \wedge \\
& (\forall v \in C \ (deg_D^+(v) = 0)) \wedge \\
& (\forall v \in V \ (v \in X \Leftrightarrow (deg_D^+(v) = 0 \wedge v \notin C)))
\end{aligned}
$$

The term $deg_D^+(v) = 0$ can be expressed in MSO logic in a similar way to the more general term of $deg_D^+(v) \leq 1$, which we define formally as

$$deg_D^+(v) \leq 1 \equiv (\exists u \in V' \ (v,u) \in D) \Rightarrow \neg(\exists w \in V' \ ((v,w) \in D \wedge w \neq u)).$$

Concerning the construction of $\phi(D, X)$, the second line expresses the fact that D is a union of disjoint directed stars so as to enforce Claim 1. Anyone with out-degree equal to one within D either delegates to some other voter or abstains (i.e. delegates to vertex a), whereas those with out-degree equal to zero in D cast a vote themselves. The third line of $\phi(D, X)$ also enforces Claim 2 (part (i)) so that members of C always cast a vote. The fourth line expresses the fact that the vertices of X are dissatisfied voters. By Claim 2 (part (ii)), the only way to make a voter v dissatisfied is by asking her to cast a ballot when $v \notin C$. Indeed, voters who are not asked to cast a ballot, have out-degree equal to one in D, so they either abstain or delegate. This means that either $(v, a) \in D$ or $(v, u) \in D$ for some $u \in V$. In the former case, v is satisfied because $v \in A$ (if $v \notin A$ then the edge (v, a) would not exist in E' and could not have been selected in D). In the latter case, v approves u (otherwise the edge (v, u) would not exist) and u casts a vote since D contains only stars. Hence v is again satisfied.

The final step is to perform optimization w.r.t. $|X|$ subject to $\phi(D, X)$ being true. To that end, we can assign a weight $w(v)$ to every vertex v such that $w(a) = 0$ and $w(v) = 1, \forall v \in V' \setminus \{a\}$. Hence $\sum_{v \in X} w(v) = |X|$. Using the result of [3], we can find a delegation function $d(\cdot)$, as given by the edges in D, that minimizes the number of dissatisfied voters within the feasible solutions. □

4.2 Adding Secondary Objectives

We continue with exhibiting that MSO frameworks can be useful for tackling other related problems as well. For the cases when we can solve MIN-SC (and MAX-SG) optimally, we are investigating whether we can find such a solution with additional properties (whenever the optimal is not unique). Motivated by questions studied in [12,13] and [15] we consider the following problems:

1. Among the optimal solutions to MIN-SC (or MAX-SG), find one in which a given voter v casts a vote, or answer that no such solution exists.
2. Ditto, with having voter v abstain in an optimal solution.
3. Among the optimal solutions to MIN-SC (or MAX-SG), find one that minimizes the number of abstainers.
4. Among the optimal solutions to MIN-SC (or MAX-SG), find one that minimizes the maximum voting power over all gurus, i.e. the number of voters that she represents (or equivalently that minimizes the maximum in-degree).

The fourth problem is quite important in models of LD, given also the critique that often applies on such models that may accumulate excessive power on some voters. Below we start by addressing the first three problems together.

Theorem 6. *Consider an instance of* ASSL, *and let G be its corresponding graph. It is in FPT w.r.t. the treewidth of G to find an optimal solution to* MIN-SC *and* MAX-SG, *in which a given voter casts a ballot or abstains (if such a solution exists). The same holds for minimizing the number of abstainers.*

We come now to the fourth problem, which is the most challenging one. For this, we will use yet another enriched version of the MSO framework, which facilitates the addition of further constraints and helps in solving several degree-constrained optimization problems. As these problems are in general more difficult [22], the results of [26] and [20] yield polynomial time algorithms w.r.t. treewidth, but do not place them in FPT.

For the presentation we will stick to the terminology of [20]. Consider a formula $\phi(X_1, \ldots, X_r)$ with free variables $X_1, \ldots X_r$. The main idea is to add so-called global and local cardinality constraints and ask for an assignment that satisfies both ϕ and the constraints. In the simpler version that we will use here, a global cardinality constraint is of the form $\sum_{i \in [r]} a_i |X_i| \leq b$ for given rational numbers $a_i, i \in [k]$ and b (some of these numbers can be zero so that we constrain the cardinality of only some of the free variables). On the other hand, a local cardinality constraint for a vertex has to do with limiting the number of its neighbors or incident edges that belong to a set corresponding to a free variable. For example, if X_1 is a free variable of ϕ that encodes a vertex set, and X_2 is a free variable encoding an edge set, we can have constraints of the form "for each vertex v of G, the number of vertices in X_1 adjacent to v belongs to a set $a(v)$", where $a(v)$ contains the allowed values (e.g., could be an interval). Similarly, we can express that the number of edges of X_2 incident with v can take only specific values from some set $a'(v)$. A nice representative illustration for local constraints in [20] is the CAPACITATED DOMINATING SET problem, where one needs to pick a dominating set D respecting capacity constraints for every $v \in D$.

Theorem 7. *Consider an instance of* ASSL *where the associated graph G has constant treewidth. Then among the optimal solutions to* MIN-SC, *we can find in polynomial time a solution that minimizes the maximum in-degree of the gurus.*

Proof Sketch. Starting from the directed graph G, let $G' = (V', E')$ be the graph used in the proof of Theorem 5, derived from G. Our first step is to use Theorem 5 and solve MIN-SC optimally so that we know the cost of an optimal solution. Suppose that we have c unsatisfied voters in an optimal solution.

In order to proceed and utilize the extended MSO framework of [20], we need to work with an undirected graph. To this end, we create an undirected graph $H = (V'', E'')$ from the directed graph G' by having each $v \in V$ correspond to 2 vertices, v_{in} and v_{out} in V''. In this manner, out-going edges from v will correspond to edges incident with v_{out} in H whereas incoming edges to v will be incident to v_{in}. The graph H will also include the node a for the abstentions, so that in total $V'' = V_{in} \cup V_{out} \cup \{a\}$. Given this construction, it is easy to verify that if G has bounded treewidth, so does H.

The next step is to produce a formula for the undirected graph H, whose satisfying assignments will correspond to valid delegation functions on the original graph G. We will denote our formula by $\psi(D, F, X)$, with the free variables D, F, X. As in Theorem 5, the set D will be a subset of edges encoding a valid delegation function. The set F will encode the set of voters who cast a ballot themselves. Finally, the set X will encode the dissatisfied voters induced by D.

Following the framework of [20], we now add 2 classes of constraints that we want to be satisfied in addition to $\psi(D, F, X)$. The first one is a so-called global cardinality constraint to ensure that the number of dissatisfied voters is no more than the optimal. Since we have already solved MIN-SC and the solution is c, and since X expresses the set of dissatisfied voters, the constraint will be $|X| \leq c$.

Finally, we add the so-called local cardinality constraints. We will produce a set of constraints depended on a fixed number $d \in \{0, 1, \dots, n\}$ such that the constraints will ensure that the maximum degree of every vertex in F is bounded by d. By using [20], we can now decide for every d if there is an assignment to the variables D, F, X that satisfies $\psi(D, F, X)$ together with the global and local cardinality constraints. To summarize, the steps of the overall algorithm are:

(1) Use Theorem 5 to find the optimal number of dissatisfied voters.
(2) Transform G to the undirected graph H and construct the formula $\psi(D, F, X)$.
(3) For $d = 0$ to n, decide if $\psi(D, F, X)$ is satisfiable subject to the global and local cardinality constraints introduced above. Stop in the first iteration where this is true and create the delegation function from D and F. □

5 Discussion and Other Directions

We have presented a model that allows voters to express preferences over delegations via an approval set. Our main goal has been to optimize the overall satisfaction of the voters, which implies that it suffices to focus only on direct delegations to actual voters. Even under this simpler solution space, the problems we study are intractable, even when the out-degree is a small constant. On the positive side, we have exhibited constant factor approximation algorithms for graphs of constant maximum out-degree, as well as exact algorithms under the bounded treewidth condition, even when secondary objectives are also present. It is therefore interesting to see if any other parameter can play a crucial role on the problem's complexity.

All our results also hold under the generalized model where a graph G is given so that the out-neighborhood $N(v)$, of voter v, expresses the set of feasible delegations which is a (possibly strict) superset of her approved delegations. On the other hand, the case where the approved delegatees of a voter v are not necessarily neighbors seems more complex (e.g., a voter approves some other person but cannot directly delegate to her due to hierarchy constraints). Finally, the results of Sect. 3 also hold for weighted voters, whereas the results of Sect. 4 only hold if the weights are polynomially bounded in unary form.

Another worthwhile direction comes from the fact that the MSO framework primarily serves as a theoretical tool for placing a problem in a certain complexity class but yields impractical running times. One could proceed with a theoretical and/or experimental study of tailor-made dynamic programming algorithms for the problems presented in Sect. 4. Coming to our last result (Theorem 7), an interesting approach for future work would be to provide algorithms with trade-offs between the total dissatisfaction and the maximum voting power (instead of optimizing one objective and keeping the other as a secondary objective).

5.1 Towards a Game-Theoretic Analysis

We conclude our work with a preliminary game-theoretic analysis, which can serve as the basis for a more elaborate future study of these models. Motivated by the approach of [12,13], we define the following simple game: Say that in an instance of ASSL each voter v acts as a strategic player, whose strategy space is $N(v) \cup \{v, \bot\}$. The utility that she can earn from an outcome is either 1, if she is satisfied with that outcome, or 0 otherwise. The first relevant question is whether such games admit pure Nash equilibria, i.e., delegation functions under which no voter is able to unilaterally change her strategy and increase her utility. In contrast to the model of rank-based preferences of [12,13], in our case, Nash equilibria are guaranteed to exist.

Proposition 1. *In every instance of* ASSL, *there exists a pure Nash equilibrium, which can be computed in polynomial time.*

In order to evaluate the equilibria of a game (in terms of the derived social good, or similarly in terms of social cost), we can use the Price of Anarchy as a standard metric. This can be defined as the worst possible ratio between the optimal solution for the social good against the number of satisfied voters at a Nash Equilibrium. Unfortunately, we show below that strategic behavior can lead to quite undesirable solutions and we note that this could act as an argument in favor of using a centralized mechanism, as done in the previous sections, to avoid such bad outcomes.

Proposition 2. *The Price of Anarchy for the strategic games of the* ASSL *model, can be as bad as $\Omega(n)$, even when $\Delta \leq 1$.*

Finally, note that Proposition 2 raises the question of coming up with richer game-theoretic models of the delegation process (e.g. richer utility functions or repeated games) so as to understand thoroughly the effects of strategic behavior.

Acknowledgements. This work has been supported by the Hellenic Foundation for Research and Innovation (H.F.R.I.) under the "First Call for H.F.R.I. Research Projects to support faculty members and researchers and the procurement of high-cost research equipment" grant (Project Number: HFRI-FM17-3512).

References

1. Abramowitz, B., Mattei, N.: Flexible representative democracy: an introduction with binary issues. In: Proceedings of the Twenty-Eighth International Joint Conference on Artificial Intelligence, (IJCAI-19), pp. 3–10 (2019)
2. Anshelevich, E., Fitzsimmons, Z., Vaish, R., Xia, L.: Representative proxy voting. arXiv preprint arXiv:2012.06747 (2020)
3. Arnborg, S., Lagergren, J., Seese, D.: Easy problems for tree-decomposable graphs. J. Algorithms **12**(2), 308–340 (1991)
4. Bloembergen, D., Grossi, D., Lackner, M.: On rational delegations in liquid democracy. In: Proceedings of the Thirty-Third AAAI Conference on Artificial Intelligence, (AAAI-19), pp. 1796–1803 (2019)
5. Blum, C., Zuber, C.I.: Liquid democracy: potentials, problems, and perspectives. J. Polit. Philos. **24**(2), 162–182 (2016)
6. Bodlaender, H.L.: A linear-time algorithm for finding tree-decompositions of small treewidth. SIAM J. Comput. **25**(6), 1305–1317 (1996)
7. Boldi, P., Bonchi, F., Castillo, C., Vigna, S.: Viscous democracy for social networks. Commun. ACM **54**(6), 129–137 (2011)
8. Caragiannis, I., Micha, E.: A contribution to the critique of liquid democracy. In: Proceedings of the Twenty-Eighth International Joint Conference on Artificial Intelligence, (IJCAI-19), pp. 116–122 (2019)
9. Christoff, Z., Grossi, D.: Binary voting with delegable proxy: an analysis of liquid democracy. In: Proceedings of the Sixteenth Conference on Theoretical Aspects of Rationality and Knowledge, (TARK-17), pp. 134–150 (2017)
10. Colley, R., Grandi, U., Novaro, A.: Smart voting. In: Proceedings of the Twenty-Ninth International Joint Conference on Artificial Intelligence, (IJCAI-20), pp. 1734–1740 (2021)
11. Courcelle, B.: The monadic second-order logic of graphs. I. Recognizable sets of finite graphs. Inf. Comput. **85**(1), 12–75 (1990)
12. Escoffier, B., Gilbert, H., Pass-Lanneau, A.: The convergence of iterative delegations in liquid democracy in a social network. In: Proceedings of the Twelfth International Symposium on Algorithmic Game Theory, (SAGT-19), pp. 284–297 (2019)
13. Escoffier, B., Gilbert, H., Pass-Lanneau, A.: Iterative delegations in liquid democracy with restricted preferences. In: Proceedings of the Thirty-Fourth AAAI Conference on Artificial Intelligence, (AAAI-20), pp. 1926–1933 (2020)
14. Golovach, P.A., Villanger, Y.: Parameterized complexity for domination problems on degenerate graphs. In: Proceedings of the Thirty-Fourth International Workshop on Graph-Theoretic Concepts in Computer Science, pp. 195–205 (2008)
15. Gölz, P., Kahng, A., Mackenzie, S., Procaccia, A.D.: The fluid mechanics of liquid democracy. In: Proceedings of the Fourteenth International Conference on Web and Internet Economics, (WINE-18), pp. 188–202 (2018)
16. Green-Armytage, J.: Direct voting and proxy voting. Const. Polit. Econ. **26**(2), 190–220 (2015)
17. Grohe, M.: Logic, graphs, and algorithms. Log. Automata **2**, 357–422 (2008)
18. Kahng, A., Mackenzie, S., Procaccia, A.: Liquid democracy: an algorithmic perspective. J. Artif. Intell. Res. **70**, 1223–1252 (2021)
19. Kavitha, T., Király, T., Matuschke, J., Schlotter, I., Schmidt-Kraepelin, U.: Popular branchings and their dual certificates. In: Proceedings of the Twenty-First International Conference on Integer Programming and Combinatorial Optimization, pp. 223–237 (2020)

20. Knop, D., Koutecký, M., Masarík, T., Toufar, T.: Simplified algorithmic metatheorems beyond MSO: treewidth and neighborhood diversity. Log. Methods Comput. Sci. **15**(4), 1–32 (2019)
21. Kreutzer, S.: Algorithmic meta-theorems. In: Grohe, M., Niedermeier, R. (eds.) IWPEC 2008. LNCS, vol. 5018, pp. 10–12. Springer, Heidelberg (2008). https://doi.org/10.1007/978-3-540-79723-4_3
22. Masařík, T., Toufar, T.: Parameterized complexity of fair deletion problems. Discret. Appl. Math. **278**, 51–61 (2020)
23. Paulin, A.: An overview of ten years of liquid democracy research. In: The Proceedings of the Twenty-First Annual International Conference on Digital Government Research, pp. 116–121 (2020)
24. Seymour, P.: The origin of the notion of treewidth. Theoretical Computer Science Stack Exchange (2014). https://cstheory.stackexchange.com/q/27317. Accessed 16 May 2021
25. Slavık, P.: A tight analysis of the greedy algorithm for set cover. J. Algorithms **25**(2), 237–254 (1997)
26. Szeider, S.: Monadic second order logic on graphs with local cardinality constraints. ACM Trans. Comput. Log. **12**(2), 1–21 (2011)
27. Zhang, Y., Grossi, D.: Tracking truth by weighting proxies in liquid democracy. arXiv preprint arXiv:2103.09081 (2021)

Two Birds with One Stone: Fairness and Welfare via Transfers

Vishnu V. Narayan[1(✉)], Mashbat Suzuki[1], and Adrian Vetta[1,2]

[1] School of Computer Science, McGill University, Montreal, Canada
{vishnu.narayan,mashbat.suzuki}@mail.mcgill.ca

[2] Department of Mathematics and Statistics, McGill University, Montreal, Canada
adrian.vetta@mcgill.ca

Abstract. We study the question of dividing a collection of indivisible items amongst a set of agents. The main objective of research in the area is to achieve one of two goals: fairness or efficiency. On the fairness side, *envy-freeness* is the central fairness criterion in economics, but envy-free allocations typically do not exist when the items are indivisible. A recent line of research shows that envy-freeness *can* be achieved if a small quantity of a homogeneous divisible item (money) is introduced into the system, or equivalently, if transfer payments are allowed between the agents. A natural question to explore, then, is whether transfer payments can be used to provide high *welfare* in addition to envy-freeness, and if so, how much money is needed to be transferred.

We show that for general monotone valuations, there always exists an allocation with transfers that is envy-free and whose Nash social welfare (NSW) is at least an $e^{-1/e}$-fraction of the optimal Nash social welfare. Additionally, when the agents have additive valuations, an envy-free allocation with negligible transfers and whose NSW is within a constant factor of optimal can be found in polynomial time. Consequently, we demonstrate that the seemingly incompatible objectives of fairness and high welfare can be achieved simultaneously via transfer payments, even for general valuations, when the welfare objective is NSW. On the other hand, we show that a similar result is impossible for utilitarian social welfare: *any* envy-freeable allocation that achieves a constant fraction of the optimal welfare requires non-negligible transfers. To complement this result we present algorithms that compute an envy-free allocation with a given target welfare and with bounded transfers.

Keywords: Fair division · Welfare · Transfers

1 Introduction

The question of how to divide a collection of items amongst a group of agents has remained of central importance to society since antiquity. Real-world examples of this problem abound, ranging from the division of land and inherited estates, border settlements, and partnership dissolutions, to more modern considerations

I. Caragiannis and K. A. Hansen (Eds.): SAGT 2021, LNCS 12885, pp. 376–390, 2021.
https://doi.org/10.1007/978-3-030-85947-3_25

such as the division of the electromagnetic spectrum, distribution of computational resources, and management of airport traffic. The predominant objective of research in this area is to study the existence of allocations that achieve one of two broad goals: *fairness* or *efficiency*. At a high level, the fairness goal is to ensure that each agent receives its due share of the items, and the efficiency goal is to distribute the items in a way that maximizes the aggregate utility achieved by all of the agents.

The study of fair division burgeoned in the decades following its formal introduction by Banach, Knaster and Steinhaus [29], and most of the early literature focused on the *divisible* setting, where a single heterogeneous divisible item (conventionally, a cake) is to be fairly shared among a set of agents with varying preferences over its pieces. The second half of the last century saw the creation of precise mathematical definitions for various fairness notions, and *envy-freeness*, where every agent prefers its piece to any piece received by another agent, has since emerged as the dominant fairness criterion in economics. More recently, research has focused on the *indivisible* setting, where each item in a collection must be allocated as a whole to some agent. It appears at first that envy-freeness cannot be achieved in this setting; consider the simple example of two agents and one item, where one agent is left envying the other in any allocation. Consequently, a common theme in the indivisible setting is the study of weaker fairness guarantees such as *EF1* and *approximate-MMS* [14,26,27].

But is it necessary to restrict ourselves to these weaker guarantees? A recent line of research shows, rather surprisingly, that it is possible to achieve canonical envy-freeness even in the indivisible setting simply by adding to the system a small quantity of a *divisible item*, akin to money [13,24], or equivalently by allowing the agents to make transfer payments between themselves. These transfer payments can always be made alongside an allocation of the indivisible items such that the result is envy-free. In this work, we ask and answer a natural follow-up question: can this tool be made to do more? Can we use it to simultaneously guarantee full envy-freeness while also achieving high *welfare*, and if so, how much in total transfer payments do we need for this?

1.1 Related Work

The formal origin of fair division dates back to the 1940s, when Banach, Knaster and Steinhaus [29] devised the Last Diminisher procedure to fairly divide a cake among n agents. Their fairness objective was *proportionality*, in which each agent receives a piece of value at least $\frac{1}{n}$ of the value of the entire cake to that agent. The pursuit of proportional cake divisions in different settings led to the creation of popular algorithmic paradigms for cake-cutting such as the moving-knives procedures [18,30]. In the following decades, *envy-freeness* (Gamow and Stern [20], Foley [19]) emerged as the canonical fairness solution. Early non-constructive results proved that, under mild assumptions, envy-free allocations always exist in the divisible setting (Stromquist [30], Woodall [35], Su [31]), and ensuing work produced finite and bounded protocols for computing these allocations (Brams and Taylor [12], Aziz and Mackenzie [5]).

The research efforts of the fair division community have undergone two major shifts in recent years. The first of these is an increased focus on *economic efficiency*. The most common type of economic efficiency is *Pareto efficiency*, in which no agent's allocation can be improved without making some other agent worse off. A classical result of Varian [34] shows that in the divisible setting there always exists an allocation that is both envy-free and Pareto efficient. A different notion of efficiency arises when we maximize a *welfare function* that measures the aggregate utility of all agents. The most common welfare functions studied in the associated literature are the *utilitarian* social welfare (or simply the social welfare), which measures the sum of the agents' valuations, and the *Nash* social welfare, which measures the geometric mean of these valuations. In the divisible setting, Bei et al. [8] and Cohler et al. [17] study the problem of maximizing social welfare under proportionality and envy-freeness constraints.

The second shift is towards the study of the indivisible setting, where m items are to be integrally divided amongst n agents. Since neither envy-freeness nor proportionality can now be guaranteed, a natural alternative is to provide relaxations or approximations of them. One such relaxation is the *EFk* guarantee. An allocation is *envy-free up to k items*, or EFk, if no agent envies another agent's bundle provided some k items are removed from that bundle. The EF1 guarantee is particularly notable, as EF1 allocations exist and can be computed in polynomial time if the valuation functions are monotone [27]. Two similar relaxations exist for proportionality, namely the *Prop1* guarantee and the *maximin share* guarantee, the latter of which is a natural extension of the two-agent cut-and-choose protocol [14]. A large body of research produced over the last decade aims to achieve these guarantees or approximations thereof (see e.g. [21,26]), including many results that show that these fairness guarantees can be achieved alongside Pareto efficiency [7] or high Nash social welfare [7,16].

The problem of achieving high utilitarian social welfare under fairness constraints was formally introduced by Caragiannis et al. [15]. The *price of fairness* (that is, of envy-freeness, EF1, or any other fairness criterion) of an instance is defined as the ratio of the social welfare of an optimal allocation without fairness constraints, to the social welfare of the best fair allocation. Intuitively, it measures the necessary worst-case loss in efficiency when we add fairness constraints. Caragiannis et al. [15] present bounds on the price of fairness (proportionality, envy-freeness and equitability) in both the divisible and indivisible settings; we remark, however, that their results for the indivisible case only consider the special set of instances for which the associated fair allocations exist. For the divisible setting, Bertsimas et al. [11] showed that the bounds of [15] are tight. Followup work on the price of fairness in the indivisible setting by Bei et al. [9] and Barman et al. [6] considers only the relaxed fairness guarantees (such as EF1 and $\frac{1}{2}$-MMS) that are always achievable in the indivisible setting.

In now classical work, Svensson [32], Maskin [28], and Tadenuma and Thomson [33] studied the indivisible item setting and asked if it is always possible to achieve an envy-free allocation simply by introducing a small quantity of a divisible item, akin to money, alongside the indivisible items. Their positive results were mirrored in followup work by Alkan et al. [1], Aragones [2], Klijn [25] and Haake et al. [23] which showed for various settings the existence of an envy-free *allocation with subsidy*. However, all of the above papers considered the restricted case where the number of items, m, is at most the number of agents n (or where the items were grouped into n fixed bundles). It was only recently that Halpern and Shah [24] extended these results to the general m-item setting, showing that an envy-free allocation with subsidy always exists in general. Brustle et al. [13] followed this up with upper bounds on the amount of money sufficient to support an envy-free allocation in all instances. Surprisingly, when the valuation functions are scaled so that the marginal value of an item is at most one dollar to any agent, at most $n-1$ dollars in the additive case and at most $O(n^2)$ dollars in the general monotone case are always sufficient to eliminate envy [13]. Note that the maximum required subsidy is *independent* of the number m of items, an observation of particular relevance to our work. Several recent papers study the problem of achieving envy-freeness alongside other properties via subsidies and transfers, including Aziz [4], Goko et al. [22].

1.2 Results and Contributions

A salient question is whether the two ideas exposited in the prior discussion can be combined: is it possible to find an allocation with subsidy that is simultaneously envy-free *and* guarantees high welfare? If so, how much subsidy is sufficient to achieve this? These questions are the focus of this paper.

Thus, one contribution of our work is to extend the literature on subsidies and their application. However, rather than subsidies, we analyze the related concept of *transfer payments* between the agents for two reasons. First, a subsidy is an external source of added utility which, in the context of welfare, would bias any subsequent comparisons with the welfare-maximizing allocation *without* subsidies. A transfer payment is neutral in this regard. Second, subsidies require an external agent willing to fund the mechanism – a typically unrealistic hope. In contrast, transfer payments require the consent only of the agents who are already willing participants in the mechanism. Provided the cost of the payments are outweighed by the benefits of participation then giving consent is reasonable. We remark that subsidies and transfers are in a sense interchangeable. Given an envy-free allocation with subsidies, subtracting the average subsidy from each agent's individual payment gives payments which sum to zero, that is, transfer payments. Conversely, given transfer payments, adding an appropriate fixed amount to each payment induces non-negative subsidy payments.[1]

A second contribution is to extend the research on the price of fairness. Specifically, we impose no *balancing constraint* on the valuation functions of the agents. To understand this, note that a common assumption in the price of fairness literature is that the valuation function of each agent is scaled so that the value

[1] Of course, whilst the correspondence between subsidies and transfers is simple, the switch to transfer payments does have a technical drawback: because transfer payments do not provide an (unnatural) external boost to welfare, obtaining welfare guarantees for the case of transfers is generally harder than for the case of subsidies.

of the grand bundle of items is *equal* for all agents. In the context of fairness, this scaling is benign because it has no affect on the most widely used measures of fairness. For example, it does not change the (relative) envy between any pair of agents. However, in the context of efficiency or welfare, this scaling can dramatically alter the welfare of any allocation by restricting attention to balanced instances, where agents are of essentially equal importance in generating welfare. This is important because it is the elimination of unbalanced instances that allows non-trivial bounds on the price of fairness to be obtainable [6,9]. Indeed, as will be seen in this paper, it is the unbalanced instances that are typically the most problematic in obtaining both fairness and high welfare.

We do, for simplicity, make the standard assumption in the literature on subsidies [13,24], and assume that the maximum marginal value for an item for any agent is always at most one dollar. We emphasize that this assumption is benign with respect to both fairness and welfare: it does not affect the relative envy between agents, and it does not affect the welfare of an allocation (as all valuations can be scaled down uniformly). Expressing the transfers in dollar amounts allows for a consistent comparison with earlier work on the topic, and equivalent bounds for the original instance can be recovered by multiplying these expressions by the maximum marginal value of an item for any agent.

We now present the main results in the paper. We study the trade-off between fairness and efficiency in the presence of transfer payments for the class of ρ-mean welfare functions, with particular focus on the two most important special cases, namely the Nash social welfare and utilitarian social welfare functions. An allocation is *envy-freeable* if it can be made envy-free with the addition of subsidies (or, equivalently, transfer payments). Our first observation is that to achieve both fairness and high welfare, it is not sufficient to simply find an envy-freeable allocation – making transfer payments is necessary. In fact, no non-zero welfare guarantee is achievable for all ρ without considering transfers in the computation of the welfare. Letting W^{ρ} denote ρ-mean welfare, we have:

Observation 1. *For any $\epsilon > 0$, there exist instances where the welfare of every envy-freeable allocation A satisfies $\frac{\mathrm{W}^{\rho}(A)}{\mathrm{W}^{\rho}(A^*)} \leq \epsilon$.*

Here A^* is the welfare-maximizing allocation. The observation applies even in the case of additive valuations with Nash social welfare functions. Consequently, the focus on allocations with transfers is justified. For ρ-mean welfare functions, we show that positive welfare guarantees are achievable with transfers.

Corollary 1. For subadditive valuations, there exists an envy-free allocation with transfers (A, t) such that $\frac{\mathrm{W}^{\rho}(A,t)}{\mathrm{W}^{\rho}(A^*)} \geq \frac{1}{n}$ and with a total transfer $\sum_i |t_i|$ of at most $2n^2$. This allocation can be computed in polynomial time.

Here n is the number of agents. Note that the total transfer is independent of the number m of items. This implies, as m grows, that the transfer payments are negligible in terms of the number of items (and of total welfare). In particular, our ultimate objective is to obtain both envy-freeness and high welfare using negligible transfers. Of course, the welfare guarantee of $\frac{1}{n}$ does not signify high welfare. So we investigate whether improved bounds can be obtained for the important special cases of $\rho = 0$ (Nash social welfare) and $\rho = 1$ (utilitarian social welfare). Strong guarantees on welfare can be obtained for the former. Specifically, there exists an envy-free allocation with transfers with a Nash social welfare that is at least an $e^{-1/e} \approx 0.6922$ fraction of the optimal welfare.

Theorem. *For general valuations, there exists an envy-free allocation with transfers (A, t) such that $\frac{\text{NSW}(A,t)}{\text{NSW}(A^*)} \geq e^{-1/e}$.*

Furthermore, for additive valuations, such constant factor welfare guarantees can be obtained with negligible transfer payments.

Theorem. *For additive valuations, given an α-approximate allocation to maximum Nash social welfare, there exists a polynomial time computable envy-free allocation with transfers (A, t) such that $\frac{\text{NSW}(A,t)}{\text{NSW}(A^*)} \geq \frac{1}{2}\alpha \cdot e^{-1/e}$ with a total transfer $\sum_i |t_i|$ of at most $2n^2$.*

In sharp contrast, for utilitarian social welfare, the factor $\frac{1}{n}$ welfare threshold is tight. To achieve any welfare guarantee greater than $\frac{1}{n}$ requires non-negligible transfer payments. Specifically, we show

Corollary. *For any $\alpha \in \left[\frac{1}{n}, 1\right]$, there exists an instance with additive valuations such that any envy-free allocation with transfers (A, t) satisfying $\frac{\text{SW}(A,t)}{\text{SW}(A^*)} \geq \alpha$ requires a total transfer $\sum_{i \in N} |t_i|$ of at least $\frac{1}{4}\left(\alpha - \frac{1}{n}\right)^2 m$.*

In fact, there exist instances for which any EFk allocation with $k = o(m)$ has a welfare guarantee of at most $\frac{1}{n} + o(1)$ (Lemma 3). This implies that EFk allocations cannot provide higher welfare with moderate transfers.

On the positive side, we can design algorithms to produce envy-free allocations with welfare guarantee α whose total transfer payment is comparable to the minimum amount possible, quantified in terms of the maximum value $\max_i v_i(A_i^*)$ any agent has in the welfare-maximizing allocation.

Theorem. *For additive valuations, for any $\alpha \in (0, 1]$, there is a polynomial time computable envy-free allocation with transfers (A, t) such that $\frac{\text{SW}(A,t)}{\text{SW}(A^*)} \geq \alpha$ with total transfer $\sum_{i \in N} |t_i| \leq n(\alpha \max_i v_i(A_i^*) + 2)$.*

Theorem. *For general valuations, for any $\alpha \in \left(0, \frac{1}{3}\right]$, there is an envy-free allocation with transfers (A, t) such that $\frac{\text{SW}(A,t)}{\text{SW}(A^*)} \geq \alpha$ with total transfer $\sum_{i \in N} |t_i| \leq 2n^2 \left(3\alpha \max_i v_i\left(A_i^*\right) + 2\right)$.*

In Sect. 2, we present our model of the fair division problem with transfers. Section 3 contains an exposition of the prior results in the literature that will be useful, along with our preliminary results on the ρ-mean welfare of envy-free allocations with transfers. In Sect. 4, we present our results on Nash social welfare, and in Sect. 5 we present our results on utilitarian social welfare. Due to length restrictions, the proofs are deferred to the full paper.

2 The Model and Preliminaries

Let $M = \{1, \cdots, m\}$ be a set of m indivisible items and let $N = \{1, \cdots, n\}$ be a set of agents. Each agent i has a *valuation function* $v_i : 2^M \rightarrow \mathbb{R}$, where

$v_i(\emptyset) = 0$. We make the standard assumption that each valuation function is *monotone*, satisfying $v_i(S) \leq v_i(T)$ whenever $S \subseteq T$. Additionally, following previous work on subsidies (see e.g. [13,24]), without loss of generality we uniformly scale the valuation functions by the same factor for each agent so that the maximum marginal value of any item is at most 1. Besides general monotone valuations, we are also interested in well-known classes of valuation function, in particular, *additive* (linear) valuations where $v(S) = \sum_{g \in S} v(g)$ for each $S \subseteq M$, and *subadditive* (complement-free) valuations where $v(S \cup T) \leq v(S) + v(T)$ for all $S, T \subseteq M$. We use $[n]$ to denote the set $\{1, \cdots, n\}$.

2.1 Fairness and Welfare

An allocation $A = (A_1, A_2, \cdots, A_n)$ is a partition of the items into n disjoint subsets, where A_i is the set of items allocated to agent i. Our aim is to obtain envy-free allocations with high welfare.

Definition 1. *An allocation $A = (A_1, \cdots, A_n)$ is envy-free if for each $i, j \in N$, $v_i(A_i) \geq v_i(A_j)$.*

In other words, an allocation is envy-free if each agent i prefers its own bundle A_i over any the bundle A_j of any other agent j. If agent i prefers the bundle of agent j then we say i envies j. Unfortunately, envy-free allocations do not always exist with indivisible item. This is evident even with two agents and one item, since the agent without an item will always envy the other. Moreover, even with two players and with identical additive valuations, determining whether an envy-free allocation exists is NP-complete. Consequently weaker notions of fairness have been introduced [14], most notably envy-freeness up to one item.

Definition 2. *An allocation A is envy-free up to one item (EF1) if for each $i, j \in N$, $v_i(A_i) \geq v_i(A_j \backslash g)$ for some $g \in A_j$.*

Rather than approximate fairness, however, our focus is on obtaining envy-freeness by adding one divisible item (money). Thus we have an *allocation with payments*; in addition to the bundle A_i, agent i has a payment p_i.

Definition 3. *An allocation with payments (A, p) is envy-free if for each $i, j \in N$, $v_i(A_i) + p_i \geq v_i(A_j) + p_j$.*

Furthermore, we say that an allocation A is *envy-freeable* if there exist payments p such that (A, p) is envy-free. An important fact is that, in contrast to envy-free allocations, envy-freeable allocations always exist for monotone valuations [24]. There are two natural types of payment. First, we have *subsidy payments* if $p_i \geq 0$. Second, we have *transfer payments* if $\sum_{i \in N} p_i = 0$, To distinguish these, we denote a subsidy payment to agent i by s_i and a transfer payment by t_i. We define the *total transfer* of an allocation as the sum $\sum_i |t_i|$.

We measure the welfare of an allocation A using the general concept of *ρ-mean welfare*, $\mathrm{W}^\rho(A) = \left(\frac{1}{n}\sum_{i \in N} v_i(A_i)^\rho\right)^{\frac{1}{\rho}}$. This class of welfare functions, introduced by Arunachaleswaran et al. [3], encompasses a range of welfare functions including the two most important cases: $\rho \to 0$, the *Nash social welfare*, is the geometric mean of the values of the agents, denoted by $\mathrm{NSW}(A) = \left(\prod_{i \in N} v_i(A_i)\right)^{\frac{1}{n}}$, and $\rho = 1$, the *utilitarian social welfare* or simply *social welfare* (scaling by the number of agents), denoted by $\mathrm{SW}(A) = \sum_{i \in N} v_i(A_i)$. With transfer payments, our interest lies in utilities rather than simply valuations. In particular, the ρ-mean welfare of an allocation with transfers (A, t) is $\mathrm{W}^{\rho}(A, t) = \left(\frac{1}{n} \sum_{i \in N} (v_i(A_i) + t_i)^{\rho}\right)^{\frac{1}{\rho}}$.

2.2 Fair Division with Transfer Payments

In this paper, we study the following question.

> Is there an allocation with transfers that simultaneously satisfies (i) envy-freeness, (ii) high welfare, and (iii) a negligible total transfer?

We have seen that envy-freeable allocations always exist. Thus, with transfer payments, we can obtain the property of envy-freeness. The reader may ask whether transfers are necessary. Specifically, given the guaranteed existence of envy-freeable allocation, can such allocations provide high welfare? The answer is NO. Even worse, no positive guarantee on welfare can be obtained without transfers. This is true even for the case of additive valuations. To see this, consider the following simple example for Nash social welfare.

Example 1. Take two agents and two items $\{a, b\}$. Let the valuation functions be additive with $v_{1,a} = 1, v_{1,b} = \frac{1}{2}$ for agent 1 and $v_{2,a} = \frac{1}{2}, v_{2,b} = \epsilon$ for agent 2. Observe there are only two envy-freeable allocations: either agent 1 gets both items or agent 1 gets item a and agent 2 gets item b. For both these envy-freeable allocations the corresponding Nash social welfare is at most $\sqrt{\epsilon}$. In contrast, the optimal Nash social welfare is $\frac{1}{2}$ when agent 1 gets b and agent 2 gets a.

It follows that to find envy-free solutions with non-zero approximation guarantees for welfare we must have transfer payments. At the outset, if we restrict ρ to be equal to 1, the result of Halpern and Shah [24] implies that the allocation that maximizes utilitarian welfare can be made envy-free with transfer payments. However, we show that this allocation can require arbitrarily large transfers relative to the number of agents. The main point of concern in using transfer payments to achieve envy-freeness is that it may be difficult for the participants to include a substantial quantity of money in the system in order to implement this solution. Consequently, this creates a third requirement, i.e. to bound the total transfers. Thus the holy grail here is to obtain high welfare using only *negligible transfers*: formally, we desire transfers whose sum (of absolute values) is independent of the number of items m. In particular, we want an allocation with transfers (A, t) such that the welfare of A is at least α times the welfare of the welfare-maximizing allocation A^* (for some large $\alpha \in [0, 1]$) and $\sum_{i \in N} |t_i| = O(f(n))$ for some function f. Specifically, the payments are negligible in the number of items (and thus in the total welfare) as m grows.

At first glance, this task seems impossible. If envy-freeable solutions cannot themselves ensure non-zero welfare guarantees, how could negligible transfer payments then induce high welfare? Very surprisingly, this is possible for some important classes of valuation functions. However, it is indeed not always possible for other classes. Investigating how and where the boundary of this dichotomy lies is the purpose of this paper.

3 Transfer Payments and ρ-Mean Welfare

In this section we familiarize the reader with the structure of envy-freeable allocations and transfer payments, and introduce our preliminary results. We begin with the general case of ρ-mean welfare.

Lemma 1. *For subadditive valuations, any envy-free allocation with transfers* (A, t) *satisfies* $\mathrm{W}^\rho(A, t) \geq \frac{1}{n}\mathrm{W}^\rho(A^*)$.

The resultant welfare guarantee of $\alpha = \frac{1}{n}$ is not particularly impressive. But it is a strictly positive guarantee, which was unachievable without transfer payments. The bound is also tight as shown by the following simple example.

Example 2. Take $m = n$ items and n agents. Let the valuation functions be additive with $v_{ii} = 1$ and $v_{ij} = 0$ for $j \neq i$. Consider the allocation assigning the grand bundle to agent 1. This is envy-freeable with transfer payments $t_1 = -\frac{n-1}{n}$ and $t_i = \frac{1}{n}$, for any agent $i \neq 1$. For social welfare ($\rho = 1$) the corresponding welfare guarantee is $\alpha = \frac{1}{n}$.

But how expensive is it to obtain this welfare guarantee? To answer this, we provide a short review concerning the computation of transfer payments. Recall that an allocation A is envy-freeable if there exist payments p such that (A, p) is envy-free. Furthermore, there is a very useful graph characterization of envy-freeability. Given an allocation A we build an envy-graph, denoted G_A. The envy-graph is directed and complete. It contains a vertex for each agent $i \in N$. For any pair of agents $i, j \in N$, the weight of arc (i, j) in G_A is the envy agent i has for agent j under the allocation A, that is, $w_A(i, j) = v_i(A_j) - v_i(A_i)$. The envy-graph induces the following characterization.

Theorem 2 ([24]). *The following statements are equivalent.*

i) The allocation A is envy-freeable.
ii) The allocation A maximizes (utilitarian) welfare across all reassignments of its bundles to agents: for every permutation π of N, we have $\sum_{i \in N} v_i(A_i) \geq \sum_{i \in N} v_i(A_{\pi(i)})$.
iii) The envy graph G_A contains no positive-weight directed cycles.

In addition, we can use the envy-graph to compute the transfer payments. It is known [24] how to find, for any envy-freeable allocation A, the minimum *subsidy* payments s such that (A, s) is envy-free. Let $l(i)$ be weight of a maximum weight path from node i to any other node in G_A. Setting $s_i = l(i)$ for each agent

i gives an envy-free allocation with minimum subsidy payments. We do not wish to subsidize the mechanism, so we convert these subsidies into transfer payments. To do this, let $\bar{s} = \frac{1}{n}\sum_{i \in N} s_i$ be the average subsidy. Then setting $t_i = s_i - \bar{s}$ for each agent gives a valid set of transfer payments, which we dub the *natural* transfer payments. We remark that the natural transfer payments do not always minimize the total transfer, but they will be sufficient for our purposes. We are now ready to compute transfer payments for subadditive valuations in the ρ-mean welfare setting. We begin with a theorem of Brustle et al. [13].

Theorem 3 ([13]). *For monotone valuations there is a polytime algorithm to find an envy-free allocation with subsidies (A, s) with $s_i \leq 2(n-1)$ for all i.*

Observe that any bound on the maximum subsidy for each agent also applies to the maximum natural transfer for each agent. Combining this observation with the previous result gives us the following corollary.

Corollary 1. *For subadditive valuations, there exists an envy-free allocation with transfers (A, t) such that $\frac{W^\rho(A,t)}{W^\rho(A^*)} \geq \frac{1}{n}$ and with a total transfer $\sum_i |t_i|$ of at most $2n^2$. This allocation can be computed in polytime.*

Thus, we can quickly obtain an envy-free allocation with transfers whose total transfer is negligible, i.e., independent of m. But, as stated, we only have a low welfare guarantee for this general ρ-mean welfare class. In the next section, we will show that high welfare and negligible transfers are achievable for the special case of $\rho = 0$, that is, NSW. First, we conclude this section by presenting a generalization of Theorem 3 that will later be useful. We say that an allocation B has *b-bounded envy* if $v_i(B_j) - v_i(B_i) \leq b$ for every pair $i, j \in N$.

Lemma 2. *Given an allocation B with b-bounded envy there is a polytime algorithm to find an envy-free allocation with transfers (A, t) with $\sum_{i \in N} |t_i| \leq 2bn^2$.*

4 Transfer Payments and Nash Social Welfare

In the following two sections, we present our main results concerning Nash social welfare and utilitarian social welfare. Here we show that, with transfers, excellent welfare guarantees can be obtained for NsW. Conversely, in Sect. 5, we will see that only much weaker guarantees can be obtained for utilitarian welfare.

4.1 NSW with General Valuation Functions

Now, recall from Example 1 that no positive welfare guarantee can be obtained in the case of Nash social welfare for even the basic case of additive valuations. Our first result for Nash social welfare is therefore somewhat surprising. With transfer payments, constant factor welfare guarantees can be obtained for general valuations. That is, envy-freeness and high welfare are simultaneously achievable.

Theorem 4. *For general valuations, there exists an envy-free allocation with transfers* (A, t) *such that* $\frac{\text{NSW}(A,t)}{\text{NSW}(A^*)} \geq e^{-1/e}$.

This theorem is rather noteworthy; for general valuation functions, with transfers, it allows us to simultaneously obtain high Nash social welfare and envy-freeness. But what of our third objective, negligible transfer payments? The approach applied in the proof of Theorem 4 cannot guarantee negligible transfers. Specifically, simply reallocating the bundles of the allocation A^* that maximizes Nash social welfare can require large transfers. In particular, the following example shows this method may require transfers as large as $\Omega(\sqrt{m})$.

Example 3. Take an instance with two agents and m items. Assume the first agent has a valuation function given by $v_1(S) = |S|$, for each $S \subseteq M$; assume the second agent has a valuation function given by $v_2(S) = \sqrt{|S|}$, for each $S \subseteq M$. The reader may verify that the Nash welfare maximizing allocation A^* is to give the first agent $\frac{2m}{3}$ items and the second agent $\frac{m}{3}$ items. This allocation is also the allocation that maximizes utilitarian social welfare by reassigning the bundles of A^*. Thus $A = A^*$. However, to make the allocation envy-free requires a minimum transfer payment of $\Omega(\sqrt{m})$, from the first agent to the second agent.

Of course, this example does not rule out the possibility that, for general valuation functions, an envy-free allocation with transfers that has high welfare and negligible payments exists. In particular, simply allocating each agent half the items requires no transfer payments at all, and gives high Nash social welfare. So simultaneously obtaining high Nash social welfare and envy-freeness via negligible transfers for general valuation functions remains an open question. Fortunately, we can show that these three properties are simultaneously achievable for important special classes of valuation function.

4.2 NSW Guarantees with Negligible Transfers

Here we prove that for (i) additive valuations, and (ii) matroid rank valuations, it is always possible to obtain envy-free allocations with high Nash social welfare and negligible transfers. Furthermore, for additive valuations we can do this using polynomial time algorithms.

Theorem 5. *For additive valuations, given an* α*-approximate allocation to maximum Nash social welfare, there exists a polynomial time computable envy-free allocation with transfers* (A, t) *such that* $\frac{\text{NSW}(A,t)}{\text{NSW}(A^*)} \geq \frac{1}{2}\alpha \cdot e^{-1/e}$ *with a total transfer* $\sum_i |t_i|$ *of at most* $2n^2$.

We remark that, for additive valuations, polytime algorithms to find allocations that α-approximate the maximum NSW do exist. Specifically, Barman et al. [7] present an algorithm with an approximation guarantee of $\alpha = \frac{1}{1.45}$. Together with Theorem 5, we thus obtain in polytime an envy-free allocation with negligible transfers and a Nash social welfare guarentee of $\frac{1}{2.9}e^{-1/e}$.

Better existence bounds can be obtained for the additive case if we remove the requirement of a polytime algorithm. A well-known result of Caragiannis et al. [16] states that for additive valuations, the Nash welfare maximizing allocation is EF1. In fact, a recent result of Benabbou et al. [10] provides a similar result for the case of *matroid rank* valuation functions, a sub-class of submodular functions. A valuation function is matroid rank if it is submodular, and the marginal value of any item is binary (i.e. for any set S of items and any item x not in S, $v_i(S \cup \{x\}) - v_i(S) \in \{0,1\}$). Here, a NSW-maximizing allocation is EF1[10]. Combining this with Lemma 2, the corresponding envy-free allocation with transfers (A,t) has transfers satisfying $\sum_i |t_i| \le 2n^2$. Further, by Theorem 4, we have $\frac{\text{NSW}(A,t)}{\text{NSW}(A^*)} \ge e^{-1/e}$ as desired.

Theorem 6. *For matroid rank valuations, there exists an envy-free allocation with transfers (A,t) with $\frac{\text{NSW}(A,t)}{\text{NSW}(A^*)} \ge e^{-1/e}$ and $\sum_i |t_i| \le 2n^2$.* □

5 Transfer Payments and Utilitarian Social Welfare

To begin, recall that an allocation B has *b-bounded envy* if $v_i(B_j) - v_i(B_i) \le b$ for every pair of agents $i, j \in N$. Without transfers, allocations with *b-bounded envy* may have very low welfare.

Lemma 3. *For utilitarian social welfare, there exist instances with additive valuation functions such that any allocation with b-bounded envy has a welfare guarantee of at most $2\sqrt{\frac{b}{m}} + \frac{1}{n}$.*

Lemma 3 implies that any EFk allocation in the given example, with $k = o(m)$, cannot provide a welfare guarantee that is significantly higher than $\frac{1}{n}$. The natural question to ask, now, is whether the problem inherent in Lemma 3 can be rectified with a small quantity of transfers. On the positive side, the result of Brustle et al. [13] shows that a small quantity of subsidy independent of the number of items is always sufficient to eliminate envy. A similar result also extends to the corresponding natural transfer payments. Combining this result with Lemma 1 tells us that a utilitarian welfare guarantee of $\frac{1}{n}$ can be achieved alongside envy-freeness with a negligible total transfer. Unfortunately, for the above example, the Iterated Matching Algorithm of [13] returns an allocation whose social welfare is only a $\frac{1}{n}$-fraction of the optimal welfare. The following corollary shows that this was inevitable: unlike for NSW, in order to make any improvement above this threshold, non-negligible transfers are required.

Corollary 2. *For any $\alpha \in \left[\frac{1}{n}, 1\right]$, there exists an instance with additive valuations such that any envy-free allocation with transfers (A,t) satisfying $\frac{\text{SW}(A,t)}{\text{SW}(A^*)} \ge \alpha$ requires a total transfer $\sum_{i \in N} |t_i| \ge \frac{1}{4}\left(\alpha - \frac{1}{n}\right)^2 m$.*

So, for utilitarian social welfare, non-negligible transfers are required to ensure both envy-freeness and high welfare. Recall, though, that balancing constraints on the valuation functions have been used in the literature to circumvent impossibility bounds on welfare. The reader may wonder if such constraints could be used to bypass the result in Corollary 2: are negligible transfer payments sufficient to obtain high welfare when the valuation functions are constant-sum? The answer is NO, as we shall see in the subsequent theorem.

In recent work, Barman et al. [6] considered the case of subadditive valuations with the constant-sum condition, and gave a polynomial-time algorithm that finds an EF1 allocation with social welfare at least $\Omega(\frac{1}{\sqrt{n}})$ of the optimal welfare. Applying the algorithm of Lemma 2 to the resulting allocation gives us an envy-free allocation with negligible transfers and welfare ratio $\Omega(\frac{1}{\sqrt{n}})$. Once again, we show that this threshold cannot be crossed without non-negligible transfers.

Theorem 7. *There exist instances with constant-sum additive valuations such that any envy-free allocation with transfers (A, t) satisfying $\frac{\mathrm{SW}(A,t)}{\mathrm{SW}(A^*)} \geq \alpha$ has a total transfer $\sum_{i \in N} |t_i| \geq (\alpha - \frac{2}{\sqrt{n}})\frac{m}{\sqrt{n}}$, for any $\alpha \in [\frac{2}{\sqrt{n}}, 1]$.*

So non-negligible transfer payments are required even assuming constant-sum valuations. This adds to our collection of negative results for utilitarian social welfare. Are any positive results possible? Specifically, can we at least match the lower bounds on transfer payments inherent in the these negative results. We will now show this can indeed be approximately achieved.

5.1 Upper Bounds on Transfer Payments

To conclude the paper, we present results that upper bound the total transfer required to obtain an envy-free allocation with a utilitarian social welfare guarantee. We give upper bounds for additive and general valuation functions. In both cases, the bound we obtain is a function of the maximum value that an agent receives in the welfare-optimal allocation. In particular, while the lower bounds are obtained as functions of m, the upper bounds we get are functions of the product of n and $\max_i v_i(A_i^*)$. In allocations that distribute utility uniformly among the agents these expressions are comparable; even in the worst case, since $v_i(A_i^*) \leq m$ for any i, they differ by some function of only n, and this difference is independent of the number of items. We begin with the additive case.

Theorem 8. *For additive valuations, for any $\alpha \in (0, 1]$, there is an envy-free allocation with transfers (A, t) such that $\frac{\mathrm{SW}(A,t)}{\mathrm{SW}(A^*)} \geq \alpha$ with total transfer $\sum_{i \in N} |t_i| \leq n(\alpha \max_i v_i(A_i^*) + 2)$.*

Finally, we show how to upper bound the transfer payments in the case of general valuation functions. Here, the welfare target is limited to the constant factor $\frac{1}{3}$, and the gap between our lower and upper bounds widens by a factor of n, but once again, this gap is independent of m.

Theorem 9. *For general valuations, for any $\alpha \in \left(0, \frac{1}{3}\right]$, there is an envy-free allocation with transfers (A, t) such that $\frac{\mathrm{SW}(A,t)}{\mathrm{SW}(A^*)} \geq \alpha$ with total transfer $\sum_{i \in N} |t_i| \leq 2n^2 \left(3\alpha \max_i v_i\left(A_i^*\right) + 2\right)$.*

Acknowledgments. We would like to thank anonymous referees for their comments and suggestions for improvements to the presentation.

References

1. Alkan, A., Demange, G., Gale, D.: Fair allocation of indivisible goods and criteria of justice. Econometrica **59**(4), 1023–1039 (1991)
2. Aragones, E.: A derivation of the money Rawlsian solution. Soc. Choice Welfare **12**(3), 267–276 (1995)
3. Arunachaleswaran, E.R., Barman, S., Kumar, R., Rathi, N.: Fair and efficient cake division with connected pieces. In: Caragiannis, I., Mirrokni, V., Nikolova, E. (eds.) WINE 2019. LNCS, vol. 11920, pp. 57–70. Springer, Cham (2019). https://doi.org/10.1007/978-3-030-35389-6_5
4. Aziz, H.: Achieving envy-freeness and equitability with monetary transfers (2020)
5. Aziz, H., Mackenzie, S.: A discrete and bounded envy-free cake cutting protocol for any number of agents. In: 2016 IEEE 57th Annual Symposium on Foundations of Computer Science (FOCS), pp. 416–427 (2016)
6. Barman, S., Bhaskar, U., Shah, N.: Optimal bounds on the price of fairness for indivisible goods. In: Chen, X., Gravin, N., Hoefer, M., Mehta, R. (eds.) WINE 2020. LNCS, vol. 12495, pp. 356–369. Springer, Cham (2020). https://doi.org/10.1007/978-3-030-64946-3_25
7. Barman, S., Krishnamurthy, S., Vaish, R.: Finding fair and efficient allocations. In: Proceedings of the 2018 ACM Conference on Economics and Computation, pp. 557–574 (2018)
8. Bei, X., Chen, N., Hua, X., Tao, B., Yang, E.: Optimal proportional cake cutting with connected pieces. In: Proceedings of the Twenty-Sixth AAAI Conference on Artificial Intelligence (2012)
9. Bei, X., Lu, X., Manurangsi, P., Suksompong, W.: The price of fairness for indivisible goods. In: Proceedings of the Twenty-Eighth International Joint Conference on Artificial Intelligence, IJCAI-19, pp. 81–87 (2019)
10. Benabbou, N., Chakraborty, M., Igarashi, A., Zick, Y.: Finding fair and efficient allocations when valuations don't add up. In: Harks, T., Klimm, M. (eds.) SAGT 2020. LNCS, vol. 12283, pp. 32–46. Springer, Cham (2020). https://doi.org/10.1007/978-3-030-57980-7_3
11. Bertsimas, D., Farias, V., Trichakis, N.: The price of fairness. Oper. Res. **59**(1), 17–31 (2011)
12. Brams, S., Taylor, A.: An envy-free cake division protocol. Am. Math. Mon. **102**(1), 9–18 (1995)
13. Brustle, J., Dippel, J., Narayan, V., Suzuki, M., Vetta, A.: One dollar each eliminates envy. In: Proceedings of the 21st ACM Conference on Economics and Computation, pp. 23–39 (2020)
14. Budish, E.: The combinatorial assignment problem: approximate competitive equilibrium from equal incomes. J. Polit. Econ. **119**(6), 1061–1103 (2011)
15. Caragiannis, I., Kaklamanis, C., Kanellopoulos, P., Kyropoulou, M.: The efficiency of fair division. In: Leonardi, S. (ed.) WINE 2009. LNCS, vol. 5929, pp. 475–482. Springer, Heidelberg (2009). https://doi.org/10.1007/978-3-642-10841-9_45
16. Caragiannis, I., Kurokawa, D., Moulin, H., Procaccia, A., Shah, N., Wang, J.: The unreasonable fairness of maximum nash welfare. ACM Trans. Econ. Comput. 7(3), 12:1–32 (2019)

17. Cohler, Y., Lai, J., Parkes, D., Procaccia, A.: Optimal envy-free cake cutting. In: Proceedings of the Twenty-Fifth AAAI Conference on Artificial Intelligence, AAAI 2011, San Francisco, California, USA, 7–11 August 2011 (2011)
18. Dubins, L.E., Spanier, E.H.: How to cut a cake fairly. Am. Math. Mon. **68**(1), 1–17 (1961)
19. Foley, D.: Resource allocation and the public sector. Yale Econ Essays **7**(1), 45–98 (1967)
20. Gamow, G., Stern, M.: Puzzle-Math. Viking Press (1958)
21. Ghodsi, M., Taghi Hajiaghayi, M., Seddighin, M., Seddighin, S., Yami, H.: Fair allocation of indivisible goods: improvements and generalizations. In: Proceedings of the 2018 ACM Conference on Economics and Computation, Ithaca, NY, USA, 18–22 June 2018, pp. 539–556 (2018)
22. Goko, H., et al.: Fair and truthful mechanism with limited subsidy (2021)
23. Haake, C.J., Raith, M., Su, F.: Bidding for envy-freeness: a procedural approach to n-player fair-division problems. Soc. Choice Welfare **19**(4), 723–749 (2002)
24. Halpern, D., Shah, N.: Fair division with subsidy. In: Fotakis, D., Markakis, E. (eds.) SAGT 2019. LNCS, vol. 11801, pp. 374–389. Springer, Cham (2019). https://doi.org/10.1007/978-3-030-30473-7_25
25. Klijn, F.: An algorithm for envy-free allocations in an economy with indivisible objects and money. Soc. Choice Welfare **17**, 201–215 (2000)
26. Kurokawa, D., Procaccia, A., Wang, J.: Fair enough: guaranteeing approximate maximin shares. J. ACM **65**(2), 1–27 (2018)
27. Lipton, R., Markakis, E., Mossel, E., Saberi, A.: On approximately fair allocations of indivisible goods. In: Proceedings of the 5th ACM Conference on Electronic Commerce (EC), pp. 125–131 (2004)
28. Maskin, E.: On the fair allocation of indivisible goods, pp. 341–349 (1987)
29. Steinhaus, H.: The problem of fair division. Econometrica **16**(1), 101–104 (1948)
30. Stromquist, W.: How to cut a cake fairly. Am. Math. Mon. **87**(8), 640–644 (1980)
31. Su, F.: Rental harmony: Sperner's lemma in fair division. Am. Math. Mon. **106**(10), 930–942 (1999)
32. Svensson, L.G.: Large indivisibles: an analysis with respect to price equilibrium and fairness. Econometrica **51**(4), 939–954 (1983)
33. Tadenuma, K., Thomson, W.: The fair allocation of an indivisible good when monetary compensations are possible. Math. Soc. Sci. **25**(2), 117–132 (1993)
34. Varian, H.: Equity, envy, and efficiency. J. Econ. Theory **9**(1), 63–91 (1974)
35. Woodall, D.R.: Dividing a cake fairly. J. Math. Anal. Appl. **78**(1), 233–247 (1980)

Pirates in Wonderland: Liquid Democracy has Bicriteria Guarantees

Jonathan A. Noel[1], Mashbat Suzuki[2(✉)], and Adrian Vetta[2]

[1] University of Victoria, Victoria, Canada
noelj@uvic.ca
[2] McGill University, Montreal, Canada
mashbat.suzuki@mail.mcgill.ca, adrian.vetta@mcgill.ca

Abstract. Liquid democracy has a natural graphical representation, the delegation graph. Consequently, the strategic aspects of liquid democracy can be studied as a game over delegation graphs, called the liquid democracy game. Our main result is that this game has bicriteria approximation guarantees, in terms of both rationality and social welfare. Specifically, we prove the price of stability for ϵ-Nash equilibria is exactly ϵ in the liquid democracy game.

1 Introduction

Liquid democracy is a form of direct and representative democracy, based on the concept of *delegation*. Each voter has the choice of voting themselves or transferring (transitively) its vote to a trusted proxy. Recent interest in liquid democracy, from both practical and theoretical perspectives, was sparked by the Pirate Party in Germany and its Liquid Feedback platform [2]. Similar initiatives have subsequently been undertaken by the Demoex Party in Sweden, the Internet Party in Spain, and the Net Party in Argentina.

There are many potential benefits of a transitive delegation mechanism. Participation may improve in quantity for several reasons. The system is easy to use and understand, induces low barriers to participation, and is inherently egalitarian: there is no distinction between voters and representatives; every one is both a voter and a delegator. Participation may also improve in quality due to the flexibility to choose different forms of participation: voters can chose to be active participants on topics they are comfortable with or delegate on topics they are less comfortable with. Accountability may improve due to the transparent nature of the mechanism and because there is a demonstrable line of responsibility between a delegated proxy and its delegators. The quality of decision making may improve via a specialization to delegated experts and a reduction in induced costs, such as the duplication of resources.

Our objective here is not to evaluate such claimed benefits, but we refer the reader to [1,2,4,11,13] for detailed discussions on the motivations underlying liquid democracy. Rather, our focus is to quantitatively measure the performance of liquid democracy in an idealized setting. Specifically, can equilibria in these

I. Caragiannis and K. A. Hansen (Eds.): SAGT 2021, LNCS 12885, pp. 391–405, 2021.
https://doi.org/10.1007/978-3-030-85947-3_26

voting mechanisms provide high social welfare? That is, we study the *price of stability of liquid democracy.*

1.1 Background

As stated, vote delegation lies at the heart of liquid democracy. Furthermore, vote delegation in liquid democracy has several fundamental characteristics: optionality, retractability, partitionability, and transitivity. So let us begin by defining these concepts and tracing their origins [1,2].

The notion of *optional* delegation proffers voters the choice of direct participation (voting themselves/choosing to abstain) or indirect participation (delegating their vote). This idea dates back over a century to the work of Charles Dodgson on parliamentary representation [9].[1]

Miller [17] proposed that delegations be *retractable* and *partitionable.* The former allows for delegation assignments to be time-sensitive and reversible. The latter allows a voter to select different delegates for different policy decisions.[2]

Finally, *transitive* delegation is due to Ford [11]. This allows a proxy to themselves delegate its vote *and* all its delegated votes. This concept is central to liquid democracy. Indeed, if an agent is better served by delegating her vote to a more informed proxy it would be perverse to prohibit that proxy from re-delegating that vote to an even more informed proxy. Moreover, such transitivity is necessary should circumstances arise causing the proxy to be unable to vote. It also reduces the duplication of efforts involved in voting.

As noted in the sixties by Tullock [18], the development of the computer opened up the possibility of large proxy voting systems. Indeed, with the internet and modern security technologies, liquid democracy is inherently practical; see Lumphier [16].

There has been a flurry of interest in liquid democracy from the AI community. This is illustrated by the large range of recent papers on the topic; see, for example, [5–8,12–15,19]. Most directly related to our work is the game theoretic model of liquid democracy studied by Escoffier et al. [10]. (A related game-theoretic model was also investigated by Bloembergen et al. [3].) Indeed, our motivation is an open question posed by Escoffier et al. [10]: are price of anarchy type results obtainable for their model of liquid democracy? We will answer this question for a generalization of their model.

1.2 Contributions

In Sect. 2, we will see that vote delegation has a natural representation in terms of a directed graph called the *delegation graph.* If each agent i has a utility of $u_{ij} \in$

[1] Dodgson was a parson and a mathematician but, as the author of "Alice in Wonderland", is more familiarly known by his *nom de plume*, Lewis Carroll.

[2] This option is particularly useful where potential delegates may have assorted competencies. For example, Alice may prefer to delegate to the Hatter on matters concerning tea-blending but to the Queen of Hearts on matters concerning horticulture.

$[0, 1]$ when agent j votes as her delegate then a game, called the *liquid democracy game*, is induced on the delegation graph. We study the *welfare ratio* in the liquid democracy game, which compares the social welfare of an equilibrium to the welfare of the optimal solution.

Pure strategy Nash equilibria need not exist in the liquid democracy game, so we focus on mixed strategy Nash equilibria. Our main result, given in Sect. 3 is that bicriteria approximation guarantees (for social welfare and rationality) exist in the game.

Theorem 1. *For all $\epsilon \in [0, 1]$, and for any instance of the liquid democracy game, there exists an ϵ-Nash equilibrium with social welfare at least $\epsilon \cdot \text{OPT}$.*

Theorem 1 is tight: the stated bicriteria guarantees cannot be improved.

Theorem 2. *For all $\epsilon \in [0, 1]$, there exist instances such that any ϵ-Nash equilibrium has welfare at most $\epsilon \cdot \text{OPT} + \gamma$ for any $\gamma > 0$.*

Theorems 1 and 2 imply that the *price of stability* for ϵ-Nash equilibria is ϵ. An important consequence of Theorem 1 is that strong approximation guarantees can *simultaneously* be obtained for both social welfare and rationality. Specifically, setting $\epsilon = \frac{1}{2}$ gives factor 2 approximation guarantees for each criteria.

Corollary 3. *For any instance of the liquid democracy game, there exists a $\frac{1}{2}$-Nash equilibrium with welfare at least $\frac{1}{2} \cdot \text{OPT}$.*

2 A Model of Liquid Democracy

In this section, we present the liquid democracy game. This game generalizes the game-theoretic model studied by Escoffier et al. [10].

2.1 The Delegation Graph

In liquid democracy each agent has three strategies: she can abstain, vote herself, or delegate her vote to another agent. So we can represent an instance by a directed network $G = (V, A)$ called the *delegation graph.* There is a vertex in $V = \{1, \cdots, n\}$ for each agent. To define the sets of arcs, there are three possibilities. First, if agent i votes herself the delegation graph contains a self-loop (i, i). Second, if agent i delegates her vote to agent $j \neq i$ then there is an arc (i, j) in G. Third, if agent i abstains then the vertex i has out-degree zero.

Now, because the out-degree of each vertex is at most one, the delegation graph G is a 1-forest. That is, each component of G is an arborescence plus at most one arc. In particular, each component is either an arborescence and, thus, contains no cycle or contains exactly one directed cycle (called a *delegation cycle*). In the former case, the component contains one sink node corresponding to an abstaining voter. In the latter case, if the delegation cycle is a self-loop the component contains exactly one voter called a *guru*; if the delegation cycle

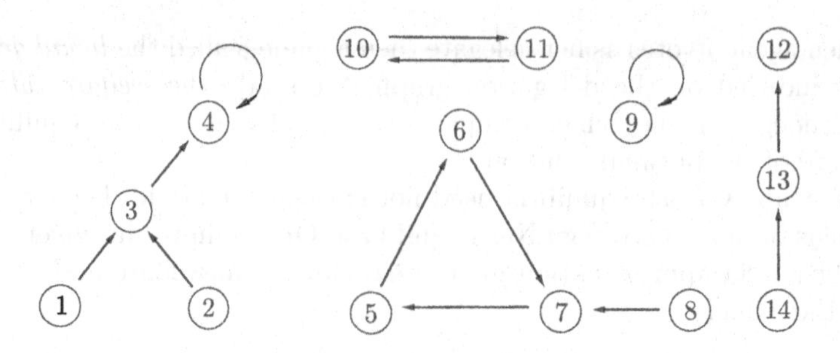

Fig. 1. A delegation graph.

has length at least two then the component contains no voters. An example of a delegation graph is shown in Fig. 1.

Observe that, by the transitivity of delegations, if an agent i is in a component containing a guru g then that guru will cast a vote on i's behalf. On the other hand, if agent i is in a component without a guru (that is, with either a sink node or a cycle of length at least two) then no vote will be cast on i's behalf. We denote the guru j representing agent i by $g(i) = j$ if it exists (we write $g(i) = \emptyset$ otherwise). Furthermore, its easy to find $g(i)$: simply apply path traversal starting at vertex i in the delegation graph.

For example, in Fig. 1 two components contain a guru. Agent 4 is the guru of agents 1, 2, 3 and itself; agent 9 is the guru only for itself. The vertices in the remaining three components have no gurus. There are two components with *delegation cycles*, namely $\{(10, 11), (11, 10)\}$ and $\{(5, 6), (6, 7), (7, 5)\}$. The final component also contains no guru as agent 12 is a sink node and thus abstains.

2.2 The Liquid Democracy Game

The game theoretic model we study is a generalization of the model of Escoffier et al. [10]. A pure strategy s_i for agent i corresponds to the selection of at most one outgoing arc. Thus we can view s_i as an n-dimensional vector $\mathbf{x}_i$. Specifically, $\mathbf{x}_i$ is a single-entry vector with entry $x_{ij} = 1$ if i delegates to agent j and $x_{ij} = 0$ otherwise. Note that if $x_{ii} = 1$ then i votes herself ("delegates herself") and that $\mathbf{x}_i = \mathbf{0}$ if agent i abstains.

It immediately follows that there is a unique delegation graph $G_\mathbf{x}$ associated with a pure strategy profile $\mathbf{x} = (\mathbf{x}_1, \mathbf{x}_2 \cdots, \mathbf{x}_n)$. To complete the description of the game, we must define the payoffs corresponding to each pure strategy profile $\mathbf{x}$. To do this, let agent i have a utility $u_{ij} \in [0, 1]$ if she has agent j as her guru. Because of the costs of voting in terms of time commitment, knowledge acquisition, etc., it may be that $u_{ii} < u_{ij}$ for some $j \neq i$ trusted by i.[3] We denote the utility of agent i in the delegation graph $G_\mathbf{x}$ by $u_i(\mathbf{x}) = u_{i,g(i)}$. If agent i

[3] Indeed, if this is not the case then liquid democracy has no relevance.

has no guru then it receives zero utility.[4] It follows that only agents that lie in a component of $G_{\mathbf{x}}$ containing a guru can obtain positive utility.

For example, in Fig. 1 agent 9 is a guru so receives utility $u_{9,9}$. Each agent $i \in \{1,2,3,4\}$ has agent 4 as its guru so receives utility $u_{i,4}$. All the remaining agents have no guru and so receive zero utility.

An agent i is playing a best response at a pure strategy $\mathbf{x} = (\mathbf{x}_1, \mathbf{x}_2, \cdots, \mathbf{x}_n)$ if he cannot increase his utility by selecting a different or no out-going arc. The strategy profile is a pure Nash equilibrium if every agent is playing a best response at $\mathbf{x}$. Our interest is in comparing the social welfare of equilibria to the optimal welfare in the game. To do this, let the social welfare of $\mathbf{x}$ be $\text{SW}(\mathbf{x}) = \sum_{i \in V} u_i(\mathbf{x})$ and let $\text{OPT} = \max_{\mathbf{x}} \sum_{i \in V} u_i(\mathbf{x})$ be the optimal welfare over all strategy profiles. The *price of stability* is the worst ratio over all instances between the *best* welfare of a Nash equilibrium and the optimal social welfare.

The reader may ask why could an equilibrium have low social welfare. The problem is that delegation is transitive, but *trust is not.* Agent i may delegate to an agent j where u_{ij} is large but j may then re-delegate to an agent k where u_{ik} is small. Worse, agent i may receive no utility if the transitive delegation of its vote leads to a delegation cycle or an abstaining voter. Unfortunately, not only can pure Nash equilibria have low social welfare in the liquid democracy game they need not even exist!

Lemma 4. *There exist liquid democracy games with no pure strategy Nash equilibrium.*

Proof. Let there be three voters with $u_1 = (\frac{1}{2}, 1, 0)$, $u_2 = (0, \frac{1}{2}, 1)$ and $u_3 = (1, 0, \frac{1}{2})$. (A similar instance was studied in [10].) Assume $\mathbf{x}$ is a pure Nash equilibrium and let S be the set of gurus in $G_{\mathbf{x}}$. There are two cases. First, if $|S| \leq 1$ then there exists an agent i with zero utility. This agent can deviate and vote herself to obtain utility of $\frac{1}{2} > 0$, a contradiction. Second, if $|S| \geq 2$ then one of the gurus can delegate its vote to another guru to obtain a utility of $1 > \frac{1}{2}$, a contradiction. Therefore, no pure Nash equilibrium exists. □

2.3 Mixed Strategy Equilibria

Lemma 4 tells us that to obtain performance guarantees we must look beyond pure strategies. Of course, as liquid democracy is a finite game, a mixed strategy Nash equilibrium always exists. It follows that we can always find a mixed Nash equilibrium in the liquid democracy game and compare its welfare to the optimal welfare.

Here a mixed strategy for agent i is now simply a non-negative vector $\mathbf{x}_i$ where the entries satisfy $\sum_{j=1}^{n} x_{ij} \leq 1$. Note that agent i then abstains with probability $1 - \sum_{j=1}^{n} x_{ij}$. However, because abstaining is a weakly dominated strategy, we may assume no voter abstains in either the optimal solution or the best Nash equilibria. Hence, subsequently, each $\mathbf{x}_i$ will be unit-sum vector.

[4] We remark that our results hold even when agents who abstain or have no guru obtain positive utility.

What is the utility $u_i(\mathbf{x})$ of an agent for a mixed strategy profile $\mathbf{x} = (\mathbf{x}_1, \mathbf{x}_2, \dots, \mathbf{x}_n)$? It is simply the expected utility given the probability distribution over the delegation graphs generated by $\mathbf{x}$. There are several equivalent ways to define this expected utility. For the purpose of analysis, the most useful formulation is in terms of directed paths in the delegation graphs. Denote by $\mathcal{P}(i, j)$ be the set of all directed paths from i to j in a complete directed graph on V. Let $\mathbb{P}_{\mathbf{x}}(g(i) = j)$ be the probability that j is the guru of i given the mixed strategy profile $\mathbf{x}$. Then

$$\begin{aligned} u_i(\mathbf{x}) &= \sum_{j \in V} \mathbb{P}_{\mathbf{x}}(g(i) = j) \cdot u_{ij} \\ &= \sum_{j \in V} \mathbb{P}_{\mathbf{x}}(\exists \text{ path from } i \text{ to } j) \cdot \mathbb{P}_{\mathbf{x}}(\exists \text{ arc } (j, j)) \cdot u_{ij} \\ &= \sum_{j \in V} \left(\sum_{P \in \mathcal{P}(i,j)} \prod_{a \in P} x_a \right) \cdot x_{jj} \cdot u_{ij} \end{aligned}$$

To understand this recall that $\mathbf{x}_i$ is a unit-sum vector. It follows that each delegation graph generated by the mixed strategy $\mathbf{x}$ has uniform out-degree equal to one. This is because we generate a delegation graph from $\mathbf{x}$ by selecting exactly one arc emanating from i according to the probability distribution $\mathbf{x}_i$. Consequently, j can be the guru of i if and only if the delegation graph contains a self-loop (j, j) *and* contains a unique path P from i to j. The second equality above then holds. Further, the choice of outgoing arc is independent at each vertex i. This implies the third equality.

Regrettably, however, no welfare guarantee is obtainable even with mixed strategy Nash equilibria.

Lemma 5. *The price of stability in liquid democracy games is zero.*

Proof. Consider an instance with three agents whose utility is given as $u_1 = (\delta, 1, 0), u_2 = (0, \delta, 1)$ and $u_3 = (1, 0, \delta)$. The optimal welfare is $\text{OPT} = 1 + 2\delta$ obtained by the first and second agent voting and the third agent delegates to the first agent.

By a similar argument to Lemma 4, we know that no pure strategy Nash equilibrium exists. It is straightforward to verify that this game has a unique mixed strategy Nash equilibria $\mathbf{x}$, where $x_1 = (\delta, 1 - \delta, 0)$, $x_2 = (0, \delta, 1 - \delta)$ and $x_3 = (1 - \delta, 0, \delta)$.

The expected utility of agent 1 is then

$$\begin{aligned} u_1(\mathbf{x}) &= \sum_{j \in V} \left(\sum_{P \in \mathcal{P}(i,j)} \prod_{a \in P} x_a \right) \cdot x_{jj} \cdot u_{ij} \\ &= x_{11} \cdot u_{11} + x_{12} \cdot x_{22} \cdot u_{12} \\ &= \delta^2 + (1 - \delta) \cdot \delta \\ &= \delta \end{aligned}$$

The case of agents 2 and 3 are symmetric, hence the social welfare of this equilibria is 3δ. Thus, the price of stability is $\frac{3\delta}{1+2\delta}$ which tends to zero as $\delta \to 0$. □

Lemma 5 appears to imply that no reasonable social welfare guarantees can be obtained for liquid democracy. This is not the case. Strong performance guarantees can be achieved, provided we relax the incentive constraints. Specifically, we switch our attention to approximate Nash equilibria. A strategy profile $\mathbf{x}$ is an *ϵ-Nash equilibrium* if, for each agent i,

$$u_i(\mathbf{x}) \geq (1-\epsilon) \cdot u_i(\hat{\mathbf{x}}_i, \mathbf{x}_{-i}) \qquad \forall \hat{\mathbf{x}}_i$$

Above, we use the notation $\mathbf{x}_{-i} = \{\mathbf{x}_i\}_{j \neq i}$. Can we obtain good welfare guarantees for approximate Nash equilibria? We will prove the answer is YES in the remainder of the paper. In particular, we present tight bounds on the price of stability for ϵ-Nash equilibria.

3 The Price of Stability of Approximate Nash Equilibria

So our task is to compare the social welfare of the best ϵ-Nash equilibrium with the social welfare of the optimal solution. Let's begin by investigating the optimal solution.

3.1 An Optimal Delegation Graph

By the linearity of expectation, there is an optimal solution in which the agents use only pure strategies. In particular, there exists an *optimal* delegation graph maximizing social welfare. Moreover, this optimal graph has interesting structural properties. To explain this, we say that agent i is *happy* if she has strictly positive utility in a delegation graph G, that is $u_{i,g(i)} > 0$. A component Q in G is *jolly* if every vertex in Q (except, possibly, the guru) is happy.

The key observation then is that there is an optimal delegation graph in which every component is a jolly star.

Lemma 6. *There is an optimal delegation graph that is the disjoint union of jolly stars.*

Proof. Let Q be a component in an optimal delegation graph G. We may assume Q contains a guru. To see this, suppose Q contains an abstaining sink node j. Then the graph $\hat{G} = G \cup (j, j)$ where j votes herself is also optimal. On the other hand, suppose Q contains a cycle C of length at least 2. Take an agent $j \in C$. Then the graph $\hat{G}$ obtained by j voting herself instead of delegating her vote is also optimal.

So we may assume each component contains a guru. Further, we may assume each component is a star. Suppose not, take a component Q with guru j containing an agent i that does not delegate to j. But then the graph $\hat{G}$ obtained by i delegating her vote directly to j is optimal.

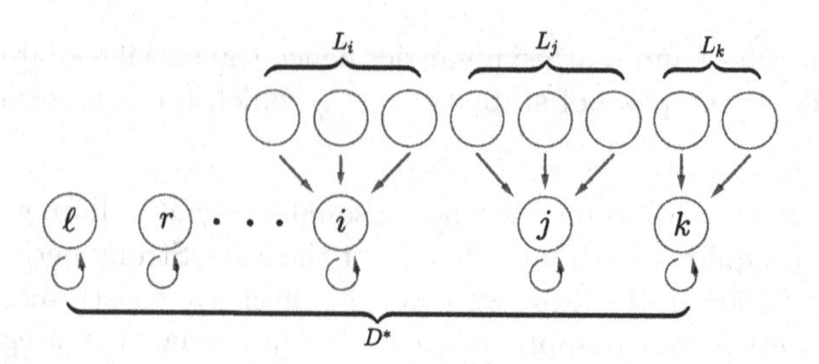

Fig. 2. Welfare optimal delegation graph that is disjoint union of jolly stars.

Finally, we may assume each star is *jolly*. Suppose i is not a happy agent in a star Q with guru j. Thus $u_{ij} = 0$. But then if i changes her delegation and votes herself we again obtain an optimal solution. In this case i will form a new singleton component which is trivially a jolly star. □

Lemma 6 states that the optimal solution can be obtained by a pure strategy $\mathbf{x}^*$ whose delegation graph is union of jolly stars. The centre of each star is a guru in the optimal solution and the leaves are the corresponding happy agents who delegated to the guru. Denote the set of gurus in the optimal solution by $D^* = \{i \in V \ : \ x^*_{ii} = 1\}$, and let $L_j = \{i \in V \setminus D^* \ : \ x^*_{ij} = 1\}$ be the agents who delegate to the guru j as illustrated in Fig. 2. It follows that the optimal solution has welfare

$$\text{OPT} = \sum_{j \in D^*} \left(u_{jj} + \sum_{i \in L_j} u_{ij} \right)$$

3.2 A Stable Solution

In order to study the best ϵ-Nash equilibrium we now show the existence of a "potentially" stable solution $\mathbf{x}$ whose definition is inspired by the set D^* of gurus in the optimal solution. In Sect. 3.3 we will prove that $\mathbf{x}$ is indeed an ϵ-Nash equilibrium and also has high social welfare.

To obtain $\mathbf{x}$ we require the following definitions. Denote the standard set of feasible mixed strategies for agent i as

$$\mathbb{S}_i = \{\mathbf{x}_i \in \mathbb{R}^n_+ \ : \ \sum_{j \in V} x_{ij} = 1 \ \}$$

Given a fixed strategy profile $\mathbf{x}_{-i} = \{\mathbf{x}_j\}_{j \neq i}$ for the other agents, let the corresponding best response for agent i be

$$B_i(\mathbf{x}_{-i}) = \arg\max_{\hat{\mathbf{x}} \in \mathbb{S}_i} u_i(\hat{\mathbf{x}}, \mathbf{x}_{-i})$$

For each $i \in D^*$, we denote a restricted set of mixed strategies

$$\mathbb{S}^R_i = \{\mathbf{x}_i \in \mathbb{R}^n_+ \ : \ \sum_{j \in V} x_{ij} = 1 \ , x_{ii} \geq \epsilon\}$$

Then for a fixed strategy profile $\mathbf{x}_{-i}$, let

$$B_i^R(\mathbf{x}_{-i}) = \arg\max_{\hat{\mathbf{x}} \in \mathbb{S}_i^R} u_i(\hat{\mathbf{x}}, \mathbf{x}_{-i})$$

be the best response for the agent i from amongst the restricted set of feasible strategies. Next recall Kakutani's Fixed Point Theorem.

Theorem 7 (Kakutani's Fixed Point Theorem). *Let K be a non-empty, compact and convex subset of $\mathbb{R}^m$, and let $\Phi : K \to 2^K$ be a set-valued function on K such that:*

(i) $\Phi(x)$ is non-empty and convex for all $x \in K$, and
(ii) Φ has a closed graph.

Then Φ has a fixed point, that is, there exists an $x^* \in K$ with $x^* \in \Phi(x^*)$.

Here a set-valued function Φ has a *closed graph* if $(x^k, y^k) \to (x, y)$ and $y^k \in \Phi(x^k)$ implies that $y \in \Phi(x)$.

Theorem 8. *There exists a strategy profile $\mathbf{x}$ such that:*

(a) For all $i \in D^*$, we have $\mathbf{x}_i \in B_i^R(\mathbf{x}_{-i})$, and
(b) For all $j \notin D^*$, we have $\mathbf{x}_j \in B_j(\mathbf{x}_{-j})$.

Proof. Let the feasible set of strategy profiles be $\Xi = \prod_{i \in D^*} \mathbb{S}_i^R \times \prod_{j \notin D^*} \mathbb{S}_j$, a subset of Euclidean space. Without loss of generality, let $D^* = \{1, 2, \ldots, k\}$. Now define a set valued function $\Phi \colon \Xi \longrightarrow 2^{\Xi}$ by

$$\mathbf{x} \mapsto (\underbrace{B_1^R(\mathbf{x}_{-1}), \cdots, B_k^R(\mathbf{x}_{-k})}_{D^*}, \underbrace{B_{k+1}(\mathbf{x}_{-(k+1)}), \cdots, B_n(\mathbf{x}_{-n})}_{V \setminus D^*})$$

That is, for each $\mathbf{x} \in \Xi$ we have $\Phi(\mathbf{x}) \subseteq \Xi$. Note the statement of the theorem is equivalent to showing that Φ has a fixed point.

Observe that Φ satisfies the conditions of Kakutani's Fixed Point Theorem. Indeed Ξ is nonempty, compact and convex, since it is a product of non-empty, compact and convex sets $\mathbb{S}_i^R$ and $\mathbb{S}_j$.

Next let's verify that $\Phi(\mathbf{x}) \neq \emptyset$. This holds since, for each agent i, we have $B_i^R(\mathbf{x}_{-i}) \neq \emptyset$ or $B_i(\mathbf{x}_{-i}) \neq \emptyset$ by the continuity of $u_i(\,\cdot\,, \mathbf{x}_{-i})$ and the Weierstrass Extreme Value Theorem.

Furthermore, for all $\mathbf{x} \in \Xi$ the set $\Phi(\mathbf{x}) \subseteq \Xi$ is convex. This is because, for each $i \in D^*$ and $j \in V \setminus D^*$, the sets $B_i^R(\mathbf{x}_{-i})$ and $B_j(\mathbf{x}_{-j})$ are convex, and thus $\Phi(\mathbf{x})$ is Cartesian product of convex sets. We must now show that both $B_j(\mathbf{x}_{-j})$ and $B_i^R(\mathbf{x}_{-i})$ are convex. The convexity of $B_j(\mathbf{x}_{-j})$ follows immediately by the multilinearity of u_i. Next take an agent $i \in D^*$. If $\mathbf{y}_i, \mathbf{z}_i \in B_i^R$ then, for all $\lambda \in [0, 1]$ and any $\hat{\mathbf{x}}_i \in \mathbb{S}_i^R$, we have

$$\begin{aligned} u_i(\lambda \mathbf{y}_i + (1-\lambda)\mathbf{z}_i, \mathbf{x}_{-i}) &= \lambda u_i(\mathbf{y}_i, \mathbf{x}_{-i}) + (1-\lambda) u_i(\mathbf{z}_i, \mathbf{x}_{-i}) \\ &\geq u_i(\hat{\mathbf{x}}_i, \mathbf{x}_{-i}) \end{aligned}$$

Observe $\lambda \mathbf{y}_i + (1-\lambda)\mathbf{z}_i \in \mathbb{S}_i^R$ since $\lambda y_{ii} + (1-\lambda) z_{ii} \geq \lambda\epsilon + (1-\lambda)\epsilon = \epsilon$ and $\lambda \sum_{j\in V} y_{ij} + (1-\lambda)\sum_{j\in V} z_{ij} = 1$. Thus $\lambda \mathbf{y}_i + (1-\lambda)\mathbf{z}_i \in B_i^R(\mathbf{x}_{-i})$ for any $\lambda \in [0,1]$, which implies $B_i^R(\mathbf{x}_{-i})$ is convex.

Finally, Φ has a closed graph because each $u_i(\mathbf{x}_i, \mathbf{x}_{-i})$ is a continuous function of $\mathbf{x}_i$ for any fixed $\mathbf{x}_{-i}$, and both sets $\mathbb{S}_i^R$ and $\mathbb{S}_i$ are compact. Thus, by Kakutani's Fixed Point Theorem, Φ has a fixed point $\mathbf{x}$. Hence (a) and (b) hold. □

3.3 Bicriteria Guarantees for Liquid Democracy

We will now prove that the fixed point $\mathbf{x}$ from Theorem 8 gives our main result: there is an ϵ-Nash equilibrium with welfare ratio at least ϵ. First let's show the incentive guarantees hold.

Lemma 9. *The fixed point* $\mathbf{x}$ *is an* ϵ*-Nash equilibrium.*

Proof. Take an agent $j \notin D^*$. Then for any $\hat{\mathbf{x}}_j \in \mathbb{S}_i$ we have

$$u_j(\mathbf{x}_j, \mathbf{x}_{-j}) \;\geq\; u_j(\hat{\mathbf{x}}_j, \mathbf{x}_{-j}) \;\geq\; (1-\epsilon)\cdot u_i(\hat{\mathbf{x}}_i, \mathbf{x}_{-i})$$

The incentive guarantee for j follows immediately.

Next consider an agent $i \in D^*$. Take any $\hat{\mathbf{x}}_i \in \mathbb{S}_i$ and define a new strategy $\mathbf{y}_i$ as follows:

$$y_{ij} = \begin{cases} \epsilon + (1-\epsilon)\cdot \hat{x}_{ii} & \text{if } j = i \\ (1-\epsilon)\cdot \hat{x}_{ij} & \text{if } j \neq i \end{cases}$$

Observe that $\mathbf{y}_i \in \mathbb{S}_i^R$ because $y_{ii} \geq \epsilon$ and $\sum_{j\in N} y_{ij} = \epsilon + (1-\epsilon)\cdot \sum_{j\in V} \hat{x}_{ij} = 1$.

Now for any path $P = \{a_1, a_2, \cdots, a_k\}$ from i to j where a_i are the arcs, the probability of obtaining this path in the delegation graph generated by the strategy profile $\{\mathbf{y}_i, \mathbf{x}_{-i}\}$ is exactly $y_{a_1} \cdot \prod_{a\in P\setminus a_1} x_a$. Thus

$$\begin{aligned}
u_i(\mathbf{y}_i, \mathbf{x}_{-i}) &= \sum_{j\in V} \mathbb{P}_{\mathbf{y}_i, \mathbf{x}_{-i}}(g(i) = j)\cdot u_{ij} \\
&= u_{ii} y_{ii} + \sum_{j\in V\setminus i} u_{ij} x_{jj} \cdot \left(\sum_{P\in\mathcal{P}(i,j)} y_{a_1} \prod_{a\in P\setminus a_1} x_a \right) \\
&\geq (1-\epsilon)\cdot u_{ii}\hat{x}_{ii} + (1-\epsilon)\cdot \sum_{j\in V\setminus i} u_{ij} x_{jj} \cdot \left(\sum_{P\in\mathcal{P}(i,j)} \hat{x}_{a_1} \prod_{a\in P\setminus a_1} x_a \right) \\
&= (1-\epsilon)\cdot \left(u_{ii}\hat{x}_{ii} + \sum_{j\in V\setminus i} u_{ij} x_{jj} \cdot \left(\sum_{P\in\mathcal{P}(i,j)} \hat{x}_{a_1} \prod_{a\in P\setminus a_1} x_a \right)\right) \\
&= (1-\epsilon)\cdot u_i(\hat{\mathbf{x}}_i, \mathbf{x}_{-i})
\end{aligned}$$

But $\mathbf{x}_i \in B_i^R(\mathbf{x}_{-i})$. Hence, $u_i(\mathbf{x}_i, \mathbf{x}_{-i}) \geq u_i(\mathbf{y}_i, \mathbf{x}_{-i})$ because $\mathbf{y}_i \in \mathbb{S}_i^R$. It follows that $u_i(\mathbf{x}_i, \mathbf{x}_{-i}) \geq (1-\epsilon)\cdot u_i(\hat{\mathbf{x}}_i, \mathbf{x}_{-i})$ and so the incentive guarantee for i. Thus $\mathbf{x}$ is an ϵ-Nash equilibrium. □

Next let's prove the social welfare guarantee holds for $\mathbf{x}$.

Theorem 10. *For all $\epsilon \in [0,1]$, and for any instance of the liquid democracy game, there exists an ϵ-Nash equilibrium with social welfare at least $\epsilon \cdot \mathrm{OPT}$.*

Proof. It suffices to prove that $\mathbf{x}$ has social welfare at least $\epsilon \cdot \mathrm{OPT}$. We have

$$\begin{aligned}
\mathrm{SW}(\mathbf{x}) &= \sum_{j \in V} u_j(\mathbf{x}) \\
&= \sum_{j \in D^*} u_j(\mathbf{x}) + \sum_{j \in D^*} \sum_{i \in L_j} u_i(\mathbf{x}) \\
&\geq \sum_{j \in D^*} u_{jj} \cdot x_{jj} + \sum_{j \in D^*} \sum_{i \in L_j} u_{ij} \cdot x_{jj} \\
&\geq \sum_{j \in D^*} \left(u_{jj} \cdot \epsilon + \sum_{i \in L_j} u_{ij} \cdot \epsilon \right) \\
&= \epsilon \cdot \sum_{j \in D^*} \left(u_{jj} + \sum_{i \in L_j} u_{ij} \right) \\
&= \epsilon \cdot \mathrm{OPT}
\end{aligned}$$

The first inequality follows since each agent $i \in L_j$ satisfies $u_i(\mathbf{x}_i, \mathbf{x}_{-i}) \geq u_i(\hat{\mathbf{x}}_i, \mathbf{x}_{-i})$ for all $\hat{\mathbf{x}}_i \in \mathbb{S}_i$. In particular, the deviation $\hat{\mathbf{y}}_i$ of delegating to the guru j with probability 1 implies $u_i(\mathbf{x}_i, \mathbf{x}_{-i}) \geq u_i(\hat{\mathbf{y}}_i, \mathbf{x}_{-i}) \geq u_{ij} x_{jj}$. Finally, the second inequality holds as we have $\mathbf{x}_j \in \mathbb{S}_j^R$, for each $j \in D^*$. Therefore $x_{jj} \geq \epsilon$ and the welfare guarantee holds. □

We can deduce from Theorem 10 that strong approximation guarantees can *simultaneously* be obtained for both social welfare and rationality. In particular, setting $\epsilon = \frac{1}{2}$ gives factor 2 approximation guarantees for both criteria.

Corollary 11. *For any instance of the liquid democracy game, there exists a $\frac{1}{2}$-Nash equilibrium with welfare at least $\frac{1}{2} \cdot \mathrm{OPT}$.*

3.4 A Tight Example

We now prove upper bounds on the welfare guarantee obtainable by *any* ϵ-Nash equilibrium. In particular, we show that the bicriteria guarantee obtained in Theorem 10 is tight.

Theorem 12. *For all $\epsilon \in [0,1]$, there exist instances such that any ϵ-Nash equilibrium has welfare at most $\epsilon \cdot \mathrm{OPT} + \gamma$ for any $\gamma > 0$*

Proof. Take an instance with $n+2$ agents. Let agents $\{1,2\}$ have identical utility functions with $u_{ij} = \delta$ if $j = 1$ and $u_{ij} = 0$ otherwise. The remaining agents $i \in \{3, \cdots, n+2\}$ have utilities

$$u_{ij} = \begin{cases} 1 & \text{if } j = 2 \\ 0 & \text{otherwise} \end{cases}$$

Observe that $\mathrm{OPT} = \delta + n$ which is obtained by agents 1 and 2 voting while the remaining agents delegate to agent 2. Now let $\mathbf{x}$ be an ϵ-Nash equilibrium. We claim that $x_{22} \leq \epsilon$. To see this, note that $x_{11} \geq (1-\epsilon)$. If $x_{11} < (1-\epsilon)$ then $u_1(\mathbf{x}) < (1-\epsilon)\delta$. But this contradicts the fact that $\mathbf{x}$ is an ϵ-Nash equilibrium, as agent 1 can deviate and vote herself to obtain a utility of δ. Furthermore, $x_{11} \geq (1-\epsilon)$ implies $x_{21} \geq (1-\epsilon)$ by a similar argument. Since $\sum_{j \in N} x_{2j} = 1$, we do have $x_{22} \leq \epsilon$. The social welfare of $\mathbf{x}$ is then

$$\mathrm{SW}(\mathbf{x}) \ \leq \ \delta + (1-\epsilon)\delta + \epsilon n \ = \ 2(1-\epsilon)\delta + \epsilon\,\mathrm{OPT}$$

Letting $\gamma = 2(1-\epsilon)\delta$ gives the desired bound. Since δ can be made arbitrarily small, γ can also be made arbitrarily small. □

Together, Theorems 10 and 12 imply that the price of stability of ϵ-Nash equilibrium is exactly ϵ in the liquid democracy game.

4 Extensions and Computational Complexity

4.1 Model Extensions

Our bicriteria results extend to the settings of repeated games, weighted voters, and multiple delegates.

- *Repeated Games.* Recollect that an underlying motivation for liquid democracy is that it can be applied repeatedly over time with agents having the option of delegating to different agents at different times. Evidently, by repeating the delegation game over time, with different utility functions for the topics considered in different time periods, the same bicriteria guarantees holds.
- *Weighted Voters.* In some electoral systems, different agents may have different voting powers. That is, the voters are weighted. In this case, the number of votes a guru casts is simply the sum of the weights of all the votes to which it was delegated. Our bicriteria guarantees then follow trivially.
- *Multiple Delegates.* In some settings it may be the case that an agent is allowed to nominate more than one delegate. That are two natural models for this. First, an agent nominates multiple delegates but the mechanism can use only one of them. Second, an agent splits its voting weight up and assigns it to multiple delegates. In both cases our techniques can be adapted and applied to give the same performance guarantees.

4.2 Computational Aspects

Let us conclude by discussing computation aspects in the liquid democracy game. Recall, to find an ϵ-Nash equilibrium with a provably optimal social welfare guarantee we solved a fixed point theorem. Moreover, solving the fixed point theorem requires knowledge of the optimal set D^* of gurus. But obtaining D^* is equivalent to finding an optimal solution to the liquid democracy game and this problem is hard.

Theorem 13. *It is NP-hard to find an optimal solution to the liquid democracy game.*

Proof. We apply a reduction from *dominating set in a directed graph.* Given a directed graph $G = (V, A)$ and an integer k: is there a set $S \subseteq V$ of cardinality k such that, for each $i \notin S$, there exists an arc $(i, j) \in A$ for some $j \in S$. This problem is NP-complete. We now give a reduction to the liquid democracy game.

Given the directed graph $G = (V, A)$ let the set of agents in the game be V. Define the utility function of an agent by:

$$u_{ij} = \begin{cases} 1 & \text{if } (i,j) \in A \\ 0 & \text{otherwise} \end{cases}$$

It immediately follows that there is a dominating set of cardinality at most k if and only if the liquid democracy game has a solution with social welfare at least $n - k$. This completes the proof. □

So is it possible for the agents to compute a good bicriteria solution in polynomial time? If we allow for sub-optimal approximation guarantees then this is achievable. The idea is simple: the characteristics required of the agents to ensure reasonable performance guarantees are *narcissism* and *avarice.*[5] First, since D^* is unknown, each agent i narcissistically assumes he himself is an optimal guru. Consequently, he will vote with probability p. Thus, he will delegate his vote with probability $(1 - p)$. This he will do avariciously, by greedily delegating to the agent i^* that gives him largest myopic utility, that is $i^* = \arg\max_{j \in V} u_{ij}$.

Theorem 14. *For any $\epsilon \in [\frac{3}{4}, 1]$, the narcissistic-avaricious algorithm is linear time and produces an ϵ-Nash equilibrium with social welfare at least* $(1-\epsilon)\cdot\text{OPT}$.

Proof. Consider the incentive guarantee for agent i for the strategy profile $\mathbf{z}$ induced by the algorithm. As i votes with probability $z_{ii} = p$ and delegates to i^* with probability $z_{ii^*} = (1 - p)$, his utility is

$$\begin{aligned} u_i(\mathbf{z}_i, \mathbf{z}_{-i}) &\geq z_{ii}u_{ii} + z_{ii^*}z_{i^*i^*}u_{ii^*} \\ &= p \cdot u_{ii} + (1-p)\cdot p \cdot u_{ii^*} \\ &\geq (1-p)\cdot p \cdot u_{ii^*} \\ &\geq (1-p)\cdot p \cdot u_i(\hat{\mathbf{z}}_i, \mathbf{z}_{-i}) \qquad \forall \hat{\mathbf{z}}_i \in \mathbb{S}_i \end{aligned}$$

Hence $\mathbf{z}$ is an ϵ-Nash equilibrium for

$$\epsilon = (1 - p(1-p)) = 1 - p + p^2 = \frac{3}{4} + (p - \frac{1}{2})^2$$

It follows that the narcissistic-avaricious algorithm can provide incentive guarantees only for $\epsilon \in [\frac{3}{4}, 1]$. Further, by solving the corresponding quadratic

[5] This is not an unreasonable assumption for both pirates and many of the inhabitants of Wonderland!.

equation, to obtain such an ϵ-Nash equilibrium we simply select $p = \frac{1}{2}\left(1+\sqrt{1-4(1-\epsilon)}\right)$.

Now, let's evaluate the social welfare guarantee for the narcissistic-avaricious algorithm. As above, we have

$$\begin{aligned}
\text{SW}(\mathbf{z}) &= \sum_{i\in V} u_i(\mathbf{z}_i, \mathbf{z}_{-i}) \\
&\geq \sum_{i\in V}(1-p)\cdot p\cdot u_{ii^*} \\
&= (1-p)\cdot p\cdot \sum_{i\in V} u_{ii^*} \\
&\geq (1-p)\cdot p\cdot \text{OPT}
\end{aligned}$$

Since $p = \frac{1}{2}\left(1+\sqrt{1-4(1-\epsilon)}\right)$ this gives

$$\text{SW}(\mathbf{z}) \geq \frac{1}{4}\Big(1-(1-4(1-\epsilon)\Big)\cdot \text{OPT} = (1-\epsilon)\cdot \text{OPT}$$

Therefore, as claimed, the narcissistic-avaricious algorithm outputs a solution whose welfare is at least $(1-\epsilon)$ times the optimal welfare.

Finally, observe that implementing the narcissistic-avaricious strategy requires that each agent i simply computes $i^* = \arg\max_{j\in V} u_{ij}$. This can be done for every agent in linear time in the size of the input. □

We emphasize two points concerning Theorem 14. One, it only works for weaker incentive guarantees, namely $\epsilon \in [\frac{3}{4}, 1]$. Unlike the fixed point algorithm it does not work for the range $\epsilon \in (0, \frac{3}{4})$. Two, the social welfare guarantee is $(1-\epsilon)$. This is a constant but, for the valid range $\epsilon \in [\frac{3}{4}, 1]$, it is much worse than the ϵ guarantee obtained by the fixed point algorithm. A very interesting open problem is to find a polynomial time algorithm that matches the optimal bicriteria guarantees provided by Theorem 10 and applies for all $\epsilon > 0$.

Acknowledgements. We thank Bundit Laekhanukit for interesting discussions. We are also grateful to the reviewers for suggestions which greatly improved the paper.

References

1. Behrens, J.: The origins of liquid democracy. Liq. Democr. J. 190–220 (2017)
2. Behrens, J., Kistner, A., Nitsche, A., Swierczek, B.: The Principles of LiquidFeedback. Interaktive Demokratie e. V., Berlin (2014)
3. Bloembergen, D., Gross, D., Lackner, M.: On rational delegations in liquid democracy. In: Proceedings of the 33rd Conference on Artificial Intelligence (AAAI), pp. 1796–1803 (2019)
4. Blum, C., Zuber, C.: Liquid democracy: potentials, problems, and perspectives. J Polit Philos **24**(2), 162–182 (2017)
5. Brill, M.: Interactive democracy. In: Proceedings of the AAMAS (2018)

6. Brill, M., Talmon, N.: Pairwise liquid democracy. In: Proceedings of the IJCAI, pp. 137–143 (2018)
7. Christoff, Z., Grossi, D.: Binary voting with delegable proxy: an analysis of liquid democracy. In: Proceedings of 16th Conference on Theoretical Aspects of Rationality and Knowledge (TARK), pp. 134–150 (2017)
8. Cohensius, G., Mannor, S., Meir, R., Meirom, E., Orda, A.: Proxy voting for better outcomes. In: Proceedings of the AAMAS, pp. 858–866 (2017)
9. Dodgson, C.: The Principles of Parliamentary Representation. Harrison and Sons, High Wycombe (1884)
10. Escoffier, B., Gilbert, H., Pass-Lanneau, A.: The convergence of iterative delegations in liquid democracy in a social network. In: Proceedings of the 12th International Symposium on Algorithmic Game Theory (SAGT), pp. 284–297 (2019)
11. Ford, B.: Delegative democracy (2002). www.brynosaurus.com/deleg/deleg.pdf
12. Gölz, P., Kahng, A., Mackenzie, S., Procaccia, A.: The fluid mechanics of liquid democracy. In: Proceedings of the 14th Conference on Web and Internet Economics (WINE), pp. 188–202 (2018)
13. Green-Armytage, J.: Direct voting and proxy voting. Const. Polit. Econ. **26**(2), 190–220 (2015)
14. Kahng, A., Mackenzie, S., Procaccia, A.: The fluid mechanics of liquid democracy. In: Proceedings of the 32nd Conference on Artificial Intelligence (AAAI), pp. 1095–1102 (2018)
15. Kotsialou, G., Riley, L.: Incentivising participation in liquid democracy with breadth first delegation. In: Proceedings of the AAMAS, pp. 638–644 (2020)
16. Lanphier, R.: A model for electronic democracy? (1995). robla.net/1996/steward/
17. Miller, J.: A program for direct and proxy voting in the legislative process. Public Choice **7**(1), 107–113 (1969)
18. Tolluck, G.: Towards a Mathematics of Politics. U of M Press, Minneapolis (1967)
19. Zhang, B., Zhou, H.: Brief announcement: statement voting and liquid democracy. In: 36th Symposium on Principles of Distributed Computing (PODC), pp. 359–361 (2017)

Abstracts

On Reward Sharing in Blockchain Mining Pools

Burak Can[1], Jens Leth Hougaard[2], and Mohsen Pourpouneh[2](✉)

[1] Department of Data Analytics and Digitalisation, Maastricht University, Maastricht, the Netherlands
b.can@maastrichtuniversity.nl

[2] Department of Food and Resource Economics (IFRO), University of Copenhagen, Copenhagen, Denmark
{jlh,mohsen}@ifro.ku.dk

Abstract. This paper provides, for the first time, a rich mathematical framework for *reward sharing schemes* in mining pools through an economic design perspective. We analyze and design schemes by proposing a comprehensive axiomatic approach. We depart from existing literature in various ways. First, our axiomatic framework is not on the consensus protocols but on the mining pools in any of these protocols. Second, our model is not restricted to a static single block, since various schemes in practice pay the miners repetitively over time in various blocks. Third, we propose reward sharing schemes and allocations not on the miners in a pool but instead on the shares submitted by these miners.

We demonstrate the flexibility of this space by formulating several desirable axioms for reward sharing schemes. The first condition ensures a *fixed total reward* that the fee charged by the pool manager is the same for any two rounds in a history. The second condition, *ordinality*, requires that time-shifts should not affect the reward distribution, so long as the order of shares is preserved. The third condition, *budget limit*, requires the pool manager to charge a nonnegative fee. The fourth condition, *round based rewards*, requires that the distribution of the rewards in a round only depends on that round. Finally, we introduce two axioms concerning fairness, *absolute redistribution* and *relative redistribution*, which demonstrates how the rewards should be redistributed when the round is extended by an additional share. We show that, together with other axioms, each of these fairness axioms, characterize two distinct classes of reward sharing schemes. Thereafter, we characterize the generalized class of proportional reward schemes, i.e., k-pseudo proportional schemes, which satisfy both of these axioms simultaneously. We introduce a final condition, *strict positivity*, which guarantees positive rewards for all shares, for any history. Imposing this additional condition single outs the well-known proportional reward scheme. The full article is available at: https://arxiv.org/abs/2107.05302.

Keywords: Blockchain · Fairness · Mining pools · Mechanism design

---

Can Gratefully Acknowledges Financial Support by the Netherlands Organisation for Scientific Research (NWO) Under the Grant with Project No. 451-13-017 (VENI, 2014) and Fonds National de la Recherche Luxembourg. Hougaard and Pourpouneh Gratefully acknowledge the Support by the Center for Blockchains and Electronic Markets Funded by the Carlsberg Foundation Under Grant No. CF18-1112

I. Caragiannis and K. A. Hansen (Eds.): SAGT 2021, LNCS 12885, p. 409, 2021.
https://doi.org/10.1007/978-3-030-85947-3

On Submodular Prophet Inequalities and Correlation Gap (Abstract)

Chandra Chekuri and Vasilis Livanos(✉)

University of Illinois at Urbana-Champaign, Urbana, IL 61801, USA

Abstract. We present a general framework for submodular prophet inequalities in the model introduced by Rubinstein and Singla [1], in which the objective function is a submodular function on the set of potential values instead of a linear one, via greedy Online Contention Resolution Schemes and correlation gaps. The framework builds upon the existing work of [1], yielding substantially improved constant factor competitive ratios for both monotone and general submodular functions, for various constraints beyond a single matroid constraint. As an additional improvement, it can be implemented in polynomial time for several classes of interesting constraints.

Along the way, we strengthen the notion of correlation gap for non-negative submodular functions introduced in [1], and provide a fine-grained variant of the standard correlation gap. For both cases, our bounds are cleaner and tighter. Furthermore, we present a refined analysis of the Measured Continuous Greedy algorithm for polytopes with small coordinates and general non-negative submodular functions, showing that, for these cases, it yields a bound that matches the bound of Continuous Greedy for the monotone case.

A full version of this paper is available at https://arxiv.org/abs/2107.03662.

Keywords: Combinatorial prophet inequality · Submodularity · OCRS

Reference

1. Rubinstein, A., Singla, S.: Combinatorial prophet inequalities. In: Proceedings of the Twenty-Eighth Annual ACM-SIAM Symposium on Discrete Algorithms, SIAM (2017), pp. 1671–1687. longer ArXiv version is at http://arxiv.org/abs/1611.00665

I. Caragiannis and K. A. Hansen (Eds.): SAGT 2021, LNCS 12885, p. 410, 2021.
https://doi.org/10.1007/978-3-030-85947-3

Vote Delegation and Misbehavior

Hans Gersbach, Akaki Mamageishvili, and Manvir Schneider(✉)

CER-ETH, ETH Zürich, Zürichbergstr. 18, 8092 Zürich, Switzerland
{hgersbach,amamageishvili,manvirschneider}@ethz.ch

In this paper we study vote delegation and compare it with conventional voting. Typical examples for vote delegation are validation or governance tasks on blockchains and liquid democracy. Specifically, we study vote delegation with well-behaving and misbehaving agents under three assumptions. First, voting is costly for well-behaving agents. That means, if a well-behaving individual abstains or delegates his/her vote, s/he is better off than with voting as long as his/her action does not affect the voting outcome. Second, the minority—composed of misbehaving voters—always votes. The rationale is that this minority is composed of determined agents who have either a strong desire to disrupt the functioning of the system, or to derive utility from expressing their minority view. Third, the preferences of agents are assumed to be private information. We evaluate vote delegation and conventional voting regarding the chance that well-behaving agents win.

Results: We provide three insights. First, if the number of misbehaving voters is high, both voting methods fail to deliver a positive outcome. Second, if the number of misbehaving voters is moderate, conventional voting delivers a positive outcome, while vote delegation fails with probability one. Third, with numerical simulations, we show that if the number of misbehaving voters is low, delegation delivers a positive outcome with a higher probability than conventional voting. Formally, we find that for any cost of voting c, there are thresholds $f^*(c)$ and $n^*(f)$ such that for any number of misbehaving voters f above f^* and an expected number of well-behaving agents above n^*, misbehaving voters will have the majority of votes and will win.

Our results have immediate implications for blockchains, i.e. they infer that vote delegation should only be allowed if it is guaranteed that the absolute number of misbehaving agents is below a certain threshold. Otherwise, vote delegation increases the risk for negative outcomes. Our results can also help assess the performance of vote delegation in democracy, a form of democracy known as "liquid democracy". Indeed, for a liquid democracy, our result is the worst-case result when delegating agents cannot trust those to whom they delegate. In the context of liquid democracy, we can view misbehaving voters as a determined minority who will vote no matter the costs. Well-behaving agents are a majority and balance costs of voting and impact on the outcome. Our result implies that if the size of the determined minority is not too small, vote delegation can lower the likelihood that the majority wins.

A Full Version of the Paper Can Be Found at https://arxiv.org/abs/2102.08823.
Partly Supported by the Zurich Information Security and Privacy Center (ZISC)

I. Caragiannis and K. A. Hansen (Eds.): SAGT 2021, LNCS 12885, p. 411, 2021.
https://doi.org/10.1007/978-3-030-85947-3

Efficiency of Equilibria in Games with Random Payoffs

Matteo Quattropani(✉) and Marco Scarsini

Luiss, Viale Romania 32, 00197 Rome, Italy
{mquattropani,marco.scarsini}@luiss.it

We consider normal-form games with n players and two strategies for each player, where the payoffs are i.i.d. random variables with some distribution F. For each strategy profile, we consider the (random) average payoff of the players, called average social utility (ASU). Most of the literature on games with random payoffs deals with the number of pure (or mixed) equilibria and its dependence on the payoffs distribution. Here we consider a different issue, i.e., efficiency of equilibria.

We first show that the optimal ASU converges in probability to a deterministic value that can be characterized in terms of the large deviation rate of F. Then we move to examine the asymptotic ASU of the pure Nash equilibrium (PNE). We start by considering the case in which F has no atoms. In this case, it is well known that asymptotically the number of PNE has a Poisson distribution with mean 1. This implies that we typically do not have many equilibria. We show that, when equilibria exist, in the limit they all share the same ASU. We then consider the case in which F has some atoms. Amiet et al. [1] show that the presence of atoms in the distribution F dramatically changes the existence issue: in this case, with probability converging to 1 as the number of players grows to infinity, there will be exponentially many PNE. We show that in this case the ASU of the best and the worst pure equilibrium converge in probability to two values, which we call x_{beq} and x_{weq}. Studying the best and worst PNE is standard in algorithmic game theory, which is often preoccupied with worst-case scenarios. The unusual phenomenon in our asymptotic framework is the high number of PNE, so that it is also important to study the efficiency of "most" equilibria. In this respect, we show that asymptotically all but a vanishingly small fraction of equilibria share the same ASU, x_{typ}, which lies between the two extrema x_{beq} and x_{weq}. In other words, most PNE have the same asymptotic ASU, but there exist also PNE having a quite different efficiency.[1]

Reference

1. Amiet, B., Collevecchio, A., Scarsini, M., Zhong, Z.: Pure Nash equilibria and best-response dynamics in random games. Math. Oper. Res., forthcoming (2021b)

[1] The full version of this paper is available at: https://arxiv.org/abs/2007.08518.

I. Caragiannis and K. A. Hansen (Eds.): SAGT 2021, LNCS 12885, p.412, 2021.
https://doi.org/10.1007/978-3-030-85947-3

Author Index